Fiber-Reinforced Polymers and Fiber-Reinforced Concrete in Civil Engineering

Fiber-Reinforced Polymers and Fiber-Reinforced Concrete in Civil Engineering

Guest Editors

Jianzhe Shi
Lili Hu
Hao Zhou
Liangliang Wei
Weiqiang Wang
Qin Zhang
Yaqiang Yang
Haitao Wang
Zheng Huang
Xing Zhao
Xu-Yang Cao

MDPI

Basel • Beijing • Wuhan • Barcelona • Belgrade • Novi Sad • Cluj • Manchester

Editors

Jianzhe Shi
Hohai University
Nanjing
China

Lili Hu
Shanghai Jiao Tong University
Shanghai
China

Hao Zhou
Central South University
Changsha
China

Liangliang Wei
Dongguan University of Technology
Dongguan
China

Weiqiang Wang
Hohai University
Nanjing
China

Qin Zhang
Hohai University
Nanjing
China

Yaqiang Yang
Jiangsu University of Science and Technology
Zhenjiang
China

Haitao Wang
Hohai University
Nanjing
China

Zheng Huang
Nantong University
Nantong
China

Xing Zhao
Nanjing University of Aeronautics and Astronautics
Nanjing
China

Xu-Yang Cao
Hohai University
Nanjing
China

Editorial Office
MDPI AG
Grosspeteranlage 5
4052 Basel, Switzerland

This is a reprint of the Special Issue, published open access by the journal *Buildings* (ISSN 2075-5309), freely accessible at: https://www.mdpi.com/journal/buildings/special_issues/SZ17047836.

For citation purposes, cite each article independently as indicated on the article page online and as indicated below:

Lastname, A.A.; Lastname, B.B. Article Title. *Journal Name* **Year**, *Volume Number*, Page Range.

ISBN 978-3-7258-4697-9 (Hbk)
ISBN 978-3-7258-4698-6 (PDF)
https://doi.org/10.3390/books978-3-7258-4698-6

Contents

buildings

MDPI

Editorial

Mechanical Behavior and Innovative Applications of Fiber-Reinforced Cementitious Composite and Fiber-Reinforced Polymer in Infrastructures

Jianzhe Shi, Haitao Wang * and Xuyang Cao *

College of Civil and Transportation Engineering, Hohai University, Nanjing 210098, China; sjz@hhu.edu.cn
* Correspondence: cewht@hhu.edu.cn (H.W.); caoxy@hhu.edu.cn (X.C.)

Abstract: Fiber-reinforced cementitious composite (FRCC) and fiber-reinforced polymer (FRP) have been widely applied in infrastructures. Their mechanical behavior and innovative applications are examined in this article. Firstly, the mechanical behaviors of FRCC and the corresponding improving methods are elaborated. The bond behavior of the FRP–concrete interface, which has a significant effect on the strengthening effect of FRP, is reviewed. A proposed method to enhance the bond behavior is also introduced. In addition, the effectiveness of FRP is demonstrated in terms of improving the load-bearing capacity, stiffness, crack resistance, fatigue resistance, and other behaviors of existing structures. Furthermore, the feasibility of fibers or FRPs in new constructions is also validated. Finally, the future prospects of the research and applications of FRCC and FRP are discussed.

Keywords: fiber-reinforced cementitious composite (FRCC); fiber-reinforced polymer (FRP); infrastructure; mechanical behavior; application

1. Introduction

The method of adding straw to delay the cracking of earthen walls has been adopted in the construction of houses since ancient times. Based on this, the addition of chopped fibers was proposed to address problems such as low tensile strength and susceptibility to cracking in cement-based materials. Now, various types of fiber-reinforced cementitious composites (FRCCs) such as fiber-reinforced concrete (FRC) [1], engineered cementitious composite (ECC) [2], and ultra-high-performance concrete (UHPC) [3] have been developed. These materials have been studied and applied by numerous researchers and engineers. These types of material can not only be used as efficient materials for structural strengthening, but also provide new methodologies for the revolution of infrastructures.

Compared to cement-based materials reinforced with chopped fibers, fiber-reinforced polymer (FRP) or plastic composed of continuous fibers and a resin matrix has light weight, high tensile strength, and obvious advantages such as good durability and fatigue resistance. The commonly used FRPs in infrastructures include carbon FRP (CFRP), glass FRP (GFRP), aramid FRP (AFRP), and basalt FRP (BFRP). The application of FRP in civil engineering consists of two major fields. One is the structural strengthening of existing structures [4], e.g., external bonding FRP sheets/grids, externally prestressing FRP tendons, and near-surface mounted (NSM) FRP. Structural strengthening is currently the mainstream field of the application of FRP, which can improve the load-bearing capacity, stiffness, seismic performance, fatigue resistance, durability, or other performance of existing structures. Besides structural strengthening, FRPs are also employed in new constructions. With the continuous development of large-span structures that are lightweight, highly durable, have a long service life, and are controllable in terms of damage and intelligence, conventional materials have become unable to meet the above requirements [5]. Therefore, FRPs are widely used in new structures such as concrete structures reinforced with FRP bars, FRP–concrete composite structures, bridges with FRP cables, and all-FRP structures. Combining various forms of FRP products with conventional materials to construct high-performance structures such as buildings, bridges, tunnels, and roads is the future direction for promoting the application of FRP in infrastructures.

In summary, researchers have achieved significant progress in studies of FRP and high-performance concrete with fibers in infrastructures with several decades of exploration. The Special Issue "Fiber-Reinforced Polymers and Fiber-Reinforced Concrete in Civil Engineering" was launched to exhibit the achievements in the mechanical behavior and innovative applications of FRP and high-performance concrete with fibers in infrastructures.

2. Mechanical Behavior of Fiber-Reinforced Cementitious Composite

The fiber types used in cement-based materials usually comprise steel fiber, carbon fiber, glass fiber, polypropylene fiber, etc. Among these types of fibers, steel fiber may corrode in harsh environments. Glass fiber can erode in the alkaline environment of concrete. Carbon fiber has stable physicochemical characteristics, but its high price is the main restriction to its application. In comparison, polypropylene fiber does not react with concrete and exhibits satisfactory bond behavior with concrete. Furthermore, basalt fiber has attracted researchers' attention recently due to its environmentally friendly production process and relatively high cost performance. Two articles focusing on polypropylene FRC and basalt FRC are included in this Special Issue. Furthermore, ultra-high-performance concrete (UHPC) has been widely studied and applied in recent decades because of its outstanding mechanical properties. However, it remains necessary to reduce its cost and address the problem of shrinkage cracking during early pouring in mass concrete construction. An improvement method is proposed in one article of this Special Issue.

2.1. Polypropylene FRC [6]

Recycled coarse aggregate (RCA) concrete reinforced with recycled polypropylene fiber (RPF), i.e., RPFRCA concrete, was proposed to effectively reuse waste petroleum products and construction waste. The effect of RCA and RPF on the tensile behavior of concrete was studied. The constitutive relationship of the specimens under axial tension and splitting tension were further investigated, and the results show that the maximum value of axial tensile strength occurred when the fiber volume content was 1.5%, and the strength increased by 21% compared with that of RCA concrete. The axial tensile strength decreased by only 10% when the replacement rate of RCA increased from 0 to 100%. Similarly, the maximum of peak strain occurred at the fiber volume content of 1.5%, which increased by 28% compared with that of RCA concrete. Furthermore, the maximum peak strain occurred when the length–diameter ratio of the fiber equaled 48, and the peak strain increased as the replacement rate of RCA increased. These results are expected to contribute to promoting recycling and sustainable concrete.

2.2. Basalt FRC [7]

Basalt fiber-reinforced pea gravel concrete (BFRPGC) has remarkable potential for applications in retrofitting covers for masonry walls. Thus, a quantitative understanding of the mechanical properties of the BFRPGC material is necessary. The effect of basalt fibers in pea gravel concrete was evaluated using compressive strength tests and splitting tensile tests. Fiber length (6 and 12 mm) and volume fraction ranging from 0.3% to 0.8% were selected as the experimental variables. The coupling effect of the bridging of the basalt fibers on concrete microcracks and the fiber agglomeration in concrete was analyzed. The best performance of the BFRPGC material was achieved by incorporating 12 mm length fibers with a volume fraction of 0.4%. A significant increase of up to 45% in compressive strength occurred in this case compared to that of the reference group. Furthermore, the failure mechanism of basalt fibers in pea gravel concrete was revealed via scanning electron microscope observations. In addition, the uniaxial compressive stress–strain model of the

BFRPGC material was proposed. The study may contribute to promoting the application of this novel material in retrofitting covers for masonry walls and other fields.

2.3. Improvement of UHPC [8]

In order to control the cost of UHPC and reduce the number of cracks caused by early pouring, volcanic rock was used as a new type of coarse aggregate, while metakaolin (MK) was also added. The volcanic rock particle size group and the MK dispersion ratio were selected as the experimental variables. Scanning electron microscopy (SEM), X-ray diffraction (XRD), and thermogravimetric (TG) microanalysis methods were used to reveal the effect of the microstructure, phase composition, material composition, and crystallinity of the mineral composition on the compressive properties of the UHPC cubes. The results show that the mechanical "lock-in effect" of the structure formed by the volcanic rock holes and mortar can effectively improve the shear resistance of the UHPC–volcanic rock interface. The compressive strength of the UHPC cubes increased with the particle size of the volcanic rock. When the MK dispersion ratio was less than 4%, the cube compressive strength of the UHPC and the contents of $CaCO_3$ crystals, C-S-H gel, and travertine in the UHPC increased with an increasing MK dispersion ratio. At an age of 28 days, compared with the addition of 1% MK, the addition of 4% MK increased the production of C-S-H gel and travertine in the UHPC matrix by 25%. Adding MK at a 4% dispersion ratio can promote the crystallization of limestone into a large amount of calcite, which can increase the strength of the UHPC. On the one hand, the addition of volcanic coarse aggregate resulted in the increased retention of free water and bound water; on the other hand, it also made it difficult to crystallize $CaCO_3$. The combined action of MK at a 4% dispersion ratio and volcanic rock significantly inhibits $CaCO_3$ crystallization. The results in this study may contribute to improving the mechanical behavior of UHPC.

3. Structural Strengthening with FRP

Reinforced concrete (RC) structures have been widely strengthened using FRP, and numerous relevant studies have emerged, mainly regarding flexural strengthening, shear strengthening, or confinement. The substantial strengthening effects of FRP have been validated. Two articles in this Special Issue focus on the bond behavior between FRP and concrete, which plays an important role in the strengthening of concrete structures using FRP. The effect of prestressing FRP laminate in the strengthening of a concrete bridge and a steel bridge was evaluated in another article, by means of distributed sensing. In addition, an interesting study, which is slightly different from other studies on FRP strengthening, is published in the Special Issue. It focuses on the cracking behavior of RC beams strengthened with FRP. Moreover, compared with structural elements at a monotonic loading, the fatigue behavior of RC beams strengthened with FRP are experimentally studied in this Special Issue. Furthermore, masonry structure strengthening, which differs from the strengthening of RC structures, is also discussed.

3.1. Dynamic Bond Behavior of FRP–Concrete Interface [9]

The majority of existing studies on the bond behavior of the CFRP–concrete interface focus on static loading conditions. This study clarified the effect of dynamic load, with strain rate as the main experimental variable. The results reveal that loading rate has no effect on the damage pattern of the interface. The ultimate bearing capacity of the CFRP–concrete interface increased as the strain rate increased. In addition, as the strain rate increased, both the maximum strain of CFRP and the maximum interfacial shear stress demonstrated a corresponding increase, with respective increase rates of 60% and 20%. The effective bond length decreased by about 20% when the strain rate increased from 10^{-5}/s to 10^{-2}/s. Finally, a formula calculating the dynamic effective bond length of CFRP laminate based on an existing model was proposed and its accuracy was verified. The study contributes to the understanding of the bond behavior of the FRP–concrete interface, under exploration or earthquake.

3.2. Improvement of Bond Behavior of FRP–Concrete Interface [10]

Bonding a ±45° biaxial GFRP under CFRP laminate can significantly improve the load-carrying capacity and deformability of the CFRP–concrete interface. However, the effect of the bonding sequence on the bond behavior remains unclear. In this study, three types of CFRP–concrete interfaces with CFRP bonded under, above, and between GFRP layers were tested under single-shear loading. Digital image correlation was used to measure the deformation of the interface during the test. The failure mode, load–displacement behavior, and principal strain distribution were analyzed. The results show that the main failure mode was the combined cohesion failure mode within the concrete and GFRP delamination, regardless of the bonding sequence. Compared to conventional CFRP–concrete interfaces, bonding the CFRP laminate above, under, and between the GFRP layers achieved 158%, 175%, and 177% increases in load-carrying capacity, respectively. Accordingly, the ultimate deformations also increased by 83%, 104%, and 86%. However, the bonding sequence showed only a slight influence on the initial stiffness of the load–displacement curve, which could be attributed to the differences in the strength and stiffness between the CFRP–concrete and the CFRP–GFRP–concrete interfaces. The results provide an effective method for improving the bond behavior of the FRP–concrete interface.

3.3. Evaluation of the Effect of Prestressing FRP Laminate on Structures

3.3.1. Concrete Bridge [11]

A method evaluating the effect of prestressed carbon-fiber-reinforced polymer (CFRP) laminate on bridges was proposed considering vehicle–bridge coupling, based on distributed sensor monitoring. Firstly, based on the principle of distributed monitoring, the strain influence line equation for bridges prestressed with CFRP laminate under moving vehicle loads was derived. Numerical simulation experiments were conducted for the verification of the method, in which a simulation method for solving the vehicle–bridge coupling vibration response based on the displacement contact method was studied. Relevant example verifications were carried out to prove the correctness of this method. On this basis, a dynamic analysis of the bridge before and after strengthening under vehicle–bridge coupling was performed. Finally, the impact of prestress levels, vehicle speeds, and vehicle axle load on the accuracy of evaluation was analyzed. The results show that prestressed CFRP laminate can effectively reduce the mid-span strain and displacement of bridges under vehicle loads. The bridge reinforcement evaluation method based on macro-strain amplitude can effectively assess the effect of prestressed CFRP laminate. Finally, vehicle speed and axle load cannot affect the accuracy of evaluation.

3.3.2. Steel Bridge [12]

Based on a distributed long-gauge fiber bragg grating (FBG) sensor, a macro-strain influence line (MSIL) formula was derived for a simply supported beam bridge under a moving vehicle load. The changes in the MSIL at the bottom of the beam were studied, with prestressing CFRP laminate. Furthermore, a rapid evaluation method was proposed for the strengthening effect based on the amplitude of the MSIL as the evaluation index for the strengthening effect. Finally, a prestressed CFRP-strengthened steel beam was tested under a moving vehicle load. The theoretical analysis and the experimental results confirm that the amplitude of the macro-strain–time history of the strengthened bridge is reduced compared to that without strengthening, and the local strengthening effect of the bridge can be monitored by the amplitude change in a single sensor. The change in global stiffness can be evaluated by monitoring the MSIL obtained from multiple long-gauge strain sensors. This study provides an effective methodology for evaluating the effectiveness of steel bridge strengthening.

The above two articles involve an efficient structural monitoring method using distributed sensing, which has great significance in the evaluation of FRP strengthening effects.

3.4. Cracking Behavior of RC Beam Strengthened with FRP [13]

The cracking behavior of RC beams strengthened by externally bonded fiber-reinforced polymer was studied, considering the effect of the reinforcement ratio. Beams strengthened with two layers of externally bonded wet carbon fiber sheets were tested under four-point bending mechanical tests up to failure. Digital image correlation was used to obtain the crack opening along the beam during loading. The results exhibited a significant effect of FRP on decreasing crack opening. At the same level of loads, the crack opening can be reduced by 20 to 50%, depending on reinforcement ratio. The crack width was compared with the theoretical values obtained based on the model proposed by Eurocode 2 (EC2), which shows the necessity of proposing a modified model. Subsequently, an empirical model was proposed as a modification of EC2, considering the effect of CFRP. The study can provide reference for the serviceability limit state design of concrete structures strengthened with CFRP.

3.5. Fatigue Model of RC Beam Strengthened with FRP [14]

Externally bonded FRP laminate is widely used in structural strengthening. Further enhancement of its strengthening effect can be achieved by inducing prestress. However, FRP debonding remains the main issue of this strengthening method, especially intermediate crack-induced debonding (IC debonding) under fatigue load. To better understand the impact of FRP debonding on the strengthening effect, a parameter analysis was conducted based on a fatigue life prediction model. The model involves the fatigue damage accumulation of the components of a strengthened beam, including concrete, steel reinforcements, and FRP laminate. The model also considers the mutual interaction among the components, and the impact of the fatigue debonding of FRP. As a result, a stress threshold for preventing the fatigue debonding of FRP was proposed, which contributes to avoiding the safety hazards caused by IC debonding in practical engineering.

3.6. Masonry Strengthened by FRP [15]

Although it is easy to construct masonry structures, the complexity of the construction materials, especially their mechanical properties, poses challenges for the numerical analysis of their structural behavior. Damage might occur in walls when they are bearing both horizontal and vertical lateral forces. Carbon-fiber-reinforced polymer (CFRP) emerges as a viable solution for strengthening unreinforced masonry walls subjected to shear loads. Experimental tests were conducted on the bond behavior of different configurations of CFRP-strengthened masonry triplets, which comprised three bricks and two mortar joints. The study affirms that masonry triplets strengthened with CFRP under shear loads exhibited strength levels at least four to six times greater than those without CFRP. Furthermore, the bond stress–slip relationship was predicted with equations proposed based on the test results. These equations were involved in a finite element analysis using ANSYS to simulate the behavior of CFRP-strengthened masonry triplets. The numerical results indicate good agreement with the test results. The outcome of this research improves the current knowledge on the use of CFRP to strengthen masonry walls with brick and mortar.

4. New Construction with Fiber or FRP

Besides structural strengthening, FRP can also be applied in new constructions. A large number of studies have been conducted on new constructions with FRP, such as concrete structures reinforced with FRP bars, concrete columns with FRP tubes, and trusses constructed using FRP profiles. This Special Issue includes three articles regarding new constructions. One article involves a concrete structure reinforced with both macro-synthetic fibers and steel reinforcements. Another article focuses on textile-reinforced concrete, which has application prospects in both new constructions such as wallboard, and in strengthening. Apart from buildings, FRP also has prospects in long-span bridges, such as in stay cables in cable-stayed bridges or main cables in suspension bridges, due to its high strength and light weight. An analysis of a long-span cable-stayed bridge with FRP cables is described in one article in this Special Issue.

4.1. Concrete Structure Reinforced with Fibers and Steel Bars [16]

Macro-synthetic fibers are increasingly used in concrete as secondary reinforcements to control temperature and shrinkage cracks, improving the durability of concrete. However, their impact on the shear strength of structural elements remains underexplored, particularly when used in combination with traditional steel reinforcement. This study developed and calibrated a nonlinear numerical model to simulate the shear response of macro-synthetic fiber-reinforced concrete (PFRC) elements. The model was calibrated with experimental data from PFRC panels subjected to pure shear loading, incorporating a custom concrete tension-softening model to capture the contribution of fibers. Validation against a broad range of PFRC beam experiments from the literature demonstrated satisfactory accuracy. The average predicted-to-experimental shear strength ratio equaled 0.99. Additionally, the model successfully replicated key response characteristics such as deformation patterns, crack propagation, and residual strength. The proposed modeling approach provides valuable insights into the interaction between fiber volume and transverse reinforcement. It also serves as a powerful tool for future numerical study, addressing the existing data gap on PFRC behavior and exploring the synergistic effects of macro-synthetic fibers and steel reinforcement on shear strength.

4.2. Enhancement of TRC [17]

To examine the flexural toughness characteristics of textile-reinforced lightweight aggregate concrete (TRLC), a four-point bending test was conducted. The effect of varying numbers of textile layers and the inclusion of hydroxypropyl methylcellulose (HPMC) on the ultimate load-bearing capacity and deformation capacity of TRLC thin plates was evaluated. The findings revealed that increasing the number of textile layers improved the initial cracking load, initial cracking deflection, ultimate load, ductility, and flexural toughness of the thin plates. The TRLC thin plate containing HPMC exhibited a reduced initial cracking load, but it displayed a strong matrix consistency and good flexural toughness. Moreover, the enhancement in the ultimate load of TRLC thin plates with HPMC was more pronounced with an increased number of textile layers, resulting in significantly more cracks compared to that without HPMC and an 11.40-fold increase in the flexural toughness index. These results validate the application prospects of TRLC, and provide references for the design of TRLC thin plates.

4.3. Long-Span Bridge with FRP Cables [18]

Based on the cable-stayed beam model, the nonlinear coupled vibration behavior of basalt fiber-reinforced polymer (BFRP) cables on long-span cable-stayed bridges was studied under parametric excitation. Considering the sag, damping of cables, and the coupled interactions between stayed cables and the main girder, the nonlinear coupling vibration model of the BFRP cable–beam composite structure was established. Taking the longest cable of Sutong Bridge as a case study, the nonlinear coupled vibration behavior of BFRP cables under parametric excitation was numerically analyzed using a finite difference method. The analysis results indicate that under parametric excitation, the large amplitude nonlinear vibration of the BFRP cable can be induced with an evident "beat" phenomenon. Under the same parametric excitation, the nonlinear coupling vibration response and the beta frequency of the BFRP cable were both smaller than that of the conventional steel cable. Furthermore, the nonlinear coupling vibration response of the BFRP cable increased with an increment in excitation amplitude and a decrement in cable force. With the increase in the excitation frequency, self-weight, and axial stiffness, the nonlinear vibration response of the BFRP cable increased first and then decreased. Finally, the damping ratio of the BFRP cable had no significant influence on the nonlinear coupling vibration. The study may contribute to the dynamic analysis of long-span bridges with FRP cables.

5. Summary

The articles published in this Special Issue, "Fiber-Reinforced Polymers and Fiber-Reinforced Concrete in Civil Engineering", involve fiber-reinforced cementitious composites (FRCCs), structures strengthened with FRPs, and new constructions using fibers or FRPs. At the material level, the mechanical behavior of FRCC was introduced, and the authors proposed corresponding improving methods. The bond behavior of the FRP–concrete interface, which has a significant effect on the strengthening effect of FRP, was also studied. Meanwhile, a method enhancing the bond behavior was proposed, based on additional FRP layers. The effectiveness of FRP was demonstrated to improve the load-bearing capacity, stiffness, crack resistance, fatigue resistance, and other behaviors of existing structures. Furthermore, the feasibility of the addition of fibers or FRPs in new constructions was also validated via experimental or numerical studies. In summary, fibers and FRPs have exhibited excellent prospects in infrastructures. In the future, researchers may concentrate on a deep understanding of the fundamental mechanical properties and the long-term behaviors of FRP and FRCC. They may also focus on the design theory and methodology, and the maintenance and diagnosis of structures with fibers and FRP.

Author Contributions: All authors have made a substantial, direct, and intellectual contribution to the work, and approved it for publication. All authors have read and agreed to the published version of the manuscript.

Funding: The authors appreciate the financial support provided by the Natural Science Foundation of Jiangsu Province (Grant Nos. BK20220985 and BK20220984) and the National Natural Science Foundation of China (Nos. 52378233, 52478165, and 52208164).

Acknowledgments: The authors would like to thank all of the anonymous reviewers for their constructive comments and suggestions.

Conflicts of Interest: The authors declare no conflicts of interest.

References

1. Cui, D.; Wang, L.; Zhang, C.; Xue, H.; Gao, D.; Chen, F. Dynamic Splitting Performance and Energy Dissipation of Fiber-Reinforced Concrete under Impact Loading. *Materials* **2024**, *17*, 421. [CrossRef]
2. Shabakhty, N.; Karimi, H.R.; Bakhtiary, A.Y. Statistical Evaluation of Fracture and Mechanical Performance of Engineered Cementitious Composites (ECC), Containing Different Percentages of Glass, Polypropylene, Polyvinyl-Alcohol Fibers, and Fly Ash. *Constr. Build. Mater.* **2024**, *417*, 135247. [CrossRef]
3. Wakjira, T.G.; Alam, M.S. Peak and Ultimate Stress-Strain Model of Confined Ultra-High-Performance Concrete (UHPC) Using Hybrid Machine Learning Model with Conditional Tabular Generative Adversarial Network. *Appl. Soft Comput.* **2024**, *154*, 111353. [CrossRef]
4. Wang, H.; Wu, Q.; Zhu, C.; Mao, Y.; Guo, C. An Innovative Prestressing System of Prestressed Carbon Fiber Sheets for Strengthening RC Beams under Flexure. *Constr. Build. Mater.* **2024**, *411*, 134409. [CrossRef]
5. Cao, X.; Shi, J.; Xu, J.; Ji, E.; She, Y.; Wang, Z. The Combined Influence of Bond–Slip and Joint-Shear in the Seismic Upgrading via Externally-Attached BFRP-Bar Reinforced Precast Sub-Frames. *J. Build. Eng.* **2023**, *80*, 107984. [CrossRef]
6. Wang, J.; Liang, J.; Li, Y.; Hou, W. Tensile Mechanical and Stress-Strain Behavior of Recycling Polypropylene Fiber Recycled Coarse Aggregate Concrete. *Buildings* **2024**, *14*, 1116. [CrossRef]
7. Li, J.; Wang, B.; Zhang, P.; Wang, Z.; Wang, M. Parametric Study on Mechanical Properties of Basalt Fiber-Reinforced Pea Gravel Concrete. *Buildings* **2024**, *14*, 380. [CrossRef]
8. Yin, Y.; Ma, Z. The Effect of Volcanic Stone and Metakaolin on the Compressive Properties of Ultrahigh-Performance Concrete Cubes. *Buildings* **2024**, *14*, 2024. [CrossRef]
9. Li, D.; Wang, X.; Zhang, J.; Jin, L.; Du, X. An Experimental Study and Result Analysis on the Dynamic Effective Bond Length of a Carbon Fiber-Reinforced Polymer Sheet Attached to a Concrete Surface. *Buildings* **2024**, *14*, 3245. [CrossRef]
10. Zhou, H.; Zhao, J.; Yang, Y.; Tan, F.; Ou, Y.; Wang, Y.; Li, C. Bond Behavior of CFRP–Concrete Bonded Joints with Additional GFRP Layer: Effect of Bonding Sequence. *Buildings* **2024**, *14*, 2936. [CrossRef]
11. Wu, B.; Fu, S.; Yan, G.; Shao, Y.; Zhou, Z.; Wang, H. Evaluation of Prestressed CFRP Plate Strengthened Bridges Using Distributed Sensing under Vehicle-Bridge Coupling. *Buildings* **2024**, *to be published*.
12. Wu, B.; Xia, Q.; Gong, Y.; Fu, S.; Wang, H.; Guo, Z. Evaluation of the Strengthening Effects on Prestressed Carbon-Fiber-Reinforced-Polymer-Strengthened Steel Beam Bridges Using Macro-Strain Influence Lines. *Buildings* **2024**, *14*, 2535. [CrossRef]
13. Armonico, A.; Michel, L.; Saidi, M.; Ferrier, E. Experimental Study on the Effect of Steel Reinforcement Ration on the Cracking Behavior of FRP-Strengthened RC Elements. *Buildings* **2024**, *14*, 950. [CrossRef]

14. Min, X.; Yang, D.; Song, S.; Li, X. Study on the Stress Threshold of Preventing Interfacial Fatigue Debonding in Concrete Beams Strengthened with Externally-Bonded FRP Laminates. *Buildings* **2024**, *14*, 430. [CrossRef]
15. Hashemi, S.M.; Ayoub, A. Improved Bond Stress-Slip Relationships for Carbon Fibre-Reinforced Polymer-Strengthened Masonry Triplets. *Buildings* **2024**, *14*, 257. [CrossRef]
16. Farag, B.F.; Thonstad, T.; Calvi, P.M. Numerical Modeling of Distributed Macro-Synthetic Fiber and Deformed Bar Reinforcement to Resist Shear. *Buildings* **2024**, *14*, 3247. [CrossRef]
17. Wang, J.; Yu, D.; Zeng, C.; Ji, X.; Ye, L.; Zhou, P.; Zhao, S. Effect of Textile Layers and Hydroxypropyl Methylcellulose on Flexural Behavior of TRLC Thin Plates. *Buildings* **2024**, *14*, 924. [CrossRef]
18. Yang, Y.; Zhou, Z.; Guan, Y.; Shi, J.; Zhan, Q.; Fahmy, M.F.M.; Wu, B. Nonlinear Coupled Vibration Behavior of BFRP Cables on Long-Span Cable-Stayed Bridges under Parametric Excitation. *Buildings* **2023**, *13*, 3082. [CrossRef]

buildings MDPI

Editorial

Fiber-Reinforced Polymers and Fiber-Reinforced Concrete in Civil Engineering

Jianzhe Shi

College of Civil and Transportation Engineering, Hohai University, Nanjing 210098, China; sjz@hhu.edu.cn

1. Introduction

Concrete has become one of the most widely used structural materials in civil engineering. However, the bottlenecks of low tensile strength and poor crack resistance are inevitable in concrete. At the beginning of the 20th century, researchers proposed the idea of mixing short fiber to improve the crack resistance of concrete, which was inspired by the method of delaying wall cracking by mixing straw or wheat straw into soil in ancient times. From the 1910s to the 1930s, patents were applied in countries such as the United Kingdom, the United States, and France to mix steel fibers for concrete modification. The "fiber crack resistance mechanism" was first proposed in the 1960s, promoting the development of steel fiber reinforced concrete. Since the 1970s, steel fiber reinforced concrete has been used in countries such as the United Kingdom, the United States, and Japan. After a development period of more than 50 years, fiber reinforced concrete (FRC) has been widely studied and applied [1–3].

Unlike short fiber reinforced composites, fiber reinforced polymers (FRPs), which are composed of continuous fibers and resin matrix, are lightweight and improve the strength/mass ratio significantly. Additionally, they possess advantageous durability and fatigue resistance. The commonly used FRPs in civil engineering include carbon FRP (CFRP), glass FRP (GFRP), aramid FRP (AFRP), and basalt FRP (BFRP). The development of FRPs began in the 1940s, when glass fiber was applied in the aviation industry. Subsequently, carbon fiber, aramid fiber, and basalt fiber were successively developed. Since the 1980s, safety and durability issues have emerged in engineering structures in developed countries such as the United States and Japan. Noting the opportunity, strengthening using FRPs was applied in those countries owing to the advantages of FRPs mentioned above. After the Hanshin earthquake in Japan, rapid and effective strengthening/repair methods were urgently necessary for the bridges and buildings damaged in the earthquake. The externally bonded carbon fiber sheet [4,5] has played an important role in post-earthquake strengthening and repair. Since then, FRPs have gradually been promoted and applied in civil engineering, receiving widespread attention from engineers. Nowadays, FRPs are widely applied both in the strengthening of existing structures and in new constructions, which demonstrates their advantages in improving the serviceability, bearing capacity, durability, and fatigue life of engineering structures [6,7].

2. Frontiers and Challenges of FRP and FRC

2.1. FRP

2.1.1. Fibers

With their rapid development, the types of fiber materials tend to be diverse. Carbon fiber, glass fiber, and basalt fiber have been widely used in civil engineering. The future development is aimed at developing fibers with higher strength, higher elastic modulus, and high-temperature resistance. Furthermore, with the increasing awareness of environmental protection, studies on green and recyclable types of fiber [8] are also of great significance in practical engineering.

2.1.2. Resin Matrix

Aging problems exist in resin matrix [9]. Therefore, the development of high-durability resin products is of great significance in improving the durability of FRP materials. In addition, FRPs rely mainly on the deformation of the resin matrix to dissipate energy under fatigue loads. Therefore, the research and development of resins with high toughness and fatigue resistance is crucial to improving the fatigue resistance of FRPs. On the other hand, the majority of FRP products used in engineering are manufactured using thermosetting resins, which cannot be reprocessed after curing. Thus, the research and development of reusable thermoplastic resins [10] is also an important direction for future development.

2.1.3. FRPs and Their Application in Construction

At present, civil engineering is developing in the direction of large span, light weight, long life, high durability, and damage controllability. With the complexity of structural forms and the harsh service environment, demands have also been put forward for FRPs. The development trends and challenges are listed as follows: first, the material system of FRPs needs to be optimized, and the fracture mechanism of FRPs under complex service environments and stress should be revealed in order to maximize the excellent performance of the fibers; second, manufacturing technology and equipment for FRPs should be improved, as should the quality and stability of FRPs; third, FRP application technologies need to be developed for the longevity, damage control, and light weight of engineering structures; furthermore, the concept of green energy conservation of FRPs should be implemented, and research on the full life cycle assessment of FRPs should be promoted; finally, further improvements to the standard system with respect to the design, production, testing, and construction of FRPs and structures using FRPs as reinforcement are necessary.

2.2. FRC

2.2.1. Mechanical Property and Durability

The mechanism of the strength and toughness enhancement of concrete via short fibers has been revealed [11], but there remain numerous issues that require further study; first, the state of fibers under load in FRC needs further exploration; second, the corrosion problems of alkaline cement on glass fibers, basalt fibers, and polyester fibers are noteworthy; in addition, the morphology of fibers and fabrics in FRC under long-term loading remains uncertain; moreover, the impacts of chemicals in the service environment on the performance of FRC [12] need to be studied; finally, there is a lack of research on the fatigue behavior of FRC, and studies on long fibers and their fabrics in concrete are rarely reported.

2.2.2. Manufacturing Technology

In terms of the manufacturing of FRC, suitable products should be designed and manufactured based on the special demands of FRC performance, e.g., impermeability, durability, stability. The flowability and operability during the production process should also be considered, and the contradiction between the final performance of FRC and the characteristics and manufacturing process of fibers is yet to be resolved; in addition, contradictions also exist between the reinforcing effects of length-diameter ratio and fiber content of fibers on FRC, and the difficulty of mixing and operation of the mixture during the manufacturing process. On the other hand, due to the high specific surface area and susceptibility to damage, it is difficult to mix short fibers in concrete, and effective solutions have not been proposed.

3. Purposes of the Special Issue

Building on the above perspectives, this Special Issue "Fiber-Reinforced Polymers and Fiber-Reinforced Concrete in Civil Engineering", seeks to contribute to the latest advancements in the research and application of FRPs and FRC. Potential topics include but, are not limited to, new materials/systems/techniques, durability/long-term performance, bond behavior, strengthening of concrete/metallic/timber/masonry structures, concrete structures reinforced/prestressed with FRPs, hybrid structures, all FRP structures, structural health monitoring and intelligent sensing, codes/standards/guidelines, field applications/case studies, and the high performance/longevity/sustainability of structures.

Funding: The author appreciates the financial support provided by the Natural Science Foundation of Jiangsu Province (Grant No. BK20220985), China Postdoctoral Science Foundations (Grant No. 2021M690882), and the Shuangchuang Doctor Program of Jiangsu Province, China (Grant No. JSSCBS20210253).

Acknowledgments: The author would like to thank all the anonymous referees for their constructive comments and suggestions.

Conflicts of Interest: The author declares no conflict of interest.

References

1. Uddin, M.N.; Ye, J.; Deng, B.; Li, L.; Yu, K. Interpretable machine learning for predicting the strength of 3D printed fiber-reinforced concrete (3DP-FRC). *J. Build. Eng.* **2023**, *72*, 106648. [CrossRef]
2. Watts, M.J.; Amin, A.; Gilbert, R.I. Time-dependent deformation and cracking behaviour of FRC beams. *Eng. Struct.* **2022**, *268*, 114741. [CrossRef]
3. Brandt, A.M.; Glinicki, M.A.; Potrzebowski, J. Application of FRC in construction of the underground railway track. *Cem. Concr. Compos.* **1996**, *18*, 305–312. [CrossRef]
4. Mohammadi, A.; Barros, J.A.O.; Sena-Cruz, J. A new model for predicting the shear strength of RC beams strengthened with externally bonded FRP sheets. *Compos. Struct.* **2023**, *319*, 117081. [CrossRef]
5. Nikraftar, A.S.A.; Dehestani, M. Reliability assessment of IC debonding in FRP strengthened RC beams using neural network. *Structures* **2022**, *46*, 1896–1917. [CrossRef]
6. Sharma, A.; Joshi, S.C. Enhancement in fatigue performance of FRP composites with various fillers: A review. *Compos. Struct.* **2023**, *309*, 116724. [CrossRef]
7. Zhang, B.; Zhu, H.; Dong, Z.; Yang, Z. Mechanical properties and durability of FRP-reinforced coral aggregate concrete structures: A critical review. *Mater. Today Commun.* **2023**, *35*, 105656. [CrossRef]
8. Shayanfar, M.; Shahrabadi, H. Flexural behavior of green RC beams with disposable glasses fibers in a marine environment. *Case Stud. Constr. Mater.* **2023**, *18*, e01972. [CrossRef]
9. Zhuang, X.; Ma, J.; Liu, F.; Dan, Y.; Huang, Y.; Jiang, L. Characterization of hydrothermal aging induced voids in carbon fiber reinforced epoxy resin composites using micro-computed tomography. *Polym. Degrad. Stab.* **2022**, *206*, 110198. [CrossRef]
10. Kazemi, M.E.; Shanmugam, L.; Dadashi, A.; Shakouri, M.; Lu, D.; Du, Z.; Hu, Y.; Wang, J.; Zhang, W.; Yang, L.; et al. Investigating the roles of fiber, resin, and stacking sequence on the low-velocity impact response of novel hybrid thermoplastic composites. *Compos. Part B* **2021**, *207*, 108554. [CrossRef]
11. Tarabin, M.; Leblouba, M.; Barakat, S.; Zahri, M. Experimental and probabilistic analysis of the crack propagation in fiber reinforced concrete. *Eng. Fail. Anal.* **2023**, *151*, 107388. [CrossRef]
12. Youssari, F.Z.; Taleb, O.; Benosman, A.S. Towards understanding the behavior of fiber-reinforced concrete in aggressive environments: Acid attacks and leaching. *Constr. Build. Mater.* **2023**, *368*, 130444. [CrossRef]

buildings

MDPI

Article

Experimental Study on Bond Fatigue Between Carbon Fiber-Reinforced Polymer Bars and Seawater–Sea Sand Concrete Under Seawater Immersion and Dry–Wet Cycle Conditions

Fu Yang [1], Weiquan Gan [1], Bingsan Chen [1], Jiangchuan Wu [2,*] and Shiwen Yuan [2]

[1] Lianyungang Zhiyuan Electric Power Design Co., Ltd., Lianyungang 222000, China; yangf@js.sgcc.com.cn (F.Y.); ganwq@js.sgcc.com.cn (W.G.); chenbs@js.sgcc.com.cn (B.C.)
[2] School of Civil and Ocean Engineering, Jiangsu Ocean University, Lianyungang 222000, China; 2024220429@jou.edu.cn
* Correspondence: wujch@jou.edu.cn; Tel.: +86-13-7051-36540

Abstract: The durability of carbon fiber-reinforced polymer (CFRP) bars in marine environments is essential for their application in seawater–sea sand concrete (SWSSC), especially under cyclic loading conditions. While previous studies primarily focused on static bonding performance, the effects of seawater immersion and dry–wet cycles on bond fatigue behavior at CFRP–SWSSC interfaces remain underexplored. This study investigated the bond fatigue performance of CFRP bars and SWSSC under seawater immersion and dry–wet cycling conditions. Eighteen CFRP bar-SWSSC bond specimens were divided into three categories and prepared for static and fatigue pull-out tests. The effects of varying stress levels (fatigue upper load/static bond ultimate load) after seawater immersion and dry–wet cycling on fatigue failure modes, bond–slip behavior, and fatigue characteristics were evaluated. The results show that seawater immersion and dry–wet cycling significantly degrade the performance of bonds between CFRP bars and SWSSC, with an average bond strength reduction of 10.31%. These conditions reduce fatigue cycles and stiffness while increasing bond–slip (relative displacement at the bar–concrete interface) and residual–slip (displacement after unloading). Moreover, dry–wet cycling has a greater negative impact on fatigue bond performance than seawater immersion. Higher fatigue stress levels exacerbate damage and crack propagation at the CFRP–SWSSC interface, leading to significant increases in both bond–slip and residual-slip. Under similar conditions, higher stress levels enhance bond stiffness. However, excessively high stresses may lead to bond fatigue failures. Using experimental data and existing fatigue bond–slip constitutive models, a customized model for CFRP bars in SWSSC was developed. These findings highlight that marine environments and fatigue loading severely impair bond performance, thereby emphasizing the importance of careful design for marine applications. The proposed model offers a reliable framework for predicting bond–slip behavior under fatigue conditions, enhancing the understanding of CFRP–SWSSC interactions and supporting the design of durable marine infrastructure.

Keywords: fatigue behavior; CFRP bars; bond; seawater–sea sand concrete; seawater immersion; dry–wet cycles

1. Introduction

The corrosion of steel reinforcements remains a pivotal issue constraining the durability and service life of marine engineering structures, as it compromises mechanical integrity and diminishes the bond strength with concrete, thereby reducing the longevity of reinforced concrete systems [1]. To mitigate this challenge, researchers have advocated for substituting steel reinforcements with fiber-reinforced polymer (FRP) bars in concrete systems. Compared to GFRP and BFRP bars, CFRP bars exhibit superior strength and stiffness, enhanced fatigue performance, and excellent corrosion resistance. In recent years, the declining cost of CFRP materials has significantly expanded their applications in civil engineering and related fields [2–6]. Additionally, substituting traditional concrete with seawater–sea sand concrete (SWSSC) addresses critical challenges, including river sand depletion, escalating environmental stress, and freshwater scarcity, while effectively leveraging abundant marine resources [4–11]. The synergy between the corrosion-resistant properties of FRP bars and the resource efficiency of SWSSC has led to the emergence of FRP-reinforced SWSSC structures, offering viable solutions to marine engineering challenges such as resource underutilization, elevated transportation costs, and steel corrosion. Consequently, the integration of FRP bars and SWSSC demonstrates immense potential for developing sustainable and durable solutions for marine infrastructure [2,12,13].

Research has shown that the bonding mechanisms of FRP bars and concrete fundamentally diverge from those of steel reinforcements, exhibiting pronounced differences in their fatigue bond behavior, particularly under the fatigue conditions commonly encountered in offshore wind power projects and cross-sea bridge engineering [3,4,6,13–20]. Consequently, ensuring the fatigue bond performance of FRP bars with concrete requires tailored approaches, differing from conventional methodologies used for steel-reinforced concrete. Studies on the fatigue bond performance of FRP bars in concrete are still in their infancy, requiring further exploration.

Zou et al. [21] evaluated the fatigue bond behavior of basalt fiber-reinforced polymer (BFRP) bars with recycled concrete and found that fatigue loading reduced bonding defects, improved bond stiffness, and elevated bond strength. A constitutive model was formulated to characterize the bond–slip behavior under fatigue loading conditions for BFRP bars and recycled concrete, integrating experimental findings and existing theoretical frameworks. Shen et al. [22] examined the bond interaction of BFRP bars with concrete under cyclic loading and determined that escalating loading cycles led to diminished bond strength due to progressive interface deterioration and mechanical degradation. Furthermore, the bond–slip displacement exhibited an upward trend with increasing loading cycles, reflecting increased relative motion at the bonding interface. Moreover, the reduced hysteresis loop area signified a progressive decline in bonding efficacy over repeated cycles. Lee et al. [23] explored the bond strength characteristics of glass fiber-reinforced polymer (GFRP) bars under monotonic and cyclic loading conditions, emphasizing the stark discrepancies in interfacial performance relative to traditional steel reinforcement. Under cyclic loading conditions, the GFRP bars demonstrated more pronounced strength degradation compared to their behavior under monotonic loading. Mai et al. [24] assessed the bond durability of BFRP bars embedded in recycled aggregate seawater–sea sand concrete (RASWSSC) during freeze–thaw cycles. The findings indicated a marked reduction in bond strength with escalating freeze–thaw cycles, whereas thicker concrete covers mitigated this degradation by enhancing bond integrity. Xiong et al. [25] analyzed the fatigue bond dynamics between BFRP bars and seawater–sea sand concrete, concluding that total slip was governed primarily by intrinsic bond performance rather than the loading protocol. The fatigue-induced bond failure modes exerted a substantial influence on fatigue bond stiffness, peak bond stress, and residual bond capacity. Xiong et al. [26] systematically investigated the influence of incorporating glass fibers (GFs) and expansive agents (EAs) into seawater–sea sand concrete (SWSSC) on the interfacial bond behavior of glass fiber-reinforced polymer (GFRP) and basalt fiber-reinforced polymer (BFRP) reinforcement bars. The findings revealed that the addition of GFs and EA significantly improved bond stiffness but resulted in only marginal improvements in bond strength. Neither bond stiffness nor bond strength exhibited consistent improvement trends with variations in fiber length, GF content, or EA content. Sun J. et al. [16] examined the mechanical characteristics of basalt fiber-reinforced seawater–sea sand recycled aggregate concrete (BFSSRAC), composed of recycled coarse aggregate (RCA), seawater, and sea sand, alongside its bond interaction with BFRP bars. The results indicated that the utilization of RCA reduced bond strength, whereas concrete of higher strength grades facilitated increased maximum slip, reflecting enhanced energy dissipation during the bond–slip mechanism. Liao et al. [27] evaluated the interfacial bond strength between glass fiber-reinforced polymer (GFRP) bars and high-strength, as well as ultra-high-strength, sea sand concrete (SWSSC). The findings demonstrated that replacing conventional concrete with SWSSC exerted no significant effect on short-term bond strength. The incorporation of polyethylene (PE) fibers prolonged the micro-slip phase within the bond stress–slip relationship, effectively delaying crack initiation and propagation. However, excessive PE fiber content induced defects, subsequently diminishing bond strength. Increased concrete strength mitigated slope-softening phenomena during the slip phase. He et al. [28] analyzed the bond performance and determinant factors associated with ribbed basalt fiber-reinforced polymer (BFRP) bars embedded in seawater–sea sand concrete (SWSSC). The results indicated that bond strength diminished with increasing rib and groove widths, whereas greater rib height contributed to enhanced bond strength. The ultimate slip exhibited a strong correlation with rib spacing. Wang et al. [29] conducted a comprehensive investigation into the bond behavior between FRP bars and concrete under fatigue loading conditions. Their findings revealed that the inclusion of fibers improved bond stiffness and load-bearing capacity demonstrated brittle failure characteristics, and mitigated the rate of fatigue-induced bond deterioration. Rensheng et al. [30] performed pull-out experiments to evaluate the bond characteristics between spiral-ribbed GFRP bars and ultra-high-performance concrete. The results revealed that augmented fiber content within a specific range significantly enhanced bond strength, achieving an increase of 74.81% to 92.22% from 0% to 2%; further increases from 2% to 3% yielded negligible changes. Deng et al. [31] executed four-point bending tests to assess the fatigue bond performance of CFRP-reinforced seawater–sea sand concrete (SWSSC) beams. The findings suggested that excessive stirrup spacing modified the static failure mode of FRP-reinforced SWSSC beams, significantly increasing fatigue crack width and deflection under fatigue conditions. The load magnitude had a profound influence on fatigue life.

Current research primarily focuses on the static bond performance of FRP bars in concrete, with limited studies exploring the fatigue bond behavior of CFRP bars in concrete, with limited exploration into the fatigue bond behavior of CFRP bars in seawater-sea sand concrete (SWSSC). Notably, research on the fatigue bond performance of CFRP bars in SWSSC under seawater immersion and wet-dry cycles is scarce. Consequently, this study investigates the fatigue bond characteristics of CFRP bar-SWSSC specimens under varying stress levels, conducting both static and fatigue tests after seawater immersion and dry–wet cycling. We systematically evaluate fatigue failure modes, bond–slip curves, bond stiffness, and residual slip characteristics under fatigue conditions. The experiment results are combined with those of the existing literature to formulate a bond–slip constitutive model for fatigue loading. This research establishes a theoretical foundation for applying CFRP bars in SWSSC structures.

2. Materials and Methods

2.1. Materials

The coarse aggregate used in the experiment was crushed granite with particle sizes of 5–25 mm, and its gradation parameters are shown in Table 1. The fine aggregate consisted of natural sea sand, while the mixing water was natural seawater, both obtained from Lianyungang City, Jiangsu Province, China. The cement used was P.O 42.5-grade ordinary Portland cement, and the water-reducing agent was a powdered polycarboxylate-based superplasticizer.

Table 1. Grading of coarse aggregates.

Aperture (mm)	Residual	2.36	4.75	9.5	16	19	26.5
Sieve Residue (g)	5.1	1.4	282.8	2601.3	1181.1	927.9	0.0
Sieve Retention Percentage (%)	0.10	0.03	5.66	52.03	23.62	18.56	0.00
Cumulative Sieve Residue (%)	/	99.90	99.87	94.21	42.18	18.56	0.00
Specification Upper Limit (%)	/	100	100	60	10	0	0
Specification Lower Limit (%)	/	95	85	30	0	0	0

The natural seawater used for concrete mixing was analyzed using ion chromatography, revealing a chloride ion concentration of 1.40% and a sulfate ion concentration of 0.17%. As shown in Figure 1, the natural sea sand was sieved through a 5 mm mesh to remove larger shells and impurities. The fineness modulus of the sand was determined to be 2.37, classifying it as medium sand according to the "Sand for Construction" standard (GB/T 14684-2011) [32]. After drying, ion chromatography analysis indicated that the surface-adhered chloride ion concentration of the natural sea sand was 0.13%, while the sulfate ion concentration was 0.03%.

(a) (b) (c)

Figure 1. Stages of sea sand processing: (**a**) sea sand collection site; (**b**) drying process of sea sand; (**c**) sieving to remove shell fragments.

The seawater and sea sand concrete, designed for a strength grade of C50, was prepared in accordance with the "Design Regulations for the Mix Ratio of the Ordinary Concrete" (JGJ 55-2011) [33] and the "Technical code for application of sea sand concrete" (JGJ 206-2010) [34]. The mix proportions of seawater and sea sand concrete are detailed in Table 2.

Table 2. Mix proportions of seawater and sea sand concrete (kg/m^3).

Strength Grade	Seawater	Sea Sand	Cement	Coarse Aggregate	Water-Reducing Admixture (% of Cement Content)	Total Mass
C50	199	613	570	1012	0.30%	2394

A total of nine cubic specimens of seawater–sea sand concrete, each with dimensions of 150 × 150 × 150 mm, were prepared for this experiment. Following 28 days of standard curing, compressive strength tests were conducted on three specimens. The remaining specimens were stored in seawater immersion tanks and dry–wet cycle test chambers for 120 days before undergoing compressive strength tests. The experimental setup is shown in Figure 2, and the corresponding test data are provided in Table 3.

(a) (b)

Figure 2. Diagrams of seawater–sea sand concrete specimens: (**a**) loading diagram of specimens; (**b**) failure diagram of specimens.

Table 3. Compressive strength of seawater–sea sand concrete.

Specimen Number	Ultimate Load (kN)	Average Ultimate Load (kN)	Compressive Strength (MPa)	Average Compressive Strength (MPa)	Corrosive Environment	Corrosion Age (d)
C-P-1	1031.1		45.83		Standard	0
C-P-2	990.6	1022.9 ± 29.1	44.03	45.46 ± 1.29	Standard	0
C-P-3	1047.0		46.53		Standard	0
C-J120-1	1236.5		54.96		Seawater Immersion	120
C-J120-2	1010.8	1163.8 ± 132.6	44.92	51.73 ± 5.89	Seawater Immersion	120
C-J120-3	1244.2		55.30		Seawater Immersion	120
C-G120-1	1106.3		49.17		Dry–Wet Cycling	120
C-G120-2	928.7	1057.8 ± 113.0	41.28	47.01 ± 5.02	Dry–Wet Cycling	120
C-G120-3	1138.5		50.60		Dry–Wet Cycling	120

The CFRP bars used in this experiment were produced by Haining Anjie Composite Materials Co., Ltd., Haining City, Zhejiang Province, China, using a pull-winding production process. The mechanical performance of the CFRP bars was tested in accordance with the "Technical Standard for Fiber Reinforced Polymer (FRP) in Construction" (GB 50608-2020) [35]. The testing apparatus consisted of a 300 kN electronic universal testing machine, operated under displacement control conditions with a loading rate of 0.2 mm/min. The loading schematic is illustrated in Figure 3. The basic mechanical properties of the CFRP bars are detailed in Table 4.

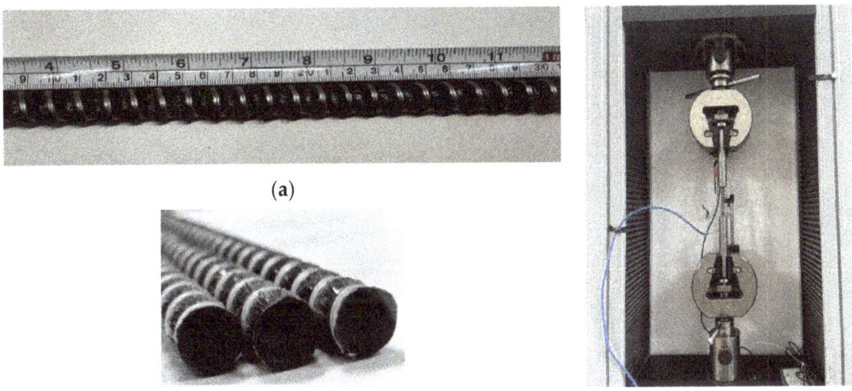

Figure 3. Test setup of CFRP bar. (**a**) CFRP bar; (**b**) cross-sectional profile; (**c**) coupon test of CFRP bar.

Table 4. Basic parameters of CFRP bar.

Reinforcement Type	Diameter (mm)	Elongation (%)	Modulus of Elasticity (GPa)	Tensile Strength (MPa)
CFRP	10	1.35	150.32	1845.63

2.2. Test Specimens

This study investigated the effects of varying stress levels (low, medium, and high) during seawater immersion and dry–wet cycling on the bond fatigue performance between CFRP bars and seawater–sea sand concrete. A total of 18 pull-out specimens were prepared and grouped into three groups: standard environment, seawater immersion, and dry–wet cycling. Each category contained three static pull-out specimens and three fatigue specimens, with stress levels assigned at 50%, 60%, and 70%. The stress ratio for all fatigue specimens was fixed at 0.2. Static bond tests were performed on nine specimens across the three categories, and their average ultimate loads were determined. The fatigue upper and lower loads for the specimens were calculated based on the average ultimate loads. Stress levels, defined as the ratio of fatigue upper load to static bond ultimate loads, were set at 50%, 60%, and 70%, with a stress ratio (fatigue lower load to fatigue upper loads) of 0.2. The detailed design of the experiment is presented in Table 5.

Both seawater immersion and dry–wet cycling utilized natural seawater, with a dry–wet cycle consisting of six hours of immersion followed by six hours of drying at room temperature. To ensure the consistent concentrations of chloride and sulfate ions in the seawater, the natural seawater was replaced weekly. Each batch of seawater was analyzed using ion chromatography, revealing chloride ion concentrations of between 1.26% and 1.44% and sulfate ion concentrations of between 0.09% and 0.18%.

Table 5. Bonding fatigue test scheme.

Specimen ID	Corrosive Environment	Corrosion Age (d)	Stress Level	Fatigue Upper Limit	Fatigue Lower Limit
S-P-1	Standard	0	-	Ultimate load in static bond tests, F_{u1}	
S-P-2			-		
S-P-3			-		
F0.5-P			50%	0.5 F_{u1}	0.1 F_{u1}
F0.6-P			60%	0.6 F_{u1}	0.12 F_{u1}
F0.7-P			70%	0.7 F_{u1}	0.14 F_{u1}
S-J120-1	Seawater Immersion	120	-	Ultimate load in static bond tests, F_{u2}	
S-J120-2					
S-J120-3					
F0.5-J120			50%	0.5 F_{u2}	0.1 F_{u2}
F0.6-J120			60%	0.6 F_{u2}	0.12 F_{u2}
F0.7-J120			70%	0.7 F_{u2}	0.14 F_{u2}
S-G120-1	Dry–Wet Cycling	120	-	Ultimate load in static bond tests, F_{u3}	
S-G120-2					
S-G120-3					
F0.5-G120			50%	0.5 F_{u3}	0.1 F_{u3}
F0.6-G120			60%	0.6 F_{u3}	0.12 F_{u3}
F0.7-G120			70%	0.7 F_{u3}	0.14 F_{u3}

In this context, "S/F" indicates static load or fatigue conditions; "P/J/G" represents standard environmental, seawater immersion, or dry–wet cycling. "F0.5/F0.6/F0.7" specifies fatigue loading at stress levels of 50%, 60%, and 70%, respectively; "J120/G120" refers to 120 days of seawater immersion or 120 days of dry–wet cycling. For example, "F0.5-J120" indicates a pull-out specimen of CFRP bar embedded in seawater–sea sand concrete and subjected to a 50% fatigue stress level after 120 days of seawater immersion.

Both the static and fatigue bond specimens in this study were designed with an eccentric pull-out configuration to closely replicate actual stress conditions. The pull-out specimens measured 150 mm × 150 mm × 150 mm. The bond length of the anchorage was set to 5 d_a (where d_a is the diameter of the reinforcement) and was achieved using PVC pipes filled with foam adhesive. The eccentric cover thickness was specified to be 35 mm. After casting, the specimens were cured in a standard curing room for one day prior to demolding and were left to continue curing in the same environment for 28 days. The seawater immersion and dry–wet cycling specimens were then transferred to a natural seawater curing pool for further curing (see Figure 4).

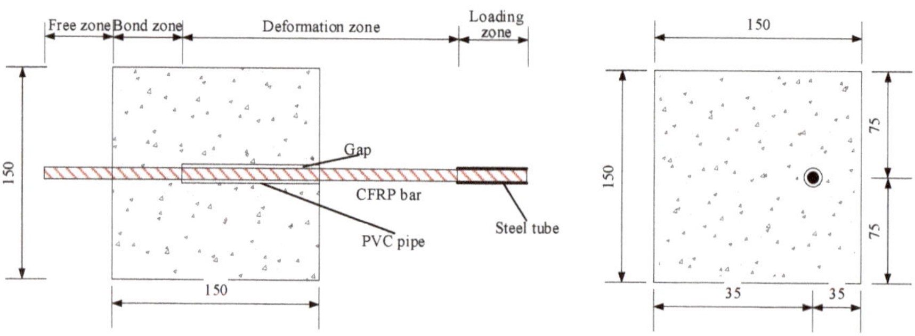

(a)

Figure 4. *Cont.*

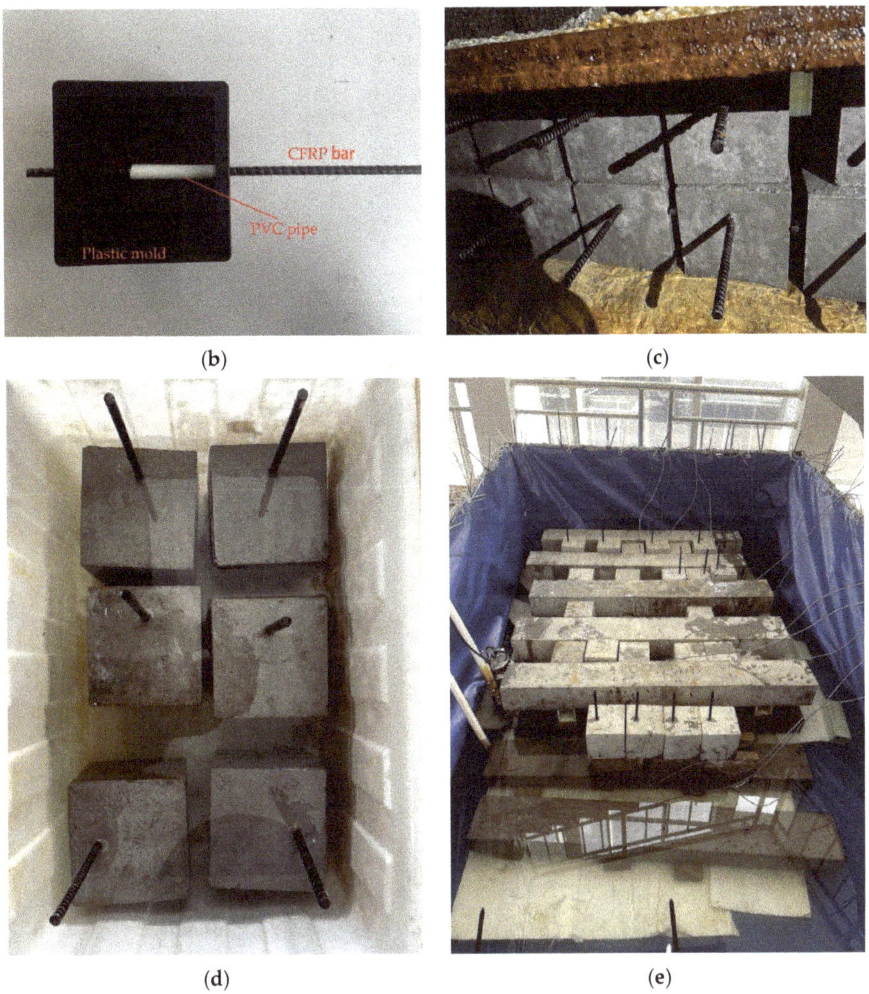

Figure 4. Pull-out specimens: (**a**) schematic of pull-out specimens; (**b**) pull-out specimen mold; (**c**) standard environment; (**d**) seawater immersion; (**e**) dry–wet cycling.

2.3. Test Methods

Static pull-out tests were carried out using an electronic universal testing machine, under displacement-controlled loading at a rate of 0.3 mm/min. The fatigue upper and lower loads were calculated based on the ultimate loads obtained from static pull-out tests for each group: standard environment, seawater immersion, and dry–wet cycling. Bond fatigue tests were conducted using the SDS-50 Electro-Hydraulic Servo Fatigue Testing Machine. Related studies indicate that when stress levels remain below 75% of the static ultimate strength, variations in loading frequency between 1 Hz and 15 Hz have negligible effects on the material fatigue strength. A loading frequency of 4 Hz was chosen for this study, and the loading schematic is shown in Figure 5.

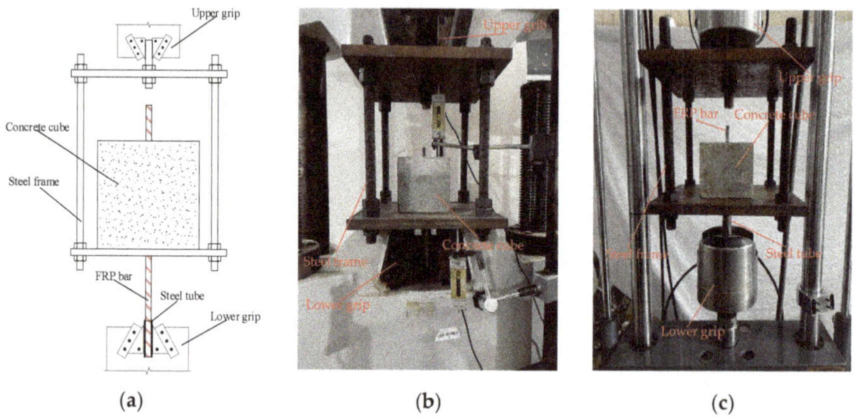

Figure 5. Test setup of static and fatigue pull-out test: (**a**) schematic drawing; (**b**) static loading apparatus; (**c**) fatigue loading apparatus.

3. Results

The bond segment length of the pull-out specimens in this experiment is defined as 5 d_a, indicating a relatively short embedded length. It is generally assumed that the bond stress is evenly distributed along the embedded length of the rebar [16]. The bond strength, denoted by τ, is calculated using Equation (1):

$$\tau = \frac{P}{(\pi d_a l_a)} \tag{1}$$

In Equation (1), τ represents the bond stress, P is the ultimate pull-out load, d_a denotes the CFRP bar diameter, and l_a is the bond length.

3.1. Static Pull-Out Test Results

The static pull-out tests revealed two failure modes: reinforcement pull-out failure and concrete splitting failure (see Figure 6). The test results are presented in Table 6. These findings align with prior studies on the bond behavior at the FRP bar–concrete interface [31–34]. Figure 7 shows the static bond–slip curves for the three specimen groups. The bond–slip curve of CFRP bars embedded in seawater–sea sand concrete during pull-out failure consists of four stages: initial micro-slip, progressive slip, descending phase, and residual phase. In the micro-slip stage, despite minimal relative slip, the bond stress rises rapidly. At this stage, pull-out resistance mainly stems from the strong adhesive forces between the CFRP bars and seawater–sea sand concrete. In the progressive slip stage, as the pull-out force increases, the slip magnitude grows rapidly, producing a nonlinear ascending trend in the bond–slip curve. The pull-out resistance in this stage is mainly attributed to mechanical interlocking and frictional forces between the transverse ribs of CFRP bars and seawater–sea sand concrete. In the pull-out phase, the bond at the CFRP bar–concrete interface is fully disrupted, causing the complete extraction of the CFRP bars. The curve slope sharply decreases and gradually flattens near peak stress. In the residual phase, after reaching peak bond stress, a rapid decline occurs, followed by stabilization at a residual stress level. This residual stress is marked by a secondary peak in the curve. At this stage, bond force mainly arises from additional resistance due to CFRP bar ribs embedding into fractured seawater–sea sand concrete during pull-out. For concrete splitting failure, the bond–slip curve of CFRP bars in seawater–sea sand concrete shows only the first three stages of the pull-out failure curve and lacks a distinct residual phase. After splitting failure, although some bond forces remain between the CFRP bars and the seawater–sea sand concrete, they sharply decrease to less than 20%, indicating severe bond degradation.

S-P-3 S-J120-1 S-J120-3

(**a**)

S-P-1 S-G120-2 S-G120-3

(**b**)

Figure 6. Failure diagram of static pull-out specimens: (**a**) concrete splitting failure; (**b**) reinforcement pull-out failure.

Table 6. Test results of static pull-out specimens.

Specimen ID	Ultimate Load (kN)	Average Ultimate Load (kN)	Bond Strength (MPa)	Average Bond Strength (MPa)	Failure Mode
S-P-1	30.52	28.60 ± 1.82	19.43	18.21 ± 1.16	RP (Reinforcement Pull-Out)
S-P-2	28.39		18.07		RP
S-P-3	26.89		17.12		CS (Concrete Splitting)
S-J120-1	33.40	32.49 ± 2.08	21.26	20.68 ± 1.32	CS
S-J120-2	30.11		19.17		CS
S-J120-3	33.96		21.62		CS
S-G120-1	28.81	28.99 ± 1.63	18.34	18.46 ± 1.04	RP
S-G120-2	27.47		17.49		RP
S-G120-3	30.71		19.55		RP

Notes: "RP" refers to reinforcement pull-out failure; "CS" refers to concrete splitting failure.

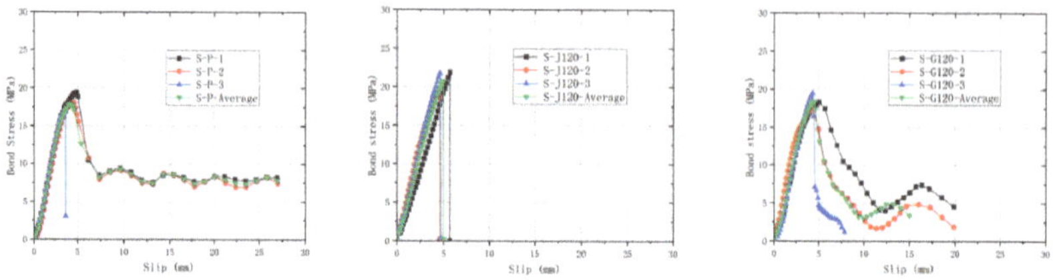

(**a**) (**b**) (**c**)

Figure 7. Bond–slip curves for static pull-out specimens: (**a**) S-P; (**b**) S-J120; (**c**) S-G120.

3.2. Fatigue Pull-Out Test Results

The fatigue failure modes of the CFRP bar and seawater–sea sand concrete specimens are categorized as no failure after 2 million fatigue cycles, pull-out failure, and splitting failure (see Figure 8). The fatigue test results of CFRP bar and seawater–sea sand concrete specimens are summarized in Table 7. Under a 50% stress level, all three groups of the CFRP bar and seawater–sea sand concrete specimens successfully endured 2 million fatigue cycles with no failure. The surface conditions remained intact, with only minor debris observed at the interface from repeated wear. After completing 2 million cycles, fatigue loading ceased, and static failure tests were conducted. The failure modes are shown in Figure 8a. At a 60% stress level, CFRP bar and seawater–sea sand concrete specimens in a standard environment achieved the maximum of 2 million fatigue cycles without failure. However, specimens under seawater immersion and dry–wet cycling conditions exhibited significantly reduced fatigue cycles. After 120 days of dry–wet cycling, the fatigue life dropped to 198,999 cycles, with pull-out failure as the predominant mode (see Figure 8b). At a 70% stress level, fatigue life declined sharply, and none of the three groups reached 2 million cycles. Specimens exposed to 120 days of dry–wet cycling showed the lowest fatigue life, falling below 10,000 cycles. In a standard environment, fatigue life was limited to 58,783 cycles, with an average of significantly less than 50,000 cycles. Specimens exposed to dry–wet cycling and seawater immersion exhibited splitting failure (Figure 8c), with significant cracks forming on the eccentric side, which caused the testing machine to halt fatigue loading.

(a) (b) (c)

Figure 8. Fatigue failure mode: (**a**) static failure after fatigue loading; (**b**) reinforcement pull-out failure; (**c**) concrete splitting failure.

Table 7. Test results of fatigue pull-out specimens.

Specimen ID	Stress Level	Fatigue Upper Load (kN)	Fatigue Lower Limit (kN)	Number of Fatigue Cycles	Failure Mode
F0.5-P	50%	14.30	2.86	2 million	SA (Static Failure)
F0.6-P	60%	17.16	3.43	2 million	SA

Table 7. *Cont.*

Specimen ID	Stress Level	Fatigue Upper Load (kN)	Fatigue Lower Limit (kN)	Number of Fatigue Cycles	Failure Mode
F0.7-P	70%	20.02	4.00	58,783	RP (Pull-Out)
F0.5-J120	50%	16.25	3.25	2 million	SA
F0.6-J120	60%	19.50	3.90	684,521	RP
F0.7-J120	70%	22.74	4.55	22,482	CS (Splitting)
F0.5-G120	50%	14.50	2.90	2 million	SA
F0.6-G120	60%	17.40	3.48	198,999	RP
F0.7-G120	70%	20.29	4.06	7234	CS

Notes: "SA" represents static failure after fatigue loading; "RP" represents reinforcement pull-out failure; "CS" represents concrete splitting failure.

3.3. Fatigue Bond–Slip Curves

Figure 9 shows the fatigue bond–slip curves of CFRP bar and seawater–sea sand concrete specimens under stress levels of 50%, 60%, and 70%, corresponding to the standard environment, seawater immersion, and dry–wet cycling conditions. The hysteresis loop area increases gradually, and the curve shifts to the right as stress levels and fatigue cycles increase. This suggests that irreversible residual slip occurs between CFRP bars and seawater–sea sand concrete under fatigue loading, with the slip magnitude increasing with higher stress levels and more load cycles.

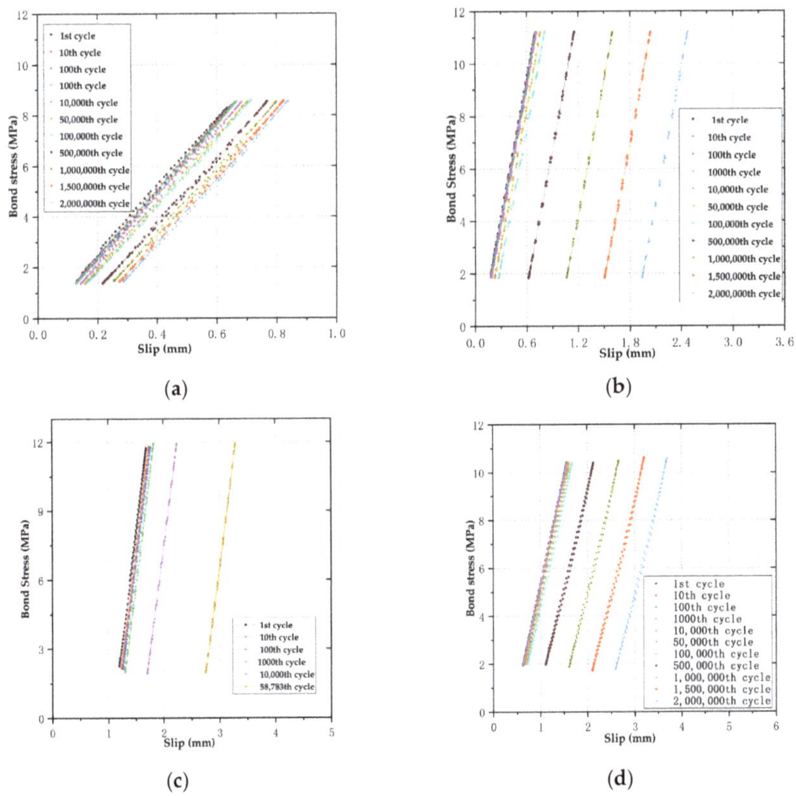

Figure 9. *Cont.*

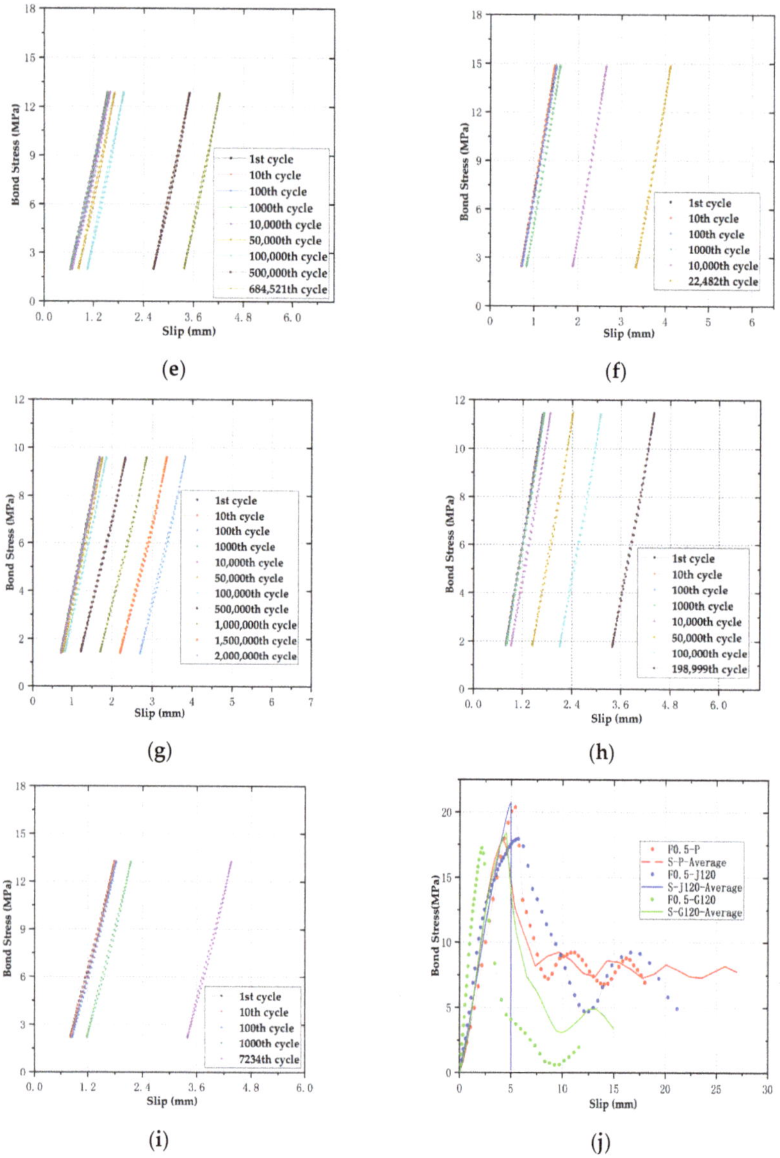

Figure 9. Fatigue bond–slip curves: (**a**) F0.5-P; (**b**) F0.6-P; (**c**) F0.7-P; (**d**) F0.5-J120; (**e**) F0.6-J120; (**f**) F0.7-J120; (**g**) F0.5-G120; (**h**) F0.6-G120; (**i**) F0.7-G120; (**j**) bond–slip curves for static loading after fatigue and direct static loading.

Under a 50% fatigue stress level, none of the CFRP bar and seawater–sea sand concrete specimens in any condition (standard environment, seawater immersion, or dry–wet cycling) failed after 2 million fatigue cycles. Subsequent monotonic pull-out tests under static loading reveal pull-out failure of the CFRP bars, exhibiting failure modes identical to those observed in specimens subjected to direct static loading. Figure 9j illustrates a comparative bond–slip curve for static loading after fatigue and direct static loading. At fatigue stress levels of 60% and 70%, specimens in seawater immersion and dry–wet cycling conditions failed during fatigue loading, exhibiting pull-out and splitting failures, while those in the standard environment did not fail. Comparative analysis shows that higher stress levels cause greater increases in slip and residual slip between CFRP bars and seawater–sea sand concrete during the initial phase. This suggests that fatigue-induced cumulative damage at higher stress levels significantly affects bond performance between CFRP bars and seawater–sea sand concrete. Additionally, the slope of the fatigue bond–slip curve (bond stiffness) slightly decreases with more fatigue cycles but increases with higher stress levels.

After fatigue loading, the bond–slip curve trends of specimens under standard conditions and after 120 days of dry–wet cycling were consistent with pre-fatigue trends; however, those after 120 days of seawater immersion showed significant differences. The consistent curves followed the previously described four stages. Moreover, stiffness in the micro-slip stage degraded by an average of 12.31% compared to pre-fatigue conditions. After 2 million fatigue cycles, the bond strength between CFRP bars and seawater–sea sand concrete decreased by an average of 10.31%, highlighting the significant degradation effect of fatigue loading on bond strength. This degradation is mainly caused by surface wear on the CFRP bars due to fatigue loading and the formation of microcracks in the seawater–sea sand concrete between the ribs. The tips of these cracks created stress concentration zones at the interface, which were exacerbated by fatigue cycles and led to progressive crack growth [36].

A comparison of fatigue bond–slip curves showed that dry–wet cycling caused more severe damage to CFRP bars and seawater–sea sand concrete specimens than seawater immersion. This is primarily attributed to the accelerated convection and diffusion of chloride ions during wet-dry cycles, resulting in more pronounced damage. After 2 million fatigue cycles, slip caused by fatigue stress under dry–wet cycling conditions exceeded that in standard environments. The hysteresis loop area of the fatigue bond–slip curves increased progressively from standard conditions to seawater immersion and then to dry–wet cycling. This was mainly due to corrosive damage from seawater immersion and dry–wet cycling at the bonding interfaces, resulting in slip phase failure. Compared to standard conditions, the fatigue bond–slip curves under corrosive environments shifted horizontally to the right, resulting in greater slip at the same stress levels. The hysteresis loop of fatigue bond–slip curves under dry–wet cycling conditions exhibited the most significant rightward shift compared to standard conditions.

3.4. Fatigue Bond Stiffness

The fatigue bond stiffness curves in Figure 10 show a clear decreasing trend in stiffness between CFRP bars and seawater–sea sand concrete as fatigue cycles increase. At a 50% stress level, after 2 million fatigue cycles, the bond stiffness under normal conditions decreased from 13.51 MPa/mm to 11.95 MPa/mm, representing an 11.55% degradation. In corrosive environments, bond stiffness degraded by 11.80% after 120 days of seawater immersion and 16.04% after 120 days of dry–wet cycling.

At a 60% stress level, the bond stiffness under normal conditions degraded from 15.20 MPa/mm to 13.60 MPa/mm until failure, representing a 10.53% reduction in stiffness. In corrosive environments, the degradation rates were 14.81% after 120 days of seawater immersion and 10.29% after 120 days of dry–wet cycling.

At a 70% stress level, the bond stiffness under normal conditions degraded from 21.28 MPa/mm to 18.68 MPa/mm until failure, showing a 12.22% reduction in stiffness. The degradation rates in other conditions were lower, but they were still significant.

A comparative analysis shows that higher stress levels and corrosive environments intensify damage between CFRP bars and seawater–sea sand concrete, accelerating interfacial crack propagation. This crack propagation leads directly to the rapid degradation of interfacial bond stiffness. This confirms the detrimental impact of high stress levels on bond fatigue performance.

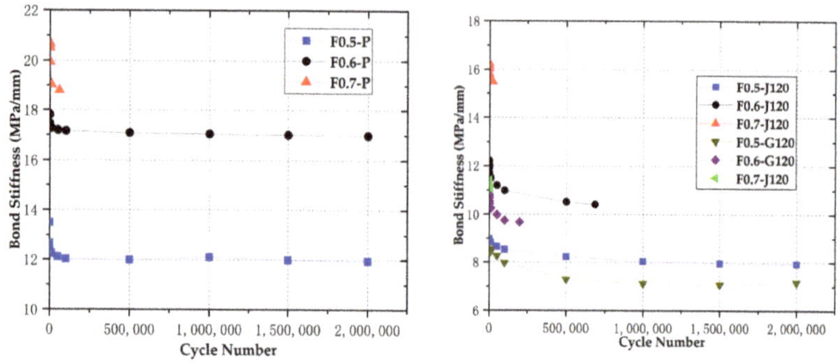

(**a**) (**b**)

Figure 10. Relationship between fatigue bond stiffness and fatigue times: (**a**) standard environment; (**b**) corrosive environment.

3.5. Residual Slip

The slip and residual slip curves in Figure 11 show progressive accumulation as fatigue cycles increase. During the first 100,000 fatigue cycles, slip and residual slip increase rapidly. However, their growth rate slows in later cycles. Thus, the evolution of slip and residual slip during fatigue cycles can be divided into two phases. The first phase involves crack initiation and rapid propagation at the interface between CFRP bars and seawater–sea sand concrete, primarily occurring during the first 100,000 fatigue cycles. Fatigue damage accumulation during this phase is significantly greater than that in the second phase. In the second phase, starting around the 100,000th fatigue cycle, crack propagation stabilizes, and the growth of slip and residual slip slows and eventually stabilizes. The slip growth trend closely matches that of residual slip, indicating that residual slip accumulation primarily drives total slip increases, reflecting fatigue-induced interfacial bond damage.

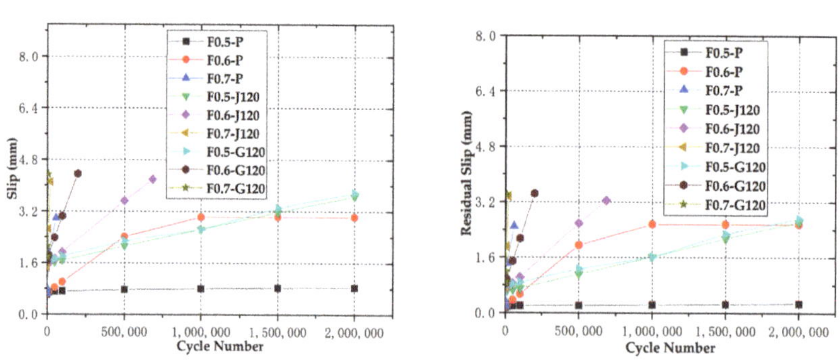

(**a**) (**b**)

Figure 11. Relationship between slip/residual slip and the number of cycles: (**a**) slip; (**b**) residual slip.

Figure 11 shows that, under standard conditions, higher stress levels significantly worsen bond performance degradation between CFRP bars and seawater–sea sand concrete. At a 50% stress level, the slip increased by 0.23 mm from the initial loading to 2 million cycles. For 60% and 70% stress levels, slip increased by 2.41 mm and 2.35 mm, respectively. After 120 days of seawater immersion, the slip increase trend became more pronounced. At a 50% stress level, slip increased by 1.6 mm, while for 60% and 70% stress levels, it increased by 2.12 mm in both. Similarly, after 120 days of dry–wet cycling, the slip increase trend remained significant. At a 50% stress level, slip increased by 1.59 mm, and for 60% and 70% stress levels, slip increased by 2.37 mm and 2.23 mm, respectively.

Regardless of the condition—standard, seawater immersion, or dry–wet cycling—high-stress fatigue loading significantly reduces bond performance, as evidenced by large increases in slip and residual slip.

4. Analysis

4.1. Influence of Bonding Fatigue Performance

4.1.1. Corrosive Environment

Figure 12 illustrates the influence of corrosive environments on the fatigue bond–slip behavior between CFRP bars and seawater–sea sand concrete. As shown in Figure 12, all specimens at a 50% stress level endured up to 2 million fatigue cycles. The fatigue bond performance trend remains consistent, regardless of whether seawater immersion or dry–wet cycling. As fatigue cycles increase, slip and residual slip follow the previously described pattern of rapid initial growth followed by gradual deceleration, accompanied by a decline in bond stiffness. Higher fatigue stress levels result in greater slip and residual slip, more significant bond stiffness reductions, and accelerated bond performance degradation caused by fatigue loading. This observation aligns with Zheng et al. [37], who studied the fatigue bond performance of FRP and concrete in humid and thermal environments. Under seawater immersion and dry–wet cycling conditions, CFRP bars and seawater–sea sand concrete experience some degradation, but bond strength variations remain within 5%. However, these conditions significantly affect fatigue loading cycles and reduce the service life of fatigue-prone structural components.

4.1.2. Stress Level

Figure 13 shows the effect of stress levels on fatigue bond performance between CFRP bars and seawater–sea sand concrete. As stress levels increase, the slope of the fatigue bond–slip curve (bond stiffness) exhibits a significant upward trend. Simultaneously, slip increases consistently with rising stress levels. At 2 million fatigue cycles, the slip in the fatigue bond–slip curve is minimal, and the hysteresis loop is at its smallest. However, as stress levels increase further, the slip increases significantly, and the hysteresis loop expands accordingly.

4.2. Bond Fatigue Test Results Characteristics

4.2.1. Bond Stiffness

Figure 14 shows the effects of various factors on bond stiffness between CFRP bars and seawater–sea sand concrete. Under prolonged seawater immersion and dry–wet cycling, bond stiffness exhibits a significant downward trend in both scenarios. At a 50% stress level, 120 days of seawater immersion reduced the bond stiffness by 33.53%. In contrast, dry–wet cycling caused more severe bond deterioration, resulting in a 36.79% reduction in stiffness. At a 60% stress level, seawater immersion reduced bond stiffness by 31.39%, while dry–wet cycling caused a 39.42% reduction. At a 70% stress level, seawater immersion reduced bond stiffness by 21.51%, while dry–wet cycling caused a significantly greater reduction of 44.80%. This trend is mainly due to the corrosive effects of seawater immersion and dry–wet cycling, which weaken the bond strength between CFRP bars and seawater–sea sand concrete. These effects increase displacement during the initial slip phase, thus reducing bond stiffness during fatigue bond–slip cycles.

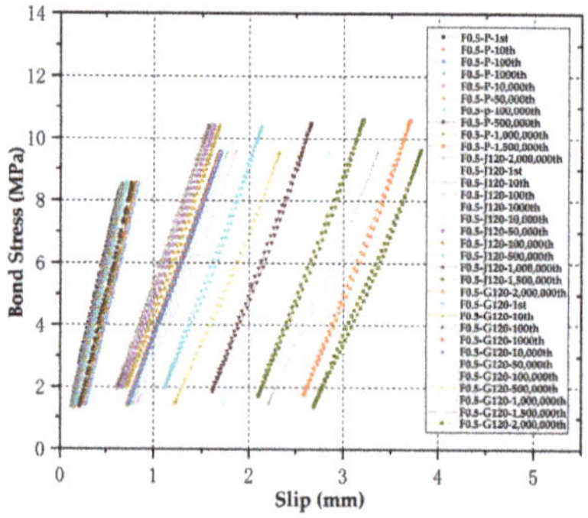

(**a**)

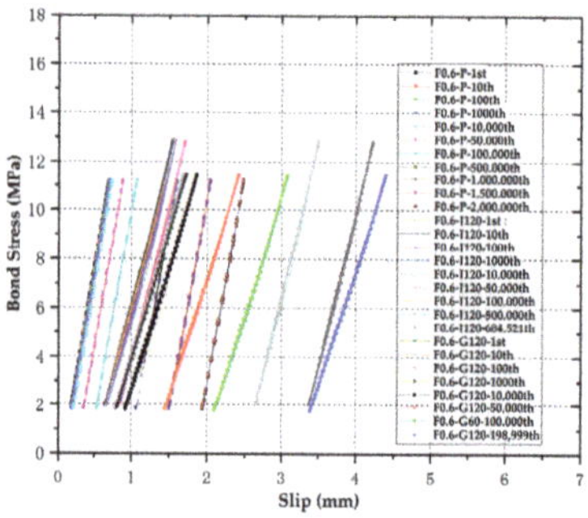

(**b**)

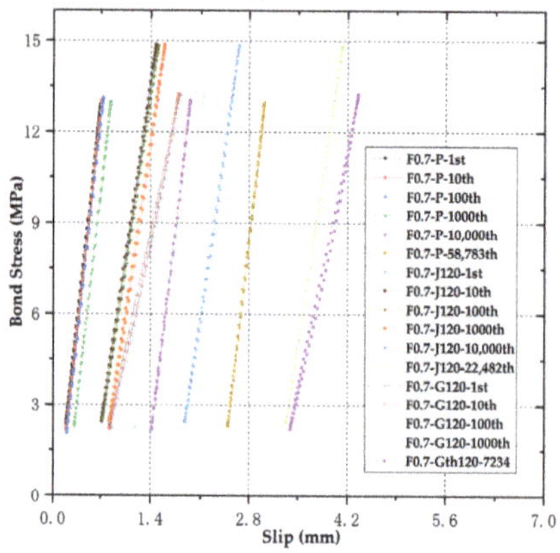

(**c**)

Figure 12. Effect of corrosive environment on fatigue bond–slip curve: (**a**) 50% fatigue stress level; (**b**) 60% fatigue stress level; (**c**) 70% fatigue stress level.

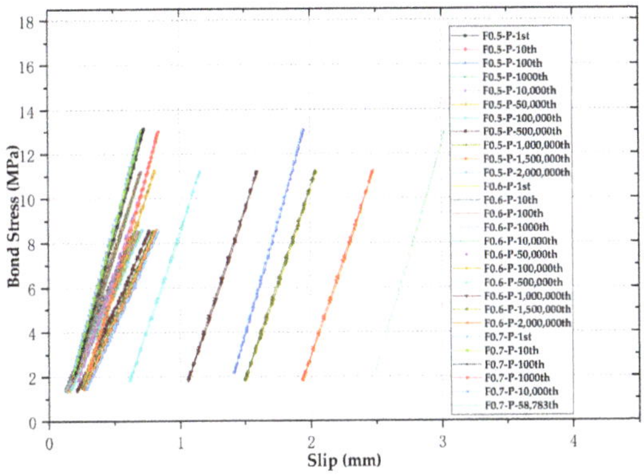

(**a**)

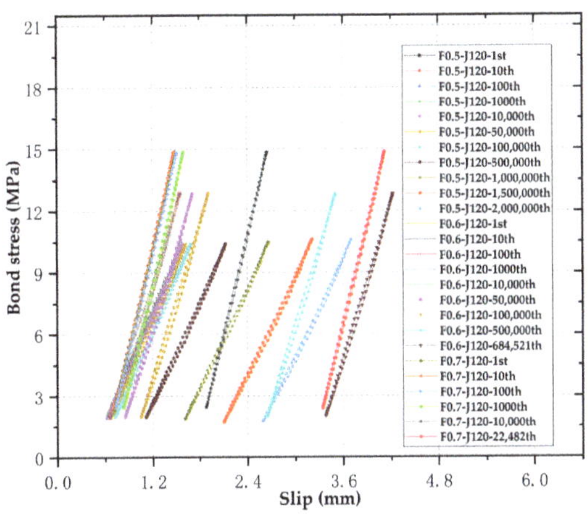

(**b**)

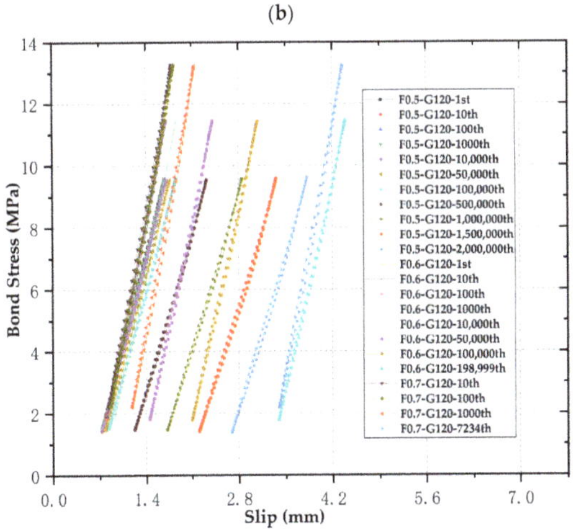

(**c**)

Figure 13. Effect of stress level on fatigue bond–slip curve: (**a**) standard environment; (**b**) seawater immersion environment; (**c**) dry–wet cycling environment.

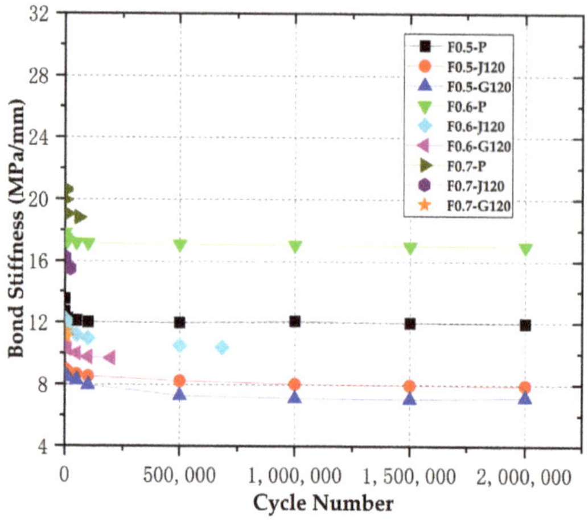

Figure 14. Bond stiffness under different factors.

Figure 14 also shows that under identical conditions, the bond stiffness increases notably with rising stress levels. In standard environments, bond stiffness increased by 31.83% and 53.15% at 60% and 70% stress levels, respectively, compared to 50%. Similarly, after 120 days of seawater immersion and dry–wet cycling, bond stiffness at higher stress levels improved by 36%, 81.85%, 26.35%, and 33.72%, respectively, compared to 50%. This is mainly because higher stress levels enhance defect mitigation at the bond interface, allowing the bond–slip process to stabilize more quickly.

4.2.2. Residual Slip

Figure 11 shows the effects of various factors on slip and residual slip of CFRP bars in seawater–sea sand concrete. As fatigue cycles increase, the growth trends of slip and residual slip exhibit similar patterns under seawater immersion and dry–wet cycling, both before and after corrosion. Notably, dry–wet cycling exhibits a more pronounced effect on slip and residual slip than seawater immersion.

At a 50% stress level, residual slip and slip increase in corrosive environments compared to standard conditions. After 2 million fatigue cycles in seawater immersion, the slip increased by 2.11 mm and residual slip by 1.99 mm. In dry–wet cycling conditions, residual slip increased by 1.59 mm and slip by 2.14 mm. This pattern persists at higher stress levels. At 60% stress level, seawater immersion increased residual slip and slip by 2.13 mm and 2.69 mm, whereas dry–wet cycling increased them by 2.37 mm and 2.71 mm, respectively. At 70% stress level, seawater immersion increased residual slip and slip by 2.11 mm and 2.65 mm, while dry–wet cycling increased them by 2.23 mm and 2.58 mm, respectively. Figure 15 also shows that both slip and residual slip increase as stress levels rise. In standard environments, slip increased by 2.15 mm and 2.19 mm at 60% and 70% stress levels, while residual slip increased by 2.22 mm and 2.32 mm. In seawater immersion, the slip increased by 0.53 mm and residual slip by 0.74 mm compared to 50% stress levels. In dry–wet cycling, slip increased by 0.58 mm and residual slip by 0.72 mm compared to 50% stress levels. The effects of 60% and 70% stress levels on slip and residual slip are relatively similar. This is mainly because higher stress levels lead to increased slip and residual slip compared to lower levels. However, excessively high stress levels make the CFRP–seawater–sea sand concrete interface prone to fatigue failure, thereby limiting slip and residual slip and reducing differences in their impacts across stress levels.

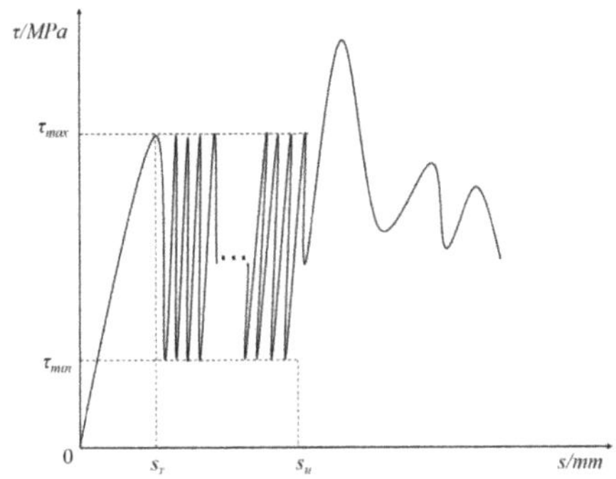

Figure 15. Fatigue bond–slip model of CFRP bars and seawater–sea sand concrete.

The analysis reveals that corrosive environments profoundly influence the bond performance between CFRP bars and seawater—sea sand concrete, with dry-wet cycles causing a more pronounced reduction in specimen fatigue life compared to seawater immersion. With the progressive increase in stress levels, the slope of the fatigue bond-slip curve (representing bond stiffness) and the slip magnitude show a marked upward trend, accompanied by a substantial reduction in fatigue life.

5. Bond–Slip Constitutive Model

Extensive research has been conducted worldwide on the bond fatigue performance between steel bars and concrete [38–40], resulting in significant experimental data and the development of fatigue bond–slip constitutive models. However, compared to steel bars and concrete, research on the bond fatigue performance of FRP bars and concrete remains relatively limited, especially regarding fatigue bond–slip constitutive models. To address this, this study uses existing experimental data and fatigue bond-slip constitutive models for steel bars and concrete to develop a bond fatigue–slip constitutive model for CFRP bars and seawater–sea sand concrete, as shown in Figure 15. The fitted model is described by the following equations, as shown below:

$$\tau = \begin{cases} 0 & s \le s_r(n) \\ \tau_{max}\left(\frac{s-s_r(n)}{s_u(n)-s_r(n)}\right)^{\alpha_n(n)} & s_r(n) < s \le s_u(n) \\ \tau_{max}\left(\frac{s-s_r(n)}{s_u(n)-s_r(n)}\right)^{\beta} & s > s_u(n) \end{cases} \tag{2}$$

$$\tau_{max} = 2f_c^{0.6} \tag{3}$$

$$\alpha_n(n) = \frac{0.24}{n^{(a\tau_{max}/\tau_u)}}(bk_c - 1) \tag{4}$$

$$\beta = -0.6(c/d)^{-0.24} \tag{5}$$

In the above equation, τ_u denotes the fatigue bond strength, τ_{max} indicates the peak bond strength, f_c refers to the compressive strength of seawater–sea sand concrete, $s_u(n)$ signifies the total peak slip after the n-th fatigue cycle under static conditions, and $s_r(n)$ represents the residual slip after the n-th fatigue cycle. Parameters $\alpha_n(n)$ and β are determined using Equations (4) and (5), where n represents the number of fatigue cycles, k_c denotes the corrosion parameter of CFRP bars, c/d indicates the ratio of seawater–sea sand concrete cover thickness to CFRP bar diameter, and a and b are correction coefficients derived from experimental data fitting.

This constitutive model integrates corrosion parameters k_c to assess the impact of seawater immersion and dry–wet cycles on the bond fatigue performance at the interface between CFRP bars and seawater sea-sand concrete. These parameters are inevitably affected by corrosive conditions, including seawater immersion and dry–wet cycles. The degradation of bond fatigue performance, however, arises from the combined effects of physical and chemical factors, such as solution composition, temperature, protective layer thickness, and corrosion duration. Adequate consideration should be given to these factors when utilizing this model.

6. Conclusions

Seawater immersion and dry–wet cycles significantly influence the bonding performance between CFRP bars and seawater–sea sand concrete. The main conclusions derived from this experimental study are as follows:

(1) The performance of CFRP bars and seawater–sea sand concrete degrades under seawater immersion and dry–wet cycling, but bond strength variation is minimal, remaining below 5%. Seawater has a limited effect on fatigue bond strength; however, fatigue loading is the primary factor. Higher stress levels reduce bond stiffness but increase slip and hysteresis loop areas.

(2) After fatigue loading, the bond–slip curve trend is consistent in standard environments and under dry–wet cycling but differs under seawater immersion. The curve shows a four-stage variation, with the micro-slip stiffness degrading by 12.31% in standard environments. After 2 million fatigue cycles, the bond strength decreases by 10.31% due to CFRP bar surface abrasion and microcracks in seawater–sea sand concrete. Stress concentration at crack tips and cumulative fatigue cycles accelerate crack propagation.

(3) Fatigue cycles significantly reduce bond stiffness between CFRP bars and seawater–sea sand concrete. The stiffness degradation rates are 11.55%, 10.53%, and 12.22% at 50%, 60%, and 70% stress levels in standard environments; 11.80% and 14.81% under prolonged immersion conditions; and 16.04% and 10.29% under dry–wet cycling conditions. Higher stress levels intensify damage and crack propagation. The slip increments at 50%, 60%, and 70% stress levels are 0.23 mm, 2.41 mm, and 2.35 mm, respectively, in standard environments. Under prolonged immersion and dry–wet cycling conditions, high-stress fatigue loading severely reduces bond performance, leading to significant increases in slip and residual slip.

(4) Seawater immersion and dry–wet cycling significantly reduce bond stiffness due to corrosion impairing bond strength and increasing initial slip displacement. Under similar conditions, higher stress levels improve bond stiffness. For specimens subjected to 120 days of immersion and dry–wet cycling, bond stiffness increases are more pronounced under high stress levels. As fatigue cycles increase, the effects of immersion and dry–wet cycling on slip and residual slip are similar; dry–wet cycling, however, has a greater impact. High stress levels increase slip and residual slip, but excessive stress levels may cause failure and reduce differences between stress levels.

(5) A fatigue bond–slip constitutive model for CFRP bars and seawater–sea sand concrete was developed based on experimental data and existing models for steel bars and concrete. This model provides a theoretical foundation for engineering professionals in the design and practical implementation of such structural systems.

Author Contributions: Conceptualization, F.Y. and J.W.; methodology, W.G.; software, S.Y.; validation, W.G., B.C. and S.Y.; formal analysis, B.C.; investigation, W.G.; resources, S.Y.; data curation, S.Y.; writing—original draft preparation, S.Y.; writing—review and editing, J.W.; visualization, F.Y.; supervision, F.Y.; project administration, F.Y.; funding acquisition, F.Y. All authors have read and agreed to the published version of the manuscript.

Funding: This research was funded by the Jiangsu Coastal Power Infrastructure Intelligent Engineering Research Center Technology Project and State Grid Jiangsu Electric Power Co., Ltd., Nanjing, China, Province Managed Industry 2024 Science and Technology Project, grant number JC2024089.

Data Availability Statement: The original data will be made available upon request.

Conflicts of Interest: Authors Fu Yang, Weiquan Gan, Bingsan Chen were employed by the company Lianyungang Zhiyuan Electric Power Design Co., Ltd., Shenzhen, China. The remaining authors declare that the research was conducted in the absence of any commercial or financial relationships that could be construed as a potential conflict of interest. The authors declare that they have no known competing financial interests or personal relationships that could have appeared to influence the work reported in this paper. The funder had no role in the design of the study; in the collection, analysis, or interpretation of data, in the writing of the manuscript, or in the decision to publish the results.

References

1. Triantafyllou, G.G.; Rousakis, T.C.; Karabinis, A.I. Effect of Patch Repair and Strengthening with EBR and NSM CFRP Laminates for RC Beams with Low, Medium and Heavy Corrosion. *Compos. Part B Eng.* **2018**, *133*, 101–111. [CrossRef]
2. Wang, D.; Gong, Q.; Yuan, Q.; Luo, S. Review of the Properties of Fiber-Reinforced Polymer-Reinforced Seawater–Sea Sand Concrete. *J. Mater. Civ. Eng.* **2021**, *33*, 04021285. [CrossRef]
3. Ahmed, A.; Guo, S.; Zhang, Z.; Shi, C.; Zhu, D. A Review on Durability of Fiber Reinforced Polymer (FRP) Bars Reinforced Seawater Sea Sand Concrete. *Constr. Build. Mater.* **2020**, *256*, 119484. [CrossRef]
4. Başaran, B.; Kalkan, İ.; Beycioğlu, A.; Kasprzyk, I. A Review on the Physical Parameters Affecting the Bond Behavior of FRP Bars Embedded in Concrete. *Polymers* **2022**, *14*, 1796. [CrossRef] [PubMed]
5. Raman, R.; Guo, F.; Al-Saadi, S.; Zhao, X.-L.; Jones, R. Understanding Fibre-Matrix Degradation of FRP Composites for Advanced Civil Engineering Applications: An Overview. *Corros. Mater. Degrad.* **2018**, *1*, 27–41. [CrossRef]
6. Nepomuceno, E.; Sena-Cruz, J.; Correia, L.; D'Antino, T. Review on the Bond Behavior and Durability of FRP Bars to Concrete. *Constr. Build. Mater.* **2021**, *287*, 123042. [CrossRef]
7. Zhao, Y.; Hu, X.; Shi, C.; Zhang, Z.; Zhu, D. A Review on Seawater Sea-Sand Concrete: Mixture Proportion, Hydration, Microstructure and Properties. *Constr. Build. Mater.* **2021**, *295*, 123602. [CrossRef]
8. Ge, L.; Feng, Z.; Sayed, U.; Li, H. Research on the Performance of Seawater Sea-Sand Concrete: A Review. *Constr. Build. Mater.* **2023**, *409*, 133921. [CrossRef]
9. Li, T.; Liu, X.; Zhang, Y.; Yang, H.; Zhi, Z.; Liu, L.; Ma, W.; Shah, S.P.; Li, W. Preparation of Sea Water Sea Sand High Performance Concrete (SHPC) and Serving Performance Study in Marine Environment. *Constr. Build. Mater.* **2020**, *254*, 119114. [CrossRef]
10. Xiao, J.; Qiang, C.; Nanni, A.; Zhang, K. Use of Sea-Sand and Seawater in Concrete Construction: Current Status and Future Opportunities. *Constr. Build. Mater.* **2017**, *155*, 1101–1111. [CrossRef]
11. Sun, Z.; Li, X.; Liu, Q.; Tang, Q.; Lin, X.; Fan, X.; Huang, X.; Gan, M.; Chen, X.; Ji, Z. Recent Advances in Alkali-Activated Materials with Seawater and Sea Sand. *Materials* **2023**, *16*, 3571. [CrossRef] [PubMed]
12. Tănase, M.; Diniță, A.; Popovici, D.R.; Portoacă, A.I.; Călin, C.; Sirbu, E.-E. Comprehensive Bibliometric Review on the Sustainability and Environmental Impact of Fiber-Reinforced Polymers. *Fibers* **2024**, *12*, 104. [CrossRef]
13. Wang, T.; Razaqpur, A.G.; Chen, S. Durability of GFRP and CFRP Bars in the Pore Solution of Calcium Sulfoaluminate Cement Concrete Made with Fresh or Seawater. *Polymers* **2023**, *15*, 3306. [CrossRef]
14. Hu, X.; Peng, G.; Niu, D.; Wang, J. Experimental Study on Bond Properties between Early-Age Concrete and Deformed Steel Bars. *Constr. Build. Mater.* **2020**, *236*, 117593. [CrossRef]
15. Xiong, Z.; Wei, W.; He, S.; Liu, F.; Luo, H.; Li, L. Dynamic Bond Behaviour of Fibre-Wrapped Basalt Fibre-Reinforced Polymer Bars Embedded in Sea Sand and Recycled Aggregate Concrete under High-Strain Rate Pull-out Tests. *Constr. Build. Mater.* **2021**, *276*, 122195. [CrossRef]
16. Sun, J.; Ding, Z.; Li, X.; Wang, Z. Bond Behavior between BFRP Bar and Basalt Fiber Reinforced Seawater Sea-Sand Recycled Aggregate Concrete. *Constr. Build. Mater.* **2021**, *285*, 122951. [CrossRef]

17. Azarkerdar, A.; Hejazi, F. Development of Analytical Model for Bonding of CFRP Rod in Concrete Subjected to Cyclic Loads. *Appl. Sci.* **2024**, *14*, 1134. [CrossRef]
18. Xian, G.; Guo, R.; Li, C.; Wang, Y. Mechanical Performance Evolution and Life Prediction of Prestressed CFRP Plate Exposed to Hygrothermal and Freeze-Thaw Environments. *Compos. Struct.* **2022**, *293*, 115719. [CrossRef]
19. Xian, G.; Zhou, P.; Li, C.; Dong, S.; Du, H.; Tian, J.; Guo, R.; Peng, Z.; Zhang, Z.; He, T. Mechanical Properties Evaluation of Glass Fiber Reinforced Thermoplastic Composite Plate under Combined Bending Loading and Water Immersion. *Constr. Build. Mater.* **2024**, *440*, 137470. [CrossRef]
20. Bae, Y.-J.; Sim, V.; Lee, S.-M.; Jung, W.-Y. Fatigue Performance Criteria for CFRP Reinforcements Based on RC Behavior in Civil Construction. *J. Asian Archit. Build. Eng.* **2024**, *ahead of print*. 1–15. [CrossRef]
21. Zou, R.; Liu, F.; Xiong, Z.; He, S.; Li, L.; Wei, W. Experimental Study on Fatigue Bond Behaviour between Basalt Fibre-Reinforced Polymer Bars and Recycled Aggregate Concrete. *Constr. Build. Mater.* **2021**, *270*, 121399. [CrossRef]
22. Shen, D.; Wen, C.; Zhu, P.; Li, M.; Ojha, B.; Li, C. Bond Behavior between Basalt Fiber-Reinforced Polymer Bars and Concrete under Cyclic Loading. *Constr. Build. Mater.* **2020**, *258*, 119518. [CrossRef]
23. Lee, J.Y.; Kim, K.H.; Kim, S.W.; Chang, M. Strength Degradation of Glass Fiber Reinforced Polymer Bars Subjected to Reversed Cyclic Load. *Strength Mater.* **2014**, *46*, 235–240. [CrossRef]
24. Mai, G.; Li, L.; Lin, J.; Wei, W.; He, S.; Zhong, R.; Xiong, Z. Bond Durability between BFRP Bars and Recycled Aggregate Seawater Sea-Sand Concrete in Freezing-Thawing Environment. *J. Build. Eng.* **2023**, *70*, 106422. [CrossRef]
25. Xiong, Z.; Mai, G.; Qiao, S.; He, S.; Zhang, B.; Wang, H.; Zhou, K.; Li, L. Fatigue Bond Behaviour between Basalt Fibre-Reinforced Polymer Bars and Seawater Sea-Sand Concrete. *Ocean Coast. Manag.* **2022**, *218*, 106038. [CrossRef]
26. Xiong, Z.; Zeng, Y.; Li, L.G.; Kwan, A.K.H.; He, S.H. Experimental Study on the Effects of Glass Fibres and Expansive Agent on the Bond Behaviour of Glass/Basalt FRP Bars in Seawater Sea-Sand Concrete. *Constr. Build. Mater.* **2021**, *274*, 122100. [CrossRef]
27. Liao, J.; Zeng, J.-J.; Bai, Y.-L.; Zhang, L. Bond Strength of GFRP Bars to High Strength and Ultra-High Strength Fiber Reinforced Seawater Sea-Sand Concrete (SSC). *Compos. Struct.* **2022**, *281*, 115013. [CrossRef]
28. He, S.; Li, L.; Lin, J.; Chen, X.; Weng, C.; Xiao, H.; Liu, P.; Qiao, S.; Xiong, Z. Bond Performance between Ribbed BFRP Bar and Seawater Sea-Sand Concrete: Influences of Rib Geometry. *Structures* **2024**, *65*, 106660. [CrossRef]
29. Wang, H.; Belarbi, A. Static and Fatigue Bond Characteristics of FRP Rebars Embedded in Fiber-Reinforced Concrete. *J. Compos. Mater.* **2010**, *44*, 1605–1622. [CrossRef]
30. Pan, R.; Zou, J.; Liao, P.; Dong, S.; Deng, J. Effects of Fiber Content and Concrete Cover on the Local Bond Behavior of Helically Ribbed GFRP Bar and UHPC. *J. Build. Eng.* **2023**, *80*, 107939. [CrossRef]
31. Deng, J.; Pan, Z.; Mai, G.; Long, Y.; Kuang, B.; Zhu, J.; Guo, Q.; Liang, J.; Huang, F.; Qiao, S.; et al. Fatigue Behaviour of CFRP Bar-Reinforced Seawater Sea Sand Concrete Beams: Deformation Analysis and Prediction. *Buildings* **2023**, *13*, 2273. [CrossRef]
32. *GB/T14684-2011*; Sand for Construction, General Administration of Quality Supervision. Inspection and Quarantine: Beijing, China, 2011.
33. *JGJ 55-2011*; Design Regulations for the Mix Ratio of the Ordinary Concrete. Ministry of Housing and Urban-Rural Development of the People's Republic of China: Beijing, China, 2011.
34. *JGJ 206-2010*; Technical Code for Application of Sea Sand Concrete. Ministry of Housing and Urban-Rural Development of the People's Republic of China: Beijing, China, 2011.
35. *GB 50608-2020*; Technical Standard for Fiber Reinforced Polymer (FRP) in Construction. Ministry of Housing and Urban-Rural Development of the People's Republic of China: Beijing, China, 2020.
36. Triller, P.; Kwiecień, K.; Kwiecień, A.; Bohinc, U.; Zając, B.; Tekieli, M.; Szumera, M.; Rousakis, T.; Vanian, V.; Akyildiz, A.T.; et al. Efficiency of FRPU Strengthening of a Damaged Masonry Infill Wall under In-Plane Cyclic Shear Loading and Elevated Temperatures. *Eng. Struct.* **2024**, *317*, 118652. [CrossRef]
37. Zheng, X.H.; Huang, P.Y.; Chen, G.M.; Tan, X.M. Fatigue Behavior of FRP–Concrete Bond under Hygrothermal Environment. *Constr. Build. Mater.* **2015**, *95*, 898–909. [CrossRef]
38. Lin, H.; Zhao, Y.; Ožbolt, J.; Hans-Wolf, R. The Bond Behavior between Concrete and Corroded Steel Bar under Repeated Loading. *Eng. Struct.* **2017**, *140*, 390–405. [CrossRef]
39. Zhang, W.; Zhang, Y.; Li, H.; Gu, X. Experimental Investigation of Fatigue Bond Behavior between Deformed Steel Bar and Concrete. *Cem. Concr. Compos.* **2020**, *108*, 103515. [CrossRef]
40. Zheng, Y.; Fan, C.; Ma, J.; Wang, S. Review of Research on Bond–Slip of Reinforced Concrete Structures. *Constr. Build. Mater.* **2023**, *385*, 131437. [CrossRef]

buildings

MDPI

Article

In Situ Testing Evaluation and Numerical Simulation of CFRP-Strengthened Reinforced Concrete Two-Way Slab with Initial Defect

Yepu Sheng and Yu Gong *

School of Civil Engineering, Shandong Jianzhu University, Jinan 250101, China; shengyepu@live.com
* Correspondence: gysdjzu@163.com

Abstract: Carbon fiber-reinforced polymer (CFRP) composites, renowned for their high strength-to-weight ratio, are increasingly utilized in the strengthening of structural components. The application of CFRP for strengthening concrete components notably improves the cracking moment and substantially elevates the ultimate load-bearing capacity. This study focuses on a reinforced concrete (RC) two-way slab with an initial defect, specifically an initial deflection. To avert deformations or damages that could break structural integrity during service, the slab was strengthened by adhering CFRP to its underside. An in situ multi-stage loading test was conducted to evaluate the load-bearing capacity of the CFRP-strengthened slab, and the findings revealed that the mid-span deflection of the two-way slab incrementally reached 1.64 mm after the loading stages, with no observable signs of concrete cracking, debonding, or tearing of the CFRP-strengthened slab. The failure modes indicated a transition from concrete compression damage to CFRP anchorage stress concentrations, highlighting the effective stress distribution and load-sharing synergy provided by CFRP-strengthening. Additionally, a numerical model based on the finite element (FE) method was developed using ABAQUS to simulate the component's performance during the loading process. A comparison between the measured mid-span deflection of the strengthened slab and the numerically simulated values confirmed the high accuracy and rationality of the simulation method. Utilizing the validated numerical model, an analysis of the slab's ultimate load capacity was conducted, demonstrating that the CFRP strengthening technique effectively increased the load-bearing capacity of the initially imperfect RC two-way slab by nearly 50%.

Keywords: two-way slab; initial deflection; loading experiment; bearing capacity; CFRP

1. Introduction

Slabs serve as crucial horizontal support and connect elements in a floor system, bridging walls and columns. As structural components are primarily designed to bear and transfer vertical loads from the floor, reinforced concrete (RC) slabs must satisfy stringent criteria for load-bearing capacity, deformation, and serviceability throughout their operational lifespan [1]. The predominant bending stresses in RC slabs can intensify load effects, substantially reducing the structural capacity and durability when faced with initial defects such as deflection or cracking. These defects, which may emerge during construction or regular use, pose a significant threat to the integrity and safety of high-rise building structures.

Currently, the application and research on the strengthening of RC slabs both domestically and internationally primarily draw on the operational modes of structural components such as beams and columns. Common strengthening methods include the use of shear studs, fiber-reinforced polymers (FRPs), and textile-reinforced mortar (TRM) [2,3]. To predict the bearing performance and strengthening effects of the strengthened RC two-way slabs, in situ loading tests under specific conditions and sophisticated numerical simulation techniques are necessary. Unlike the common strengthening schemes for beams and columns, such as increasing cross-sectional areas, external reinforcement, and adding energy-dissipating components, the strengthening and retrofit design of two-way slab structures often employ materials with a higher elastic modulus and tensile strength, such as bonded steel plates and carbon fiber-reinforced polymer (CFRP). This strengthening method can be naturally completed in situ without causing damage to the original structure and without significantly increasing the size of the component [4,5]. The strengthening method using metal plates has good applicability and workability, providing effective strengthening for various structural component forms. However, during the load-bearing process, high demands are placed on the anchorage of the steel plates and the quality of their bonding. For components with initial defects or irregular, uneven sections, this method may struggle to achieve the reliable adhesion of steel or aluminum plates, leading to suboptimal strengthening effects and underutilization of the performance advantages of metal materials [6–8]. In contrast, the use of high-performance carbon fiber materials significantly improves and enhances the bearing performance of two-way slabs [9,10]. CFRP, known for its high strength and efficiency, excellent impact and fatigue resistance, and superior corrosion resistance [11–14], has been shown in studies to achieve good adhesion to concrete with reliable load-sharing synergy. However, due to the high strength and extremely thin profile of CFRP sheets, they are prone to brittle damage during strengthening and construction operations [15–17].

Recent studies have significantly contributed to the understanding of the CFRP-strengthened effects on RC slabs. Al-Rousan et al. [18] demonstrated that CFRP-strengthening, particularly in unidirectional sheet form, enhances flexural strength while maintaining ductility. Santos et al. [19] showed that CFRP systems, including the stitch and embedding through section (ETS) methods, effectively increase punching shear strength. Breveglieri and Czaderski [20] found that CFRP strips' long-term performance is resilient to thermal cycles and sustained loads, with adhesive viscoelasticity being a key service factor. Al-Rousan et al. [21] emphasized the importance of anchorage systems in CFRP strengthening to prevent debonding and optimize load-carrying capacity. Ghayeb et al. [22] confirmed that CFRP sheets significantly boost two-way RC slabs' punching shear resistance, with their analytical model aligning with existing standards, aiding in the design and assessment of strengthened slabs. Lastly, Zhou et al. [23] introduced a self-locking hybrid anchoring technique for CFRP sheets in one-way RC slabs, showing improved debonding resistance and strengthening efficiency, which is crucial for enhancing the reliability and durability of CFRP-strengthened RC elements. In parallel, the referenced articles [24–28] explore the specific advancements and challenges associated with the use of CFRP and glass fiber-reinforced polymer (GFRP) in strengthening structures. The study by Li [24] introduced a wedge-shaped bond anchorage system for pultruded CFRP plates, emphasizing its effectiveness in uniformly distributing shear stress and reducing stress concentration, significantly enhancing anchorage efficiency and the overall performance of CFRP-strengthening systems. As for the influence factors, Ke et al. [25] provided a comprehensive review of the influence of environmental factors such as temperature, moisture, and corrosion on bond performance, as well as underscoring the importance of adhesive properties and joint geometry in enhancing durability and effectiveness. Complementing the body of research on CFRP-strengthening, Xian et al. [26,27] provided a thorough evaluation of the mechanical properties of GFRP composites across a range of conditions. It reported the long-term performance of GFRP over thermosetting composites. These studies, through comparative analysis and leveraging their strengths, collectively highlight the multifaceted advantages of CFRP in RC slab strengthening, from immediate strength gains to long-term durability and performance under various conditions.

In summary, research outcomes have been achieved based on the in situ testing of reinforced concrete, full-scale component testing, and load-bearing capacity tests of scaled components for strengthening effects. However, to study and evaluate the strengthening effects throughout the entire load-bearing process, sophisticated numerical simulation based on the finite element (FE) method is required. Previous studies [29–33] have provided a new perspective for strengthening research and the effective evaluation of structures. A study by Shu et al. [34] analyzed two-way RC slabs under bending through nonlinear finite element analysis, examining the impact of various modeling parameters on the prediction accuracy of load-carrying capacity, load–deflection curve response, and crack patterns. Xu et al.'s theoretical calculations [35], complemented by experimental evidence, quantified the influence of bonding layers on the flexural strengthening of RC slabs with CFRP sheets. Sharaky et al. [36] conducted a FE model to predict the slabs' behavior, offering insights into the effects of bond length, material properties, and strengthening area on the strengthening outcomes. Zou et al. [37] simulated the fatigue behavior of RC beams strengthened with side-bonded CFRP, demonstrating its effectiveness in predicting the fatigue performance of CFRP-strengthened beams and providing a basis for predicting their fatigue life under different loading conditions.

The subject of this study is an RC two-way slab component located in an existing 34-story residential building. The component exhibited an initial deflection due to the influence of curing, environmental factors, and construction loads. To prevent unexpected damage and interaction effects during subsequent operations, the authors' team conducted in situ strengthening measures prior to the normal use phase of the structure. This paper investigated the application and performance enhancement achieved by using bonding CFRP in situ as a strengthening method. A multi-stage loading test, designed to the permissible load-bearing capacity level, was employed to measure the performance of the strengthened component and to study the distribution regulation of stress and strain. A FE model was established and calibrated based on field measurement data, and the model was applied to analyze and predict the ultimate load-bearing capacity of the element.

2. Project Overview and Test Plan

2.1. Project Information

This study concerns a high-rise residential building, designed and completed in 2016, that is located in northern China and features a structural system of frames and shear walls. The focus is on a RC slab that exhibits mid-span deflection, with dimensions of 4100 mm × 5900 mm × 160 mm. The slab is reinforced with double-layer, two-way CRB600H rebars of Φ 6 at 150 mm spacing, and the concrete grade is C25. The building was classified with a seismic fortification intensity of 7 degrees (0.15 g) and a designed service life of 50 years. Post-construction, it was observed that the slab's deflection negatively impacted the comfort and load-bearing safety of the residential building. After a comparative analysis of strengthening schemes, the application of CFRP bonding to the underside of the RC two-way slab was selected for strengthening.

CFRP strips were uniformly bonded along the transverse and longitudinal directions on the underside of the slab using an epoxy resin adhesive. The CFRP grade used was I-300, which corresponds to a design tensile strength greater than 3400 MPa (Grade I) and a weight of 300 g/m^2. The CFRP strengthening consisted of a single layer with a thickness of 0.167 mm, a width of 100 mm, and a clear spacing of 100 mm between the strips. The testing of the CFRP material showed that the average tensile strength was 3500 MPa, complemented by an average elastic modulus of 254 GPa. For the installation of CFRP, the concrete surfaces were meticulously prepared following a standardized procedure. Initially, the surfaces were roughened to enhance adhesion, creating a mechanical key that would secure the epoxy adhesive, followed by thorough cleaning to ensure a robust bond at the interface with a designed bond strength of 4.0 MPa. Subsequently, the CFRP sheets, each 100 mm in width with a clear spacing of 100 mm between strips, were installed with precision. Between 15 and 30 min post-initial application, a second layer of epoxy was applied to foster significant chemical bonding, a critical method for withstanding shear stresses that are inherent to the structure's service life. The curing period for the strengthened RC slabs, inclusive of the CFRP sheets and the epoxy resin, spanned a full seven days after the strengthening system's application. Further, this curing process necessitated the maintenance of specific temperature and humidity conditions to guarantee the proper curing of the resin. The implementation of the strengthening scheme and the in situ arrangement are captured in Figure 1.

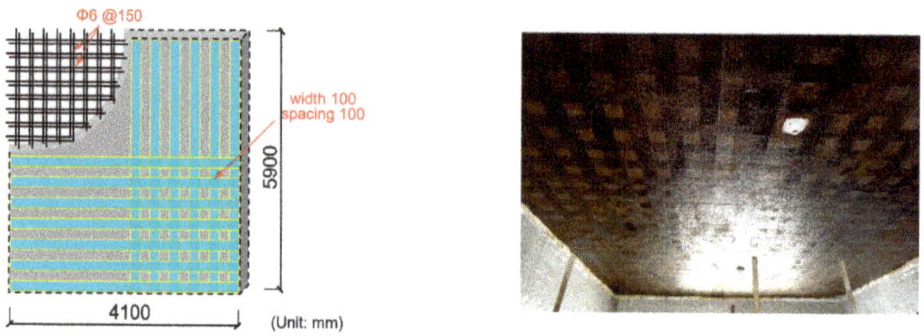

(**a**) (**b**)

Figure 1. CFRP-strengthening on the underside of the two-way slab; (**a**) Construction details; (**b**) Post-construction condition.

The two-way slab specimens involved in the experimental analysis exhibited initial defects in deflection. Each slab's initial defect was marked by mid-span deflection. Being precast composite slabs, they were noted to have frequent water ponding at the center of the upper surface following the placement of a site-cast layer and the curing process. Using a gradienter to measure the deformation of each slab's upper or lower surfaces, it was confirmed that the mid-span deflection reached a stable value of 12 mm. To ensure that their load-carrying and deformation capacities meet the requirements during the subsequent normal use phase of the structure, the slabs were strengthened by evenly spacing and fully adhering CFRP strips to the lower surface. Considering that the RC two-way slab had already been cast and that it was necessary to conduct performance testing of structural components in accordance with the requirements of the "Code for Acceptance of Construction Quality of Concrete Structures" [38], a stepwise loading test was conducted at a safe level to assess and evaluate the effectiveness of the strengthening measures. The test response of the two-way slab under load was utilized to appraise the structural strengthening outcomes.

2.2. Testing Method

The load test was conducted to assess the safety of the slab under the quasi-permanent combination (QC) of loads, including the detection of cracks at the underside of the slab and mid-span deflections, as well as under the basic combination (BC) of loads (design loads). For the purpose of crack detection, a thorough flashlight inspection was performed on the underside of the slab, focusing on critical areas such as the central region, the vicinity of diagonals, and the perimeter to scrutinize the spaces between CFRP strips for the occurrence or propagation of concrete cracks. During the loading test, scaffolding was erected at the underside of the slab, and five dial gauges were installed in groups to collect vertical deformation values at different locations, as shown in Figure 2. To ensure the consistency of data monitoring throughout the prolonged experimental period, mechanical dial gauges were utilized. Meticulous attention was devoted to minimizing any disruptions to the gauges during both data acquisition and observational phases, thereby guaranteeing the precision and dependability of the data collected.

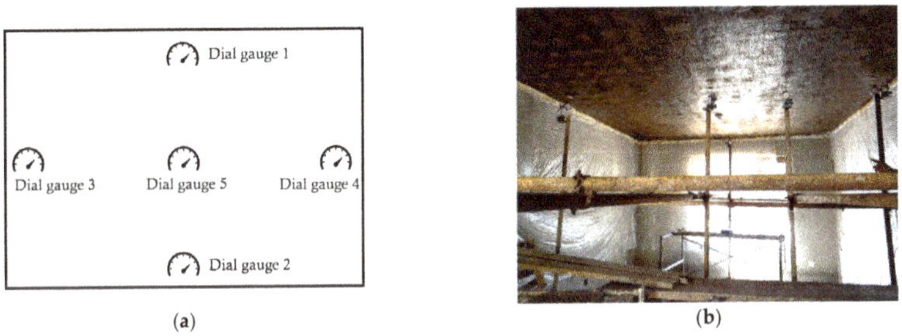

Figure 2. Dial gauges layout under the RC slab; (**a**) Schematic diagram; (**b**) Construction layout.

The self-weight of the two-way slab, G_{1k}, is 4.0 kN/m^2; the self-weight of the concrete post-pouring layer, G_{2k}, is 1.2 kN/m^2; and the live load, Q_k, is 2.0 kN/m^2 [39].

(1) Load on QC stage:

The total load of the slab is:

$$Q_s = G_{1k} + G_{2k} + 0.4Q_k = 6.0 \text{ kN/m}^2 \tag{1}$$

The loading weight for the quasi-permanent load combination is determined by deducting the self-weight G_{1k} from the load value:

$$W_{QC} = (Q_s - G_{1k})S_0 = 47.4 \text{ kN} \tag{2}$$

In the formula, S_0 represents the net area of the slab component, and it is calculated that S_0 is 23.69 m^2.

(2) Load on BC stage:

$$Q_d = 1.2(G_{1k} + G_{2k}) + 1.4Q_k = 9.04 \text{ kN/m}^2 \tag{3}$$

$$W_{BC} = (Q_d - G_{1k})S_0 = 119.40 \text{ kN} \tag{4}$$

Combined with Equations (3) and (4), it is calculated that W_{BC} is 119.40 kN.

2.3. Multi-Stage Loading

Upon completing the load combination calculations, it was ascertained that the loading process would be bifurcated into two distinct phases, designated QC and BC, encompassing a total of nine distinct levels, culminating in a cumulative load of 119.4 kN. In the execution of in situ testing and loading procedures, 25 kg sandbags were strategically employed to incrementally apply the load. At the onset of each loading stage, the precise number of sandbags was determined in accordance with the pre-calculated load specifications. Throughout the loading phase, a lightweight engineer carefully placed the sandbags across the floor, ensuring a uniform and stable distribution, thus mitigating any potential adverse effects from the loading activities. The initial phase, QC, was dedicated to the scrutiny of cracks and deflections and was executed in four progressive increments. Each increment was calibrated at 0.5 kN/m^2, with the initial three increments maintained for a duration of approximately 5 to 10 min and the final increment sustained for approximately 30 min; at each loading tier, deflections and crack widths were meticulously measured. A comprehensive illustration of the loading increments is presented in Figure 3.

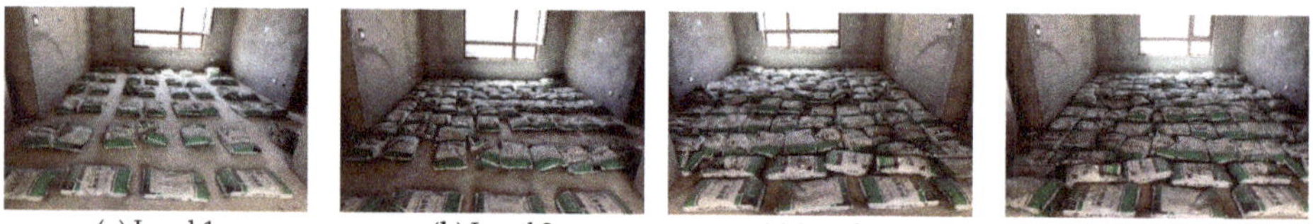

(**a**) Level 1 (**b**) Level 2 (**c**) Level 3 (**d**) Level 4

Figure 3. Load increments at stage QC.

The second phase, designated the BC stage, focuses on the inspection of cracks and deflections. This phase was carried out in five increments, with each increment loading at a determined weight of 0.6 kN/m^2. Consistent with the QC stage, the first four increments were maintained for a period of approximately 5 to 10 min, whereas the ultimate increment was held for around 30 min. During this phase, close attention was paid to any indications of structural failure. A detailed depiction of the loading increments is provided in Figure 4. The corresponding deflections of the mid-span of the slab in each load increment are listed in Table 1.

(**a**) Level 5 (**b**) Level 6 (**c**) Level 7 (**d**) Level 8 (**e**) Level 9

Figure 4. Load increments at stage BC.

Table 1. Multi-stage loading and the corresponding deflections.

Load Stage	Load Increment (kN/m^2)	Load in Total (kN)	Mid-Span Deflection (mm)
Level 1	0.50	11.85	0.11
Level 2	0.50	23.70	0.24
Level 3	0.50	35.55	0.42

Table 1. *Cont.*

Load Stage	Load Increment (kN/m^2)	Load in Total (kN)	Mid-Span Deflection (mm)
Level 4	0.50	47.40	0.53
Level 5	0.60	61.60	0.71
Level 6	0.60	75.80	0.91
Level 7	0.60	90.00	1.11
Level 8	0.60	104.25	1.39
Level 9	0.65	119.60	1.64

2.4. Testing Result

Upon completion of each increment of the load in the two stages, the underside of the slab was inspected for any signs of concrete cracking, and the CFRP was examined for signs of tensile rupture or adhesive failure. Observations revealed that after the ninth level of load application, the mid-span deflection of the slab increased gradually, yet no concrete cracking, debonding, or tearing of the CFRP-strengthened slab was observed. These experimental findings indicate that the load-bearing capacity of the two-way slab strengthened with bonded CFRP is reliable and meets the standards for normal structural use. However, the impact of the strengthening treatment on the ultimate strength of the slab requires further validation through numerical simulation.

3. FE Modelling

3.1. Simulation Set-Up

In conducting an ABAQUS FE analysis of CFRP-strengthened concrete two-way slabs with bonded CFRP, the epoxy adhesive used was shown to provide effective bonding after surface treatment, as indicated by material property tests. Consequently, it was assumed that the adhesion between the CFRP and the concrete was sound, with no relative slippage occurring, and the "tie" contact was applied to simulate the bonding effect of CFRP. Additionally, given the CFRP's negligible thickness of only 0.167 mm in the model, its thickness was disregarded in the calculations, treating it as a thin film closely adhering to the underside of the two-way slab.

The concrete material was selected as a damage plasticity model [40], with the stress–strain relationships for compression and tension modes depicted in Figure 5. To enhance computational efficiency, the constitutive relationship for the steel material was modeled using a bilinear ideal elastoplastic model. The CFRP was characterized by a linear elastic criterion, with material failure occurring when the tensile stress exceeded the tensile strength. In the FE model, the materials were defined by different values tested in a laboratory, with the parameters listed in Table 2.

Table 2. Materials' performance in the laboratory.

Material	Grade	Sectional Dimension (mm)	Strength (MPa)	SD. (MPa)	Elastic Modulus (10^5 MPa)
Concrete	C25	160.81	28.7 (compression)	3.19	0.29
Steel	CRB600H	6.02	527.3 (yield stress)	19.76	2.09
CFRP	I-300	0.167	3512.7 (tension)	9.75	2.54

For the finite element simulation, the mesh type selection was as follows: twenty-node quadratic reduced integration hexahedral element (C3D20R) was used for concrete components, three-node quadratic integration three-dimensional truss element (T3D3) was applied for CFRP-strengthened components, and eight-node reduced integration shell element (S8R) was utilized for CFRP components. After several trial calculations, the element dimensions for each component were approximately discretized to a range of 30 to 50 mm, ensuring a balance between accuracy and computational efficiency.

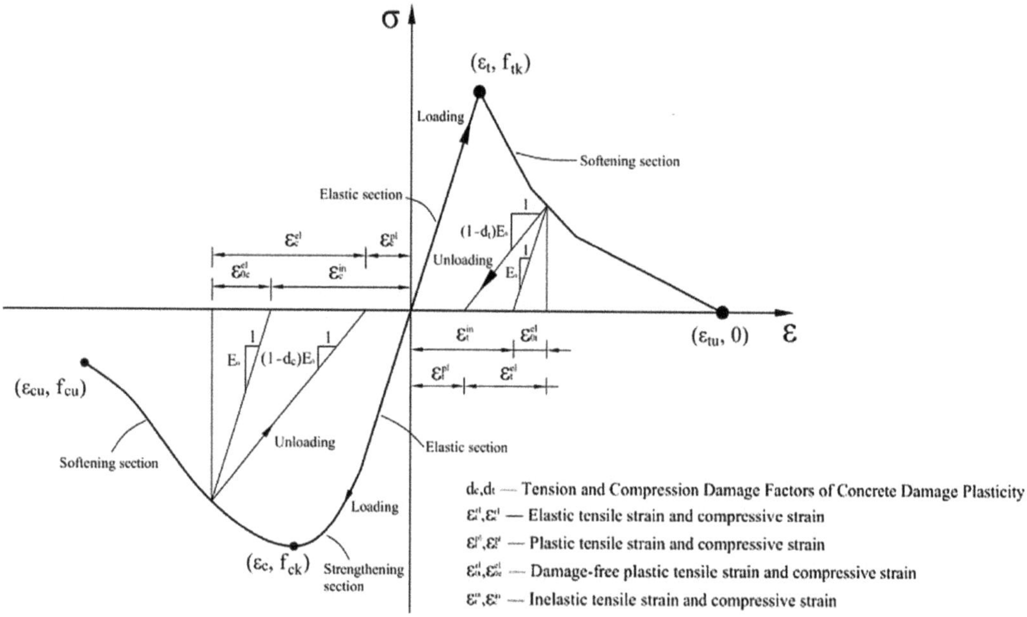

Figure 5. Constitutive curve of the concrete-damaged plasticity model.

3.2. Boundary and Loading Conditions

To accurately simulate the structural support conditions and loading characteristics that a two-way slab would experience under real-world service conditions, it was noted that negative bending moments were present at the slab's edges during the loading process. Consequently, the boundary constraints for the slab in the simulation were modeled as fully fixed to reflect these conditions.

Additionally, to enhance computational accuracy during the loading phase, several measures were implemented. These included refining the mesh to improve the resolution of the model, utilizing element types that employ quadratic interpolation functions for a more accurate representation of material behavior. While the refinements significantly increased the computational load, the structural form, boundary conditions, and loading modes of the problem exhibited ideal symmetry.

Given this symmetry, this study adopted a semi-modeling optimization method for the structure. This approach involved creating a model that represents only a portion of the structure, leveraging the symmetry to reduce the computational domain. This simplified modeling technique ensured that the computational results met the necessary accuracy requirements while also achieving a reduction in computational complexity and an increase in simulation efficiency. Figure 6 provides a visual representation of the quarter model, with the total load applied within the model constituting a quarter of the total structural load. During post-processing, to obtain the reaction forces from the quarter model that correspond to the full structure, the extracted values must be multiplied by 4 to achieve the correct scaling. The upper right portion of the figure illustrates the doubly reinforced two-way slab structure that was included in the calculations, while the lower part demonstrates the layout of the CFRP after symmetric restoration.

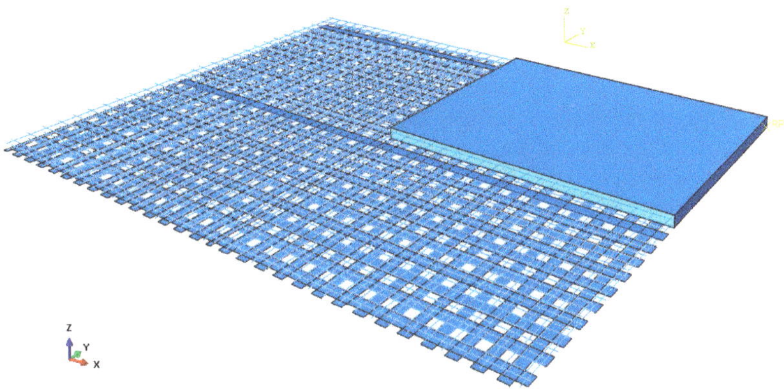

Figure 6. Quarter model simplification of CFRP-strengthened slab in ABAQUS.

The loading process was meticulously executed by configuring ten standardized analysis steps within the model. In the first analysis step, a uniform load was applied until the mid-span deflection of the slab matched the initial deflection of the original two-way slab component, simulating the initial deformation. It is noteworthy that during this step, the interaction of the CFRP elements was temporarily deactivated until the two-way slab reached a pre-defined downward deflection. During this phase, the CFRP elements were not engaged in load-bearing, allowing only the CFRP-strengthened concrete two-way slab to undergo deformation. Once the slab achieved the preset deflection, the CFRP elements were reactivated in the subsequent analysis steps to participate in the load-bearing process and to actualize their strengthening effect.

The remaining nine analysis steps systematically applied the test load values, as detailed in Figure 7, to further evaluate the slab's response to increasing loads and to assess the effectiveness of CFRP-strengthened in enhancing the load-bearing capacity and structural performance of the two-way slab. This structured approach ensured a comprehensive analysis of the slab's behavior under simulated service conditions, providing valuable insights into the CFRP-strengthened strategy's efficacy.

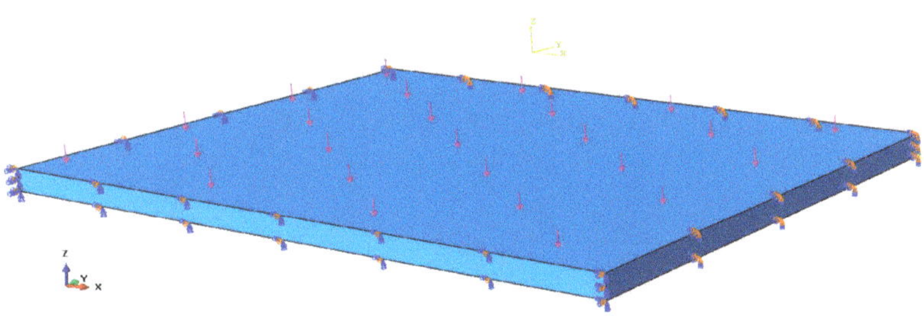

Figure 7. Constraints and loading of computing model.

4. FE Results

During the incremental loading process, observations were recorded at each stage for the lower surface of the slab. The deflection values from the in situ load test were recorded by the dial gauges, and the numerical simulation results were analyzed using data that stabilized after multiple mesh refinements. Table 3 presents a comparison of mid-span deflection values. After subjecting the slab to nine levels of loading, no concrete cracking or CFRP debonding failure was observed at the underside of the slab, and no load-bearing failure was apparent. The finite element analysis (FEA) results were consistent with the test results, and there were no component failures due to material yielding or contact failure that terminated the calculation. The deflection comparison data presented in the table reveal that the calculated values closely correspond with the measured data, which underscores the reliability and precision of the FEA method. This alignment between the numerical predictions and the empirical observations validates the FEA as a dependable tool for predicting the structural behavior of CFRP-strengthened concrete slabs strengthened with CFRP under loading conditions.

Table 3. Deflection comparison of load test and FEA of two-way slab.

Load Stage	Mid-Span Deflection (mm)	FEA Deflection (mm)	Error
Level 1	0.11	0.15	0.36
Level 2	0.24	0.29	0.21
Level 3	0.42	0.44	0.05
Level 4	0.53	0.59	0.11
Level 5	0.71	0.78	0.10
Level 6	0.91	0.97	0.07
Level 7	1.11	1.15	0.04
Level 8	1.39	1.34	−0.04
Level 9	1.64	1.53	−0.07

Figure 8 illustrates a graded statistical analysis of the deflection of the two-way slab. The simulated data accurately reflected the test results of the in situ experiment. The proximity of the two curves is related to the fact that the concrete, steel reinforcement, and CFRP materials were still in the elastic phase of loading during the testing phase. Upon comparison of the load–deflection curves of various models, it was observed that the model of the two-way slab without CFRP-strengthening showed a mid-span deflection approximately 50% greater than that of the CFRP-strengthened slab model under the same load. Although the mid-span deflections were all below 3 mm, the strengthening effect of the bonded CFRP was further demonstrated through a more detailed comparison.

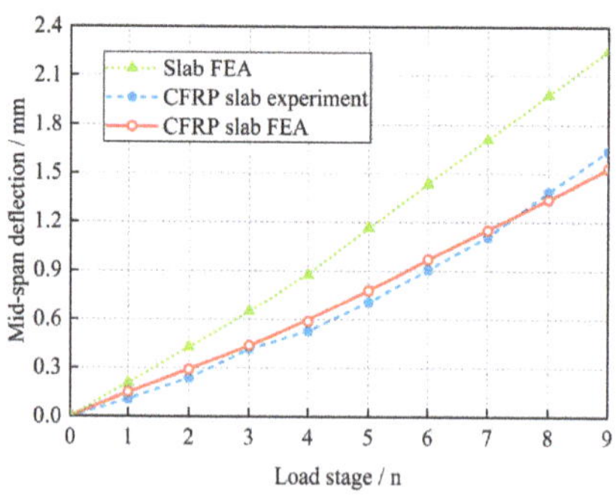

Figure 8. Load–deflection curve comparison between experimental and FEA result.

Figure 9 depicts the distribution of von Mises stress within the concrete and CFRP elements during the loading process. The stress contour plot revealed that the CFRP actively participated in load-bearing during the experiment, effectively sharing the load with the RC slab. The stress contour plot revealed that the CFRP actively participated in load-bearing throughout the test, efficiently sharing the load with the RC slab. This led to a more uniform distribution of stress across the concrete slab, with Mises stress values roughly remaining below 24.14 MPa. Notably, the stress values in the edge regions, which experienced a combination of bending and shear forces, were less than 3.45 MPa. In contrast, the stress within the CFRP was markedly higher, particularly at the anchorage zones. The surface load applied to the upper surface of the slab was effectively transferred to the CFRP, which served to strengthen and enhance the structural capacity of the component. The analysis results indicated that the stress was primarily concentrated at the anchorage edges of the CFRP and at mid-span areas where the strain was relatively high. The numerical simulation results accurately reflected the actual stress mechanisms, demonstrating the effectiveness of CFRP in reinforcing and strengthening the structural integrity of the slab.

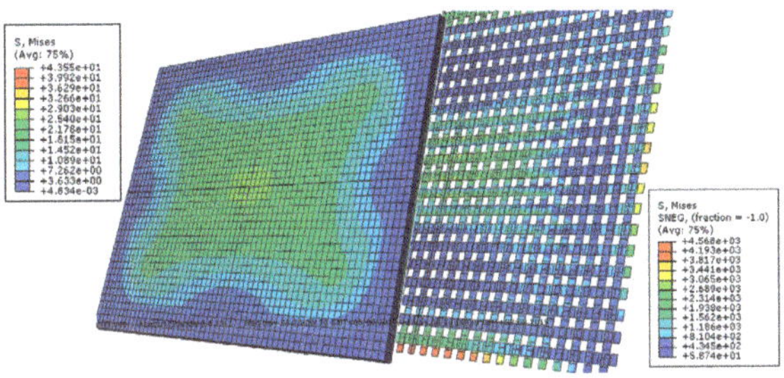

Figure 9. Stress contour plot of concrete and CFRP at loading stage Level 7.

In order to substantiate the enhancement of the load-bearing capacity of the two-way slab attributable to the CFRP-strengthened strategy, a numerical simulation was utilized to simulate an ultimate loading experiment. This simulation facilitated a comparative analysis of the stress values on both the lower and upper surfaces of the slab, with a focus on the CFRP-strengthened conditions, across the entire loading process. The objective was to evaluate the ultimate load-bearing capability and the deformation characteristics at the yield point of the specimens, as well as to ascertain the degree of improvement of the slabs strengthened with CFRP. Figure 10 depicts the stress distribution within the concrete section of the two-way slab. It is evident from the figure that in the absence of strengthening, the slab experiences elevated stress concentrations along the 45-degree diagonals at the lower surface, within the support boundary regions, and around the circumference of the central area on the upper surface. These stress concentrations are indicative of areas that are prone to cracking or crushing under load, underscoring the importance of CFRP-strengthening in such critical structural components.

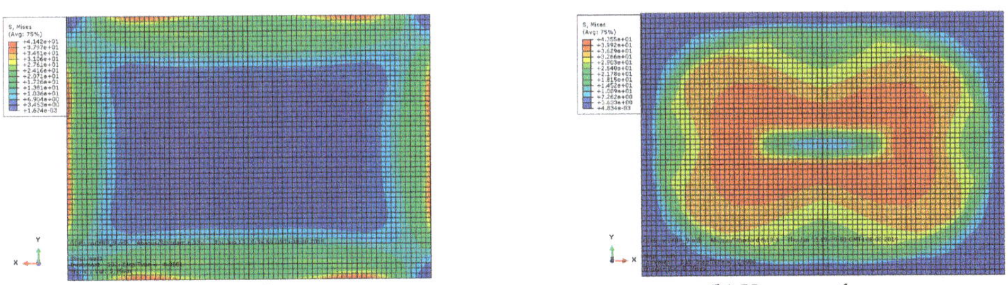

(**a**) lower surface (**b**) Upper surface

Figure 10. Stress contour plots of the unstrengthened slab.

The tensile damage contour plots of the concrete in the unstrengthened slab are depicted in Figures 11 and 12. It was observed that from the first (Level 4) to the second (Level 9) phase, the damaged areas evolved and spread, with significant tensile failure in the concrete at the mid-span underside of the slab and the constrained boundary areas on the upper surface, where cracking was notably pronounced. The distribution of tensile damage in the lower concrete surface of the unstrengthened slab, as reflected by the DAMAGET factor, escalated from a limited area exceeding 0.75 in Figure 11a to nearly 70% of the area with values exceeding 0.8 in Figure 11b. A similar consistent pattern is observed upon comparison of Figure 12a,b. This observation underscores the essential and critical role of CFRP-strengthening in preventing the propagation of cracks and potential structural failure in these regions.

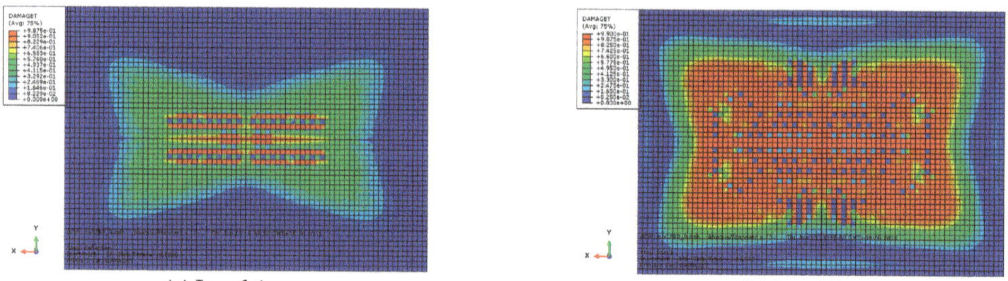

(**a**) Level 4 (**b**) Level 9

Figure 11. Concrete tensile damage contour plot of lower surface in unstrengthened slab.

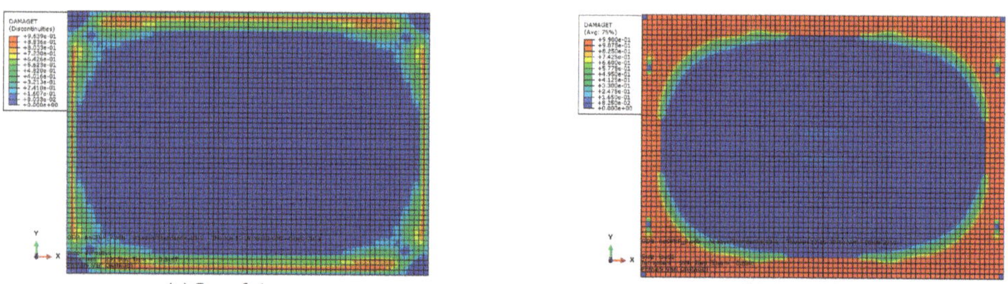

(**a**) Level 4 (**b**) Level 9

Figure 12. Concrete tensile damage contour plot of upper surface in unstrengthened slab.

Figure 13 illustrates the steel stress distribution within the doubly reinforced slab. The stress contour plot of the steel correlates with the stress and failure patterns observed in the concrete of the slab. It was analyzed that in the unstrengthened slab, CFRP-strengthening played a crucial role in bearing a portion of the load and in limiting excessive deformation at the mid-span. The stress contour revealed that the concrete stress values were notably higher at the 45-degree diagonals and the support boundary areas on the lower surface of the slab, as well as within the circumferential range at the center of the upper surface, where cracks or crushing phenomena in the concrete were likely to occur under load.

The comparison of the stress contour plots revealed that after strengthening with CFRP, the stress levels in the concrete were significantly lower under the same loading stage, as depicted in Figure 14. The lower surface of the strengthened slab displayed a notably uniform distribution of stress, reaching a peak of 19.76 MPa, whereas the upper surface experienced maximum compressive stress of 24.70 MPa, with the high-stress region notably localized to a narrow area at the slab's mid-span. Simulation during the ultimate load stage indicated that the stress in the CFRP strips (as shown in Figure 15) was approximately an order of magnitude higher than the tensile stress in the concrete. The synergistic action of CFRP and steel reinforcement enhanced the ultimate load-bearing capacity and deformation capacity of the two-way slab, thereby increasing the safety margin of the structural component. The comparison of the stress data between the strengthened and unstrengthened slabs demonstrated the substantial enhancement in load-bearing capacity achieved through the application of CFRP-strengthening. CFRP-strengthening not only increased the overall strength of the slab but also helped to distribute the stress more evenly, reducing the concentration of stress in critical areas and thereby delaying the onset of cracking or crushing.

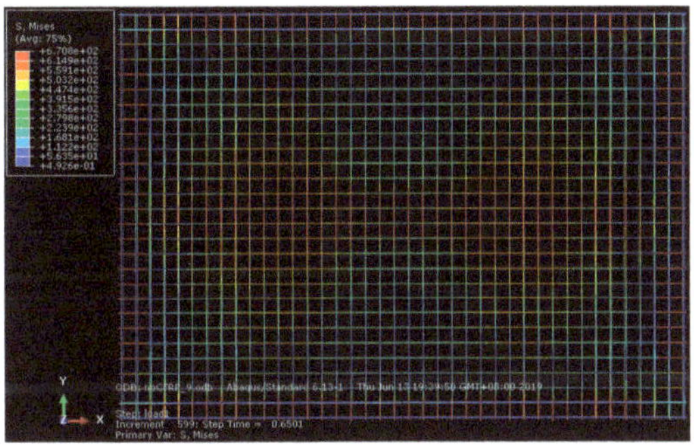

Figure 13. Mises stress contour plot of steel in the unstrengthened slab.

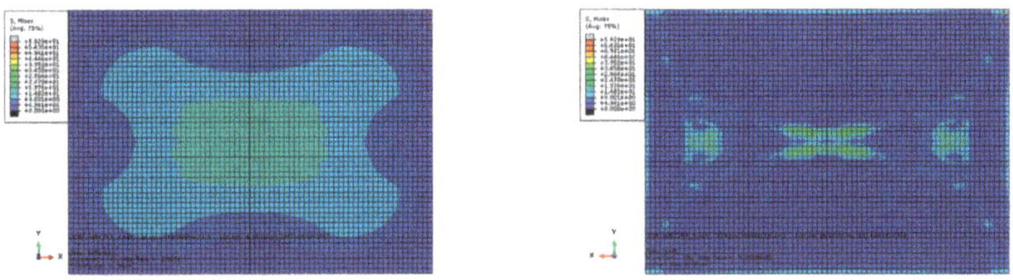

(**a**) lower surface (**b**) Upper surface

Figure 14. Stress contour plot of the CFRP-strengthened slab.

Figure 15. Stress contour plot of CFRP of the strengthened slab.

The yielding of the component was defined by the experimental deflection reaching 1/50 of the span (approximately 120 mm). Figure 16 illustrates the load–displacement curve of the slab component. At the yielding point of the unstrengthened two-way slab, the load capacity was 95 kN/m^2. Post-yielding, the component's load capacity could not increase further, and the deflection rapidly increased, leading to complete failure of the slab. In contrast, the slab strengthened with CFRP could withstand a load of 135 kN/m^2 when the deflection reached 120 mm, and the load capacity could further increase to 140~150 kN/m^2 after yielding. The ultimate load capacity was approximately 50% higher than that of the unstrengthened contrast model. The comparative results indicate that the CFRP-strengthened scheme effectively ensured an adequate level of structural safety and security.

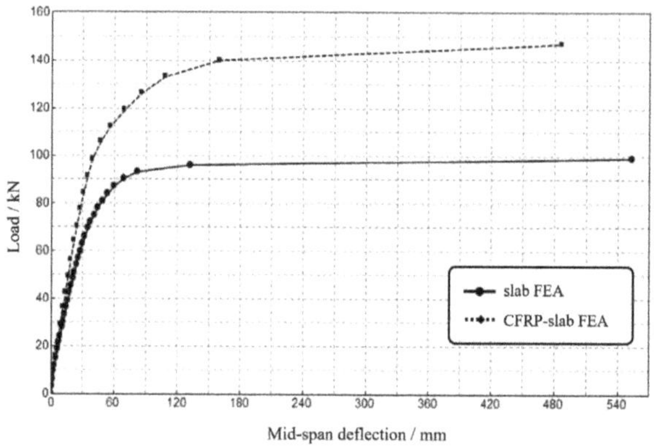

Figure 16. Curve comparison of two-way slabs under ultimate loading experiment.

5. Conclusions

In this paper, an in situ test and FE simulation of the effectiveness of CFRP-strengthening on the load-bearing performance of RC two-way slabs was conducted, utilizing both in situ multi-stage loading experiments and nonlinear FEA. The findings are encapsulated in the conclusions presented below:

(1) The experimental evidence has confirmed that the utilization of CFRP for the strengthening of RC two-way slabs exhibiting an initial defect is a dependable and efficient approach. The strengthening application of CFRP proves to be not only feasible but also highly practical during the construction phase of strengthening. In devising CFRP strengthening strategies, the quantity of CFRP layers and the spacing between them can be meticulously tailored to accommodate the specific geometrical attributes and load-bearing specifications of the component. Such customization enhances the overall efficacy of the strengthening scheme, ensuring an optimized outcome aligned with structural performance objectives.

(2) After effectively simplifying the structure within ABAQUS 6.13 software, the process of CFRP and RC two-way slabs jointly sustaining loads can be accurately simulated by appropriately configuring the finite element properties, material constitutive models, and loading procedures. A comparison of the numerical simulation outcomes with the in situ test results has validated the consistency between the numerical analysis and the experimental data. The final mid-span deflection results exhibited discrepancies of −7% between the simulated calculations and the actual measurements, thereby underscoring the accuracy of the analytical methodology.

(3) Numerical simulation outcomes, including stress and damage contour plots, demonstrate a marked enhancement in the load-bearing capacity of the slab components due to the presence of reinforcement bars and bonded CFRP layers. The slab's underside

area, which initially exhibited tensile damage coefficients above 0.8 for concrete in over 70% of its area, has been changed to no signs of tensile or compressive damage on either surface. The simulation's analytical outcomes reveal that CFRP strengthening has significantly enhanced the slab's structural integrity and load-bearing capacity. Additionally, the slabs experience a more uniform distribution of stress, ably shared by the CFRP layers. Simulation calculations have confirmed that the ultimate load capacity of the CFRP-strengthened components has increased by about 50% relative to the unstrengthened component.

Author Contributions: Conceptualization, Y.S.; methodology, Y.S.; software, Y.S. and Y.G.; validation, Y.G.; formal analysis, Y.S.; investigation, Y.S. and Y.G.; resources, Y.S.; data curation, Y.G.; writing—original draft preparation, Y.S.; writing—review and editing, Y.G.; visualization, Y.S. and Y.G.; supervision, Y.G.; project administration, Y.S. and Y.G.; funding acquisition, Y.S. All authors have read and agreed to the published version of the manuscript.

Funding: This research received no external funding.

Data Availability Statement: The data presented in this study are available upon corroborated request from the corresponding author.

Conflicts of Interest: The authors declare no conflicts of interest.

References

1. Buitrago, M.; Adam, J.M.; Moragues, J.J. Load transmission between slabs and shores during the construction of RC building structures—A review. *Eng. Struct.* **2018**, *173*, 951–959. [CrossRef]
2. Mahmoud, H.S.; Hawileh, R.A.; Abdalla, J.A. Strengthening of high strength reinforced concrete thin slabs with CFRP laminates. *Compos. Struct.* **2021**, *275*, 114412. [CrossRef]
3. Mercimek, Ö.; Ghoroubi, R.; Erbaş, Y. Comparison of strengthening methods to improve punching behavior of two-way RC flat slabs. *Structures* **2022**, *46*, 1495–1516. [CrossRef]
4. Shen, J.; Dai, H.; Zhou, G. Shear behavior of pre-damaged RC beams strengthened with UHPFRC-CFRP grid layer. *Compos. Struct.* **2024**, *349–350*, 118520. [CrossRef]
5. Mansour, W.; Li, W.; Ghalla, M. Improving the punching capacity of two-way RC flat slabs via external strengthening using various configurations of aluminum sheets. *Constr. Build. Mater.* **2024**, *420*, 135611. [CrossRef]
6. Aljidda, O.; El Refai, A.; Alnahhal, W. Experimental and numerical investigation of the flexural behavior of one–way RC slabs strengthened with near–surface mounted and externally bonded systems. *Constr. Build. Mater.* **2024**, *421*, 135709. [CrossRef]
7. Hawileh, R.; Douier, K.; Gowrishankar, P. Durability of externally strengthened concrete prisms with CFRP laminates and galvanized steel mesh attached with epoxy adhesives and mortar. *Compos. Part C Open Access* **2024**, *14*, 100493. [CrossRef]
8. Sahranavard, S.; Jahangir, H.; Haji kazemi, H. An experimental and numerical investigation of RC slabs externally strengthened by perforated steel plates. *Structures* **2022**, *45*, 1239–1252. [CrossRef]
9. Oret, G.; Limam, O. Experimental and numerical analysis of RC two-way slabs strengthened with NSM CFRP rods. *Constr. Build. Mater.* **2008**, *22*, 2025–2030.
10. Limam, O.; Nguyen, V.T.; Foret, G. Numerical and experimental analysis of two-way slabs strengthened with CFRP strips. *Eng. Struct.* **2005**, *27*, 841–845. [CrossRef]
11. Jin, L.; Zhang, B.; Chen, F. CFRP-strengthened shear walls: Combined effects of CFRP and reinforcement ratio. *Int. J. Mech. Sci.* **2024**, *282*, 109634. [CrossRef]
12. Li, L.; Chen, J.; Wang, W.T.; Dai, K.Y.; Hu, W.B. Experimental and prediction model of axial compressive responses of corroded RC cylinders strengthened with CFRP. *Eng. Struct.* **2024**, *317*, 118649. [CrossRef]
13. Gholami, M.; Sam, A.R.M.; Yatim, J.M. A review on steel/CFRP strengthening systems focusing environmental performance. *Constr. Build. Mater.* **2013**, *47*, 301–310. [CrossRef]
14. Jung, D.; Kim, K.H.E.; Choi, H. Shear strengthening and damage control effects of CFRP and self-prestressing iron-based SMA for concrete columns. *Compos. Struct.* **2025**, *351*, 118550. [CrossRef]
15. Wang, H.T.; Liu, S.S.; Shi, J.W. Bonded and unbonded strengthening of RC beams with the prestressed CFRP plate system: An experimental study. *Constr. Build. Mater.* **2024**, *438*, 137118. [CrossRef]
16. Elnagar, M.E.; Makhlouf, M.H.; El-sayed, K.M. An experimental and numerical study on two-way RC slabs with openings strengthened by ferrocement technique. *Structures* **2024**, *65*, 106673. [CrossRef]

17. Joyklad, P.; Krishna Gadagamma, C.; Maneengamlert, B. Structural behavior of RC one-way slabs strengthened with ferrocement and FRP composites. *Eng. Fail. Anal.* **2024**, *161*, 108328. [CrossRef]
18. Al-Rousan, R. Performance of reinforced concrete slabs strengthened with different types and configurations of CFRP. *Compos. Part B Eng.* **2012**, *43*, 510–521. [CrossRef]
19. Santos, G.S. Punching CFRP-based strengthening solutions for reinforced concrete flat slabs. *Compos. Struct.* **2019**, *224*, 111077. [CrossRef]
20. Breveglieri, M. RC slabs strengthened with externally bonded CFRP strips under long-term environmental exposure and sustained loading. Part 2: Laboratory experiments. *Compos. Part C Open Access* **2021**, *6*, 100210. [CrossRef]
21. Al-Rousan, R.Z.; Ababneh, A.N.; Bani Hani, M.S. Effect of different anchorage schemes on the flexural behavior of RC slabs strengthened with CFRP sheets. *Constr. Build. Mater.* **2024**, *428*, 136348. [CrossRef]
22. Ghayeb, H.H.; Atea, R.S.; Al-kannoon, M.A.A. Performance of reinforced concrete flat slab strengthened with CFRP for punching shear. *Case Stud. Constr. Mater.* **2023**, *18*, e01801. [CrossRef]
23. Zhou, C. Experimental study on the flexural strengthening of one-way RC slabs with end-buckled and/or externally bonded CFRP sheets. *Eng. Struct.* **2023**, *282*, 115832. [CrossRef]
24. Li, C.; Xian, G. Novel wedge-shaped bond anchorage system for pultruded CFRP plates. *Mater. Structures.* **2018**, *51*, 162. [CrossRef]
25. Ke, L.; Li, Y.; Li, C.; Cheng, Z.; Ma, K.; Zeng, J. Bond Behavior of CFRP-Strengthened Steel Structures and Its Environmental Influence Factors: A Critical Review. *Sustain. Struct.* **2024**, *4*, 000038. [CrossRef]
26. Xian, G.; Zhou, P.; Li, C.; Dong, S.; Du, H.; Tian, J.; Guo, R.; Peng, Z.; Zhang, Z.; He, T. Mechanical Properties Evaluation of Glass Fiber Reinforced Thermoplastic Composite Plate under Combined Bending Loading and Water Immersion. *Constr. Build. Mater.* **2024**, *440*, 137470. [CrossRef]
27. Dong, S.; Zhou, P.; Ning, Z.; Wu, X.; Li, C.; Xian, G. Durability of Carbon- and Glass-Fiber Reinforced Thermoplastic Polymer Composites: A Literature Review. *J. Build. Eng.* **2024**, *98*, 111055. [CrossRef]
28. Tian, L.; Wang, L.; Xian, G. Machine Learning Prediction of Interfacial Bond Strength of FRP Bars with Different Surface Characteristics to Concrete. *Case Stud. Constr. Mater.* **2024**, *21*, e03984. [CrossRef]
29. Ghalla, M.; Badawi, M.; Hu, J.W.; Elsamak, G.; Mlybari, E.A.; Emara, M. Ultimate Performance of Two-Way Reinforced Concrete Flat Slabs Enhanced by SHCC Drop Panels Mitigating Punching Failure. *J. Build. Eng.* **2024**, *99*, 111574. [CrossRef]
30. Abdulrazzaq, M.M.; Hilo, S.J.; Sabih, S.M. Numerical investigation for the flexural load behaviour of profiled composite slab strengthened with CFRP. *Mater. Today Proc.* **2022**, *61*, 1115–1125. [CrossRef]
31. Abas Golham, M.; Al-ahmed, A.H.A. Behavior of GFRP reinforced concrete slabs with openings strengthened by CFRP strips. *Results Eng.* **2023**, *18*, 101033. [CrossRef]
32. Anas, S.M.; Alam, M. Role of shear reinforcements on the punching shear resistance of two-way RC slab subjected to impact loading. In Proceedings of the 13th International Symposium on Plasticity and Impact Mechanics, Chennai, India, 21–26 August 2022; Volume 87, pp. 43–54.
33. Anas, S.M.; Alam, M.; Shariq, M. Behavior of two-way RC slab with different reinforcement orientation layouts of tension steel under drop load impact. In Proceedings of the 13th International Symposium on Plasticity and Impact Mechanics, Chennai, India, 21–26 August 2022; Volume 87, pp. 30–42.
34. Shu, J.; Fall, D.; Plos, M. Development of modelling strategies for two-way RC slabs. *Eng. Struct.* **2015**, *101*, 439–449. [CrossRef]
35. Xu, S.; Xie, H.; Li, Q. Bending performance of RC slabs strengthened by CFRP sheets using sprayed UHTCC as bonding layer: Experimental study and theoretical analysis. *Eng. Struct.* **2023**, *292*, 116575. [CrossRef]
36. Sharaky, I.A.; Elamary, A.S.; Alharthi, Y.M. Experimental and numerical investigation on the flexural performance of RC slabs strengthened with EB/NSM CFRP reinforcement and bonded reinforced HSC layers. *Eng. Struct.* **2023**, *289*, 116338. [CrossRef]
37. Zou, C.; Ibrahim, Z.; Hashim, H. Fatigue performance simulation of reinforced concrete beams externally strengthened with side bonded CFRP sheets. *Eng. Struct.* **2024**, *316*, 118530. [CrossRef]
38. *GB50204-2015;* Code for Acceptance of Construction Quality of Concrete Structures. China Academy of Building Research: Beijing, China; China Architecture & Building Press: Beijing, China, 2015.
39. *GB 50010-2010;* Code for Design of Concrete Structures. China Ministry of Construction: Beijing, China; China Architecture & Building Press: Beijing, China, 2015.
40. Shao, S.; Wu, Y.; Fu, H.; Feng, S.; Zhang, J. Numerical Investigation on the Mechanical Properties of Vault Void Lining and Steel Plate Strengthening. *Materials* **2023**, *16*, 789. [CrossRef]

buildings

MDPI

Article

Dynamic Monitoring of CFRP Strengthened Bridges: A Distributed Sensing Approach Under Bridge–Vehicle Coupling

Bitao Wu [1], Sicheng Fu [1], Yan Gong [1], Yanchao Shao [2], Zhenwei Zhou [1,*] and Haitao Wang [3,*]

[1] School of Civil Engineering and Architecture, East China Jiaotong University, Nanchang 330013, China; wubitao@ecjtu.edu.cn (B.W.); sichengfu@ecjtu.edu.cn (S.F.); gongyan98@ecjtu.edu.cn (Y.G.)
[2] Central Research Institute of Building and Construction Co., Ltd. MCC, Beijing 100088, China; shaoyanchao@cribc.com
[3] College of Civil and Transportation Engineering, Hohai University, Nanjing 210098, China
* Correspondence: 3383@ecjtu.edu.cn (Z.Z.); cewht@hhu.edu.cn (H.W.)

Abstract: To achieve a rapid assessment of reinforced bridges without interrupting traffic, the method and influence parameters for evaluating prestressed CFRP (Carbon Fiber Reinforced Polymer) plate-strengthened bridges were investigated based on distributed sensing while considering the bridge–vehicle coupling. Firstly, based on the principle of distributed monitoring, the strain influence line equation for prestressed CFRP plate-strengthened bridges under moving vehicle loads was derived. An evaluation index and method for assessing strengthening effectiveness based on distributed strain monitoring was proposed. Numerical simulation experiments were conducted for verification, in which a simulation method for solving the bridge–vehicle coupling vibration response based on the displacement contact method was studied. Relevant example verifications were carried out to prove the correctness of this method. On this basis, a dynamic performance analysis of the bridge before and after strengthening under bridge–vehicle coupling was performed. Finally, a study on influencing parameters was conducted to investigate the impact of different prestress levels of CFRP plates, different vehicle speeds, and different vehicle axle loads on the evaluation method. The results demonstrate that the bridge–vehicle coupling calculation model based on the displacement contact method can effectively simulate the dynamic response of the bridge after reinforcement. The evaluation method based on distributed sensing monitoring is unaffected by the speed and axle load of the vehicle, enabling the assessment of the condition and stress state of the strengthened bridge.

Keywords: prestressed CFRP plate; bridge–vehicle coupling; distributed monitoring; strengthening effect; bridge evaluation

1. Introduction

CFRP exhibits numerous advantages, including lightweight, high strength, corrosion resistance, fatigue resistance, and ease of transportation and construction. It has been widely employed for the strengthening and maintenance of bridges and various building components [1–3]. The most common CFRP strengthening methods consist of direct bonding and prestressed CFRP strengthening. The direct bonding method involves adhering CFRP to the bottom of the reinforced component, significantly enhancing the component's load-bearing capacity and fatigue resistance [4–7]. However, this method still has drawbacks, such as the under-utilization of CFRP strength and limited improvement in the rigidity of the reinforced components [8,9]. Existing studies indicate that when externally bonded CFRP plates are applied for the flexural reinforcement of structures, the effectiveness of the CFRP plates often depends on the strength of the bonding interface between the plates and the structure. When the CFRP-reinforced structure fails, the usual failure mode is interface failure, where the CFRP plate peels off from the structure. At this point, the strength of the CFRP plates is only about one-fifth of their ultimate tensile strength. Due to the relatively low interfacial bonding strength, the high tensile performance of CFRP sheets is often not fully utilized [10,11].

To solve the above issues, researchers have proposed and developed the prestressed CFRP strengthening method. This method involves applying prestress to CFRP plates [12–14], which not only fully leverages the material's advantages but also enhances structural performance. In terms of experimental research, the use of mechanical anchoring for prestressed tensioning of CFRP plates has been proven to be an effective method for maintaining the synergistic work of CFRP and concrete beams in the event of premature debonding [15–17]. Research has shown that the use of prestressed reinforcement technology with mechanical anchoring at both ends can effectively improve the load-bearing capacity of the structure and the utilization rate of CFRP plates, avoiding the failure of the anchored end of CFRP and the delamination failure of CFRP plates [18,19]. The application of mechanical anchoring prestressed CFRP reinforcement technology in actual bridge-strengthening projects has achieved good strengthening effects [20,21]. Kim et al. [22] selected a method of fiber-reinforced polymer (FRP) bonded in sheet form to conduct an experimental study on the performance of the reinforced structure. Soureshjani et al. [23] studied the seismic performance of bridge piers strengthened with CFRP. The results show that CFRP reinforcement can improve the seismic performance of structures. However, with the increase in service time and the influence of the external environment, prestressed CFRP plate-strengthened bridges inevitably lead to prestress loss, which affects the strengthening effect.

To investigate the mechanism and causes of prestress loss in the mechanical anchoring reinforcement method, many scholars have conducted extensive monitoring experiments [24–26] Hu5ang et al. [27] proposed a fatigue life prediction model for reinforced concrete beams strengthened with post-tensioned CFRP plates, considering the component materials and partial delamination of CFRP. They found that the fatigue performance of prestressed CFRP-reinforced beams was significantly better than that of unreinforced beams. Peng et al. [28] studied the effects of different strengthening methods on the flexural bearing capacity and fatigue performance of seven reinforced concrete beams. In addition to the concrete test components mentioned above, studies have also been conducted on the strengthening of prestressed CFRP plates in bridges. Puurula et al. [29] established a finite element model based on a 50-year full-scale failure test of a reinforced concrete railway bridge and evaluated the bearing capacity of different strengthening schemes using the model. The bridge can withstand loads 6.5 times greater than the existing load requirements. Wu [30] studied the longitudinal and transverse stress redistribution law of continuous beam bridges reinforced with prestressed CFRP plates. For bridges composed of multiple beams, the uneven transverse stress caused by the prestressed system during strengthening cannot be ignored, and the strengthening effect between adjacent spans will affect each other and cannot be ignored. In summary, compared with other strengthening methods, the prestressed CFRP strengthening method has the advantages of improving the bending and shear bearing capacity of components, improving the stress distribution of components, suppressing deformation and crack propagation, almost not increasing the self-weight of the structure, and a high material strength utilization rate. In the existing research on the bending performance of prestressed CFRP plate-strengthened beams, the strengthening effect is usually evaluated through ultimate load tests. However, the ultimate load test is a destructive test, and it is not practical to evaluate the reinforcement performance of actual bridges by conducting ultimate load tests.

Static vehicle load testing is used to evaluate the bearing capacity of actual bridges. By applying vehicle loads equivalent to the design loads, the monitoring data is compared with the theoretical values calculated by the model to evaluate whether the deformations and stresses meet the requirements for bridge-bearing capacity. However, conducting static load tests on bridges requires interrupting normal traffic, organizing and arranging loading vehicles, and requires a significant amount of time and economic costs. In addition, in order to evaluate the effectiveness of bridge strengthening, two static load tests are required, one before and one after strengthening, to compare the results before and after strengthening, which will incur greater costs. To assess the current bridge conditions, researchers investigated the use of influence lines for evaluating bridge stiffness under moving vehicle loads [31–33]. By monitoring the dynamic response of vehicles passing over bridges, the deformation and stress conditions of the bridges can be evaluated. The most commonly used theory for this evaluation is the influence line theory, which converts moving vehicle loads into quasi-static loads. These approaches can effectively identify and assess the longitudinal distribution pattern of stress on the bridges. However, in practical engineering, the interaction between vehicles and bridges is a complex dynamic process. Simplifying the vehicle to a moving concentrated force in finite element simulations can inevitably lead to significant errors.

In addition, the existing research on strengthening monitoring methods mainly involves attaching "point" type strain gauges to the surface of CFRP plates and evaluating the stress conditions during the reinforcement process by monitoring the stress of the CFRP plates. However, if the gauge length of the strain gauge is too small, it is prone to failure due to the influence of small surface cracks on the CFRP plate. This study proposes a method and influence parameters for evaluating prestressed CFRP plate-strengthened bridges under bridge–vehicle coupling, which utilizes the regional monitoring advantages of distributed long-gauge strain sensing technology. A simulation based on the displacement contact method was studied to analyze the vibration response of the strengthened bridge–vehicle coupling, and a finite element model of a prestressed CFRP plate-strengthened multi-beam composite bridge under bridge–vehicle coupling was established. By employing numerical simulation methods, the influence of different prestressing levels of CFRP plates, variations in vehicle speeds, and different axle loads on the strengthened bridge evaluation were studied. The proposed method attempts to quickly evaluate reinforced bridges without interrupting normal traffic. Using distributed long-gauge sensors to achieve full coverage of key areas in reinforced bridges overcomes the problem of traditional strain monitoring being too "local" and unable to achieve regional monitoring. This research shows that the longitudinal strain distribution pattern before and after bridge reinforcement is well identified by moving vehicle loads and is not affected by changes in vehicle parameters.

2. Evaluation of Prestressed CFRP Plate-Strengthened Bridge Using Distributed Sensing

By secondary encapsulating fiber Bragg grating sensors, long-gauge strain (or macro-strain) sensors can be produced, enabling distributed monitoring. In this section, the formula for solving the long-gauge strain at the bottom of the concrete beam of the bridge structure after strengthening with prestressed CFRP plates is derived. The bridge structure is illustrated in Figure 1, where the span of the bridge is l, and the bottom of the bridge is strengthened with prestressed CFRP plates. The symbols E, I, A, l, y, and A_{CFRP} represent the elastic modulus of the beam structure, the moment of inertia of the cross-section about the neutral axis, the cross-sectional area, the calculated length of the beam, the height of the neutral axis, and the cross-sectional area of the prestressed CFRP plates, respectively. There are i long-gauge strain sensors arranged at the bottom of the beam, with the center of the *i-th* sensor located at a distance x_i from the support. The CFRP plates are tensioned and anchored at both ends to the bottom of the beam to work together with the concrete bridge, with a tensile force of F_p. The prestress in the CFRP plates is $\sigma_p = F_p/A_{CFRP}$.

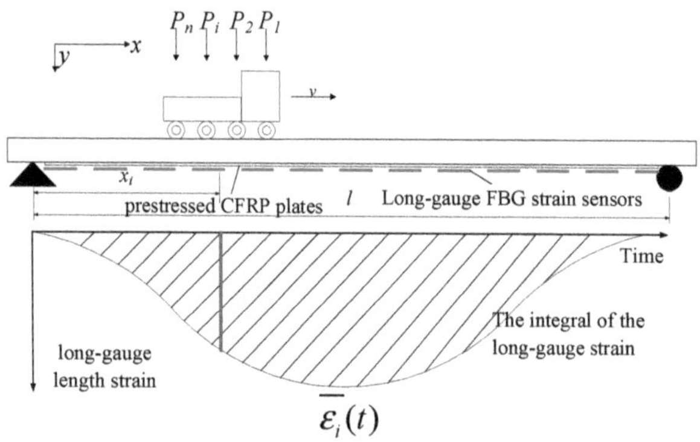

Figure 1. Schematic diagram of evaluation under moving loads.

The long-gauge strain $\overline{\varepsilon}_{pi}$ at the *i*-th sensor inside the anchor point caused by prestressed CFRP strengthening is given by Equation (1):

$$\overline{\varepsilon}_{pi} = \left(\frac{F_p}{EA} + \frac{My}{EI}\right) = \left(\frac{F_p}{EA} + \frac{F_p y^2}{EI}\right) \tag{1}$$

When the CFRP plates experience prestress loss, the corresponding tensile force and the prestress of the CFRP plates are denoted as F_l and σ_l, respectively. After the prestress loss occurs, the area of the CFRP plates A_{CFRP} remains unchanged. At this point, the long-gauge strain of the *i*-th sensor after prestress loss is as follows:

$$\overline{\varepsilon}_{li} = \left(\frac{F_l}{EA} + \frac{My}{EI}\right) = \left(\frac{F_l}{EA} + \frac{F_l y^2}{EI}\right) \tag{2}$$

Dividing $\overline{\varepsilon}_{li}$ by $\overline{\varepsilon}_{pi}$, the following is obtained:

$$\frac{\overline{\varepsilon}_{li}}{\overline{\varepsilon}_{pi}} = \frac{F_l}{F_p} = \frac{\sigma_{li}}{\sigma_{pi}} \tag{3}$$

According to Equation (3), given σ_{li} and σ_{pi}, the long-gauge strain of the *i-th* sensor can be used as an evaluation index to assess the prestress loss of the CFRP plate at position x_i. This index is applicable for the prestress evaluation of CFRP plates during the strengthening process. In the operational phase, after the completion of bridge strengthening, the strengthened bridge bears moving vehicle loads. Therefore, it is necessary to study the evaluation index for the prestress loss of the CFRP plate under traffic conditions without interruption. The strain influence line was primarily used for vehicle load identification by

Ojio and Yamada [34]. The strain influence line $f_i(x)$ at the bottom of the bridge at position x_i is expressed as follows in Equation (4):

$$\begin{cases} f_i(x) = \frac{(l-x_i)yx}{EIl} & 0 \le x \le x_i \\ f_i(x) = \frac{x_i y}{EI}\left(1 - \frac{x}{l}\right) & x_i < x \le l \end{cases} \tag{4}$$

When a vehicle with speed v passes through the bridge, the axle weight of the j-th axle is $P_j(j = 1{\sim}n)$, and the distance between wheel j and the first wheel is $d_j(j = 1{\sim}n)$, the long-gauge strain response of the i-th sensor under the moving vehicle load can be expressed as follows:

$$\bar{\varepsilon}_{Vi}(x) = \sum_{j=1}^{n} P_j f_i(x - d_j) \tag{5}$$

The integral of the long-gauge strain generated by the vehicle load along the span direction of the beam is as follows:

$$\int_0^{l+d_n} \bar{\varepsilon}_{Vi}(x)dx = \sum_{j=1}^{n} P_j \int_0^{l+d_n} f_i(x - d_j)dx = \sum_{j=1}^{n} P_j \int_0^{l} f_i(x)dx \tag{6}$$

By substituting $f_i(x)$ and $t = (l + d_n)/v$ into the above equation, the integral value S_{vi} with respect to time t for the long-gauge strain of the i-th sensor under vehicle load is obtained:

$$\int_0^{l+dn} \bar{\varepsilon}_{Vi}(x)dx = vS_{vi} = \frac{(l - x_i)yx_i \sum_{j=1}^{n} P_j}{2EI} \tag{7}$$

The integral value S_{pi} of the long-gauge strain of the i-th sensor caused by prestressing strengthening with respect to time t can be expressed as follows:

$$S_{Pi} = \int_0^t \bar{\varepsilon}_{Pi}dt \tag{8}$$

Substituting Equation (2) into Equation (8), respectively, the following is obtained:

$$S_{Pi} = \left(\frac{F_P}{EA} + \frac{F_P y^2}{EI}\right) \cdot t \tag{9}$$

Thus, the integral value S_i of the long-gauge strain with respect to time t for the bridge strengthened by prestressed CFRP plates as the vehicle passes is expressed as shown in Equation (10):

$$S_i = S_{Vi} + S_{Pi} \tag{10}$$

Substituting S_{Pi} and S_{Vi} into Equation (10) yields the expression for S_i:

$$S_i = S_{Pi} + S_{Vi} = \frac{(l - x_i)yx_i \sum_{j=1}^{n} P_j}{2EIv} - \left(\frac{F_P}{EA} + \frac{F_P y^2}{EI}\right) \cdot \frac{l + d_n}{v} \tag{11}$$

According to Equation (11), it can be concluded that the damage S_i caused by a test vehicle with known speed and axle load passing over the strengthened bridge can be used as an evaluation index to assess the prestress loss of the prestressed CFRP plates at the corresponding location x_i. This can serve as the evaluation metric for the prestress loss in the CFRP plates, leading to the proposal of a corresponding distributed static and dynamic prestress loss assessment method. The above derivation process is based on the concentrated force of movement without considering the influence of different vehicle parameters, such as the influence of different vehicle speeds, weights, and types. Therefore, this study mainly investigates the effects of different vehicle parameters and different tensile stresses on the above methods.

3. Numerical Simulation of Bridge Strengthening Under Bridge–Vehicle Coupled Vibration

3.1. Calculation of Bridge–Vehicle Coupled Vibration Based on the Displacement Contact Method

In the previous section, a theoretical analysis was conducted on the assessment method for prestressed CFRP plate-strengthened bridges. To verify the feasibility of the proposed method, a simulation method based on the displacement contact method for bridge–vehicle coupled vibration is employed. A finite element model of a prestressed CFRP plate-strengthened multi-beam composite bridge under bridge–vehicle coupled vibration is established to investigate the influence of different CFRP plate prestressing levels and vehicle parameters on the evaluation indices of bridge strengthening based on the envelope line of distributed macro-strain amplitude. The finite element model adopts a 32 m span simply supported multi-beam composite bridge consisting of five I-shaped steel beams with a main beam spacing of 2.13 m, a bridge deck width of 10.65 m, and a deck thickness of 0.20 m [35]. The deck material is C50 concrete, and the I-shaped steel beams are made of Q345 steel. The cross-sectional schematic of the bridge is shown in Figure 2a. The parameters of the main beams are presented in Figure 2b.

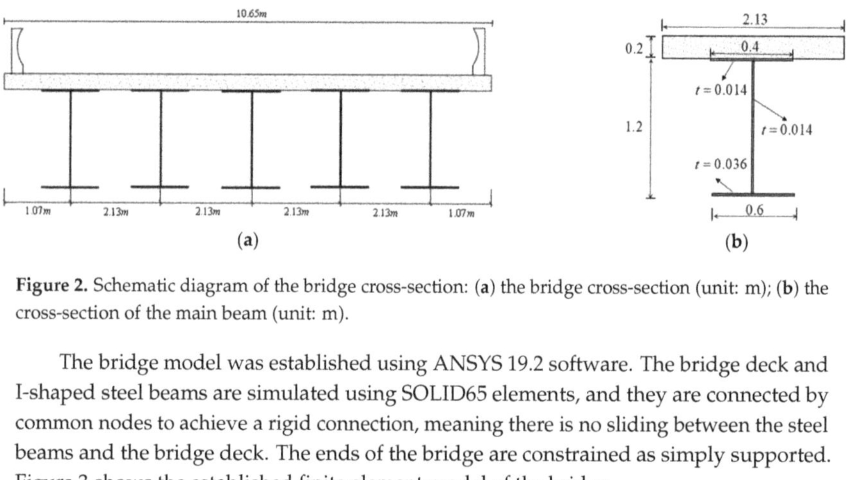

Figure 2. Schematic diagram of the bridge cross-section: (**a**) the bridge cross-section (unit: m); (**b**) the cross-section of the main beam (unit: m).

The bridge model was established using ANSYS 19.2 software. The bridge deck and I-shaped steel beams are simulated using SOLID65 elements, and they are connected by common nodes to achieve a rigid connection, meaning there is no sliding between the steel beams and the bridge deck. The ends of the bridge are constrained as simply supported. Figure 3 shows the established finite element model of the bridge.

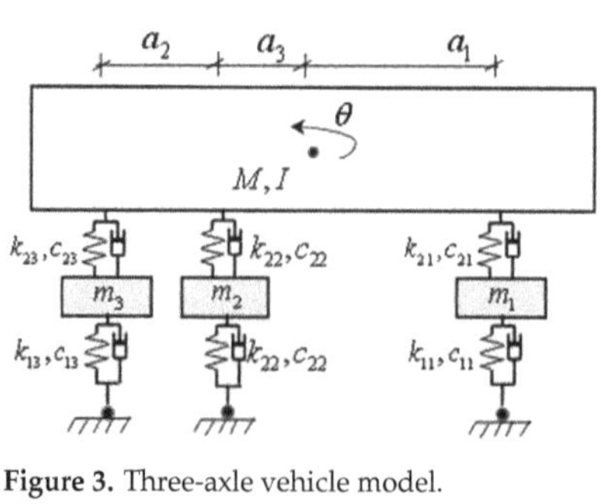

Figure 3. Three-axle vehicle model.

A 3-axle vehicle model was selected as the simulated loading vehicle, which consists of three parts: the vehicle body, suspension, and tires. The vehicle body considers degrees of freedom for vertical movement and pitching, while the three tires consider vertical movement degrees of freedom. The vehicle body and tire elements are simulated using MASS21 elements, and the spring-damping system between the vehicle body, suspension, and tires is simulated using COMBIN14 elements, as shown in Figure 3. The parameters of the model vehicle are shown in Table 1.

Table 1. Vehicle Parameters.

Parameters	Units	Value
Weight of the middle and rear wheels m_1, m_2	kg	1600
body weight M	kg	27,400
Front-wheel weight m_1	kg	1400
Vehicle pitch moment of inertia I	kg·m^2	288,000
Stiffness coefficient of the upper suspension system of the front wheel k_{21}	kN·m^{-1}	4200
Damping coefficient of front wheel upper suspension system c_{21}	kN·s·m^{-1}	1.75
Stiffness coefficient of front wheel lower suspension system k_{11}	kN·m^{-1}	4000
Damping coefficient of front wheel lower suspension system c_{11}	kN·s·m^{-1}	0.852
Stiffness coefficient of the upper suspension system of the middle and rear wheels k_{22}, k_{23}	kN·m^{-1}	15,000
Damping coefficient of the upper suspension system of the middle and rear wheels c_{22}, c_{23}	kN·s·m^{-1}	3.872
Stiffness coefficient of the lower suspension system of the middle and rear wheels k_{12}, k_{13}	kN·m^{-1}	4000
Damping coefficient of the lower suspension system of the middle and rear wheels c_{12}, c_{13}	kN·s·m^{-1}	7
Front wheelbase to vehicle weight center distance a_1	m	2.7
Distance from the center of gravity of the vehicle to the center of the middle wheelbase a_2	m	1.3
Distance between the rear wheels and the center of gravity of the vehicle $a_1 + a_2$	m	2.7

The displacement contact method was used to simulate the interaction between the vehicle and the bridge by establishing contact pairs between the bridge model and the wheel contact points. Displacement is then applied along the bridge direction to the wheel contact points to simulate the movement of vehicle loads, ultimately achieving bridge–vehicle coupling vibration, as shown in Figure 4. The advantage of this method is that the moving mass can move at any position on the bridge, not confined to the nodes of the bridge model, which facilitates the creation of a more refined bridge model.

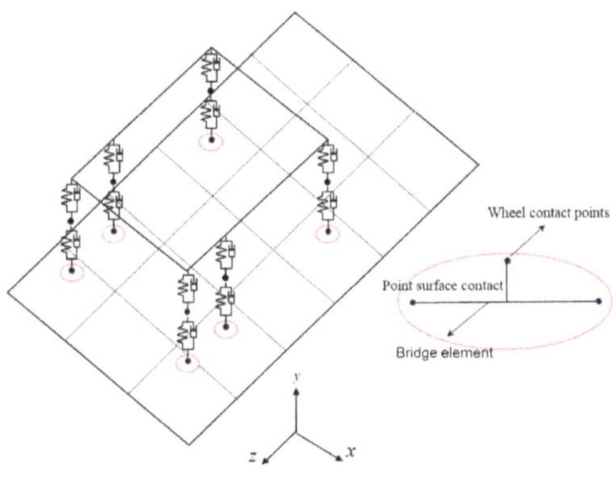

Figure 4. Schematic diagram of displacement contact method.

The bridge deck is defined as the target surface, simulated using the target element TARGE170. The wheel contact points are defined as the contact surface, simulated using the contact element CONTA175. The contact between the contact elements and target elements is set to be non-separable. The displacement coordination between the two contact surfaces is achieved through the augmented Lagrangian algorithm. Considering the structure's predominant bending deformation, the contact stiffness is set to 0.1. The effect of pavement irregularities on the bridge–vehicle coupling system is simulated by applying a pair of interaction forces at the wheel-bridge contact points.

The central displacement time history diagrams of the simply supported beam at different speeds, derived from the algorithm in this paper, are compared with the displacement time history diagrams from the original literature, as shown in Figure 5a,b. Figure 5a,b compare the mid-span displacement time histories between the displacement contact method and the reference calculation method, with vehicle speeds of 60 km/h and 120 km/h, respectively. By comparing the central displacement time history diagrams at different speeds, it is evident that the results from the literature reference align well with the results calculated by the method in this paper. This indicates that the method described in this paper can effectively and accurately simulate and analyze bridge–vehicle coupling vibration.

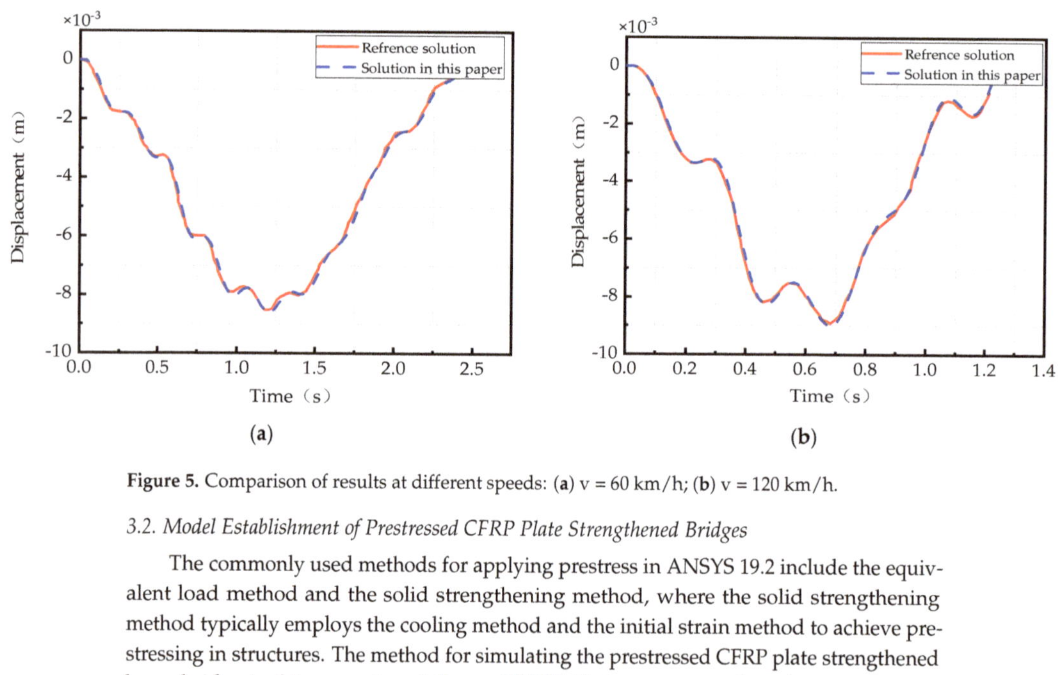

Figure 5. Comparison of results at different speeds: (**a**) v = 60 km/h; (**b**) v = 120 km/h.

3.2. Model Establishment of Prestressed CFRP Plate Strengthened Bridges

The commonly used methods for applying prestress in ANSYS 19.2 include the equivalent load method and the solid strengthening method, where the solid strengthening method typically employs the cooling method and the initial strain method to achieve prestressing in structures. The method for simulating the prestressed CFRP plate strengthened beam bridge in this paper is as follows: SHELL63 elements are selected to simulate the CFRP plates, which are arranged at the bottom of each steel beam; all degrees of freedom of the end nodes of the CFRP plates, simulated with SHELL63 elements, are coupled with the corresponding nodes of the steel beam simulated with SOLID65 elements to model the effect of the CFRP plate anchoring system in practice; the midsection nodes of the CFRP plate elements are coupled with the corresponding steel beam nodes for lateral and vertical degrees of freedom to ensure the collaborative work between the CFRP plates and the steel beams; the cooling method is used to simulate the application of prestress to the CFRP plates, where the relationship between temperature change and prestress in the CFRP plates is given by $\Delta T = -F/(E\alpha_f)$, where F is the applied prestress, E is the elastic modulus of the CFRP plates, and α_f is the linear expansion coefficient of the CFRP plates. The specific parameters of CFRP plates are shown in Table 2.

Table 2. Specific parameters of CFRP plates.

Parameters	Value
Tensile strength (MPa)	2400
Elastic modulus (GPa)	163
Thickness (m)	3×10^{-3}
Width (m)	0.1
Length (m)	26
Linear expansion coefficient	3.2×10^{-5}

The finite element model of the bridge has a total of 18,160 nodes and 25,587 elements. In order to meet the calculation accuracy, the mesh size of the bridge deck is 0.5 × 0.28 × 0.2 m, the mesh size of the steel beam web element is 0.5 × 0.014 × 0.23 m, and the mesh size of the flange plate is 0.5 × 0.125 × 0.014 m (upper flange) and 0.5 × 0.125 × 0.036 m (lower flange). The prestressed CFRP plate has 780 elements and a grid size of 0.5 × 0.043 × 0.003 m. The main beam material of the bridge is Q345, with an elastic modulus of 2.10×10^{11} Pa, Poisson's ratio of 0.3, density of 7850 kg/m^3, yield strength of 3.45×10^8 Pa, tensile strength of 5.60×10^8 Pa, and moment of inertia of a single beam of 0.011 m^4. The bridge deck is made of C50 concrete with an elastic modulus of 3.45×10^{10} Pa, Poisson's ratio of 0.2, density of 2500 kg/m^3, compressive strength of 5×10^7 Pa, and moment of inertia of 20.13 m^4.

After the bridge model is established, the coupling effect of the CFRP plates on the bridge and the relevant prestressing are considered according to the above methods. Finally, a vehicle model and the contact relationship between the vehicle and the bridge are established. The finite element model of the multi-beam composite bridge strengthened with prestressed CFRP plates under the bridge–vehicle coupling effect, established in this paper, is shown in Figure 6.

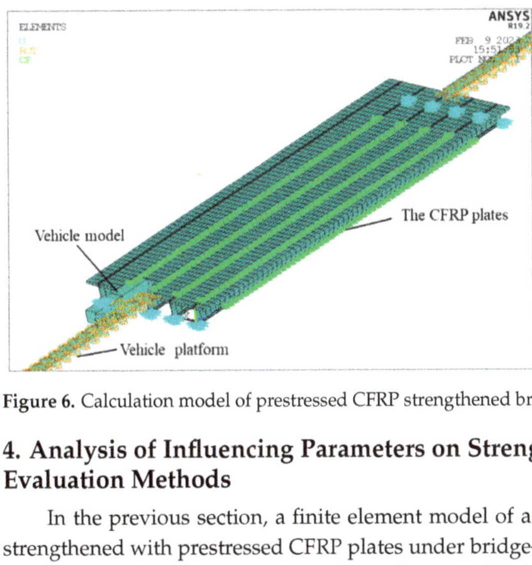

Figure 6. Calculation model of prestressed CFRP strengthened bridge with bridge–vehicle coupling.

4. Analysis of Influencing Parameters on Strengthening Effect Evaluation Methods

In the previous section, a finite element model of a multi-beam composite bridge strengthened with prestressed CFRP plates under bridge–vehicle coupling effects and a bridge–vehicle coupling calculation method was established. To investigate the influence of different strengthening levels and vehicle parameters on the evaluation results, a relevant parameter influence analysis was conducted. This analysis examines the effects of varying CFRP plate prestress levels, vehicle speeds, and vehicle axle loads on the proposed evaluation method for strengthening effects. It is assumed that 31 long-gauge fiber Bragg grating sensor (FBG) strain sensors with a gauge length of 1 m are arranged at the bottom of the beam, and the sensors are sequentially numbered as G1~G31, as shown in Figure 7.

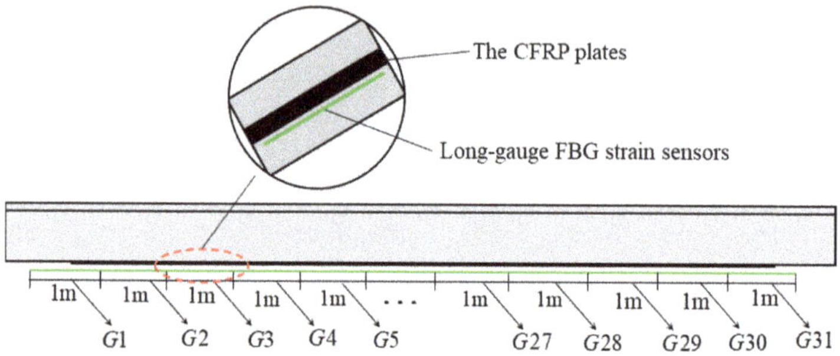

Figure 7. Sensor position layout.

4.1. The Impact of Different Prestress Levels of CFRP Plates

This section primarily considers the impact of different prestress levels of CFRP plates on the evaluation methods for strengthening effectiveness. One initial scenario, P0, and five strengthened scenarios, P1 to P5, were designed. The prestress levels of CFRP plates under each strengthened scenario are detailed in Table 3. The prestress level for 100% CFRP plates is 1200 MPa. The vehicle speed is set at 120 km/h, and the vehicle axle load and related parameters are provided in Section 3.1.

Table 3. Design of CFRP plate strengthening conditions.

Cases	Level (%)	Tensile Stress (Mpa)	CFRP Plate Strain (με)	CFRP Plate Sizes		
				Width (m)	Thickness (m)	Area (m²)
P0	/	/	/	/	/	/
P1	20	240	1472	0.1	3×10^{-3}	3×10^{-4}
P2	40	480	2945	0.1	3×10^{-3}	3×10^{-4}
P3	60	720	4417	0.1	3×10^{-3}	3×10^{-4}
P4	80	960	5890	0.1	3×10^{-3}	3×10^{-4}
P5	100	1200	7362	0.1	3×10^{-3}	3×10^{-4}

The calculated results for different cases under the bridge–vehicle coupling effect are extracted. The macro-strain time–history curves at the mid-span of the beam (covering the area of sensor G16) under different CFRP plate strengthening cases are shown in Figure 8a, and the mid-span node displacement time–history curves are shown in Figure 8b. From Figure 8a,b, it can be observed that after reinforcing the bridge with prestressed CFRP plates, the macro-strain time–history curves in the sensor coverage area and the mid-span node displacement time–history curves have an initial value, which is due to the initial stress. This initial value is related to the prestress level of the CFRP plate, but the presence of the prestress does not alter the trend of the curves or the location of the maximum amplitude. For each reinforcing case, the macro-strain amplitude and displacement amplitude of the curves both occur within the sensor span. With the increase in CFRP plate tension prestress, the amplitude of macro-strain and displacement time–history curves at the mid-span of the bridge generally decrease, indicating that prestressing strengthening has a good effect on reducing bridge deformation and stress. The results indicate that the bridge–vehicle coupling calculation method based on the displacement contact method can effectively solve the dynamic response of bridges reinforced with pre-stressed CFRP plates.

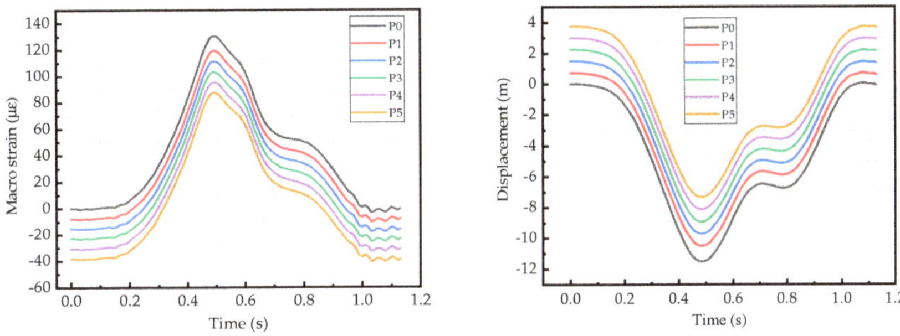

(**a**) (**b**)

Figure 8. Displacement and the macro-strain response of a strengthened bridge under bridge–vehicle coupling vibration: (**a**) macro-strain time-history curves; (**b**) displacement time–history curve.

The measured distributed macro-strain amplitudes under different strengthening cases are plotted as envelopes, as shown in Figure 9a. From the figure, it can be observed that the macro-strain amplitude envelope exhibits a staged decline trend within the anchorage zone, particularly in the coverage area of sensors G4 to G28, as the prestress level of the CFRP plate increases. Outside the anchorage zone, the effect of the initial stress gradually diminishes, and the declining trend of the measured macro-strain amplitude decreases with increasing distance from the anchorage point. Near the supports, the macro-strain amplitudes measured by the sensors remain relatively unchanged across different scenarios. The maximum values of the macro-strain amplitude envelopes for all strengthening scenarios occur at sensor G16 in the mid-span, with the values being 130.6 με, 116.0 με, 104.4 με, 92.7 με, 81.0 με, and 69.4 με, respectively. In the P5 case, where the CFRP plate is at full prestress, the macro-strain amplitude at sensor G16 is reduced by approximately 61.2 με.

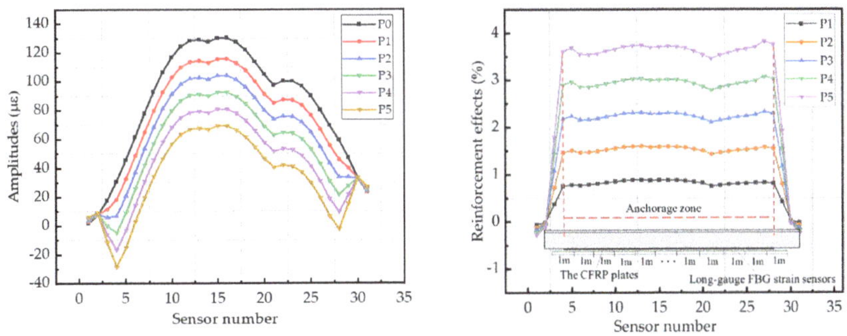

(**a**) (**b**)

Figure 9. Identification results of strengthening effect under different tensile stress conditions: (**a**) envelopes of macro-strain under different cases; (**b**) strengthening effect measured by each sensor under different cases.

Through the above analysis, it can be concluded that the prestressed CFRP plate strengthening of beam bridges effectively reduces the mid-span strain and vertical deflection of the bridges when subjected to vehicle loads. By substituting the macro-strain amplitudes measured by different sensors under each strengthening scenario into Equation (11), the strengthening effect of the prestressed CFRP plate on the beam bridge in this simulation can be determined. From Figure 9b, it can be observed that the strengthening effect distribution curves measured by the sensors exhibit an overall upward trend with increasing CFRP plate prestress. Within the anchorage zone, the strengthening effects measured by each sensor are generally consistent. Outside the anchorage zone, the strengthening effects measured by each sensor decrease with increasing distance from the anchorage zone. Under the full prestress level of the CFRP plate, i.e., the P5 strengthening case, the strengthening effects measured near the mid-span by sensors G15, G16, and G17 are 3.71%, 3.73%, and 3.72%, respectively.

4.2. Impact of Different Vehicle Speeds on the Assessment Method

To study the impact of vehicle speed on the proposed strengthening effect assessment method, the changes in the distributed macro-strain amplitude envelopes of the bridge under different vehicle speeds were analyzed. The vehicle speed scenarios were designed as 40 km/h, 60 km/h, 80 km/h, 100 km/h, and 120 km/h. The macro-strain amplitude envelopes at different speeds before and after strengthening can be seen in Figure 10.

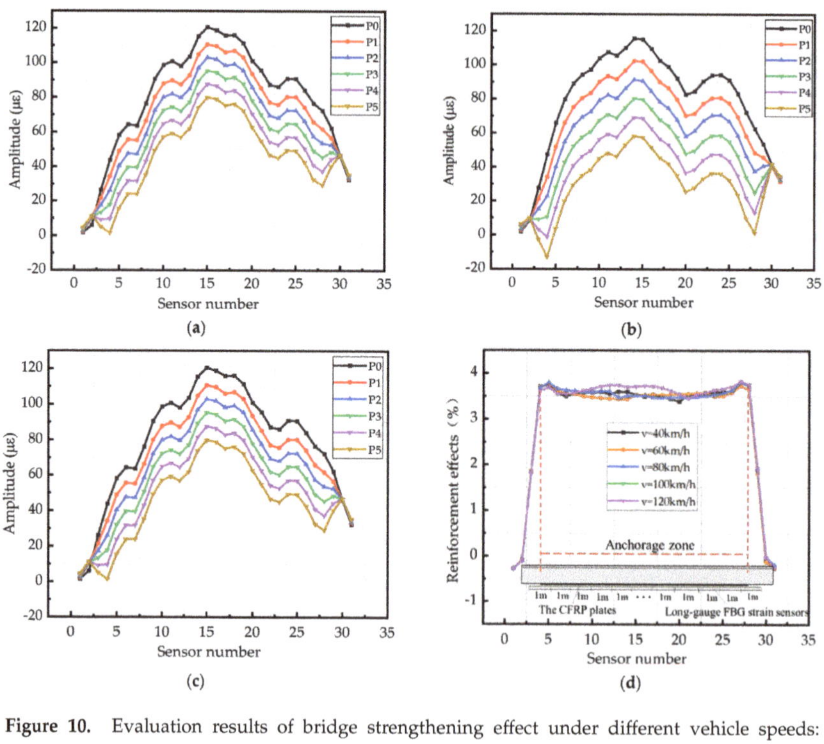

Figure 10. Evaluation results of bridge strengthening effect under different vehicle speeds: (**a**) v = 40 km/h; (**b**) v = 80 km/h; (**c**) v = 120 km/h; (**d**) envelope lines of macro-strain amplitude at different speeds.

From Figure 10a–c, it can be observed that the distributed macro-strain amplitude envelopes at the same speed show a regular decreasing trend under different strengthening cases. However, the shapes of the macro-strain amplitude envelopes measured by each sensor vary at different vehicle speeds, and the corresponding macro-strain amplitudes for each sensor also differ. To verify whether this phenomenon impacts the assessment method for the strengthening effects proposed in this study, a comparative analysis of the strengthening effects measured by each sensor under different speeds in the strengthening scenario P5 is shown in Figure 10d. It can be found that the distribution curves of the bridge strengthening effects at different speeds are largely overlapping, with the maximum error in the strengthening effect at the mid-span being approximately 6%. Thus, the vehicle speed has little effect on the assessment method for strengthening effects presented in this paper.

4.3. Impact of Different Vehicle Weights on the Assessment Method

In actual situations, the types of vehicles are random, and the axle loads of the vehicles will vary. By altering the vehicle mass, the influence of different vehicle weights on the proposed strengthening effect assessment method in this paper is simulated. Assuming the standard vehicle mass is P, 0.5P and 1.5P are used to simulate the effects of light and heavy vehicles in practice, respectively. Other vehicle parameters remain unchanged, and the vehicle speed is set to 120 km/h. Figure 11 shows the macro-strain amplitude envelopes for vehicles with different axle loads under the P0 scenario. It can be seen that with the increase in vehicle axle load, the macro-strain amplitudes measured by each sensor increase synchronously, but the shapes of the macro-strain amplitude envelopes remain largely unchanged.

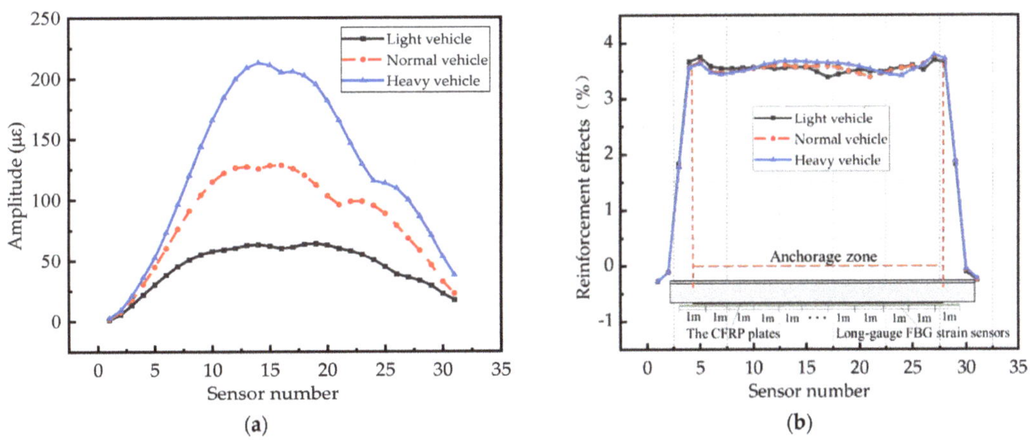

Figure 11. Impact of different vehicle weights on the assessment method: (**a**) envelope lines of macro-strain amplitude values for different vehicle weights under P0 operating conditions; (**b**) strengthening effect under different vehicle axle loads.

The evaluation method for the strengthening effect discussed in this paper is suitable for the detection of medium and small-span bridges. Under the condition of ensuring consistent vehicle parameters before and after strengthening, this method can effectively carry out assessments of the bridge strengthening effect. Outside the anchoring zone of the CFRP plates, the strengthening effect is not significant due to the reduced utility of initial stress. Through the longitudinal stress distribution of the bridge during the reinforcement process, it is known that there is a significant stress concentration and complex stress distribution at the anchorage points of the CFRP plate. However, at positions within the anchoring zone, the strengthening effect measured by each sensor is basically the same. Within the range of two anchoring points, as the tensile stress increases, the longitudinal stress at the bottom of the beam increases significantly, which can better evaluate the reinforcement effect of the bridge. However, at positions within the anchoring zone, the strengthening effect measured by each sensor is basically the same. Therefore, when conducting evaluations and monitoring the strengthening effect of prestressed CFRP plates on bridges, long-gauge FBG strain sensors should preferably be arranged in the anchoring sections of the CFRP plates.

5. Conclusions

This article mainly studies the evaluation method of reinforced bridges based on distributed monitoring technology under bridge–vehicle coupling, as well as the influence of vehicle parameter changes on tension and prestress changes. Firstly, the calculation formula for the distributed strain influence line of reinforced bridges under moving vehicle loads was derived, and the distributed strain amplitudes before and after reinforcement were used as evaluation indicators. The main conclusions of this article are as follows:

(1) The finite element computation results show that, when subjected to vehicle loads, the mid-span strain and mid-span displacement of the bridge are significantly reduced after strengthening with prestressed CFRP plates. Comparing the mid-span macro-strain values and mid-span displacement values before and after strengthening, the mid-span macro-strain value and mid-span displacement value after strengthening are reduced by 61.2 $\mu\varepsilon$ and 0.0047 m, respectively;

(2) With changes in vehicle speed, the shape and magnitude of the macro-strain envelope of the bridge also change. However, when the vehicle speeds before and after strengthening are consistent, the distribution curves of the strengthening effect measured by each sensor do not exhibit significant changes due to the differences in the shape and magnitude of the macro-strain envelope. The maximum error of the mid-span reinforcement effect is about 6%. The distribution curves of the strengthening effect at different vehicle speeds essentially overlap;

(3) The influence of vehicle load on the evaluation method for the strengthening effect is that changes in axle load cause synchronous variations in the macro-strain measured by each sensor, but the shape of the macro-strain envelope remains unchanged. When using long-gauge sensors for monitoring and evaluating the anchored prestressed CFRP plates strengthening method, it is more effective to install the sensors within the area between the two end anchorage points.

Author Contributions: Conceptualization, B.W.; Methodology, Z.Z.; Software, Y.G., Y.S. and H.W.; Validation, Z.Z. and H.W.; Writing—original draft, S.F.; Writing—review and editing, B.W.; Supervision, Z.Z.; Funding acquisition, B.W. All authors have read and agreed to the published version of the manuscript.

Funding: This study was supported by the National Natural Science Foundation of China (Grant No. 52368042), the Jiangxi Provincial Cultivation Program for Academic and Technical Leaders of Major Subjects (Grant No. 20225BCJ23025), the Natural Science Foundation of Jiangxi Province (Grant No. 20242BAB22008), and the Beijing Nova Program (20220484088, 20230484402).

Data Availability Statement: The data presented in this study are available in the article.

Conflicts of Interest: Author Yanchao Shao is employed by the Central Research Institute of Building and Construction Co., Ltd. MCC. The remaining authors declare that the research was conducted in the absence of any commercial or financial relationships that could be construed as a potential conflict of interest.

References

1. Havez, A.A.; Al-Mayah, A. Flexural strengthening of concrete structures using externally bonded and unbonded prestressed CFRP laminates: A Literature review. *J. Compos. Constr.* **2023**, *27*, 03123001. [CrossRef]
2. Shi, J.W.; Wu, Q.Q.; Li, B.; Liu, Y.; Cao, W.H.; Wang, H.T. Fatigue bond behavior of FRP-to-concrete joints with various bonding adhesives. *Eng. Struct.* **2024**, *301*, 117311. [CrossRef]
3. Duan, L.; Wang, C.; Brühwiler, E.; Wang, Q. Experiments on the flexural behavior of full-scale PC box girders with service damage strengthened by prestressed CFRP plates. *Compos. Struct.* **2023**, *312*, 116864. [CrossRef]
4. Taheri, F.; Shahin, K.; Widiarsa, I. On the parameters influencing the performance of reinforced concrete beams strengthened with FRP plates. *Compos. Struct.* **2002**, *58*, 217–226. [CrossRef]
5. Esfahani, M.R.; Kianoush, M.R.; Tajari, A.R. Flexural behaviour of reinforced concrete beams strengthened by CFRP sheets. *Eng. Struct.* **2007**, *29*, 2428–2444. [CrossRef]
6. Boukhezar, M.; Samai, M.L.; Mesbah, H.A. Flexural behaviour of reinforced low-strength concrete beams strengthened with CFRP plates. *Struct. Eng. Mech.* **2013**, *47*, 819–838. [CrossRef]
7. Gong, S.; Su, M.; Yoshitake, I.; Zhu, C.; Peng, H. Factors affecting flexural properties of RC beams strengthened with gradually prestressed NSM CFRP strips. *Eng. Struct.* **2024**, *306*, 117865. [CrossRef]
8. Wang, H.T.; Liu, S.S.; Shi, J.W.; Xu, G.W.; Chen, M.S.; Bian, Z.N. Bonded and unbonded strengthening of RC beams with the prestressed CFRP plate system: An experimental study. *Constr. Build. Mater.* **2024**, *438*, 137118. [CrossRef]
9. Zaki, M.A.; Rasheed, H.A.; Roukerd, R.R.; Raheem, M. Performance of reinforced concrete T beams strengthened with flexural CFRP sheets and secured using CFRP splay anchors. *Eng. Struct.* **2020**, *210*, 110304. [CrossRef]
10. Aram, M.R.; Czaderski, C.; Motavalli, M. Debonding failure modes of flexural FRPstrengthened RC beams. *Compos. Part B Eng.* **2008**, *39*, 826–841. [CrossRef]
11. ACI Committee 440. *Guide for the Design and Construction of Externally Bonded FRP Systems for Strengthening Concrete Structures (ACI 440.2 R-22)*; American Concrete Institute: Farmington Hills, MI, USA, 2022. [CrossRef]
12. Wang, H.T.; Bian, Z.N.; Wu, Q.; Hu, L.; Chen, M.S.; Tang, C. Development of a novel anchorage and tensioning system for strengthening steel beams with unbonded prestressed CFRP plates. *Eng. Struct.* **2024**, *312*, 118234. [CrossRef]
13. Piatek, B.; Siwowski, T.; Michalowski, J.; Blazewicz, S. Flexural strengthening of RC beams with prestressed CFRP strips: Development of novel anchor and tensioning system. *J. Compos. Constr.* **2020**, *24*, 04020015. [CrossRef]
14. Deng, J.; Rashid, K.; Li, X.; Xie, Y.; Chen, S. Comparative study on prestress loss and flexural performance of rectangular and T beam strengthened by prestressing CFRP plate. *Compos. Struct.* **2021**, *262*, 113340. [CrossRef]
15. Zaki, M.A.; Rasheed, H.A. Impact of efficiency and practicality of CFRP anchor installation techniques on the performance of RC beams strengthened with CFRP sheets. *Can. J. Civ. Eng.* **2019**, *46*, 796–809. [CrossRef]
16. Chen, L.; Qiang, X.; Jiang, X.; Bai, J. Numerical study of steel–concrete composite beams strengthened by CFRP plates with prestressed unbonded reinforcement system. *Eng. Fail. Anal.* **2024**, *157*, 107905. [CrossRef]
17. Wang, H.T.; Wu, Q.; Zhu, C.Y.; Mao, Y.H.; Guo, C.C. An innovative prestressing system of prestressed carbon fiber sheets for strengthening RC beams under flexure. *Constr. Build. Mater.* **2024**, *411*, 134409. [CrossRef]
18. Su, M.; Gong, S.; Liu, Y.; Peng, H. Flexural behavior of RC beams strengthened with fully or partially prestressed near-surface mounted FRP strips: An experimental investigation. *Eng. Struct.* **2022**, *262*, 114345. [CrossRef]
19. Lee, H.Y.; Jung, W.T.; Chung, W. Flexural strengthening of reinforced concrete beams with pre-stressed near surface mounted CFRP systems. *Compos. Struct.* **2017**, *163*, 1–12. [CrossRef]
20. Michels, J.; Staśkiewicz, M.; Czaderski, C. Prestressed CFRP strips for concrete bridge girder retrofitting: Application and static loading test. *J. Bridge Eng.* **2016**, *21*, 04016003. [CrossRef]
21. Grace, N.; Jensen, E.; Matsagar, V.; Penjendra, P. Performance of an AASHTO beam bridge prestressed with CFRP tendons. *J. Bridge Eng.* **2013**, *18*, 110–121. [CrossRef]
22. Soureshjani, O.K.; Nouri, G. Seismic assessment and rehabilitation of a RC structure under mainshock-aftershock seismic sequences using beam-column bonded CFRP strategy. *Bull. Earthq. Eng.* **2022**, *20*, 4989–5016. [CrossRef]
23. Kim, T.K.; Park, J.S. Flexural Behavior Evaluation for Seismic, Durability and Structure Performance Improvement of Aged Bridge According to Reinforcement Methods. *Int. J. Concr. Struct. Mater.* **2024**, *18*, 5. [CrossRef]
24. Ghafoori, E.; Hosseini, A.; Al-Mahaidi, R.; Zhao, X.L.; Motavalli, M. Prestressed CFRP-strengthening and long-term wireless monitoring of an old roadway metallic bridge. *Eng. Struct.* **2018**, *176*, 585–605. [CrossRef]

25. Kianmofrad, F.; Ghafoori, E.; Elyasi, M.M.; Motavalli, M.; Rahimian, M. Strengthening of metallic beams with different types of pre-stressed un-bonded retrofit systems. *Compos. Struct.* **2017**, *159*, 81–95. [CrossRef]
26. Hosseini, A.; Ghafoori, E.; Al-Mahaidi, R. Strengthening of a 19th-century roadway metallic bridge using nonprestressed bonded and prestressed unbonded CFRP plates. *Constr. Build. Mater.* **2019**, *209*, 240–259. [CrossRef]
27. Huang, H.; Wang, W.W.; Dai, J.G. Fatigue behavior of reinforced concrete beams strengthened with externally bonded prestressed CFRP sheets. *J. Compos. Const.* **2017**, *21*, 04016108. [CrossRef]
28. Peng, H.; Zhang, J.; Shang, S.; Liu, Y. Experimental study of flexural fatigue performance of reinforced concrete beams strengthened with prestressed CFRP plates. *Eng. Struct.* **2016**, *127*, 62–72. [CrossRef]
29. Puurula, A.M.; Enochsson, O.; Sas, G.; Blanksvärd, T.; Ohlsson, U.; Bernspång, L.; Täljsten, B.; Carolin, A.; Paulsson, B.; Elfgren, L. Assessment of the strengthening of an RC railway bridge with CFRP utilizing a full-scale failure test and finite-element analysis. *J. Struct. Eng.* **2015**, *141*, D4014008. [CrossRef]
30. Wu, B.; Zhou, Y.; Yin, X.; Lin, Z.; Zhou, Z. The anti-arch inhibition effect of multispan continuous girder bridge strengthened with prestressed CFRP plates. *Structures* **2022**, *35*, 845–855. [CrossRef]
31. Zheng, X.; Yi, T.H.; Yang, D.H.; Li, H.N. Stiffness estimation of girder bridges using influence lines identified from vehicle-induced structural responses. *J. Eng. Mech.* **2021**, *147*, 04021042. [CrossRef]
32. Zhang, G.W.; Cheng, Y.Y.; Xia, Q.; Zhang, J. Curvature envelope area based rapid identification method of bridge distributional element stiffness using microwave interference radar. *Mech. Syst. Signal Process.* **2023**, *197*, 110390. [CrossRef]
33. Hester, D.; Brownjohn, J.; Huseynov, F.; Obrien, E.; Gonzalez, A.; Casero, M. Identifying damage in a bridge by analysing rotation response to a moving load. *Struct. Infrastruct. Eng.* **2020**, *16*, 1050–1065. [CrossRef]
34. Ojio, T.; Yamada, K. Bridge weigh-in-motion systems using stringers of plate girder bridges. In Proceedings of the Third International Conference on Weigh-in-Motion (ICWIM3) Iowa State University, Orlando, FL, USA, 13–15 May 2002.
35. Wang, W.; Deng, L.; He, X.; Cai, C.S.; Bi, T. Truck Weight Limit for Simply Supported Steel Girder Bridges Based on Bridge Fatigue Reliability. *J. Aerosp. Eng.* **2018**, *31*, 4018079. [CrossRef]

buildings

MDPI

Article

Effect of Fiber Type and Length on Strength, Fracture Energy, and Durability Properties of Microwave-Cured Fiber-Reinforced Geopolymer Mortars

Adil Gultekin

Department of Civil Engineering, Duzce University, Duzce 81620, Turkiye; adilgultekin@duzce.edu.tr

Abstract: Microwave curing can be an alternative curing method for geopolymer production. Although many properties of microwave-cured geopolymer composites have been investigated, the effect of microwave curing on the strength and durability properties of fiber-reinforced geopolymers remains a topic that requires investigation. In this study, the effect of fiber type and length on the properties of microwave-cured metakaolin-based geopolymers was investigated. For this purpose, PVA (6, 12 mm) and polymer (15, 30 mm) fibers were utilized. Compressive and flexural strength, fracture energy, abrasion resistance, high-temperature resistance, water absorption capacity and rate of capillary water absorption tests were conducted and the microstructure was examined using scanning electron microscopy. For curing, a household microwave oven was used at a power level of 300 watts. With the fibers' inclusion, fracture energies could be increased by up to 1150%, ductility was enhanced, flexural strengths were increased and compressive strengths decreased. Moreover, the rate of capillary water absorption decreased by up to 13%, while water absorption values increased by between 5% and 12%. The results suggested that microwave curing could be an alternative curing method for the production of fiber-reinforced geopolymer composites, offering shorter curing times and lower energy consumption.

Keywords: geopolymer; PVA fiber; polymer fiber; microwave curing; fracture energy; durability

1. Introduction

Portland cement is a widely used building material due to certain advantages. However, the production of cement results in significant carbon dioxide emissions and the process consumes a considerable amount of energy, which has driven the search for alternative binders [1]. Furthermore, the production of cement contributes to environmental issues such as greenhouse gas emissions, soil erosion, noise pollution and reductions in air quality [2]. Nowadays, climate change and the increase in global temperatures caused by greenhouse gases, particularly carbon dioxide, have become pressing concerns [3]. The increase in the global population has led to a continuous increase in the demand for concrete and, consequently, Portland cement [4]. Cement production is responsible for approximately 7% of global human-made greenhouse gas emissions, highlighting the significance of the issue [5]. In response, researchers have focused on reducing the quantity of cement by exploring the substitution of cement with alternative materials or by developing cement-free construction materials [6]. Studies indicate that geopolymers, which have a lower environmental impact compared to cement, can be regarded as a potential alternative binder [7].

Geopolymers have advantages such as significantly reducing CO_2 emissions resulting from cement production, ensuring the disposal of waste materials [8], good resistance to seawater, high temperature and freeze–thaw cycles [9]. Geopolymers are binders produced by the combination of powdered materials rich in alumina and silica with alkaline solutions [10]. Numerous studies have investigated the use of natural materials such as kaolin, zeolite, laterite, as well as agricultural or industrial waste materials, including fly ash, bottom ash, ground granulated blast furnace slag, silica fume, red mud, rice husk ash and waste glass, in geopolymer production [11]. Additionally, it has been reported that the calcination of natural pozzolans like trass enhances the strength of geopolymers [12].

Despite numerous advantages, geopolymers have low tensile and flexural strength due to their brittle structure, which limits the areas of application. To address these weaknesses, the incorporation of fibers is considered an effective approach [13]. The addition of fiber reinforcement to geopolymer composites enhances tensile strength, flexural strength, ductility and toughness, while also improving crack control [14]. Research indicates that microfibers slow down the propagation of cracks by acting on microcracks, thereby increasing the peak load capacity, while macro fibers bridge larger cracks and prevent or delay further crack development [15]. The working mechanism of fibers in such composites is characterized by different events like debonding, sliding and pull-out of fibers [16]. Numerous studies have investigated the reinforcement of brittle matrices using various types of fibers. Yang et al. [17] investigated the effects of basalt, polypropylene and steel fibers on the properties of fly ash–slag-based geopolymer concretes, reporting that the appropriate dosage of fiber resulted in increased splitting tensile strength, fracture toughness and fracture energy values. The most favorable results regarding fracture energy were achieved with steel fibers; specifically, the addition of 1% steel fiber led to an approximate 1910% increase in fracture energy compared to the reference mixture produced without fiber. Wang et al. [18] investigated the effects of basalt fiber addition at various dosages on the fracture behavior of fly ash-based geopolymer concretes. Findings indicated an increase in peak load, fracture toughness and fracture energy values with the incorporation of fibers, with the most favorable dosage for fracture energy being 0.05%, resulting in a 56% increase. A similar study was conducted by Li et al. [19] utilizing polypropylene fibers of 6 mm and 12 mm lengths.

The addition of fibers is known to have positive effects not only on mechanical properties but also on some durability characteristics in geopolymer composites. Studies conducted on this subject further corroborate this finding. Alves et al. [20] examined the effects of glass fibers in slag-based geopolymers, reporting a decrease in the drying shrinkage values with the incorporation of fibers. Singh et al. [21] investigated the combined use of steel and polypropylene fibers in fly ash–slag-based geopolymers, indicating that the use of fibers increased abrasion resistance and reduced the amount of capillary absorbed water. Kuranli et al. [22] investigated the effects of adding steel, polypropylene and polyamide fibers on the strength and durability properties of blast furnace slag-based alkali-activated concretes. The researchers stated that the appropriate selection of fiber type and dosage can reduce water absorption values by 13% and porosity by 12.5%. Additionally, it was stated that the incorporation of fibers can positively affect high-temperature resistance. Zaid et al. [23] reported that both the mechanical properties and drying shrinkage performance of metakaolin-based geopolymers improved with the addition of steel fiber. Similar studies have been conducted on the strength and durability properties of fiber-reinforced geopolymer composites. Xu et al. [24], Deng et al. [25] and Cai et al. [26] investigated the effects of polyvinyl alcohol (PVA) fibers, while Bhutta et al. [27], Hussein et al. [28] and Xu et al. [29] examined the effects of polyester fiber addition. These studies demonstrated that the use of fibers at appropriate dosages and lengths positively influences the properties of geopolymer composites.

Many factors, including the type of aluminosilicate, amount of glassy phase in the aluminosilicate, type and concentration of alkali solution, alkali/aluminosilicate ratio and mix proportions significantly influence the geopolymerization process. Additionally, the curing method, curing temperature and curing duration have been reported to significantly affect geopolymerization [30]. Ambient curing and oven curing are the two traditional curing methods for geopolymer production [31]. Curing under ambient conditions in geopolymers produced with low-calcium-bearing aluminosilicates results in low strength values [32]. The application of thermal curing in geopolymer production requires external energy. Therefore, researchers have initiated efforts to minimize the energy required during the curing process [33].

According to previous studies, microwave technology has emerged as an innovative heating method utilized in various fields in recent years. The application of microwaves creates a rapid heating effect throughout the entire volume, resulting in a more uniform temperature distribution within the material [34]. Additionally, microwave heating provides energy savings [35]. Due to these advantages, microwave energy finds applications in diverse areas, including food processing, chemical treatment, sintering and composite material processing [36]. One notable application is geopolymer production. Recent studies have explored the use of microwave technology as an innovative curing method in the production of geopolymers. Refaat et al. [37] investigated the applicability of microwave technology in the production of one-part blast furnace slag-based alkali-activated materials. Graytee et al. [38] examined the use of microwaves in the production of fly ash-based geopolymer paste and mortar, reporting that a compressive strength of 72 MPa could be achieved with only 15 min of curing at 300 watts, while a compressive strength of 90 MPa could be obtained with 60 min of curing at 200 watts. In a similar study, Gultekin and Ramyar [39] explored the production of geopolymer mortars using fly ash, pumice, perlite and burnt clay with microwave technology, indicating that geopolymer mortar production is feasible within short curing times. Abdullah et al. [40] also utilized microwave technology for the production of fly ash-based geopolymer mortar. Collectively, these studies demonstrated that microwave technology holds significant potential for the production of geopolymers and alkali-activated materials. Similarly, recent studies have focused on the production of various types of microwave-cured geopolymer and alkali-activated building materials. Zheng et al. [41] investigated the properties of metakaolin-based foam pastes produced with microwave curing. Nadeem et al. [42] examined the effect of the sodium silicate/aluminosilicate ratio on the properties of microwave-cured natural soil-based geopolymer foams, reporting that the porosities of the foams obtained with different ratios varied between 38% and 77%, while compressive strengths ranged from 3.1 to 8.8 MPa. Shen et al. [35] explored the usability of microwave curing in the production of ultra-high-performance geopolymer concrete. Alqahtani et al. [43] conducted a study on the characterization and properties of microwave-cured lightweight geopolymer aggregates produced from nine different mixtures containing varying proportions of fly ash, sodium silicate, sodium hydroxide and sodium bicarbonate. The study emphasized that microwave technology holds significant potential for the production of lightweight geopolymer aggregates. In a similar study, Hanif et al. [44] examined the properties of fly ash-based geopolymer lightweight aggregates produced using microwave curing, reporting the successful production of lightweight aggregates with a loose bulk density of 443 kg/m^3, a compact bulk density of 485 kg/m^3, an oven-dry specific gravity of 0.8 and a total porosity of 55%.

Although microwave curing has been studied in recent years for the production of geopolymer and alkali-activated materials, many of these investigations have been about paste, mortar, concrete, lightweight concrete, lightweight aggregates and high-strength concrete. However, the exploration microwave-cured fiber-reinforced geopolymer composites remains an under-researched area. This study evaluates the feasibility of microwave curing in the production of fiber-reinforced metakaolin-based geopolymer mortar, focusing on the effects of fiber type, length and geometry on fracture energy, as well as strength and durability properties. For these purposes, 6 mm and 12 mm polyvinyl alcohol (PVA) fibers with a straight shape and 15 mm and 30 mm curved-wave polyester fibers were used at volume dosages of 0.375% and 0.5%, respectively. In this context, CEN standard sand was used as the aggregate and a mixture of sodium silicate and sodium hydroxide was used as the alkaline activator. Curing was conducted using a household microwave oven at 300 watts for durations of up to 30 min. With the addition of fibers, fracture energies could be increased by up to 1150%, ductility was enhanced, flexural strengths were increased and compressive strengths decreased. Moreover, the rate of capillary water absorption decreased by up to 13%, while water absorption values increased by between 5% and 12%. The results suggested that microwave curing could be an alternative curing method for the production of fiber-reinforced geopolymer composites, offering shorter curing times and lower energy consumption.

2. Materials and Methods

2.1. Materials

In the experimental study, kaolin obtained from the Bilecik region was used as the aluminosilicate source. The chemical composition and some physical properties of the raw kaolin are shown in Table 1. The kaolin was calcined at 800 °C for 5 h using a laboratory-type muffle furnace to increase its activity. A mixture of liquid sodium silicate and pellet-form sodium hydroxide with at least 98% purity was used as the alkali activator. The sodium silicate consisted of 9.1% Na_2O, 28.6% SiO_2 and 62.3% water by weight. CEN standard sand, conforming to TS EN 196-1 [45], and tap water were also used to produce mortar mixtures.

Table 1. Chemical composition and some physical properties of kaolin.

Compound	%	Property	Value
SiO_2	73.15	Specific gravity	2.60
Al_2O_3	16.55	Retain on 32 μm (%)	4.30
Fe_2O_3	0.69	Retain on 45 μm (%)	0.25
CaO	0.38	Retain on 90 μm (%)	0.15
MgO	0.31	Blaine specific surface area (cm^2/g)	8800
Na_2O	0.10		
K_2O	0.24		
SO_3	0.82		
TiO_2	0.32		
Loss on ignition	6.90		

Straight polyvinyl alcohol (PVA) fibers and curved-wave polymer (polyester) fibers were employed as reinforcement for the mortar mixtures. The PVA fibers were utilized at lengths of 6 mm and 12 mm, whereas the polymer fibers were used at lengths of 15 mm and 30 mm. Some properties of the fibers, provided by the manufacturer, are presented in Table 2. Scanning electron microscopy (SEM) images and photographs of the fibers are presented in Figure 1. SEM images showed that the PVA fibers have a more homogeneous and smooth surface texture, while the surfaces of the polymer fibers are rough and heterogeneous.

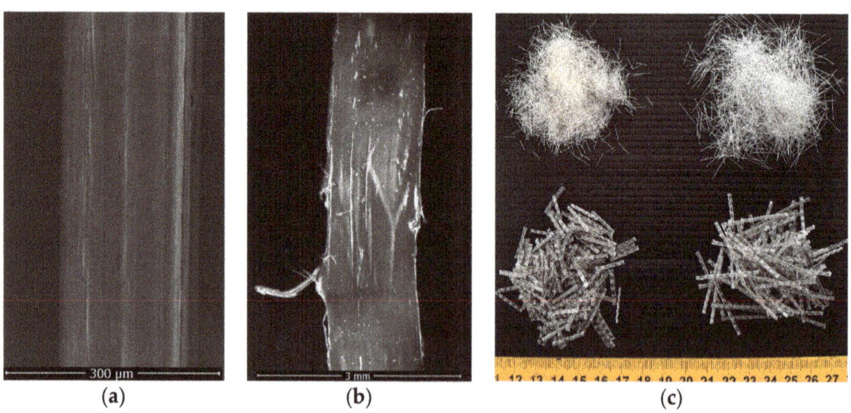

Figure 1. SEM images and photographs of fibers: (**a**) PVA fiber; (**b**) polymer fiber; and (**c**) photograph of fibers.

Table 2. Properties of fibers.

Property	Unit	PVA Fiber	Polymer Fiber
Length	mm	6 and 12	15 and 30
Diameter of fiber	µm	170	-
Thickness of fiber	mm	-	0.5
Width of fiber	mm	-	1.0–1.3
Geometry of fiber	-	Straight	Wavy
Tensile strength	MPa	1000	400–800
Specific gravity	-	1.30	1.36
Melting point	°C	>240	253

2.2. Preparation of Mixtures and Curing Regimes

Mixtures were prepared with a 5 L capacity Hobart-type mortar mixer. Sodium hydroxide pellets were dissolved in the sodium silicate solution and the resultant mixture was rested for 24 h. CEN standard sand, metakaolin, activator solution and water were added to the mixer bowl, respectively, and the mixer was operated at 62.5 RPM for 90 s. In the fiber-reinforced mixtures, the fibers were added to the bowl by hand using the sprinkling method during approximately the first 30 s.

After 90 s of mixing, the materials adhered to the sides of the mixer bowl were scraped off with a spoon and the mixing was continued at the same speed for another 90 s. Immediately after mixing, the flow diameters of the mixtures were determined in accordance with the ASTM C1437 Standard [46]. Subsequently, the mixtures were cast into molds in two layers and each layer was compacted with 25 jolts using a jolting table. To determine flexural strength, compressive strength, water absorption, high-temperature resistance and the rate of capillary absorption, $40 \times 40 \times 160$ mm^3 prism specimens were used. For abrasion resistance and fracture energy tests, $71 \times 71 \times 71$ mm^3 cube and $50 \times 50 \times 240$ mm^3 notched prism specimens were produced, respectively. The height and width of the notch were 10 mm and 3 mm, respectively. The top of the molds was covered with a polyvinyl chloride (PVC) film after the compaction and placement process.

It is known that geopolymers produced with low calcium content have long curing times under ambient conditions. Due to technical reasons, it is not possible to cure fresh mortars with metal molds in microwave ovens. Therefore, the mortars were pre-cured in a curing cabinet to ensure setting. Immediately after the placement and compaction process, the specimens in the molds were placed into a humidity cabin and pre-cured for 8 h at 30 °C. After this stage, the specimens were demolded for microwave curing. The compressive strength of the mixtures after pre-curing was approximately 2–3 MPa. The samples removed from the molds were cured using a household microwave oven at a power level of 300 watts. The curing time for the $40 \times 40 \times 160$ mm^3 prism samples was 30 min, while the curing duration for the $71 \times 71 \times 71$ mm^3 cube and $50 \times 50 \times 240$ mm^3 notched samples was limited to 15 min. In these larger samples, the increased dimensions caused fiber damage and melting of the fiber at the center of the specimen, necessitating a shorter curing time of 15 min. The first outcome of this study indicates that the melting temperatures of the fibers should be considered when using microwave curing in geopolymer production. After 30 min of curing for the $50 \times 50 \times 240$ mm^3 samples, surface temperatures measured with a laser thermometer ranged from 171 °C to 182 °C. However, it is evident that the inner parts of the samples reached higher temperatures, causing the fibers to melt. No such issue was observed in the $40 \times 40 \times 160$ mm^3 prism samples. The smaller sample size likely prevented the internal temperatures from rising to levels that would melt the fibers. A schematic summary of the curing process is presented in Figure 2.

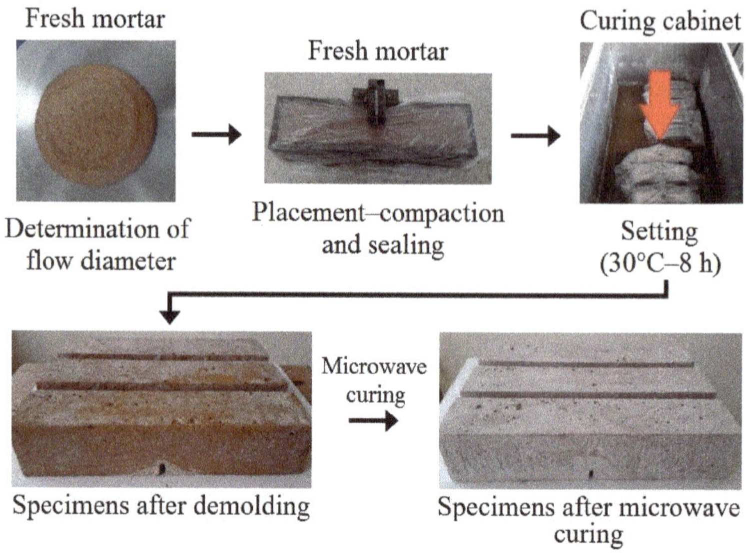

Figure 2. Curing process.

2.3. Determination of Flexural and Compressive Strength

The flexural and compressive strength tests were conducted in accordance with ASTM C348 [47] and ASTM C349 [48] Standards, respectively. Three 40 × 40 × 160 mm^3 prism specimens were used to determine flexural strength and six samples (portions of the prisms broken in flexure) were used to determine compressive strength. The averages of these values were reported.

2.4. Determination of Fracture Energy

Fracture energy values were determined using notched samples of 50 × 50 × 240 mm^3 in accordance with the Rilem [49] recommendation, with some modifications. The notch thickness and height were 3 mm and 10 mm, respectively. The test was performed using a universal device with a 3-point flexural test setup and the loading rate was set at 0.05 mm/min. A clip-on gauge was used to measure the crack mouth opening displacement (CMOD) values. This device was fixed to two metal blades adhered to the lower parts of the samples. As the CMOD in the midspan increased, the clip-on gauge took measurements and transmitted the data to the computer, allowing the load corresponding to the current CMOD to be measured. The load–CMOD graphs plotted with these data were used to calculate the fracture energy. The fracture energy tests were stopped when the maximum load was reduced by 95% (only one sample from the series containing 30 mm polymer fiber was finished earlier because the fiber took all the load and the crack opening did not progress in a time-correlated manner).

The experimental setup, details of the specimen and photographs of the test are shown in Figure 3. Two samples were used for each series and the average values were reported. Fracture energies were calculated using Equation (1). In the equation, W_0 represents the area under the load–crack opening graph (N·mm); m represents the mass of the part of the samples between the supports; g represents the gravitational acceleration; δ represents the maximum crack opening; and A represents the unnotched cross-sectional area in the middle of the supports (50 × 40 mm^2).

$$\text{Fracture energy} = (W_0 + mg\delta)/A \quad (1)$$

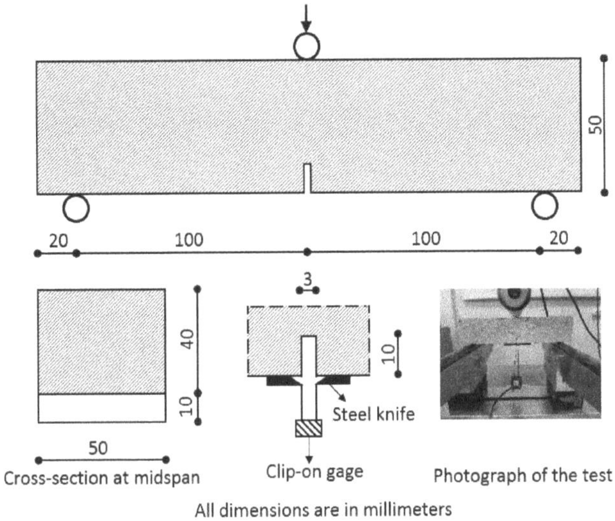

Figure 3. Geometry of specimens, notch details and test photograph.

2.5. Determination of High-Temperature Resistance

High-temperature resistance tests were conducted using the 40 × 40 × 160 mm^3 prism samples. The temperature increase rate of the muffle furnace used was approximately 10 °C per minute, while the burning temperatures were 600 °C and 900 °C. After the furnace temperature reached the burning temperature, the samples were held for 3 h and at the end of this period, the furnace was closed, allowing the samples to cool inside the furnace. When the temperature of the samples reached room temperature, the flexural and compressive strengths were determined and the strength losses were calculated.

2.6. Determination of Water Absorption and Rate of Capillary Water Absorption

Capillary water absorption tests were conducted using the 40 × 40 × 160 mm^3 prism samples in accordance with ASTM C1585 [50], with some modifications. Specimens were placed in an oven at 50 °C for 3 days. After that, the side surfaces of the samples (40 × 160 mm^2) and the upper part (40 × 40 mm^2) were covered with a waterproof material and the weight changes during the relevant standard periods were recorded. Using this information, I-s$^{0.5}$ points were plotted and rate of absorption values were determined from the slope of the best-fitting line to these points. Three samples were used for each series and the average values were reported.

In the water absorption capacity tests, the 40 × 40 × 160 mm^3 samples were kept in water for 48 h and then the saturated-surface-dry weights were measured. The samples were dried in a 105 °C oven until a constant weight was achieved and water absorption capacities were determined using these two weight measurements. For each mortar series, 3 samples were used and the average value was reported.

2.7. Determination of Abrasion Resistance

Abrasion resistance tests were carried out using a Böhme abrasion test device in accordance with TS 2824 EN 1338 Standard [51], with some modifications. Before and after the test, the dimensions of the cube samples were measured with an electronic digital caliper and the difference between the two volumes was recorded as the total volume loss. Three samples were used for each series and the average value was reported.

2.8. Mixtures

The first mixture used in this study was a reference mortar prepared without fibers, while the other four mixtures were reinforced with PVA or polymer fibers. In the preliminary study [52], fibers were used at dosages of 0.125%, 0.250%, 0.375% and 0.500% by volume and the effects of fiber dosage, fiber type and fiber length on flexural and compressive strength were determined. As a result of the preliminary study, one sample of each fiber type and fiber length was selected for further research in this study. The materials and some properties of the mixtures are shown in Table 3. The reference mixture produced within the scope of this study was coded with the abbreviation "REF", while the fiber-reinforced mixtures were named according to the fiber type and fiber length. Polymer fibers were indicated with "POL", while the numbers "6, 12, 15 and 30" represent the lengths of the fibers.

Table 3. Mixture details and flow diameter values.

Mixture	Ingredient (g)					Fiber Dosage (%)	Flow Diameter (cm)
	Metakaolin	Sand	Sodium Silicate	Sodium Hydroxide	Water		
REF	450	1350	189	24	82	-	16.3
PVA-6	450	1350	189	24	82	0.375	15.1
PVA-12	450	1350	189	24	82	0.375	15.3
POL-15	450	1350	189	24	82	0.500	14.4
POL-30	450	1350	189	24	82	0.500	14.8

3. Results and Discussion

3.1. Compressive and Flexural Strength

The effects of fiber addition on compressive and flexural strength are illustrated in Figure 4 comparatively. While the addition of fibers resulted in a decrease in compressive strength, flexural strength exhibited an increase. The compressive strength of the reference sample, which measured 40.1 MPa, decreased by 16% and 15% with the incorporation of 6 mm and 12 mm PVA fibers, respectively. The reductions in compressive strength with the use of 15 mm and 30 mm polymer fibers were 21% and 23%, respectively.

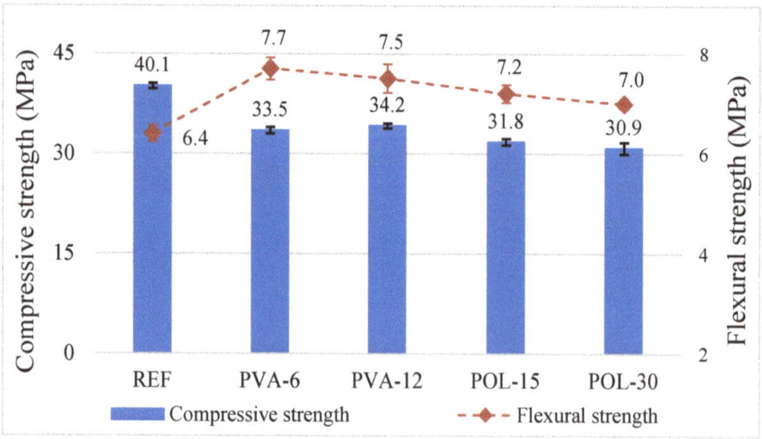

Figure 4. Compressive and flexural strength of mortars.

The flexural strengths of all fiber-reinforced series were higher than that of the reference sample. An increase of 20% and 17% in flexural strength was achieved with the incorporation of 6 mm and 12 mm PVA fibers, respectively, while the increases associated with the use of 15 mm and 30 mm polymer fibers were 13% and 9%, respectively. PVA fibers exhibited a more favorable contribution to flexural strength, likely due to their shorter lengths and smaller cross-sections. This allows for a greater number of fibers to be present on the fracture surface, resulting in a more pronounced increase in flexural strength.

Numerous studies have investigated the effect of fiber addition on the compressive strength of geopolymer/alkali-activated materials. While some of these studies concluded that fiber addition increases compressive strength, others reported opposing results. One significant finding from these studies is the crucial role of fiber dosage in this context. Kantarci [53] reported that the compressive strength of slag-based geopolymer concretes increased with the fiber inclusion at appropriate dosages and that this increase could reach 19% and 14% with the use of 6 mm and 12 mm polypropylene fibers, respectively. Prasad et al. [54] examined the effect of polypropylene fiber addition on the properties of fly ash–slag-based geopolymer concrete produced with varying concentrations of sodium hydroxide. The researchers stated that compressive strength increased with appropriate fiber dosages in concretes produced using 8 and 12 M NaOH solutions but decreased at higher fiber contents. Aygormez et al. [55] investigated the effect of polypropylene fiber (0.8% by volume) on the properties of metakaolin-based geopolymer mortars and concluded that the inclusion of fibers led to a decrease in compressive strength, which can be attributed to the combination of reduced workability, increased porosity, a lower density of fibers and potential fiber agglomeration.

Previous studies have shown that the addition of fibers positively contributes to the flexural strength of geopolymer/alkali-activated materials produced with appropriate fiber dosages and mixing ratios. Al-mashhadani et al. [56] examined the effect of steel, polypropylene and PVA fibers on geopolymer mortars produced using slag and fly ash, and reported that fiber addition increased the flexural strength by up to 49%. The researchers also noted that the addition of fibers had no significant effect on compressive strength, resulting in either very limited increases or slight strength losses. In a similar study, Murthy and Patil [57] investigated the effect of polypropylene fiber inclusion on the properties of ambient-cured fly ash–slag-based geopolymer concretes. The researchers reported that flexural strength increased with fiber addition up to an optimum dosage of 1%, but decreased beyond this point. The increases in flexural strength due to 1% fiber addition were 88% and 58% for 7 and 28 days, respectively. In this respect, it can be said that the results of flexural strength tests are consistent with similar studies in the literature.

3.2. Fracture Energy

The load–CMOD graphs obtained from the fracture energy test are presented in Figure 5. Examination of the graphs corresponding to the reference samples reveals that maximum force is reached in a short duration (at low crack opening values), followed by a rapid decrease in the load carried by the composite. Although the composite sustained the force for a longer duration up to a certain crack opening, the force exhibited a continuous decline, ultimately decreasing to 5% of the maximum load by the end of the test. With the addition of 6 mm PVA fiber, an increase in the maximum crack opening values was recorded; however, the overall behavior remained similar. In contrast, the incorporation of 12 mm PVA fibers resulted in a secondary increase phase after the initial decrease in force, where the sustained force increased to a certain level before starting to decline. This suggests that longer PVA fibers are more effective in bridging under these conditions. For polymer fibers, both 15 mm and 30 mm long fibers demonstrated greater effectiveness regarding flexural behavior compared to PVA fibers, particularly highlighting the beneficial effects of the wavy surface geometry in this context.

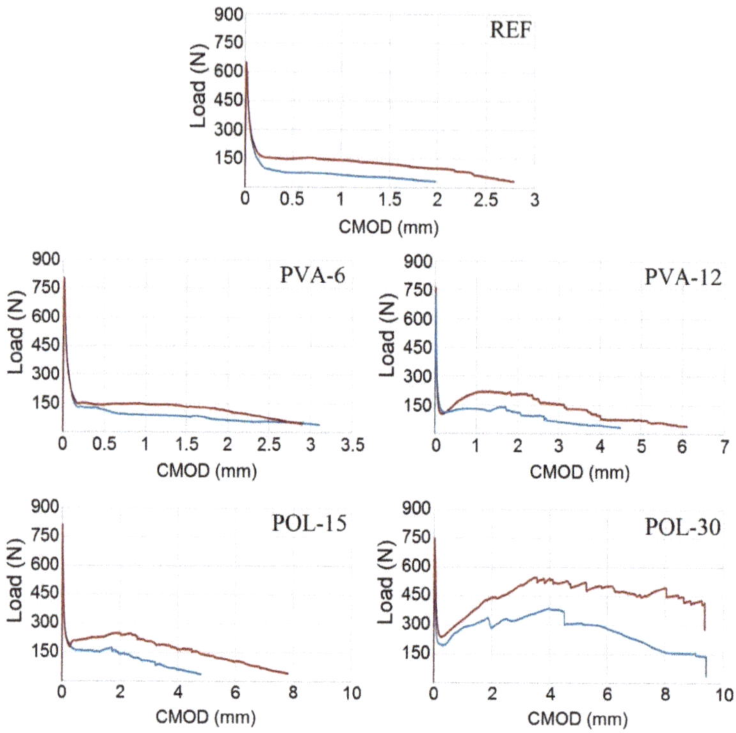

Figure 5. Load–CMOD graphs.

The peak load, CMOD and fracture energy values obtained in the fracture energy tests are presented in Figure 6. While the average peak load of the reference samples was 639.8 N, these values increased by 16–21% with the addition of fiber due to the bridging effects (Figure 7) of fiber on cracks. While the highest peak load was achieved with the use of 6 mm PVA fiber, it was observed that short fibers in the same fiber type provided higher peak loads than long fibers. This is probably due to the fact that short fibers in the same fiber type were more densely present on the fracture surface. When the CMOD values are examined, it is seen that higher values were achieved with the use of fibers, similar to the peak loads. In this context, the average value of 2.3 mm in the reference samples increased by 30–309% with the fiber inclusion. Long fibers provided higher CMOD values compared to short fibers. It is thought that this situation was due to the higher success of long fibers in bridging. The fact that higher CMOD values were obtained in the polymer fiber-reinforced mortars compared to the PVA-bearing series was due to both the larger fiber length and the wavy geometry of the fibers. When the fracture energy values are examined, it is seen that the fracture energies increased with the use of fibers, as expected. The fracture energy, which was 0.14 N/mm in the reference sample, increased by 29% and 143% to 0.18 and 0.34 N/mm with the use of 6 and 12 mm PVA fibers. These increases were 264% (0.51 N/mm) and 1150% (1.75 N/mm) with the use of 15 and 30 mm polymer fibers, respectively. It was determined that wavy and longer polymer fibers were more beneficial in this regard and had a greater effect on both toughness and ductility. It can be concluded that the fiber-reinforced geopolymer mortars produced by microwave curing demonstrated notable effectiveness in terms of fracture energy.

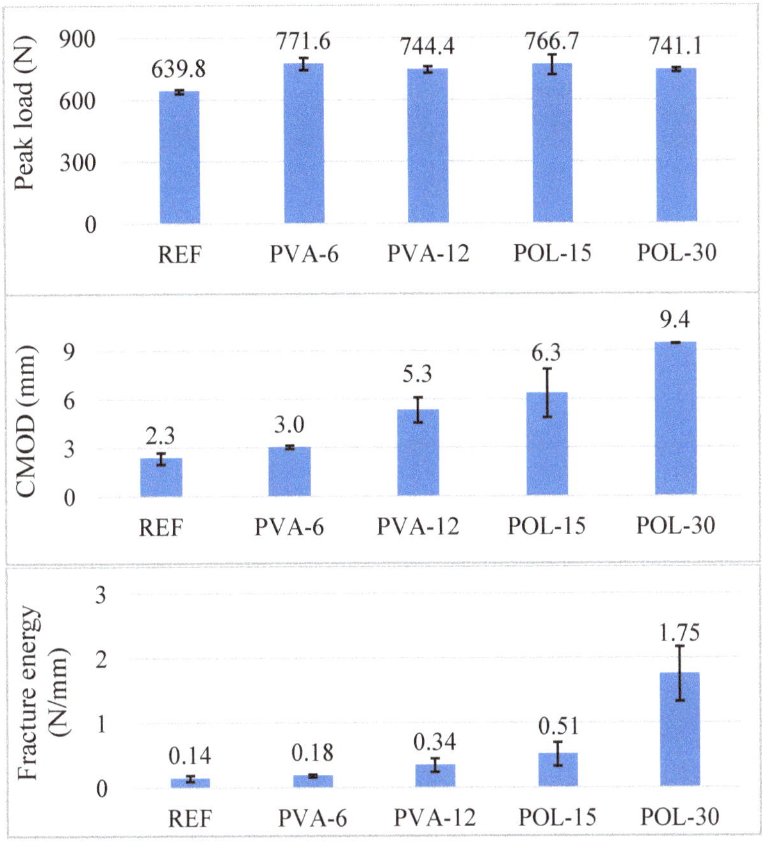

Figure 6. Peak load, CMOD and fracture energy values.

Figure 7. Bridging effects of fibers: (**a**) REF; (**b**) PVA-12; and (**c**) POL-30.

The fracture energy of fiber-reinforced geopolymer and alkali-activated materials has been the subject of various studies. Yurt [58] examined the fracture energy of fiber-reinforced blast furnace slag-based alkali-activated concretes using glass, basalt and polypropylene fibers at different dosages. The researcher used notched samples and reported that both the peak loads and the fracture energy increased with the use of fibers at appropriate dosages and the highest fracture energy values were obtained with the use of polypropylene fibers in general. Bernal et al. [59] investigated the various properties of blast furnace slag-based alkali-activated concretes with different dosages of steel fiber additions at different curing times. The researchers reported that brittle fracture was observed in samples without fibers, while fracture occurred in a more ductile manner with the addition of fibers. It was found that the critical stress intensity factor could be increased by 37% and 52% at 14 and 28 days, respectively, which was attributed to the bridging abilities of the fibers. Similar studies were conducted by Si et al. [60] with basalt fiber, Gomes et al. [61] with steel fiber and Zhang et al. [62] with polyethylene fiber and similar results were obtained.

3.3. High-Temperature Resistance

High-temperature resistance tests were conducted with exposure to temperatures of 600 °C and 900 °C for a duration of 3 h. Upon visual examination of the samples post-exposure, no signs of volumetric instability, significant cracking, shrinkage or swelling were observed. An example illustrating these findings is presented in Figure 8.

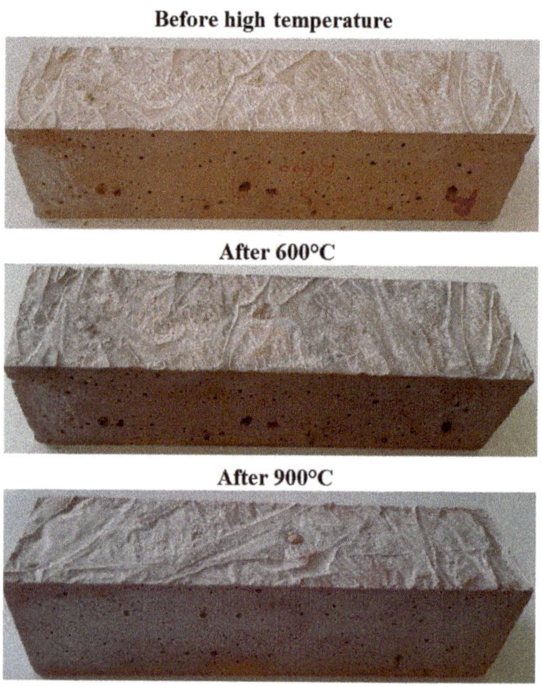

Figure 8. Effect of high temperature on reference mortar.

In the series exposed to 600 °C, no cracks or visible deformations were observed. Conversely, in the series subjected to 900 °C, small cracks developed in certain areas of some fiber-reinforced samples; however, no significant cracks or deformations were noted. It is believed that these small cracks resulted from stress concentrations around pores or in regions where the fibers melted and disappeared. A visual representation of this phenomenon following the 900 °C exposure is presented in Figure 9.

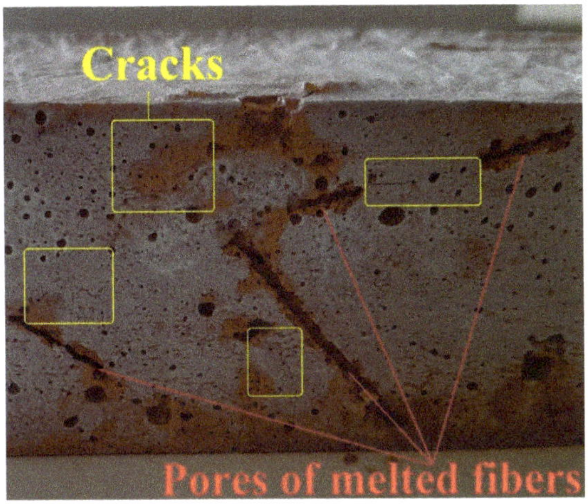

Figure 9. Cracks after exposure to 900 °C.

The high-temperature resistances of the mortars produced in this study were assessed by evaluating flexural and compressive strength losses after exposure to temperatures of 600 °C and 900 °C for a duration of 3 h. In this context, the compressive strengths and corresponding strength losses before and after high-temperature exposure are illustrated in Figures 10 and 11, respectively. As anticipated, compressive strengths decreased in all series after high-temperature exposure. Compressive strength reductions ranged from 24% to 30% in all series subjected to 600 °C, with the least strength loss observed in the series incorporating 6 mm PVA fibers. The compressive strength of the reference mixture, which measured 40.1 MPa prior to high-temperature exposure, decreased to 29.6 MPa. In the series containing 6 mm and 12 mm PVA fibers, initial strengths of 33.5 MPa and 34.2 MPa reduced to 25.6 MPa and 23.9 MPa, respectively, after exposure to 600 °C. Similarly, the strengths of the series with 15 mm and 30 mm polymer fibers, originally 31.8 MPa and 30.9 MPa, decreased to 24.8 MPa and 22.8 MPa, respectively. Upon increasing the exposure temperature to 900 °C, two distinct outcomes were observed. The compressive strengths of the PVA-bearing mortars and reference mixture exhibited slight increases relative to the 600 °C exposure, while the strengths in the polymer fiber series experienced minor decreases. Following exposure to 900 °C, the compressive strength of the reference sample increased to 31.4 MPa and the strengths of the series containing PVA fibers rose to 27.3 MPa and 26.0 MPa, respectively. In contrast, the strengths of the series containing short and long polymer fibers decreased slightly to 23.0 MPa and 22.1 MPa.

Giannopoulou et al. [63] examined the high-temperature resistance (at 600, 800 and 1050 °C) of waste brick powder- and waste ceramic tile powder-based geopolymer pastes and noted that strength losses occurred after exposure to 600 °C in geopolymers based on waste brick and at 600 and 800 °C in geopolymers based on ceramic tile. However, it was reported that the strengths increased at higher temperatures, even surpassing those of reference samples not exposed to high temperatures. The researchers stated that the observed strength losses were due to dehydration, dehydroxylation and the release of vapor pressure, while the strength gains were attributed to sintering of the matrix. In a similar study, Tahwia et al. [64] investigated the high-temperature resistance of ultra-high-performance geopolymer concretes and reported that compressive strengths after exposing to 100 and 200 °C increased. The researchers reported that the reason for the strength gains was the ongoing geopolymerization reactions. Payakaniti et al. [65] stated that the strength increases in geopolymers at high temperatures are due to further geopolymerization reactions at temperatures up to around 200 °C. At higher temperatures, newly formed crystals, such as zeolite and nepheline, in the internal structure create a filler effect. However, the researchers also noted that an increased number of these crystals may negatively affect strength due to internal stress.

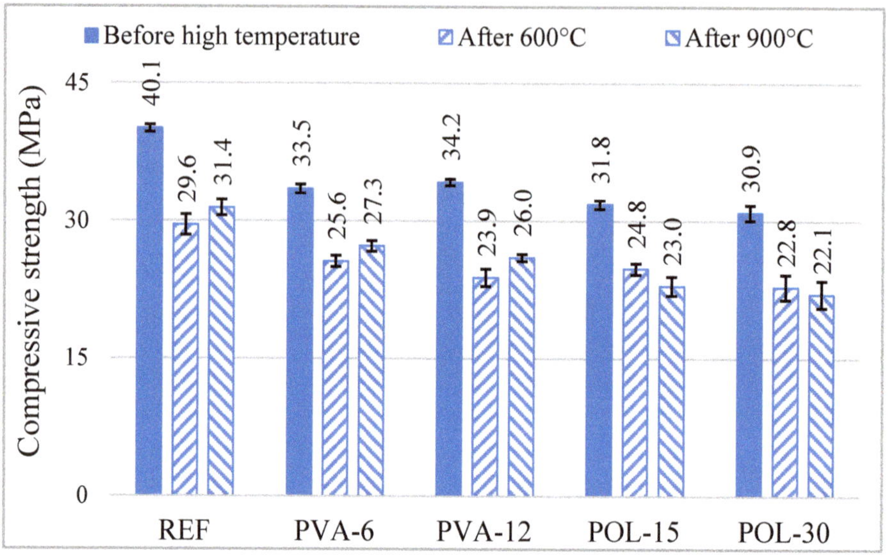

Figure 10. Effect of high temperature on compressive strength.

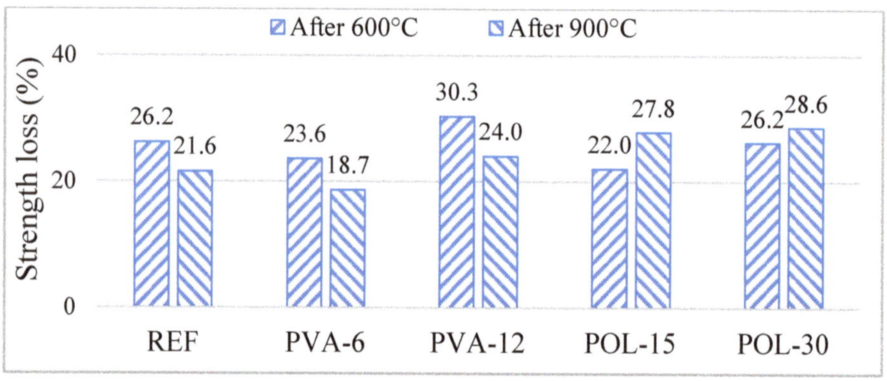

Figure 11. Compressive strength losses after high-temperature exposure.

The flexural strengths and the strength losses before and after high-temperature exposure are shown in Figures 12 and 13, respectively. Upon examination of the results, a trend similar to the compressive strength results was observed: flexural strengths decreased after exposure to high temperatures in all series, but the flexural strengths, which reduced after exposure to 600 °C, increased slightly after 900 °C. The strength losses after 900 °C showed a slight reduction with the addition of fibers, although no significant differences were observed between the fiber types.

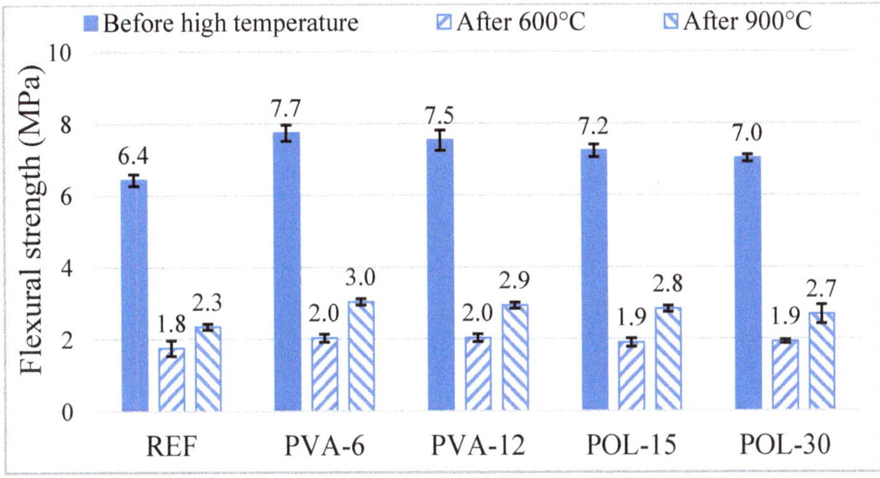

Figure 12. Effect of high temperature on compressive strength.

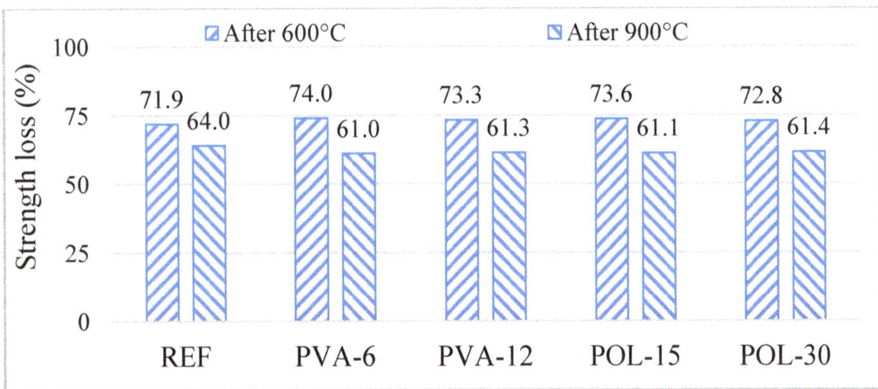

Figure 13. Flexural strength losses after high-temperature exposure.

3.4. Water Absorption and Rate of Capillary Water Absorption

The water absorption and rate of the initial water absorption values of the mortar series are shown in Figure 14. When the water absorption capacities were examined, it was observed that the values were close to each other; however, the water absorption rates increased with the fiber inclusion. The water absorption value, which was 8.7% in the reference mixture, increased to 9.1% with PVA fiber addition, while it was 9.2% and 9.7% in the series containing 15 and 30 mm polymer fibers, respectively. The reduction in workability associated with the fiber inclusion likely accounts for this situation. While the effect of fiber addition on the water absorption capacity of geopolymer and alkali-activated materials has been investigated by many researchers, the results are contradictory. While some studies report that fiber addition increases water absorption capacity, others report that it decreases it; additionally, there are also studies indicating that water absorption capacity is not seriously affected by fiber addition. Shah et al. [66] investigated the effect of PVA, basalt and micro steel fibers on fly ash–slag-based alkali-activated mortars and stated that water absorption values increased slightly with the fiber inclusion. Moradikhou and Esparham [67] examined the effect of polypropylene fiber dosage on the water absorption capacity of metakaolin-based geopolymer concretes. The researchers reported that water absorption values decreased with fiber addition and the water absorption capacity of the mixture containing 0.25% fiber was approximately 16% lower than that of the reference mixture. In this study, the increase in water absorption capacity (Figure 14) with fiber inclusion (ranged between 5% and 12%) was attributed to the loss of workability that occurs with the addition of fibers. The fact that the mixtures with lower flow diameters (Table 3) have a higher water absorption capacity also supports this idea.

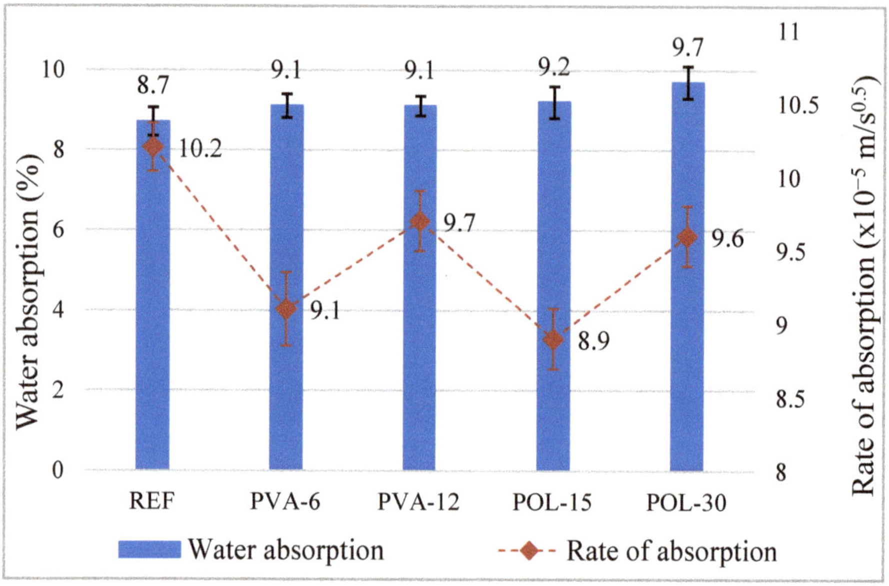

Figure 14. Water absorption and rate of capillary water absorption values.

When the rate of capillary water absorption test results are examined, it is seen that the value of 10.2×10^{-5} m/s$^{0.5}$ in the reference sample decreased by 11% and 5% with the use of 6 mm and 12 mm PVA fibers, respectively, while the decreases resulting from the inclusion of 15 mm and 30 mm polymer fibers were approximately 13% and 6%, respectively. Short fibers exhibited greater advantages than long fibers regarding the capillary water absorption rate. This is likely due to the fact that, for the same fiber type and volume, a greater number of short fibers are distributed throughout the mortar, resulting in greater effectiveness in preventing microcracks.

3.5. Abrasion Resistance

The total volume losses of the mortars in the Böhme abrasion resistance test are shown in Figure 15. The results indicated that the fiber-free reference sample exhibited the highest abrasion resistance among the five mortar mixtures. The total volume loss of reference sample was measured as 15 cm^3. Although the abrasion values increased by 11–37% with the addition of fibers, it is important to consider that the compressive strengths of the fiber-reinforced series were 15–23% lower than those of the reference mortar. Since the compressive strengths of mortars produced with the same type of fiber were relatively close to each other, a more meaningful comparison can be made between them. In this context, it can be stated that mortars containing long fibers exhibited slightly higher abrasion resistance compared to those with short fibers.

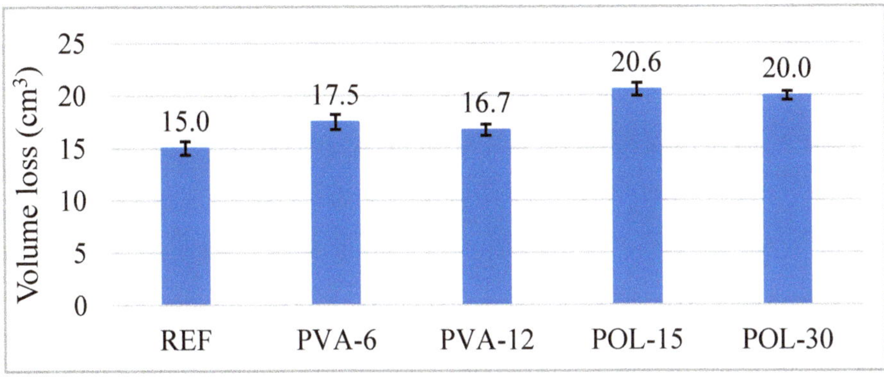

Figure 15. Total volume losses in the abrasion resistance test.

The abrasion resistance of fiber-reinforced geopolymers has been studied by many researchers and these studies have reported that the fiber inclusion positively contributes to abrasion resistance generally. For instance, Nuaklong et al. [68] investigated the effect of carbon fiber inclusion (0.1%, 0.2% and 0.3% by the weight of the binder) on the abrasion resistance of fly ash-based geopolymer mortars. It was determined that with the addition of fibers, the compressive strength increased by 3–14% and weight losses in the abrasion test conducted according to ASTM C944 decreased by 14–26%. It was stated that this effect was due to the increase in compressive strength resulting from the addition of fibers. In a similar study, Ganesan et al. [69] investigated the effect of steel fiber dosage on the abrasion resistance of fly ash-based geopolymer concretes and reported that the amount of abrasion decreased with the addition of fibers and that increasing the fiber dosage further enhanced the positive contribution to abrasion resistance. It can be concluded that the results obtained in this study (Figure 15) regarding abrasion resistance contradict the previously mentioned studies. However, it is important to note that, unlike in the mentioned studies, significant decreases in compressive strength were seen with the fiber inclusion in this study and these losses impact abrasion resistance. Ramujee and Potharaju [70] stated that, although various factors affect the abrasion resistance of concrete, the most critical factor is the compressive strength of concrete.

3.6. SEM Investigation

SEM images of the samples taken from the fracture surface after the flexural strength tests of the mortars are presented in Figure 16. To examine the bond between the fiber and the matrix, samples were selected from regions specifically containing fibers. Additionally, due to the limited sample size that can be analyzed using the scanning electron microscope, samples were chosen from series containing short fibers in both fiber types. A general examination of the images revealed that the structures contained geopolymer gel, aggregate particles, fibers, as well as voids and cracks of various sizes. The voids and cracks in the transition zone between both the aggregate particles and the matrix and between the fibers and the matrix, were clearly visible. Additionally, the fracture surfaces of the polymer fibers showed more deformation compared to those of the PVA fibers, which exhibited smoother fracture surfaces. This suggested that polymer fibers established better adhesion with the matrix due to their wavy surfaces, leading to a more ductile appearance of the fracture surface. In addition, the presence of mortar particles on the fibers can be interpreted as evidence of strong adhesion between the fibers and the matrix. Although cracks and pores were observed at the fiber–mortar interface, the presence of mortar particles on the fibers suggested that the length and thickness of the interfacial transition zone (ITZ) was somewhat smaller prior to the flexural strength test. It is thought that the reduction in fiber thickness due to tensile effects and potential debonding contributed to a slightly wider appearance of the fiber–mortar interface in SEM images. Although this study was conducted in a limited area, the absence of fiber flocculation in specific regions indicated a homogeneous distribution of fibers within the mortar.

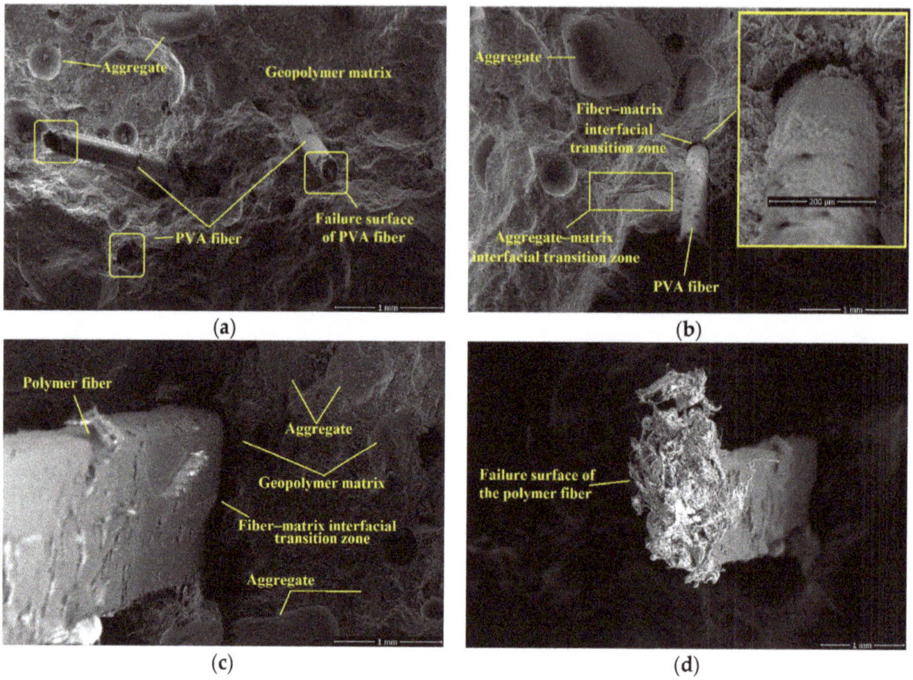

Figure 16. SEM images: (**a**,**b**) PVA-6 sample; (**c**,**d**) POL-15 sample.

4. Conclusions

In this study, the effects of fiber length and fiber type on the strength, durability and fracture energy of microwave-cured metakaolin-based geopolymer mortars were investigated. Polyvinyl alcohol (PVA) fibers of 6 mm and 12 mm length and polymer fibers of 15 mm and 30 mm length were utilized for this purpose. The curing process was conducted using a household microwave oven at a power level of 300 watts for either 15 or 30 min. Based on the materials used and the tests applied, the following conclusions were drawn:

In the production of fiber-reinforced geopolymer composites with microwave curing, an increase in sample size may cause the temperature of the sample, especially at its core, to approach the melting point of the fibers. These factors should be carefully considered in the selection of fiber type and curing time.

With the fiber inclusion, flexural strength increased while compressive strength decreased. Short fibers contributed more effectively than long fibers and PVA fibers had a greater positive impact on the flexural strength compared to polymer fibers.

In the fracture energy test, peak loads, crack mouth opening values and fracture energies increased with the addition of fibers. The positive effect of polymer fibers on fracture energy was found to be greater than that of the PVA fibers and short fibers were more advantageous at the peak load point.

While the fiber inclusion did not significantly affect high-temperature resistance, it increased the water absorption capacity of the fiber-reinforced series and decreased the rate of capillary water absorption.

Although the abrasion resistance test results indicated that abrasion resistance decreased with the fiber addition, it is important to note that the compressive strengths of the fiber-reinforced series were 15–23% lower than those of the reference mortar.

This study was conducted using only two fiber types and a single fiber dosage for each type. Additionally, only one energy level was applied during the curing process. It is evident that variations in fiber types, dosages and microwave energy levels could significantly influence the results. Therefore, the findings cannot be generalized to other fiber types, dosages or curing conditions. Further studies incorporating these variables are needed to fully understand their impact. Therefore, this study can be considered as a preliminary investigation into the use of microwave curing in the production of fiber-reinforced geopolymer/alkali-activated materials. However, the results indicate that microwave curing has potential for the production of fiber-reinforced geopolymers and alkali-activated materials and that materials with reasonable properties can be produced even with a short curing time of 30 min. However, further research is needed to examine microwave-cured geopolymers produced with different aluminosilicates and various alkaline solutions and reinforced with different fibers. Future studies should also investigate freeze–thaw, wetting–drying, shrinkage and creep properties, which were not addressed in this study.

Funding: This research was funded by "Duzce University Scientific Research Projects Coordination Center" under grant number 2023.06.05.1426.

Data Availability Statement: The raw data supporting the conclusions of this article will be made available by the author on request.

Conflicts of Interest: The author declares no conflicts of interest.

References

1. Manzoor, T.; Bhat, J.A.; Shah, A.H. Advancements in Geopolymer Concrete: A State-of-the-Art Analysis of Its Mechanical and Durability Features. *Iran. J. Sci. Technol. Trans. Civ. Eng.* **2024**, *48*, 1777–1816. [CrossRef]
2. Khalid, S.M.; Shobha, M.S. Effect of Ternary Blends on Mechanical Strength, Durability and Microstructural Properties of Geopolymer Concrete. *Iran. J. Sci. Technol. Trans. Civ. Eng.* **2024**, *48*, 1395–1411. [CrossRef]
3. Prabhakar, P.; Kumar, R. Compatibility of Sustainable Geopolymer Based on Artificial Neural Network. *Innov. Infrastruct. Solut.* **2024**, *9*, 323. [CrossRef]
4. Rekha, M.S.; Sumathy, S.R.; Arunachalam, K.P.; Avudaiappan, S.; Abbas, M.; Fernande, D.B. Effects of Alkaline Concentration on Workability and Strength Properties of Ambient Cured Green Geopolymer Concrete. *Asian J. Civ. Eng.* **2024**, *25*, 4893–4910. [CrossRef]
5. Meesala, C.R.; Verma, N.K.; Kumar, S. Critical Review on Fly-ash Based Geopolymer Concrete. *Struct. Concr.* **2020**, *21*, 1013–1028. [CrossRef]
6. Saadatmand, H.; Zehtab, B.; Najafabadi, H.G.; Tavakoli, D.; Behforouz, B. Mechanical Properties of Sustainable Metakaolin/Rockwool Based Geopolymer Mortar. *Innov. Infrastruct. Solut.* **2024**, *9*, 275. [CrossRef]
7. Singh, P.K.; Rajhans, P. Comparative Analysis of Regression and ANN Algorithm for Predicting Compressive Strength of Sustainable Geopolymer Concrete at Varying NaOH Concentration and Curing Temperature. *Iran. J. Sci. Technol. Trans. Civ. Eng.* **2024**, *48*, 1273–1298. [CrossRef]
8. Al Mohammad, A.; Çevik, A.; Jwaida, Z.; Shubbar, A. Effect of Expanded Glass Lightweight Aggregate on the Performance of Geopolymer Mortar at Elevated Temperatures. *Iran. J. Sci. Technol. Trans. Civ. Eng.* **2024**. [CrossRef]
9. Zhang, B.; Peng, H.; Xiong, T.; Zhu, H. Towards Enhancing the Durability of Seawater Coral Aggregate Concrete Under Drying-wetting Cycles with Slag-based Geopolymers. *J. Sustain. Cem.-Based Mater.* **2024**, *13*, 389–401. [CrossRef]
10. Kumar, G.B.; Selvam, R.; Govindaraj, V. Durability Assessment of Blended Slag-fly Ash Geopolymer Concrete in Field Conditions After 5 Years. *J. Sustain. Cem.-Based Mater.* **2024**, *13*, 1770–1781. [CrossRef]
11. Matsimbe, J.; Dinka, M.; Olukanni, D.; Musonda, I. Geopolymer: A Systematic Review of Methodologies. *Materials* **2022**, *15*, 6852. [CrossRef] [PubMed]
12. Gultekin, A.; Yazici, S.; Ramyar, K. Effect of Trass Calcination on Properties of Geopolymer Mixtures. *J. Aust. Ceram. Soc.* **2022**, *58*, 1623–1631. [CrossRef]
13. Haddaji, Y.; Majdoubi, H.; Mansouri, S.; Alomayri, T.S.; Allaoui, D.; Manoun, B.; Oumam, M.; Hannache, H. Microstructure and Flexural Performances of Glass Fibers Reinforced Phosphate Sludge Based Geopolymers at Elevated Temperatures. *Case Stud. Constr. Mater.* **2022**, *16*, e00928. [CrossRef]

14. Pham, K.V.A.; Nguyen, K.T.; Le, T.A.; Lee, K. Investigation of Impact Behavior of Innovative Non-curing Steel Fiber Geopolymer Composites. *Case Stud. Constr. Mater.* **2022**, *16*, e01011. [CrossRef]
15. Farhan, K.Z.; Johari, M.A.M.; Demirboga, R. Impact of Fiber Reinforcements on Properties of Geopolymer Composites: A Review. *J. Build. Eng.* **2021**, *44*, 102628. [CrossRef]
16. Ranjbar, N.; Talebian, S.; Mehrali, M.; Kuenzel, C.; Cornelis Metselaar, H.S.; Jumaat, M.Z. Mechanisms of Interfacial Bond in Steel and Polypropylene Fiber Reinforced Geopolymer Composites. *Compos. Sci. Technol.* **2016**, *122*, 73–81. [CrossRef]
17. Yang, S.; Zhao, R.; Ma, B.; Si, R.; Zeng, X. Mechanical and Fracture Properties of Fly Ash-Based Geopolymer Concrete with Different Fibers. *J. Build. Eng.* **2023**, *63*, 105281. [CrossRef]
18. Wang, Y.; Hu, S.; Sun, X. Experimental Investigation on the Elastic Modulus and Fracture Properties of Basalt Fiber–reinforced Fly Ash Geopolymer Concrete. *Constr. Build. Mater.* **2022**, *338*, 127570. [CrossRef]
19. Li, L.; Yan, C.; Zhang, N.; Farooqi, M.U.; Xu, S.; Deifalla, A.F. Flexural Fracture Parameters of Polypropylene Fiber Reinforced Geopolymer. *J. Mater. Res. Technol.* **2023**, *24*, 1839–1855. [CrossRef]
20. Alves, L.; Leklou, N.; de Souza, F.; de Barros, S. Assessment of the Effect of Fiber Percentage in Glass Fiber Reinforced Slag-based Geopolymer. *J. Asian Ceram. Soc.* **2021**, *9*, 1265–1274. [CrossRef]
21. Singh, S.; Sharma, S.K.; Akbar, M.A. A Detailed Investigation on Assessing the Introduction of Hybrid Fibres in Geopolymer Composites for Pavement Application. *Int. J. Pavement Eng.* **2024**, *25*, 2403681. [CrossRef]
22. Kuranli, O.F.; Uysal, M.; Abbas, M.T.; Cosgun, T.; Nis, A.; Aygormez, Y.; Canpolat, O.; Al-Mashhadani, M.M. Mechanical and Durability Properties of Steel, Polypropylene and Polyamide Fiber Reinforced Slag-based Alkali-activated Concrete. *Eur. J. Environ. Civ. Eng.* **2023**, *27*, 114–139. [CrossRef]
23. Zaid, O.; Martínez-García, R.; Abadel, A.A.; Fraile-Fernández, F.J.; Alshaikh, I.M.H.; Palencia-Coto, C. To Determine the Performance of Metakaolin-based Fiber-reinforced Geopolymer Concrete with Recycled Aggregates. *Arch. Civ. Mech. Eng.* **2022**, *22*, 114. [CrossRef]
24. Xu, F.; Deng, X.; Peng, C.; Zhu, J.; Chen, J. Mix Design and Flexural Toughness of PVA Fiber Reinforced Fly Ash-geopolymer Composites. *Constr. Build. Mater.* **2017**, *150*, 179–189. [CrossRef]
25. Deng, Z.; Zhang, S.; Deng, Z. PVA Fiber-reinforced Geopolymer Mortar Made With Hybrid Recycled Aggregates: Toward Thermal Insulation, Lightweight and Improved Durability. *J. Clean. Prod.* **2023**, *426*, 139200. [CrossRef]
26. Cai, J.; Jiang, J.; Gao, X.; Ding, M. Improving the Mechanical Properties of Fly Ash-Based Geopolymer Composites with PVA Fiber and Powder. *Materials* **2022**, *15*, 2363. [CrossRef]
27. Bhutta, A.; Farooq, M.; Banthia, N. Performance Characteristics of Micro Fiber-reinforced Geopolymer Mortars for Repair. *Constr. Build. Mater.* **2019**, *215*, 605–612. [CrossRef]
28. Hussein, T.A.; Mosaberpanah, M.A.; Kurda, R. Synergic Effect of Polyester Fiber and Nano Silica on Chemical Resistance of Geopolymer Mortar. *PLoS ONE* **2023**, *18*, e0289497. [CrossRef]
29. Xu, Z.; Wu, J.; Zhao, M.; Bai, Z.; Wang, K.; Miao, J.; Tan, Z. Mechanical and Microscopic Properties of Fiber-reinforced Coal Gangue-based Geopolymer Concrete. *Nanotechnol. Rev.* **2022**, *11*, 526–543. [CrossRef]
30. Singh, N.B. Fly Ash-Based Geopolymer Binder: A Future Construction Material. *Minerals* **2018**, *8*, 299. [CrossRef]
31. Li, Z.; Du, P.; Zhou, Y.; Wang, J.; Cheng, X. Synchronous Hot-pressed Metakaolin-fly ash Based Geopolymer: Compressive strength and hydration products. *J. Build. Eng.* **2024**, *97*, 110997. [CrossRef]
32. Qu, C.; Qin, Y.; Wang, T. From Cement to Geopolymers: Performances and Sustainability Advantages of Ambient Curing. *J. Build. Eng.* **2024**, *91*, 109555. [CrossRef]
33. Celikten, S.; Ozturk, Z.B.; Atabey, I.I. High-temperature Resistance of Ceramic Sanitaryware Waste and Fly Ash-based Geopolymer and Hybrid Geopolymer Mortars Produced at Ambient Curing Conditions. *Constr. Build. Mater.* **2024**, *446*, 137990. [CrossRef]
34. Shi, H.; Ma, H.; Tian, L.; Yang, J.; Yuan, J. Effect of Microwave Curing on Metakaolin-quartz-based Geopolymer Bricks. *Constr. Build. Mater.* **2020**, *258*, 120354. [CrossRef]
35. Shen, J.; Li, Y.; Lin, H.; Liu, J. The Experimental Study on Microwave-assisted Preparation of Ultra-High Performance Geopolymer Concrete (UHPGC). *Constr. Build. Mater.* **2024**, *414*, 134934. [CrossRef]
36. Singh, C.; Khanna, V.; Singh, S. Sustainability of Microwave Heating in Materials Processing Technologies. *Mater. Today Proc.* **2023**, *73*, 241–248. [CrossRef]
37. Refaat, M.; Mohsen, A.; Nasr, E.-S.A.R.; Kohail, M. Utilization of Optimized Microwave Sintering to Produce Safe and Sustainable One-part Alkali-activated Materials. *Sci. Rep.* **2023**, *13*, 4611. [CrossRef]
38. Graytee, A.; Sanjayan, J.G.; Nazari, A. Development of a High Strength Fly Ash-based Geopolymer in Short Time by Using Microwave Curing. *Ceram. Int.* **2018**, *44*, 8216–8222. [CrossRef]
39. Gultekin, A.; Ramyar, K. Effect of Curing Type on Microstructure and Compressive Strength of Geopolymer Mortars. *Ceram. Int.* **2022**, *48*, 16156–16172. [CrossRef]
40. Abdullah, A.B.M.M.; Mohd Tahir, M.F.; Hussin, K.; Binhussain, M.; Ekaputri, J.J. Effect of Microwave Curing to the Compressive Strength of Fly Ash Based Geopolymer Mortar. *Mater. Sci. Forum* **2016**, *841*, 193–199. [CrossRef]
41. Zheng, J.; Li, X.; Bai, C.; Zheng, K.; Wang, X.; Sun, G.; Zheng, T.; Zhang, X.; Colombo, P. Rapid Fabrication of Porous Metakaolin-based Geopolymer via Microwave Foaming. *Appl. Clay Sci.* **2024**, *249*, 107238. [CrossRef]
42. Nadeem, M.; Ul Haq, E.; Ahmed, F.; Asif Rafiq, M.; Hameed Awan, G.; Zain-ul-Abdein, M. Effect of Microwave Curing on the Construction Properties of Natural Soil Based Geopolymer Foam. *Constr. Build. Mater.* **2020**, *230*, 117074. [CrossRef]

43. Alqahtani, F.K.; Rashid, K.; Zafar, I.; Iqbal Khan, M. Assessment of Morphological Characteristics and Physico-mechanical Properties of Geopolymer Green Foam Lightweight Aggregate Formulated by Microwave Irradiation. *J. Build. Eng.* **2021**, *35*, 102081. [CrossRef]
44. Hanif, S.; Tahir, M.A.; Rashid, K.; Rehman, M.U.; Saleem, N.; Aslam, A.; Naeem, G. Physico-mechanical Performance of Lightweight Geopolymer foam Aggregates Developed by Geopolymerization Through Microwave-oven Irradiations. *J. King Saud Univ.-Eng. Sci.* **2023**, *35*, 311–318. [CrossRef]
45. *TS EN 196-1*; Methods of Testing Cement—Part 1: Determination of Strength. TSE: Ankara, Turkiye, 2016.
46. *ASTM C1437*; Standard Test Method for Flow of Hydraulic Cement Mortar. ASTM International: West Conshohocken, PA, USA, 2020.
47. *ASTM C348*; Standard Test Method for Flexural Strength of Hydraulic-Cement Mortars. ASTM International: West Conshohocken, PA, USA, 2021.
48. *ASTM C349*; Standard Test Method for Compressive Strength of Hydraulic-Cement Mortars (Using Portions of Prisms Broken in Flexure). ASTM International: West Conshohocken, PA, USA, 2018.
49. RILEM (50-FMC). Determination of the Fracture Energy of Mortar and Concrete by Means of Three-point Bend Tests on Notched Beams. *Mater. Struct.* **1985**, *18*, 287–290. [CrossRef]
50. *ASTM C1585*; Standard Test Method for Measurement of Rate of Absorption of Water by Hydraulic-Cement Concretes. ASTM International: West Conshohocken, PA, USA, 2020.
51. *TS 2824 EN 1338*; Concrete Paving Blocks—Requirements and Test Methods. TSE: Ankara, Turkiye, 2005.
52. Gultekin, A. Investigation of Mechanical Properties of Microwave-cured Fiber-reinforced Geopolymer Mortars. In Proceedings of the Bingol 2nd International Conference on Applied Sciences, Bingol, Turkiye, 19–21 July 2024; pp. 28–37.
53. Kantarci, F. The Effect of Fiber Length and Content on Acid Resistance of Geopolymer Concrete. *Gumushane Univ. J. Sci. Technol.* **2021**, *11*, 424–437. [CrossRef]
54. Prasad, V.B.; Anand, N.; Kiran, T.; Jayakumar, G.; Sohliya, A.; Ebenezer, S. Influence of Fibers on Fresh Properties and Compressive Strength of Geo-polymer concrete. *Mater. Today Proc.* **2022**, *57*, 2355–2363. [CrossRef]
55. Aygormez, Y.; Canpolat, O.; Al-mashhadani, M.M.; Uysal, M. Elevated Temperature, Freezing-thawing and Wetting-drying Effects on Polypropylene Fiber Reinforced Metakaolin Based Geopolymer Composites. *Constr. Build. Mater.* **2020**, *235*, 117502. [CrossRef]
56. Al-mashhadani, M.M.; Canpolat, O.; Aygormez, Y.; Uysal, M.; Erdem, S. Mechanical and Microstructural Characterization of Fiber Reinforced Fly Ash Based Geopolymer Composites. *Constr. Build. Mater.* **2018**, *167*, 505–513. [CrossRef]
57. Murthy, S.S.; Patil, N.N. Mechanical and Durability Characteristics of Fly Ash and GGBFS Based Polypropylene Fibre Reinforced Geopolymer Concrete. *Mater. Today Proc.* **2022**, *59*, 18–24. [CrossRef]
58. Yurt, U. An Experimental Study on Fracture Energy of Alkali Activated Slag Composites Incorporated Different Fibers. *J. Build. Eng.* **2020**, *32*, 101519. [CrossRef]
59. Bernal, S.; De Gutierrez, R.; Delvasto, S.; Rodriguez, E. Performance of an Alkali-activated Slag Concrete Reinforced with Steel Fibers. *Constr. Build. Mater.* **2010**, *24*, 208–214. [CrossRef]
60. Si, R.; Zhan, Y.; Zang, Y.; Sun, Y.; Huang, Y. Effect of Basalt Fiber on Fracture Properties and Drying Shrinkage of Alkali-activated Slag with Different Silicate Modulus. *J. Mater. Res. Technol.* **2023**, *25*, 552–569. [CrossRef]
61. Gomes, R.F.; Dias, D.P.; Silva, F.D.A. Determination of the Fracture Parameters of Steel Fiber-reinforced Geopolymer Concrete. *Theor. Appl. Fract. Mech.* **2020**, *107*, 102568. [CrossRef]
62. Zhang, N.; Yan, C.; Li, L.; Khan, M. Assessment of fiber factor for the fracture toughness of polyethylene fiber reinforced geopolymer. *Constr. Build. Mater.* **2022**, *319*, 126130. [CrossRef]
63. Giannopoulou, I.; Robert, P.M.; Sakkas, K.-M.; Petrou, M.F.; Nicolaides, D. High Temperature Performance of Geopolymers Based on Construction and Demolition Waste. *J. Build. Eng.* **2023**, *72*, 106575. [CrossRef]
64. Tahwia, A.M.; Ellatief, M.A.; Bassioni, G.; Heniegal, A.M.; Elrahman, M.A. Influence of High Temperature Exposure on Compressive Strength and Microstructure of Ultra-high Performance Geopolymer Concrete with Waste Glass and Ceramic. *J. Mater. Res. Technol.* **2023**, *23*, 5681–5697. [CrossRef]
65. Payakaniti, P.; Chuewangkam, N.; Yensano, R.; Pinitsoontorn, S.; Chindaprasirt, P. Changes in Compressive Strength, Microstructure and Magnetic Properties of a High-calcium Fly Ash Geopolymer Subjected to High Temperatures. *Constr. Build. Mater.* **2020**, *265*, 120650. [CrossRef]
66. Shah, S.F.A.; Chen, B.; Oderji, S.Y.; Aminul Haque, M.; Ahmad, M.R. Comparative Study on the Effect of Fiber Type and Content on the Performance of One-part Alkali-activated Mortar. *Constr. Build. Mater.* **2020**, *243*, 118221. [CrossRef]
67. Moradikhou, A.B.; Esparham, A. Water Absorption, Density, Mechanical Strengths and High-temperature Resistance of Metakaolin-based Geopolymer Concrete Reinforced with Hybrid Polyolefin and Simple Polypropylene Fibers. *Adv. Res. Civ. Eng.* **2021**, *3*, 1–15.
68. Nuaklong, P.; Wongsa, A.; Boonserm, K.; Ngohpok, C.; Jongvivatsakul, P.; Sata, V.; Sukontasukkul, P.; Chindaprasirt, P. Enhancement of Mechanical Properties of Fly Ash Geopolymer Containing Fine Recycled Concrete Aggregate with Micro Carbon Fiber. *J. Build. Eng.* **2021**, *41*, 102403. [CrossRef]

69. Ganesan, N.; Abraham, R.; Deepa Raj, S. Durability Characteristics of Steel Fibre Reinforced Geopolymer Concrete. *Constr. Build. Mater.* **2015**, *93*, 471–476. [CrossRef]
70. Ramujee, K.; Potharaju, M. Abrasion Resistance of Geopolymer Composites. *Procedia Mater. Sci.* **2014**, *6*, 1961–1966. [CrossRef]

buildings

MDPI

Article

Numerical Modeling of Distributed Macro-Synthetic Fiber and Deformed Bar Reinforcement to Resist Shear

Benedikt Fadel Farag, Travis Thonstad and Paolo Martino Calvi *

Department of Civil and Environmental Engineering, University of Washington, Seattle, WA 98195, USA; bf20@uw.edu (B.F.F.); thonstat@uw.edu (T.T.)
* Correspondence: pmc85@uw.edu

Abstract: Macro-synthetic fibers are increasingly used in concrete as secondary reinforcement to control temperature and shrinkage cracks, improving durability by limiting crack widths. However, their impact on the shear strength of structural elements remains underexplored, particularly when used in combination with traditional steel reinforcement. To address this knowledge gap, this study developed and calibrated a non-linear numerical model to simulate the shear response of macro-synthetic fiber-reinforced concrete (PFRC) elements, using finite element software VecTor2. The model was calibrated with experimental data from PFRC panels subjected to pure shear loading, incorporating a custom concrete tension-softening model to capture the contribution of fibers. Validation against a broad range of PFRC beam experiments from the literature demonstrated the model's accuracy, achieving an average predicted-to-experimental shear strength ratio of 0.99 (COV = 5.5%). Additionally, the model successfully replicated key response characteristics such as deformation patterns, crack propagation, and residual strength. The proposed modeling approach provides valuable insights into the interaction between fiber volume and transverse reinforcement. It also serves as a powerful tool for future numerical studies, addressing the existing data gap on PFRC behavior and exploring the synergistic effects of macro-synthetic fibers and steel reinforcement on shear strength.

Keywords: fiber-reinforced concrete; macro-synthetic fibers; polypropylene fibers; finite element analysis; numerical modeling

1. Introduction

Macro-synthetic fibers are often added to concrete mixtures as secondary reinforcement, designed to control shrinkage and temperature cracks. Adding fibers to concrete also enhances various concrete properties, such as tensile strength, deformation capacity, toughness, and overall performance [1].

Significant research has been conducted on replacing conventional transverse steel reinforcement bars with distributed steel fibers [2,3] (among many others). The impact of the research conducted on steel-fiber-reinforced concrete (SFRC) led to a revision of the minimum shear reinforcement requirement in ACI 318-08 [4], permitting the use of steel fibers exceeding 0.75% by volume and meeting other requirements. This code change was primarily based on research by Parra-Montesinos [5], who analyzed a database of 147 SFRC beams. The study showed that all slender SFRC beams with a fiber volume of $V_f \geq 0.75\%$ achieved a shear stress higher than $0.29\sqrt{f'_c}$ $\left(3.5\sqrt{f'_c}\ psi\right)$.

In recent years, the use of macro-synthetic fibers for structural applications has gained research interest, and significant experimental work has been performed [6–9]. The results available in the literature suggest that the inclusion of macro-synthetic fibers can enhance concrete performance. However, most of the existing experimental programs have focused on testing beam specimens subjected to flexure or combinations of shear and flexure, which makes it somewhat difficult to interpret the results and to isolate the fiber contribution to the individual resisting mechanisms.

Additionally, the combined use of macro-synthetic fibers and traditional shear reinforcement, which could be beneficial with respect to reducing fiber content, and hence improving concrete workability [10], has not been investigated thoroughly. While there is evidence of synergy between the two types of reinforcement, contrasting results prevent definitive conclusions from being drawn [11].

To address both shortcomings, a recent experimental program [12] investigated experimentally the response of concrete panel specimens reinforced with combinations of macro-synthetic fibers and steel bars, under pure shear loading. The experiments were conducted using the Panel Element Tester at University of Washington with the setup shown in Figure 1. As previously discussed by several authors, e.g., Vecchio and Collins [13], these panel specimens represent web regions of larger structural elements such as girders or RC walls (see Figure 1). Hence, the experimental data provided by Gaston et al. [12] fill an important gap and provide insight into the shear response of macro-synthetic fiber-reinforced concrete (PFRC) structures that can serve both numerical modeling calibration and design approach formulation purposes.

This paper illustrates a numerical modeling approach for PFRC elements subjected to shear that was calibrated relying on the experimental results provided by Gaston et al. [12]. The models were implemented in the finite element software VecTor2 [14] and thoroughly validated against a large database consisting of 77 PFRC beam specimens collected from the literature.

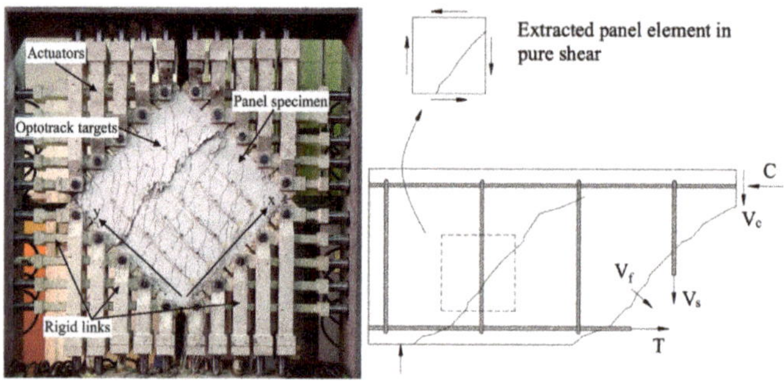

(a) (b)

Figure 1. Panel test experimental setup and illustration of panel element with respect to a beam component. (**a**) Panel Element Tester. (**b**) Schematic view of beam subject to shear.

2. Literature Review

The potential utilization of fibers in structural applications has garnered the sustained interest of researchers and engineers for over half a century. Experiments have studied the influence of concrete strength, fiber type, fiber percentage by volume, fiber aspect ratio, and longitudinal reinforcement ratio (among other factors) on the strength and deformability of FRC beams. Much of the existing research has focused on SFRC beams without transverse deformed bar reinforcement. In 2019, Lantsoghts [15] compiled a database of 488 tests of SFRC beams without stirrups from sixty-three studies available in the literature, and since the publication of this database, the number of studies investigating the behavior of SFRC beams without stirrups has steadily continued to grow (e.g., [16–18]).

A smaller number of studies have investigated the shear behavior of SFRC beams that also contain shear reinforcement (e.g., [19–25]). From the limited evidence available, the combined use of stirrups and steel fibers has been shown to have a synergistic effect on shear capacity. For example, Ding et al. [23] found that the increase in shear capacity from the addition of steel fibers was 23% larger for beams containing transverse reinforcement, in comparison to similarly reinforced beams that did not contain stirrups [23]. In addition, the inclusion of steel fibers in beams containing transverse reinforcement has been shown to increase deformation capacity and can alter the failure mode from shear to flexural failure [19]. For example, Aoude et al. [24] found that the addition of steel fibers to beams containing transverse reinforcement improved the deformation capacity relative to beams with similar reinforcement but without fibers, such that yielding of the longitudinal reinforcement could occur.

In contrast to the immense interest in SFRC beams, very few studies have investigated the shear behavior of PFRC beams ([26–34]), and only a handful of studies have investigated PFRC beams that also contain transverse shear reinforcement (e.g., [11,35,36]). A brief review of the pertinent literature is presented subsequently, highlighting relevant observations from experimental tests of PFRC beams and panel elements, and approaches for modeling FRC structural elements under shear loading.

2.1. State of the Art and Available Experimental Evidence

Several experimental programs have been conducted on structural concrete components containing macro-synthetic fibers employed as shear reinforcement [11,27,30,37]. These studies demonstrated that including macro-synthetic fibers in the concrete mix enhances the shear strength of transversely unreinforced elements. For instance, in these studies it was reported that a fiber volume of $V_f = 0.75\%$ produced shear strength increments of up to 23% and 28%, for slender and deep beams, respectively.

A handful of experimental programs have also investigated the combined use of macro-synthetic fibers and deformed bar reinforcement, to resist shear forces in structural applications, e.g., [11,35,36]. These have shown that elements containing both macro-synthetic fibers and transverse steel reinforcement can achieve ultimate shear strength that is 11% to 47% higher than their reinforced concrete (RC) counterparts.

Synergistic effects were identified between macro-synthetic fibers and transverse steel reinforcement, as their interaction produced shear strength increments that surpassed those obtained by separately adding the incremental strength contribution provided by the individual reinforcement components. For instance, a study by Majdzadeh et al. [35] indicated synergistic improvement of 10% for PFRC beams containing $V_f = 0.5\%$ macro-synthetic fibers and a transverse reinforcement ratio of $\rho_t = 0.28\%$.

Several experimental programs have also been performed on steel FRC panels subjected to pure shear loads (e.g., [2,38,39]), with few investigating the response of panel specimens containing macro-synthetic fiber reinforcement [38]. The outcome of these panel experiments indicated that macro-synthetic fiber reinforcement can improve the shear response by effectively bridging cracks, hence reducing crack widths. However, higher content of macro-synthetic fibers is required compared to steel fibers to achieve analogous shear strength. For instance, Carnovale [38] found that a macro-synthetic fiber volume $V_f = 2.0\%$ was needed to achieve the shear strength of a panel specimen containing $V_f = 0.5\%$ hooked-end steel fibers. Evidently, concrete mixes containing high fiber volumes can be affected by workability issues.

In addition to experimental activities, several studies have attempted to develop/implement numerical models that are suitable for simulating the response of FRC elements subjected to shear loads. Susetyo et al. [2] carried out the VecTor2 finite element (FE) analysis of the FRC panel specimens they previously tested. The study by Susetyo revealed that the tension-stiffening and compression-softening relationships developed for RC elements subject to biaxial stresses could not be used to accurately capture the FRC behavior; hence, alternative constitutive laws were proposed. Carnovale [38] also modeled a series of panel specimens in VecTor2 and proposed adjustments to the Simplified Diverse Embedment Model (SDEM) to improve the accuracy of the numerical models.

Chasioti [39] performed FE analysis of hybrid (i.e., containing two or more fiber types) steel FRC (HSFRC) panels in VecTor2. The study revealed that the default constitutive models available in the FE program could not lead to accurate predictions. However, improved behavior simulations could be achieved by custom-defining in the software the tensile response of the material, relying on the results of companion "dog-bone" laboratory tests.

Cankaya and Akan [40] conducted an experimental program on micro-synthetic fiber-reinforced beam specimens and carried out FE analysis of the specimens in VecTor2. The numerical results matched the results of the experiments at a qualitative level, but the yield and ultimate strength of the specimens were generally overestimated in the analysis. The authors attributed this outcome to the lack of adequate tension constitutive models for FRC under biaxial loads.

Hence, a reliable modeling approach for PFRC subjected to shear loads is still missing. Additionally, due to the lack of experimental evidence, no attempts have been made to date at calibrating behavior models that are specific to PFRC elements with transverse steel reinforcement.

2.2. Summary of a Recent Experimental Program by Gaston (2024) [12]

As discussed in the previous sections, the available experimental evidence on the benefits of macro-synthetic fibers with respect to shear strength is still limited. Hence, to improve understanding and address the lack of fundamental data pertaining to the shear behavior of elements containing both conventional transverse deformed bar and macro-synthetic fiber reinforcement, an experimental program was recently conducted by Gaston [12].

Gaston [12] tested twelve panel specimens (eight PFRC and four RC) under monotonic pure shear loading conditions, using the Panel Element Tester available in the Structural Engineering Testing Laboratory at University of Washington. The shear loads were generated via 37–267 kN (60 kips) force-controlled hydraulic actuators and three (3) rigid links. The average shear stress v_{xy} was computed by dividing the applied shear force by the area of the cross-section of the specimen, while the average strains (ε_ℓ, ε_t, ε_{45}, and ε_{135}) of the cracked composite material were computed from the corner LED target part of the 5 × 5, 540 mm × 540 mm (21.25 in × 21.25 in) grid that covered the test region. The average shear strain γ_{xy} was computed from the measured average strains using Mohr's circle transformations, as outlined in [41].

The main variables of interest of the experimental program were the transverse reinforcement ratio and the macro-synthetic fiber content (GCP Strux 90/40). The transverse reinforcement ratio ρ_t varied from 0% to 0.91%, while the fiber volume V_f ranged from 0% to 0.52%, as indicated in Table 1. The longitudinal reinforcement ratio ρ_l was maintained constant at 2.28% for all specimens, selected to be more than two times the maximum transverse reinforcement of ρ_t = 0.91%; this ensured the occurrence of a shear failure mechanism while preventing biaxial yielding failure mode. The reinforcement layout for a representative panel specimen is shown in Figure 2.

Table 1. Gaston [12] panel element test program summary.

Specimen	V_f [%]	ρ_t [%]	f'_c [MPa]	v_{exp} [MPa]	v_{pred} [MPa]	v_{pred}/v_{exp} [-]
PFRC-000-000	0.00	0.00	44.5	2.12	2.25	1.06
PFRC-000-029	0.00	0.29	37.7	4.27	4.09	0.96
PFRC-000-058	0.00	0.58	31.2	5.41	5.49	1.01
PFRC-000-091	0.00	0.91	42.2	7.82	7.46	0.95
PFRC-026-000	0.26	0.00	32.6	1.43	1.34	0.93
PFRC-026-029	0.26	0.29	38.3	4.31	4.21	0.98
PFRC-026-058	0.26	0.58	34.3	4.24	5.69	1.34
PFRC-026-091	0.26	0.91	43.7	7.43	7.70	1.04
PFRC-052-000	0.52	0.00	29.4	1.50	1.72	1.14
PFRC-052-029	0.52	0.29	45.0	5.00	4.98	1.00
PFRC-052-058	0.52	0.58	35.5	4.94	5.85	1.18
PFRC-052-091	0.52	0.91	36.1	7.47	7.78	1.04

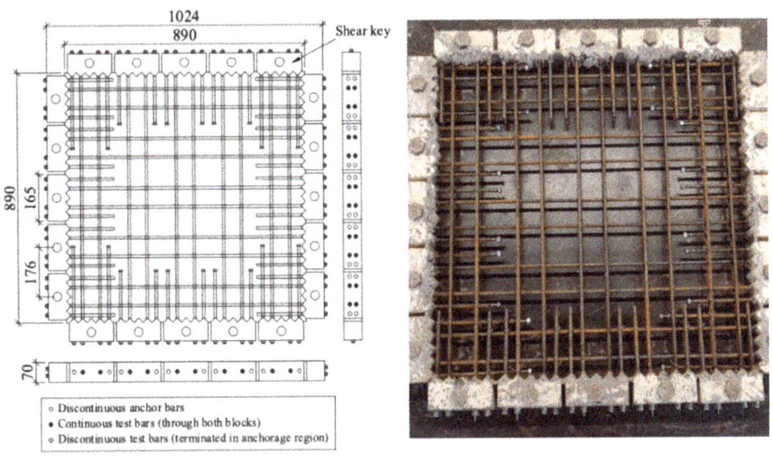

(a) (b)

Figure 2. Reinforcement layout of a representative specimen (ρ_l = 2.28%, ρ_t = 0.91%). (**a**) Reinforcement layout. (**b**) Assembled panel cage ready for cast.

Figure 3 displays the results in terms of normalized shear stress–strain response (the shear stress was normalized by $\sqrt{f'_c}$ to ensure an equitable comparison between specimens with different concrete strength). Figure 3 reports the results of all 12 tests, but the response of four representative specimens (PFRC-026-000, PFRC-026-029, PFRC-000-058, and PFRC-052-091) is highlighted to facilitate visualization and better illustrate the observed trends as a function of the variables studied.

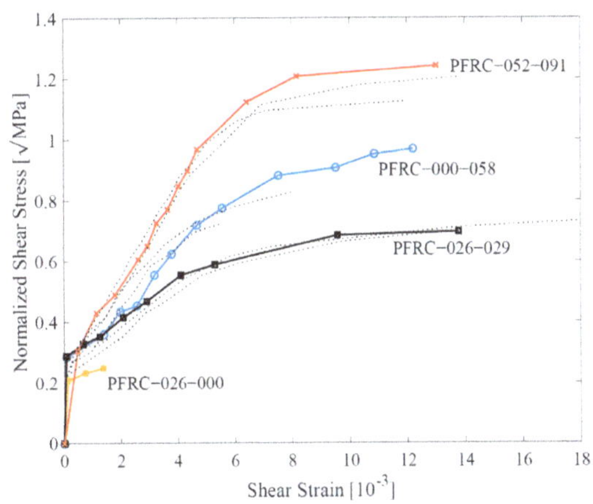

Figure 3. Normalized shear stress–strain response of panel specimens tested by Gaston [12].

The results showed that the shear strength of the specimens tested was predominantly affected by the reinforcement ratio ρ_t, while minor increases in shear strength were observed by adding macro-synthetic fibers at the addition rates used in Gaston's study. However, macro-synthetic fibers were effective in decreasing both the maximum and average crack widths during shear loading.

It should also be noted that the observed benefits of fiber addition appeared to depend on ρ_t [12]. More pronounced strength gains were observed in panels with lower reinforcement ratios (ρ_t) compared to those with higher reinforcement levels. For instance, increasing the fiber volume content from 0.26% to 0.52% resulted in strength improvements of 15% and 18% for reinforcement ratios of ρ_t = 0.29% and 0.58%, respectively. In contrast, the same increase in fiber content produced virtually no strength enhancement in panels with a reinforcement ratio of ρ_t = 0.91%.

This is consistent with the results of Navas et al. [11], where macro-synthetic fibers appeared to be more effective for ρ_t = 0.10% compared to ρ_t = 0.15%.

For a more detailed discussion of the experimental results, interested readers are invited to refer to [12].

3. Numerical Modeling Approach Calibration

Previous studies have illustrated that modeling reinforced concrete elements containing both macro-synthetic fibers and transverse steel reinforcement (and FRC elements subject to shear more broadly) is challenging because of the limited amount of experimental data available to calibrate reliable constitutive laws that capture the post-cracking biaxial response of the composite material. The experimental results obtained by Gaston [12] summarized in the previous section provide much needed information that can help fill this crucial knowledge gap.

Hence, the outcome of the experimental program by Gaston [12] was used to calibrate a modeling approach in the finite element software VecTor2 [14]. VecTor2 is based on the Modified Compression Field Theory (MCFT) [13] and the Disturbed Stress Field Model (DSFM) [42], which were specifically formulated to simulate the response of cracked reinforced concrete elements subjected to in-plane loads. In these theories, cracked concrete is treated as a new material with its own constitutive relationships, and governed by compatibility and equilibrium equations. The key equations of the MCFT are illustrated in Equations (1)–(7). The DSFM utilizes the same framework as the MCFT, but modifications to compatibility and constitutive relationships are introduced [42].

Equilibrium:

$$f_x = f_{c1} - v_{xy}\cot\theta + \rho_x f_{sx} \tag{1}$$

$$f_y = f_{c1} - v_{xy}\cot\theta + \rho_y f_{sy} \tag{2}$$

Compatibility:

$$w = s_\theta \epsilon_1 \tag{3}$$

$$s_\theta = 1/\left(\frac{\sin\theta}{s_x} + \frac{\cos\theta}{s_y}\right) \tag{4}$$

Constitutive:

$$f_{c1} = \frac{0.33\sqrt{f'_c}}{1+\sqrt{200\epsilon_1}} \tag{5}$$

$$f_{c2} = \frac{-f'_c}{0.8-0.34\epsilon_1/\epsilon_o}\left[2\frac{\epsilon_2}{\epsilon_o} - \left(\frac{\epsilon_2}{\epsilon_o}\right)^2\right] \tag{6}$$

$$v_{ci} \leq \frac{0.18\sqrt{f'_c}}{0.31+24w/(a_g+16)} \tag{7}$$

where f_x and f_y are the applied stress in x and y directions, v_{xy} is the applied shear stress, f_{c1} and ϵ_1 are the principal tensile stress and strain, respectively, f_{c2} and ϵ_2 are the principal compressive stress and strain, respectively, ρ_x and ρ_y are the reinforcement ratio in x and y directions, w is the average crack width, s_θ is the crack spacing with crack control parameters s_x and s_y in x and y directions, f'_c is the concrete compressive strength with corresponding compressive strain ϵ_o, a_g is the aggregate size, and v_{ci} is the shear stress on the crack planes.

In recent years, the MCFT/DSFM have been adapted to deal with FRC elements, primarily by implementing modifications to the principal tension/compression constitutive models based on the results of uniaxial tension/compression tests on FRC elements, in the attempt to account for the influence of fibers on the axial behavior of the material. To this end, the Variable Engagement Model (VEM) [43], the Diverse Embedment Model (DEM) [44], and the Simplified Diverse Embedment Model [45] have been implemented in VecTor2 [14] and complement the constitutive models originally developed for "plain" reinforced concrete.

The modeling strategy adopted in this study relied on the "single membrane element" approach discussed by [2,38,39], amongst others. Hence, each panel was modeled as a single four-node membrane element with smeared reinforcement in both longitudinal and transverse directions, as illustrated in Figure 4. The shear force was generated with nodal forces applied at the corners of the element, which were increased monotonically from zero load to failure. The boundary conditions were defined using a pin and a roller positioned at the two bottom corners of the element. This approach has been effectively utilized in previous studies (e.g., [2,38,39]) to accurately replicate the actual boundary conditions observed during laboratory panel testing.

The steel reinforcement stress–strain behavior was assigned based on the experimental results collected from tension tests performed on five steel reinforcement samples. Based on these coupon tests, the average properties computed and used as inputs in the VecTor2 models were as follows: yield strength, f_{sy} = 512 MPa, ultimate strength, f_{su} = 698 MPa, elastic modulus, E_s = 180,000 MPa, hardening strain, ε_{sh} = 0.013, and ultimate strain, ε_{su} = 0.18. The horizontal reinforcement ratio was fixed at ρ_l = 2.28% for all specimens and ρ_t was specified according to values in Table 1. For simplicity, the default VecTor settings (i.e., perfect bond) were applied to model the bond between PFRC and steel reinforcement. However, more refined strategies can be implemented, as discussed by Zhang et al. [46].

The membrane elements were assigned a thickness t = 70 mm, the concrete compressive strengths and elastic moduli values reported in Table 1, and a maximum crack spacing in both longitudinal and transverse directions, $s_l = s_t$ = 70 mm, since the maximum crack spacing reported in the experiments did not appear to vary with fiber content.

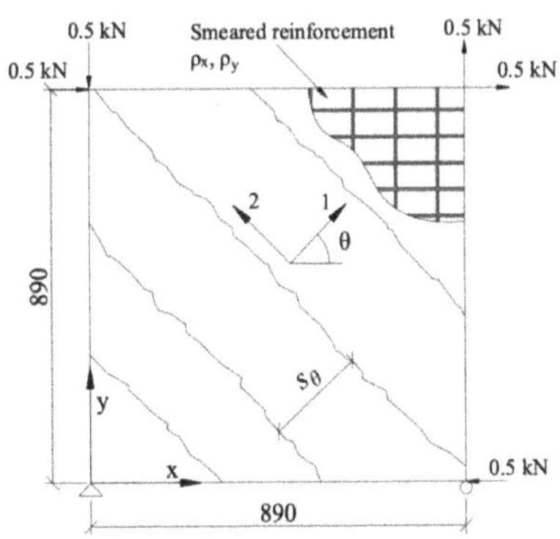

Figure 4. View of the single element model implemented in VecTor2.

Following an extensive parametric study (see [41] for details), in which tension-softening, crack slip, and tensile strength models (among many others) were varied, it was found that the most accurate predictions of the full response of the specimens tested by Gaston [12] could be achieved by selecting the material constitutive models summarized in Table 2. It should be noted that, while the tension-softening model has significant effects on the results, none of the approaches available in VecTor2 appeared suitable to lead to consistently accurate predictions of the behavior of the PFRC elements tested experimentally. Therefore, a new tension-softening model was calibrated, based on the experimental results, and implemented into the overall element model as a custom-input function.

Table 2. Concrete, reinforcement, and bond constitutive models used in the analysis.

Property	Constitutive Model
Compression pre-peak	Hognestad (Parabola)
Compression post-peak	Modified Park–Kent
Compression softening	Vecchio 1992-A (e1/e2-Form)
Tension stiffening	Modified Bentz (2005)
Tension softening	Custom input (Strain Based)
FRC tension	Not considered
Confined strength	Kupfer/Richart
Dilation	Variable-Isotropic
Cracking criterion	Mohr–Coulomb (Stress)
Crack stress calculation	Basic (DSFM/MCFT)
Crack width check	Agg/2.5 Max Crack Width
Crack slip	Not Considered
Creep and relaxation	Not Considered
Hysteretic response	Nonlinear w/plastic offsets
Hysteretic Response	Bauschinger Effect (Seckin)
Dowel action	Tassios (Crack Slip)
Buckling	Akkaya (2012)
Concrete bond	Eligehausen

Readers will note that several variables in Table 2 are labeled as "not considered", and this designation is based on the following justifications:

(i) FRC Tension: This variable was excluded from the model because a customized tension-softening input was calibrated and implemented. Activating both models would result in double-counting the tensile response of the material, leading to inaccuracies.

(ii) Creep and Relaxation: These time-dependent behaviors are critical as they significantly influence the material's long-term performance. While these phenomena should be accurately incorporated into the numerical model for assessing long-term responses, all laboratory specimens modeled were approximately 30 days old on the test day. Therefore, long-term considerations were not necessary for this study.

(iii) Crack Slip: Accounting for crack slip affects the compatibility conditions in the numerical model [42]. While including crack slip can be important in certain scenarios, such as panels with extensive reinforcement in both directions, no notable crack slips were observed in the panels used to calibrate the model, as reported by Gaston [12]. Moreover, incorporating crack slip in the numerical model yielded results that diverged from experimental findings, leading to the recommendation to disable this function in this instance. It is worth noting that crack slip may need to be considered when modeling structures with significantly different reinforcement arrangements.

To proceed with the calibration of a new tension-softening model, the principle tensile stress–strain relationships were first computed for all specimens, from the available experimental readings. This was done using the available experimental measurements of average stresses and strains, and the available experimental response of the steel reinforcement, and relying on principles of equilibrium (see [41] for details).

The computed principal tension stress–strain response was then used to calibrate a modified exponential tension-softening model that could capture the residual tensile strength observed in the tests:

$$\frac{f_{c1}}{f_{cr}} = (1-b)e^{-a\,\varepsilon_{c1}} + b \tag{8}$$

The variables in Equation (8) are defined as follows: ε_{c1} and f_{c1} are the FRC principal tensile strain and stress, f_{cr} is the cracking stress taken as maximum principal tensile stress measured during each panel test, a is a decay parameter, and b defines the residual tensile strength. The constants a and b depend on many factors, including concrete mixture design, fiber volume fraction, fiber type, fiber orientation, and fiber modulus, among many others. Because the experimental program [12] only included a single fiber type and concrete mixture, the model was formulated in terms of the fiber volume fraction alone and is to be considered specific to the characteristics of the mix and fiber types utilized in this study. However, the calibration approach used in this study can be adapted to establish analogous constitutive relationships suitable for various fiber types and concrete mixtures. As new experimental results emerge or existing data from the literature are processed, the proposed model may be refined in the future, either by maintaining the current framework or by Increasing its complexity.

A comparison between the proposed tension-softening model and the experimental results is provided in Figure 5a. Via optimization, the highest level of agreement between the measured and predicted principal tension stress–strain response was found by formulating parameters a and b as follows:

$$a = 450 - 20000\ V_f \geq 50 \tag{9}$$

$$b = 3\sqrt{V_f} \tag{10}$$

where V_f is the fiber volume fraction (i.e., 0.00%, 0.26% and 0.52% for the tested panels). The values in Equation (9) were obtained using the R-squared (R^2) method [47]. Hence, the selected model parameters (a and b) were adjusted to achieve the highest possible R^2 value, thus minimizing the difference between observed and predicted values.

This simplified model was deemed appropriate at this stage, but further research would be necessary to develop a more general approach that accounts for the complex mechanisms of fiber activation and pullout for the various macro-synthetic fibers presently available.

The tension-softening model outlined in Equations (8) and (9) was implemented into VecTor2 as a custom-input tension-softening curve, defined by specifying four key points in the tension stress–strain space, as shown in Figure 5a.

Figure 5b compares the proposed tension-softening model with several models available in VecTor, including the bilinear model, Hordijk's nonlinear model, and the exponential model by Carnovale [38]. Both the bilinear and nonlinear models do not account for the inclusion of fibers in the concrete mix [14], whereas Carnovale's exponential model was calibrated specifically for PFRC (polymer-fiber-reinforced concrete), though it does not adjust based on fiber content. Figure 5b shows that none of these existing models closely match the experimental results used for calibration in this study, and they differ significantly from the proposed model.

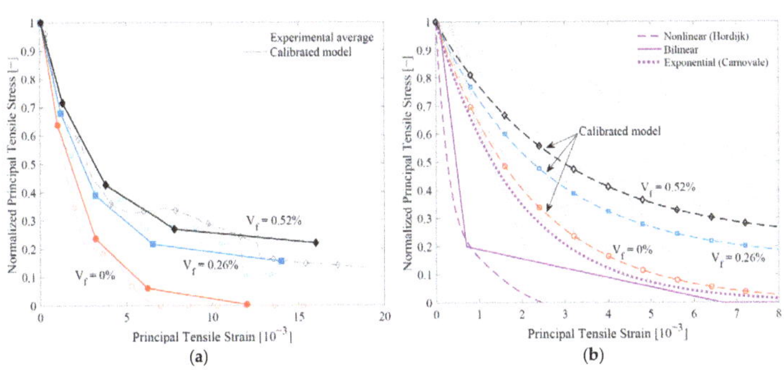

Figure 5. Calibration of the tension-softening model. (**a**) Proposed model vs. experimental results. (**b**) Proposed model vs. models available in VecTor.

Comparisons between predicted and measured responses of the FRC panel specimens are provided in Table 1, in terms of predicted-to-experimental shear strength ratios, and in Figure 6, in terms of full shear stress–strain response for four representative specimens. The average predicted-to-experimental strength ratio was 1.05 with a coefficient of variation (COV) of 11%; the whole stress–strain behavior was also accurately represented for all panels. Although not shown here, the principal stress–strain response and the progression of the principal stress and strain orientations were also captured with a high degree of accuracy [41].

Based on this outcome, the overall model calibration process was deemed successful; hence, extensive validation activities were undertaken, which are discussed in the next section of this paper.

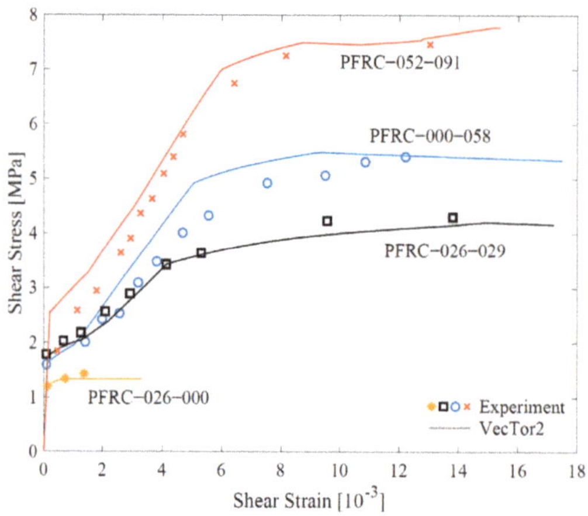

Figure 6. Predicted against experimental shear stress–strain responses for representative specimens.

4. Numerical Modeling Approach Validation

To validate the reliability of the numerical modeling approach discussed in the previous section, it is necessary to verify whether accurate response predictions can be obtained for structural components that differ in geometry, loading conditions, material properties, etc., from the experiments that were used to calibrate the key features of the model itself.

Hence, a database of suitable PFRC experiments was assembled based on data available in the literature. The primary focus was to identify experimental programs that included specimens with macro-synthetic fibers, with or without stirrups. Twelve beam experimental programs were found, and the specimen geometry and load–displacement response were collected into a database. Only beams containing macro-synthetic fibers were included, where the classification between micro- and macro-synthetic fibers was based on the criteria outlined by ACI Committee 544 [48] in terms of the fiber diameter d_f, such that:

$$\text{micro} - \text{fibers}: \ d_f < 0.3 \text{ mm}$$

$$\text{macro} - \text{fibers}: \ d_f \geq 0.3 \text{ mm}$$

A total of 77 beams were identified that possessed the desired characteristics, including 42 PFRC beams without stirrups, 10 PFRC beams with stirrups, 17 RC "control" beams without fibers, and 8 RC beams with stirrups but without fibers. An overview of the experimental variables included in the database, with the considered ranges of values, is provided in Table 3. More details pertaining to the individual specimens are provided in Appendix A, in Table A1.

The variables of the database were analyzed to investigate any correlations. Generally, the variables were not found to be correlated. However, the transverse reinforcement ratio was found to be positively correlated with the compressive strength at 0.42, meaning that beams with a higher reinforcement ratio tended to have greater concrete strength. Conversely, the compressive strength was negatively correlated with fiber volume at −0.40, indicating that higher compressive strength is associated with lower fiber volumes.

Table 3. Overview of parameters in the assembled beam database.

Parameter	Average	Max	Min
d [mm]	280	481	120
a/d [-]	3.3	5.2	1.5
ρ_l [%]	1.90	3.22	0.47
f'_c [MPa]	41.5	101	13.85
V_f [%]	0.7	3.0	0
P_u [kN]	267	1169	49

The variables in Table 3 are defined as follows: d is the effective depth, a/d is the span-to-depth ratio, ρ_l is the longitudinal reinforcement ratio, and P_u is the ultimate load; other variables have been defined. The 77 specimens comprising the assembled database were modeled and analyzed in VecTor2 to assess the reliability of the proposed numerical modeling approach. Two-dimensional non-linear finite element models were constructed using four-node quadrilateral elements, as schematically outlined in Figure 7. The quadrilateral element size was chosen such that at least 10 elements would span the height of the specimen [49]. The assigned constitutive material models are those summarized in Table 2, identified during the calibration process discussed earlier.

Due to symmetry, only one half of each beam was modeled, with appropriate boundary conditions consisting of a vertical roller assigned at the support, and a total of twelve (one per node) horizontal supports (rollers) assigned at the beam mid-span section. This modeling strategy, recommended by Wong et al. [14], amongst many others, can be used to accurately simulate actual boundary conditions while modeling only half of a symmetric beam element, as recommended by Wong et al. [14]. This approach has been successfully implemented in numerous previous studies (e.g., [49,50]).

The longitudinal reinforcement was modeled explicitly with two-node truss–bar elements, while the transverse reinforcement was included as smeared reinforcement embedded in the concrete elements.

The elastic modulus of the steel reinforcement was taken as E_s = 200,000 Mpa for all specimens. The yield and ultimate steel stress were taken as the experimental values if reported, or assigned yield and ultimate stress values, f_{sy} = 420 Mpa and f_{su} = 620 Mpa, corresponding to Grade 60 rebar.

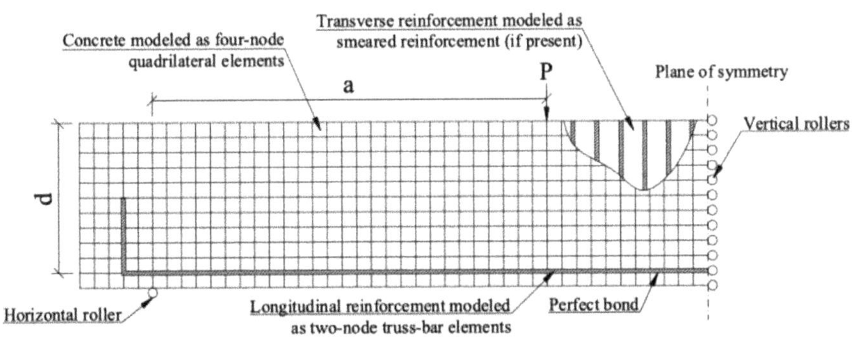

Figure 7. Finite element model of the beams analyzed in VecTor2.

The concrete strength was specified in the models based on the values reported in the various studies (see Table A1). Whenever experimental values of elastic modulus and aggregate size were available, they were specified for the concrete material. The software default values, based on the compressive strength of the concrete, were assigned for all the other concrete material properties, except for the crack spacing. The crack spacing was determined from AASHTO LRFD SBD [51] and CSA A23.3 [52] guidelines, adopting the lower of the two values. The AASHTO and CSA crack spacing equations stem from the recommendations of [13].

The AASTHO guidelines recommend that the minimum crack spacing be calculated as:

$$s_{xe} = s_x \cdot 35/\left(a_g + 16\right) \tag{11}$$

where s_{xe} must be between 300 mm and 2000 mm and s_x is taken as d_v or the maximum distance between layers of longitudinal reinforcement, whichever is less. The CSA guidelines recommend using the crack spacing parameter $s_{xe} = 300$ mm for sections with minimum transverse reinforcement; alternatively, for cross-sections without minimum transverse reinforcement, the minimum crack spacing should be computed as follows:

$$s_{xe} = s_x \cdot 35/\left(a_g + 15\right) \tag{12}$$

An upper limit on the crack spacing, equal to the member depth h, was also imposed [49]. Hence, for beams without minimum transverse reinforcement, the crack spacing was specified as:

$$s_\ell = s_t = \min(\mathrm{h},\ s_{csa},\ s_{aashto}). \tag{13}$$

where s_ℓ and s_t is the maximum crack spacing perpendicular to the longitudinal and transverse reinforcement, respectively. For beams with transverse deformed bar reinforcement, the transverse crack spacing was further limited to the stirrup spacing s, i.e., for beams with transverse reinforcement:

$$s_t = \min(\mathrm{s}, \mathrm{h},\ s_{csa},\ s_{aashto}). \tag{14}$$

The shear capacities predicted by VecTor2 are presented in Table 4, in terms of predicted-to-experimental shear strength ratios for all sets of beams analyzed. The overall experimental strength values are estimated with good accuracy in the numerical simulations, with an average numerical-to-experimental shear strength ratio of 1.062 and COV of 24.4%. An analogous-to-higher level of accuracy is observed with respect to the individual groups of beam specimens. The consistently high-accuracy prediction of strength indicates that the modeling approach outlined earlier, and the recommended material models of Table 2, are reliable and can be adopted to simulate the response of PFRC structural components.

Further, the numerical models appeared to properly capture the load–deflection behavior (Figure 8) of the experimental specimens, replicating the response quite well during all loading stages, as well as crack patterns and failure mechanisms (as shown in Figure 9 for a representative specimen).

Table 4. Predicted-to-experimental strength ratios for each beam type.

Type	Count	Average	COV (%)
All beams	77	1.062	24.4
Control beams	17	1.066	25.1
Beams with fibers	42	1.074	27.6
Beams with stirrups	8	1.088	18.2
Beams with fibers and stirrups	10	0.985	5.50

It should be noted that, given the broad range of specimens considered, comprising 77 beams from 12 independent experimental programs, the wide range of aspect ratios, reinforcement ratios, fiber volumes, and concrete compressive strengths contained within the database, and considering that the numerical modeling approach adopted was calibrated from panel experiments, the outcome of this validation study is considered remarkable.

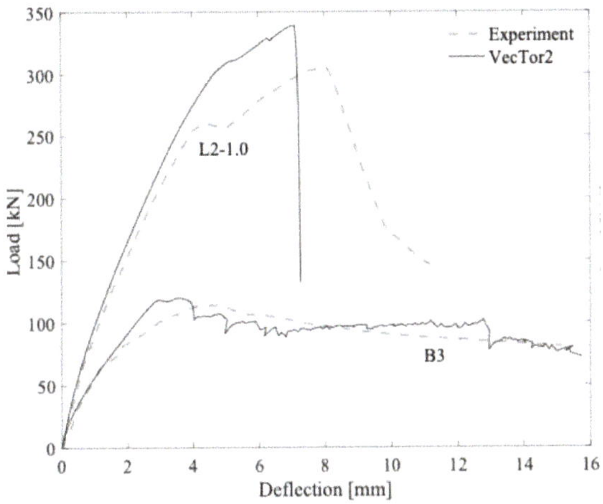

Figure 8. Predicted vs. experimental load–deflection response for two representative specimens.

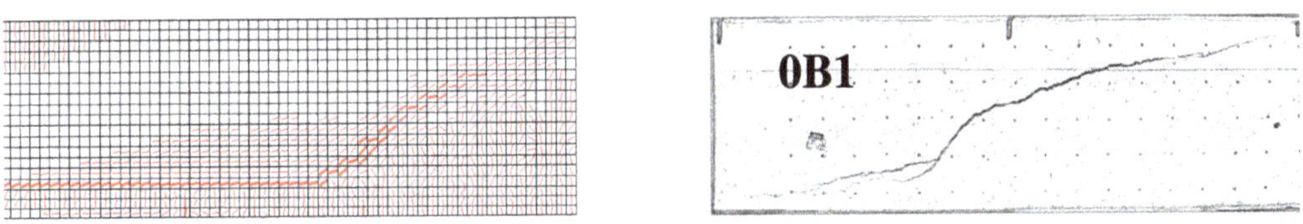

(**a**) (**b**)

Figure 9. Crack pattern and failure mechanism for representative specimen (OB1). (**a**) VecTor2 crack map. (**b**) Experimental crack map [11].

Shear strength predictions for the individual specimens are compared to the associated experimental strengths in Figure 10. The shear capacity of the beams analyzed spans from 1.0 MPa to 4.3 MPa. As discussed earlier, the proposed modeling approach led to reasonably accurate predictions of the shear strength of the full range. However, the shear strength of the beams tested by [53] and the lower aspect ratio beams tested by [14] were notably overpredicted (numerical-to-experimental shear strength ratios > 1.4). These beams had span-to-depth ratios, $a/d \leq 2.3$, putting them in the lowest quartile of the experimental database in terms of span length, which may have contributed to the overprediction, because at lower span-to-depth ratios beams tend to fail in shear rather than in flexure or shear-flexure, and the failure mechanism is more complex, involving bond phenomena and compression strut mechanisms.

It should be noted that in this study, a consistent modeling approach was applied to all beams. However, beams with lower span-to-depth ratios could benefit from modeling adjustments better suited for deep beams. Improved accuracy may be achieved by following recommendations from the literature, such as using a finer mesh and modeling steel reinforcement as discrete bars instead of smeared elements ([50,54]). Using experimental (not always available) rather than nominal stress–strain relationships may also lead to improved predictions. Nonetheless, investigating specialized modeling techniques for deep PFRC beams was beyond the scope of this work. This is a complex task that warrants a dedicated, in-depth study to be pursued in future research.

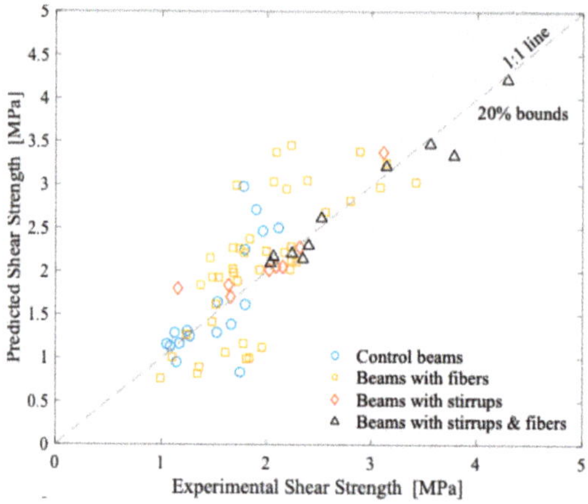

Figure 10. Predicted versus experimental beam strength.

The strength prediction accuracy appeared to improve with greater fiber volume, but the fiber volume was not observed to significantly influence the predictions. Higher average predicted-to-experimental strength ratios were observed for lower span-to-depth ratios ($a/d \leq 2.5$) than for greater span-to-depth ratios ($a/d > 2.5$), with computed average ratios of 1.13 (COV = 28%) and 1.05 (COV = 17%), respectively.

The results in Figure 10 confirm that, overall, the proposed modeling approach provides accurate strength predictions for the PFRC beams selected from the literature, which were not part of the dataset used for calibrating the model itself.

A multiple linear regression analysis was conducted to investigate relationships between key variables and the predicted-to-experimental ratio. The model considered six predictors: longitudinal reinforcement ratio, transverse reinforcement ratio, fiber volume, compressive strength, span-to-depth ratio, and yield strength of longitudinal reinforcement. The regression analysis revealed that the longitudinal reinforcement ratio was a highly significant predictor (p-value < 0.001), while the other variables were not statistically significant (p-value > 0.10). Specifically, holding other variables constant, a unit increase in the longitudinal reinforcement ratio was associated with an average increase of 0.136 in the predicted-to-experimental ratio. However, the model only explained about 28% of the variance in the predicted-to-experimental ratio, suggesting that the linear model may not be fully capable of capturing the complexity involved in the numerical modeling of the beams.

5. Conclusions

This study introduced a non-linear numerical modeling approach to simulate the shear response of concrete elements reinforced with both deformed steel bars and distributed macro-synthetic fibers. Using the finite-element software VecTor2, the model was calibrated with experimental data from PFRC (polymer-fiber-reinforced concrete) panels incorporating a user-defined concrete tension-softening model to capture the contribution of fibers. The methodology combined finite element analysis and experimental data validation, providing a simple yet robust framework for evaluating the performance of PFRC structural elements.

The results outlined in this study show that the proposed modeling approach, calibrated based on the results of panel tests, is in fact more general and suitable for simulating the behavior of fiber-reinforced concrete (FRC) beams, providing accurate estimates of peak strength, full force–displacement response, and crack maps. To this end, validation against a diverse range of beam experiments from the literature was carried out. The strong correlation between experimental and numerical results, particularly for beams with transverse reinforcement (the main focus of this study), underscores the accuracy of the model. An average predicted-to-experimental shear strength ratio of 0.99 (COV = 5.5%) was achieved, demonstrating its reliability in capturing key aspects of the response. Additionally, the model successfully predicted other response characteristics, such as deformation patterns, crack propagation, and in some cases, residual strength, further validating its applicability. It should be further noted that the strength prediction accuracy appeared to improve with higher fiber volume, although the fiber volume did not appear to significantly influence the predictions.

In conclusion, this study provides a relatively simple yet effective modeling approach for simulating the behavior of PFRC structural elements under shear, offering valuable insights into their mechanical response. The model's ability to accurately predict shear strength and other response parameters suggests its potential for further applications, including parametric studies to explore the synergistic effects of fiber volume and transverse reinforcement, which are not yet fully understood. This work contributes significantly to the limited data on macro-synthetic fiber-reinforced concrete and provides a strong foundation for future research on enhancing the structural performance of PFRC elements.

It should be noted that the existing data on the behavior of macro-synthetic fiber-reinforced concrete structural elements remain limited, and significant uncertainty should be expected when forming conclusions based on any single experimental research program. Future work should include further experimental research on PFRC members with transverse reinforcement to expand the available dataset, which currently consists of only 16 elements. Additionally, a more comprehensive PFRC tension-softening model should be developed, considering a wider range of fiber and concrete properties beyond just fiber volume and cracking strength. The model presented herein should also be validated under different loading conditions.

Author Contributions: Conceptualization, T.T. and P.M.C.; methodology, T.T. and P.M.C.; validation, B.F.F., T.T. and P.M.C.; formal analysis, B.F.F.; investigation, B.F.F.; resources, B.F.F., T.T. and P.M.C.; data curation, B.F.F., T.T. and P.M.C.; writing—original draft preparation, B.F.F., T.T. and P.M.C.; writing—review and editing, B.F.F., T.T. and P.M.C.; visualization, B.F.F.; supervision, T.T. and P.M.C.; project administration, T.T. and P.M.C.; funding acquisition, T.T. and P.M.C. All authors have read and agreed to the published version of the manuscript.

Funding: This research was funded by the Accelerated Bridge Construction University Transportation Center, grant number 69A3551747121.

Data Availability Statement: The raw data supporting the conclusions of this article will be made available by the authors on request.

Acknowledgments: The authors would also like to thank members of the project's advisory panel for their technical assistance and guidance throughout the project.

Conflicts of Interest: The authors declare no conflicts of interest.

Appendix A

Table A1. Database of PFRC beams tested in flexure.

Reference	Specimen ID	d	a/d	ρ_l	ρ_t	f_c	V_f	l_f	AR	$v_{u,exp}$	$v_{u,pred}$	Pred/Exp
		(mm)		(%)	(%)	(MPa)	(%)	(mm)		(MPa)	(MPa)	
[35]	B1	120	3.0	2.62	0.28	45.3	-	-	-	3.11	3.38	1.09
	B2	120	3.0	2.62	-	47.1	-	-	-	2.11	2.50	1.18
	B3	120	3.0	2.62	0.28	37.8	0.50	50	85	3.78	3.34	0.88
	B6	120	3.0	2.62	-	43.9	0.50	50	85	2.39	3.05	1.28
	B7	120	3.0	2.62	-	44.2	1.00	50	85	3.14	3.26	1.04
	B8	120	3.0	2.62	-	43.1	1.50	50	85	2.89	3.39	1.17

Table A1. *Cont.*

Reference	Specimen ID	d (mm)	a/d	ρ_l (%)	ρ_t (%)	f_c (MPa)	V_f (%)	l_f (mm)	AR	$v_{u,exp}$ (MPa)	$v_{u,pred}$ (MPa)	Pred/Exp
[26]	Reference mixture	265	3.0	1.78	-	39.2	-	-	-	1.25	1.31	1.05
	P-WV-50-0.5	265	3.0	1.78	-	41.9	0.50	50	63	1.47	2.15	1.47
	P-WV-50-0.75	265	3.0	1.78	-	39.0	0.75	50	63	1.79	2.21	1.24
	P-WV-50-1.0	265	3.0	1.78	-	37.9	1.00	50	63	1.68	2.27	1.35
[27]	L1-0.0	400	3.5	2.15	-	40.9	-	-	-	1.53	1.29	0.85
	L1-0.50	400	3.5	2.15	-	41.9	0.50	40	90	1.72	1.89	1.10
	L1-0.75	400	3.5	2.15	-	41.9	0.75	40	90	1.93	2.02	1.04
	L2-0.0	330	3.5	3.18	-	40.9	-	-	-	1.53	1.65	1.07
	L2-0.50	330	3.5	3.18	-	41.9	0.50	40	90	1.75	2.26	1.30
	L2-0.75	330	3.5	3.18	-	41.9	0.75	40	90	1.84	2.37	1.29
	L2-1.0	330	3.5	3.18	-	35.6	1.00	40	90	2.00	2.23	1.12
	Sh2-0.0	330	2.3	3.18	-	40.9	-	-	-	1.78	2.98	1.68
	Sh2-0.50	330	2.3	3.18	-	41.9	0.50	40	90	2.09	3.38	1.61
	Sh2-0.75	330	2.3	3.18	-	41.9	0.75	40	90	2.23	3.46	1.55
[28]	B-2	270	2.5	1.16	-	43.4	-	-	-	1.16	1.80	1.56
	Sy4.5-1	270	1.5	1.16	-	46.3	0.49	50	71	3.55	3.47	0.98
	Sy4.5-2	270	2.5	1.16	-	46.3	0.49	50	71	2.03	2.10	1.03
[55]	RC	172	5.2	2.34	0.22	36.2	-	-	-	2.09	2.05	0.98
	PFRC	172	5.2	2.34	-	37.6	1.00	12.5	25	1.49	1.93	1.30
[37]	W430PC	215	2.5	1.30	-	31.2	-	-	-	1.80	1.62	0.90
	W510PFRC	255	2.5	1.24	-	26.0	1.45	40	53.3	2.24	2.13	0.95
	W650PFRC	215	3.0	1.15	-	26.0	1.45	40	53.3	2.17	2.22	1.02
	W770PC	255	2.5	1.23	-	31.2	-	-	-	1.66	1.39	0.83
	W770MSR	255	2.5	1.23	0.10	31.2	-	-	-	2.32	2.28	0.98
	W770PFRC	255	2.5	1.23	-	26.0	1.45	40	53.3	2.29	2.11	0.92
	W890PFRC	295	2.2	1.23	-	26.0	1.45	40	53.3	2.23	2.02	0.91
[30]	B2.5R	210	2.5	1.28	-	26.5	-	-	-	1.13	1.29	1.15
	B2.5P1.0	210	2.5	1.28	-	27.0	1.00	39	51	1.52	1.62	1.06
	B2.5P2.0	210	2.5	1.28	-	13.9	2.00	39	51	1.36	0.90	0.66
	B2.5P3.0	210	2.5	1.28	-	18.5	3.00	39	51	1.78	1.17	0.66
	B3.5R	210	3.5	1.28	-	26.5	-	-	-	1.14	0.95	0.83
	B3.5P1.0	210	3.5	1.28	-	27.0	1.00	39	51	1.48	1.42	0.95
	B3.5P2.0	210	3.5	1.28	-	13.9	2.00	39	51	1.35	0.82	0.60
	B3.5P3.0	210	3.5	1.28	-	18.5	3.00	39	51	1.61	1.06	0.66
	B4.5P1.0	210	4.5	1.28	-	27.0	1.00	39	51	1.24	1.27	1.02
	B4.5P2.0	210	4.5	1.28	-	13.9	2.00	39	51	0.99	0.76	0.76
	B4.5P3.0	210	4.5	1.28	-	18.5	3.00	39	51	1.10	1.01	0.92
[29]	B1V0S0	125	2.4	3.22	-	43.1	-	-	-	1.96	2.47	1.26
	B1V3S0	125	2.4	3.22	-	44.4	0.33	40	90	2.56	2.69	1.05
	B1V5S0	125	2.4	3.22	-	45.1	0.55	40	90	2.80	2.82	1.01
	B1V7S0	125	2.4	3.22	-	45.9	0.77	40	90	3.08	2.98	0.97
[11]	OA1	473	3.9	1.68	-	41.8	-	-	-	1.08	1.14	1.05
	OA2	474	4.8	2.23	-	42.2	-	-	-	1.17	1.17	1.00
	OB1	473	3.9	2.23	-	38.3	-	-	-	1.27	1.25	0.99
	OB2	471	4.8	2.24	-	39.8	-	-	-	1.05	1.16	1.11
	A1	473	3.9	1.68	0.10	39.8	-	-	-	1.64	1.84	1.12
	A2	473	4.8	2.23	0.10	40.8	-	-	-	1.66	1.70	1.03
	B1	474	3.9	2.22	0.15	40.1	-	-	-	2.15	2.05	0.95

Table A1. *Cont.*

Reference	Specimen ID	d	a/d	ρ_l	ρ_t	f_c	V_f	l_f	AR	$v_{u,exp}$	$v_{u,pred}$	Pred/Exp
		(mm)		(%)	(%)	(MPa)	(%)	(mm)		(MPa)	(MPa)	
[11]	B2	474	4.8	2.22	0.15	41.0	-	-	-	2.02	2.02	1.00
	OAP1	473	3.9	1.67	-	43.1	1.10	48	56	1.55	1.93	1.25
	OAP2	473	4.8	2.23	-	44.9	1.10	48	56	1.69	1.98	1.18
	OBP1	471	3.9	2.24	-	42.7	1.10	48	56	1.68	2.03	1.21
	OBP2	469	4.9	2.25	-	42.0	1.10	48	56	1.38	1.84	1.34
	AP1	475	3.9	1.67	0.10	44.0	1.10	48	56	2.40	2.30	0.96
	AP2	474	4.8	2.23	0.10	44.6	1.10	48	56	2.34	2.14	0.91
	BP1	481	3.8	2.19	0.15	45.0	1.10	48	56	2.52	2.61	1.04
	BP2	475	4.8	2.22	0.15	44.2	1.10	48	56	2.24	2.21	0.98
[36]	HSC-0%-15M-0	201	3.7	1.41	-	96.5	-	-	-	1.79	2.25	1.26
	HSC-0.75%S1-15M-0	201	3.7	1.41	-	71.4	0.75	50	74	2.23	2.28	1.02
	HSC-0.75%S1-15M-S	201	3.7	1.41	0.50	57.0	0.75	50	74	2.06	2.17	1.05
	HSC-0.75%S1-20M-0	199	3.7	2.53	-	78.3	0.75	50	74	3.42	3.04	0.89
	HSC-0.75%S1-20M-S	199	3.7	2.53	0.50	89.2	0.75	50	74	3.14	3.22	1.03
	HSC-0.75%S1-No.5(HS)-S	199	3.7	1.60	-	91.6	0.75	50	74	4.28	4.22	0.98
[31]	Control	225	1.7	1.19	-	41.7	-	-	-	1.90	2.72	1.43
	P2	225	1.7	1.19	-	45.3	0.22	54	68	2.19	2.95	1.35
	P4	225	1.7	1.19	-	40.8	0.44	54	68	1.71	3.00	1.75
	P8	225	1.7	1.19	-	37.7	0.88	54	68	2.07	3.04	1.47
[34]	RC 00	225	3.1	0.47	-	57.7	-	-	-	1.75	0.84	0.48
	RC 2.5	225	3.1	0.47	-	56.3	0.27	48	56	1.81	0.99	0.55
	RC 4.0	225	3.1	0.47	-	55.3	0.44	48	56	1.84	1.00	0.55
	RC 5.5	225	3.1	0.47	-	54.8	0.60	48	56	1.96	1.13	0.58

References

1. Balaguru, P.N.; Shah, S.P. Fiber-reinforced cement composites. In *Cement Based Materials*; IntechOpen: London, UK, 1992.
2. Susetyo, J.; Gauvreau, P.; Vecchio, F.J. Steel fiber-reinforced concrete panels in shear: Analysis and modeling. *ACI Struct. J.* **2013**, *110*, 285–296.
3. You, Z.; Ding, Y.; Niederegger, C. Replacing stirrups of self-compacting concrete beams with steel fibers. *Trans. Tianjin Univ.* **2010**, *16*, 411–416. [CrossRef]
4. ACI Committee 318. *Building Code Requirements for Structural Concrete (ACI 318-19) and Commentary (ACI 318R-19)*; American Concrete Institute: Farmington Hills, MI, USA, 2019; p. 623.
5. Parra-Montesinos, G.J. Shear strength of beams with deformed steel fibers. *Concr. Int.* **2006**, *28*, 57–66.
6. Karthik, M.P.; Maruthachalam, D. Experimental study on shear behaviour of hybrid Fibre Reinforced Concrete beams. *KSCE J. Civ. Eng.* **2015**, *19*, 259–264. [CrossRef]
7. AbdelAleem, B.H.; Ismail, M.K.; Hassan, A.A. The combined effect of crumb rubber and synthetic fibers on impact resistance of self-consolidating concrete. *Constr. Build. Mater.* **2018**, *162*, 816–829. [CrossRef]
8. Joshi, S.S.; Thammishetti, N.; Prakash, S.S. Efficiency of steel and macro-synthetic structural fibers on the flexure-shear behaviour of prestressed concrete beams. *Eng. Struct.* **2018**, *171*, 47–55. [CrossRef]
9. Hossain, F.Z.; Pal, A.; Ahmed, K.S.; Bediwy, A.; Alam, M.S. Shear behavior of polypropylene fiber-reinforced concrete beams containing recycled aggregate and crumb rubber. *J. Clean. Prod.* **2023**, *412*, 137370. [CrossRef]
10. Zhao, C.; Wang, Z.; Zhu, Z.; Guo, Q.; Wu, X.; Zhao, R. Research on different types of fiber reinforced concrete in recent years: An overview. *Constr. Build. Mater.* **2023**, *365*, 130075. [CrossRef]
11. Navas, F.O.; Navarro-Gregori, J.; Herdocia, G.L.; Serna, P.; Cuenca, E. An experimental study on the shear behaviour of reinforced concrete beams with macro-synthetic fibres. *Constr. Build. Mater.* **2018**, *169*, 888–899. [CrossRef]
12. Gaston, J.P. Shear Behavior of Macro-Synthetic Fiber-Reinforced Concrete Panels. Master's Thesis, University of Washington, Seattle, WA, USA, 2023.
13. Vecchio, F.J.; Collins, M.P. The modified compression-field theory for reinforced concrete elements subjected to shear. *ACI J.* **1986**, *83*, 219–231.

14. Wong, P.S.; Vecchio, F.J.; Trommels, H. *Vector2 & Formworks User's Manual Second Edition*; University of Toronto: Toronto, ON, Canada, 2013.
15. Lantsoght, E.O. Database of shear experiments on steel fiber reinforced concrete beams without stirrups. *Materials* **2019**, *12*, 917. [CrossRef] [PubMed]
16. Chao, S.H. Size Effect on Ultimate Shear Strength of Steel Fiber-Reinforced Concrete Slender Beams. *ACI Struct. J.* **2020**, *117*, 145–158. [CrossRef]
17. Visser, D.; Boshoff, W.P. Shear Behaviour of Concrete V-Shaped Beams with and without Steel Fibres. *Mater. Struct.* **2021**, *54*, 28. [CrossRef]
18. Ismail, M.K.; Yosri, A.; El-Dakhakhni, W. A Multi-Gene Genetic Programming Model for Predicting Shear Strength of Steel Fiber Concrete Beams. *ACI Struct. J.* **2022**, *119*, 317–328.
19. Furlan, S.; Bento-de-Hanai, J.B. Shear behavior of fiber reinforced concrete beams. *Cem. Concr. Compos.* **1997**, *19*, 359–366. [CrossRef]
20. Swamy, R.N.; Bahia, H.M. The effectiveness of steel fibers as shear reinforcement. *Concr. Int.* **1985**, *7*, 35–40.
21. Narayanan, R.; Darwish, I.Y.S. Use of steel fibers as shear reinforcement. *ACI Struct. J.* **1987**, *84*, 216–227. [CrossRef]
22. Cucchiara, C.; Mendola, L.L.; Papia, M. Effectiveness of stirrups and steel fibres as shear reinforcement. *Cem. Concr. Compos.* **2004**, *26*, 777–786. [CrossRef]
23. Ding, Y.; You, Z.; Jalai, S. The composite effect of steel fibres and stirrups on the shear behaviour of beams using self-consolidating concrete. *Eng. Struct.* **2011**, *33*, 107–117. [CrossRef]
24. Aoude, H.; Belghiti, M.; Cook, W.D.; Mitchell, D. Response of steel fiber-reinforced concrete beams with and without stirrups. *ACI Struct. J.* **2012**, *109*, 359–368.
25. Amin, A.; Foster, S.J. Shear strength of steel fibre reinforced concrete beams with stirrups. *Eng. Struct.* **2015**, *111*, 323–332. [CrossRef]
26. Greenough, T.; Nehdi, M. Shear behavior of fiber-reinforced self-consolidating concrete slender beams. *ACI Mater. J.* **2008**, *105*, 468.
27. Altoubat, S.; Yazdanbakhsh, A.; Rieder, K.A. Shear behavior of macro-synthetic fiber-reinforced concrete beams without stirrups. *ACI Mater. J.* **2009**, *106*, 381.
28. Parmentier, B.; Cauberg, N.; Vandewalle, L. Shear resistance of macro-synthetic and steel fibre reinforced concrete beams without stirrups. In Proceedings of the 8th RILEM International Symposium on Fibre Reinforced Concrete: Challenges and Opportunities, Guimaraes, Portugal, 19–21 September 2012; pp. 19–21.
29. Ababneh, A.; Al-Rousan, R.; Alhassan, M.; Alqadami, M. Influence of synthetic fibers on the shear behavior of lightweight concrete beams. *Adv. Struct. Eng.* **2017**, *20*, 1671–1683. [CrossRef]
30. Arslan, G.; Keskin, R.S.O.; Ozturk, M. Shear behaviour of polypropylene fibre-reinforced-concrete beams without stirrups. *Proc. Inst. Civ. Eng. Struct. Build.* **2017**, *170*, 190–198. [CrossRef]
31. Murad, Y.; Abdel-Jabbar, H. Shear behavior of RC beams prepared with basalt and polypropylene fibers. *Case Stud. Constr. Mater.* **2022**, *16*, e00835. [CrossRef]
32. Ahmad, H.; ElNemr, A.; Ali, N.; Hussain, Q.; Chaiyasarn, K.; Joyklad, P. Finite Element Analysis of Glass Fiber-Reinforced Polymer-(GFRP) Reinforced Continuous Concrete Beams. *Polymers* **2021**, *13*, 4468. [CrossRef]
33. Hammad, N.; ElNemr, A.M.; Hassan, E.H. Flexural performance of reinforced Alkali-activated concrete beams incorporating steel and structural Macro synthetic polypropylene fiber. *Constr. Build. Mat.* **2022**, *324*, 126634. [CrossRef]
34. Koura, M.M.; Tahwia, A.M.; Matthana, M.H. Influence of macro-synthetic fibers on the flexural behavior of high strength concrete beams reinforced with GFRP bars. *Mansoura Eng. J.* **2024**, *49*, 4. [CrossRef]
35. Majdzadeh, F.; Soleimani, S.M.; Banthia, N. Shear strength of reinforced concrete beams with a fiber concrete matrix. *Can. J. Civ. Eng.* **2006**, *33*, 726–734. [CrossRef]
36. Bastami, R. Structural Performance of High-Strength Reinforced Concrete Beams Built with Synthetic Fibers. Doctoral Dissertation, University of Ottawa, Ottawa, ON, Canada, 2019.
37. Conforti, A.; Minelli, F.; Tinini, A.; Plizzari, G.A. Influence of polypropylene fibre reinforcement and width-to-effective depth ratio in wide-shallow beams. *Eng. Struct.* **2015**, *88*, 12–21. [CrossRef]
38. Carnovale, D.J. Behaviour and Analysis of Steel and Macro-Synthetic Fibre Reinforced Concrete Subjected to Reversed Cyclic Loading: A Pilot Investigation. Doctoral Dissertation, University of Toronto, Toronto, ON, Canada, 2013.
39. Chasioti, S. *Hybrid Steel Fibre Reinforced Concrete in Shear: From the Material to the Structural Level*; University of Toronto: Toronto, ON, Canada, 2017.
40. Cankaya, M.A.; Akan, C. An Experimental and Numerical Investigation on the Bending Behavior of Fiber Reinforced Concrete Beams. *Turk. J. Civ. Eng.* **2023**, *34*, 59–78. [CrossRef]
41. Farag, B.F. Predicting the Shear Strength of Macro-Synthetic Fiber-Reinforced Concrete. Master's Thesis, University of Washington, Seattle, WA, USA, 2024.
42. Vecchio, F.J. Disturbed stress field model for reinforced concrete: Formulation. *J. Struct. Eng.* **2000**, *126*, 1070–1077. [CrossRef]

43. Voo, J.Y.L.; Foster, S.J. Variable engagement model for the design of fibre reinforced concrete structures. In *Advanced Materials for Construction of Bridges, Buildings, and Other Structures III*; Vasant Mistry, P.E., Ed.; Office of Bridge Technology, Federal Highway Administration: Wahington, DC, USA, 2003; Available online: https://dc.engconfintl.org/advanced_materials/2 (accessed on 1 July 2024).
44. Lee, S.C.; Cho, J.Y.; Vecchio, F.J. Diverse embedment model for steel fiber-reinforced concrete in tension: Model development. *ACI Mater. J.* **2011**, *108*, 516.
45. Lee, S.C.; Cho, J.Y.; Vecchio, F.J. Simplified diverse embedment model for steel fiber-reinforced concrete elements in tension. *Mater. J.* **2013**, *110*, 403–412.
46. Zhang, W.; Zheng, D.; Huang, Y.; Kang, S. Experimental and simulative analysis of flexural performance in UHPC-RC hybrid beams. *Constr. Build. Mater.* **2024**, *436*, 136889. [CrossRef]
47. Montgomery, D.C.; Peck, E.A.; Vining, G.G. *Introduction to Linear Regression Analysis*, 5th ed.; John Wiley & Sons: New York, NY, USA, 2012.
48. ACI Committee 544. *Guide to Design with Fiber-Reinforced Concrete (ACI 544.4R-18)*; American Concrete Institute: Farmington Hills, MI, USA, 2018; pp. 1–39.
49. Galik, W.D.; Calvi, P.M. Shear strength of steel-concrete composite "NPS® Basic" truss beams. *Eng. Struct.* **2023**, *290*, 116362. [CrossRef]
50. Liu, J.; Mihaylov, B. Shear strength of RC deep beams with web openings based on two-parameter kinematic theory. *Struct. Concr. J.* **2020**, *21*, 349–361. [CrossRef]
51. AASHTO. *AASHTO LRFD Bridge Design Specifications*, 4th ed.; SI Units, American Association of State Highway and Transportation Officials: Washington, DC, USA, 2007.
52. CSA (Canadian Standards Association). *Design of Concrete Structures*; CSA (Canadian Standards Association): Toronto, ON, Canada, 2014.
53. Zhang, Z.; Gu, G.X. Finite-Element-Based Deep-Learning Model for Deformation Behavior of Digital Materials. *Adv. Theory Simul.* **2020**, *3*, 2000031. [CrossRef]
54. Saleh, M.; AlHamaydeh, M.; Zakaria, M. Finite element analysis of reinforced concrete deep beams with square web openings using damage plasticity model. *Eng. Struct.* **2023**, *278*, 115496. [CrossRef]
55. Sahoo, D.R.; Maran, K.; Kumar, A. Effect of steel and synthetic fibers on shear strength of RC beams without shear stirrups. *Constr. Build. Mater.* **2015**, *83*, 150–158. [CrossRef]

buildings

MDPI

Article

An Experimental Study and Result Analysis on the Dynamic Effective Bond Length of a Carbon Fiber-Reinforced Polymer Sheet Attached to a Concrete Surface

Dong Li [1], Xinrui Wang [1], Jiangxing Zhang [1,2], Liu Jin [1,*] and Xiuli Du [1]

[1] Key Laboratory of Urban Security and Disaster Engineering of Ministry of Education, Beijing University of Technology, Beijing 100124, China; dong.li@bjut.edu.cn (D.L.); wangxr@emails.bjut.edu.cn (X.W.); zhangjiangxingbjut@163.com (J.Z.); duxiuli@bjut.edu.cn (X.D.)

[2] School of Civil Engineering, Hebei University of Science and Technology, Shijiazhuang 050018, China

* Correspondence: jinliu@bjut.edu.cn

Abstract: A carbon fiber-reinforced polymer (CFRP) is a common material utilized for the enhancement in reinforced concrete (RC) constructions. Previous research indicates that the bonding performance between a CFRP sheet and concrete determines whether the bonding of CFRP material is effective. However, the majority of existing research on the bonding performance of the CFRP–concrete interface is concentrated on static loading conditions. In order to clarify the effect of dynamic load on the bonding performance of the CFRP sheet–concrete interface, this study adopts the double-sided shear test method to carry out dynamic experimental research. The test findings reveal that the damage pattern of the CFRP sheet–concrete interface remains consistent across different loading rates. The ultimate bearing capacity increases as the strain rate increases. As the strain rate increases from 10^{-5} s^{-1} to 10^{-2} s^{-1}, the effect of bond length on ultimate bearing capacity increases by about 7%. As the strain rate increases, both the maximum strain of CFRP and the maximum interfacial shear stress demonstrate a corresponding increase, with respective increase rates of 60% and 20%. The effective bond length decreases by about 20% when the strain rate rises from 10^{-5} s^{-1} to 10^{-2} s^{-1}. Finally, a formula for calculating the dynamic effective bond length of a CFRP sheet, grounded in the Chen and Teng formula, has been proposed and verified.

Keywords: CFRP sheet–concrete interaction; loading rates; CFRP strain; stripping bearing capacity; dynamic effective bond length

1. Introduction

In building structural engineering, it is challenging to build structures that are both stable and durable, while also ensuring the use of sustainable materials. Current technical tools and materials are unable to achieve further reductions in weight or improve existing strength, which has stimulated the investigation of composite materials with potential value. A fiber-reinforced polymer (FRP) exhibits a combination of high strength and light weight. The application of FRP material can effectively reinforce the existing reinforced concrete (RC) structure and repair the existing damage [1]. Among the various types of FRP materials, a carbon fiber-reinforced polymer (CFRP) has the highest strength and is the most widely used [2]. Moreover, CFRP exhibits the highest fatigue and creep resistance compared to other FRP materials [3]. CFRP sheets are frequently employed in the reinforcement and repair of RC structures due to these characteristics [4–6]. At present, many studies [7–9] have shown that the most common damage pattern in CFRP applications is the stripping damage at the CFRP–concrete interface. The CFRP–concrete interface bonding problem is a primary focus and challenge of CFRP application research.

Under static load, a substantial body of experimental research has explored the bonding properties between an FRP sheet and concrete through single shear tests (such as Lee et al. [10], Subramaniam et al. [11], Ali-Ahmad et al. [12]), double shear tests (such as Zheng et al. [13], Raoof et al. [14], Pellegrino et al. [15]), and modified beam tests (such as Wang [16], De et al. [17], Hassan et al. [18]). Numerous tests have demonstrated that the interfacial shear stress is the main bond stress at the FRP–concrete interface. Consequently, evaluations of bond performance at the FRP–concrete interface focus primarily on its ability to withstand pure shear stresses. The double shear test is preferred by many researchers for its operational simplicity, clear force state, and enhanced ability to minimize FRP force eccentricity [19]. Research has identified several parameters that influence bond behavior at the FRP–concrete interface. In a study by Xue et al. [20], the impact of concrete strength and CFRP width on interfacial bonding properties were studied by double shear tests. The test results indicate that the interfacial bond strength increases with concrete strength, while decreasing with the increase in CFRP width. The bond strength of C40 concrete and a CFRP sheet is about 30% higher than that of C30 concrete. The bond strength of 25 mm, 40 mm, and 50 mm wide CFRP sheets was reduced by 28%, 37%, and 61%, respectively, compared to a 10 mm wide CFRP sheet. Anne et al. [21] studied the bonding properties of different adhesives in detail. It was found that the combination of different adhesives and the different mixing ratios of the same adhesive would affect the bonding properties of the FRP–concrete interface. Fang et al. [22] investigated the influence of different CFRP bond lengths on the interfacial bonding performance under different concrete strengths through experiments. The findings indicate that as the concrete strength rises, the average bond strength also gradually increases, but the increase is small. The effective bond length is about 100 mm. Yuan et al. [23] analyzed the effective bond length of the AFRP–concrete interface through double shear tests, and found that as the number of layers increased from 1 to 2 and 3, the effective bond length exhibited a 29% and 43% increase, respectively. At the FRP–concrete interface, the "effective bond length" is the length near the loading end where most of the shear stress is concentrated. This region of the interface withstands the majority of the external load. Beyond the "effective bond length", the shear stress of the bonded layer is insignificant, and the contribution to the overall shear resistance of the bonded surface is negligible. In other words, when the bond length exceeds the effective bond length, the shear stress on the bond layer becomes negligible. Consequently, the contribution of shear stress to the bond surface's resistance for the entire member is also negligible [24,25]. Therefore, it is very important to study the effective bond length in order to ascertain the bond performance of the FRP–concrete interface.

In recent years, occurrences such as earthquakes, chemical explosions, and terrorist attacks have been prevalent. Dynamic load has the characteristics of short action time and large action strength. Concrete materials exhibit completely different mechanical properties under dynamic and static loads [26,27]. Consequently, the safety performance of concrete structures under dynamic loading has garnered widespread attention. However, due to the strain rate effect, the mechanical properties of the FRP–concrete interface under dynamic loading are quite different from those under static loading [28–31]. There is limited research on the dynamic bonding properties of the FRP–concrete interface, with the majority of these studies concentrating on the interface's ultimate bearing capacity and bond–slip behavior. Shi et al. [32] investigated the bonding performance of the FRP–concrete interface under dynamic loading conditions. It has been demonstrated that the interfacial ultimate bearing capacity, interfacial fracture energy, and interfacial shear stress exhibited significant strain rate effects. As the strain rate increases from 2.5×10^{-5} s^{-1} to 0.1 s^{-1}, the ultimate bearing capacity increases by 21%. Shen et al. [33] carried out a dynamic double shear test to explore the impact of the load strain rate on the dynamic bond–slip relationship and effective bond length. The findings indicate that as the load strain rate increases from 0.61×10^{-3} s^{-1} to 0.63 s^{-1}, there is a 32.8% increase in ultimate bearing capacity and a 60% increase in maximum bond stress, while the dynamic effective bond length decreases by 14.3%. Based on dynamic single shear tests and a regression analysis, Li et al. [34] derived the empirical formula of the dynamic increase factor (DIF) related to interfacial bond strength, peak shear stress, and peak slip. Li et al. [35] constructed a finite element model of FRP–concrete lap joints, and studied the interfacial bonding properties under static load and rapid load. It is found that the maximum bearing capacity increases with the loading rate. At present, the method of calculating the bond–slip relationship, stress distribution, and interfacial bearing capacity of the FRP–concrete interface under a dynamic rate is still unclear. Determining the effective bond length under dynamic loading conditions is crucial for understanding the dynamic bonding properties of the FRP–concrete interface. It is regrettable that the calculation method of effective bond length under dynamic load conditions is rarely studied, and the form is not unified. Therefore, studying the impact of the strain rate on the effective bond length at the FRP–concrete interface holds immense significance.

In this study, dynamic tensile tests were performed on CFRP–concrete double-sided shear specimens to investigate the bonding properties of the CFRP–concrete interface. The tests were conducted using different loading rates and varying bond lengths of CFRP sheets as variables. The analyses focused on how loading rates affect damage modes, FRP strain distribution patterns, and the bond–slip relationship. Based on these results, the effective bond length of CFRP sheets under various loading rates was determined, and a formula for calculating the bond length under dynamic conditions was established.

2. Test Program

2.1. Test Design and Loading Scheme

In this study, the test design was carried out according to the relevant requirements of GB 50608-2020 [36] and T/CECS 146-2022 [37]. Figure 1 presents the diagram of the double shear test setup utilized in this study. The dimensions of the concrete specimen are 300 mm × 100 mm × 100 mm. The specimen is fixed with a self-designed fixture and connected to a universal testing machine. The upper end of the fixture and its base are attached to the universal testing machine using a pin and secured with a nut. The CFRP sheet is suspended from a fixture matching the width of the concrete. The CFRP sheet fixture is connected to the upper fixture with four screws, and a vertical force is applied in the tension direction of the CFRP sheet by moving the crossbeam of the universal testing machine. To prevent stress concentration from severing the CFRP sheet at the contact point during pulling, the fixture is chamfered at this contact point. The concrete specimen is secured to the lower base using two steel plates and four bolts. To prevent wedge damage at the concrete loading end, a 50 mm unbonded zone is created at this end, ensuring a 150 mm distance from the unbonded zone's lower edge to the loading fixture. A CFRP sheet with a width of 50 mm is symmetrically bonded to both sides of the concrete specimen along the central axis, and the bond lengths are 40 mm, 60 mm, 80 mm, 100 mm, 120 mm, 160 mm, and 200 mm, respectively.

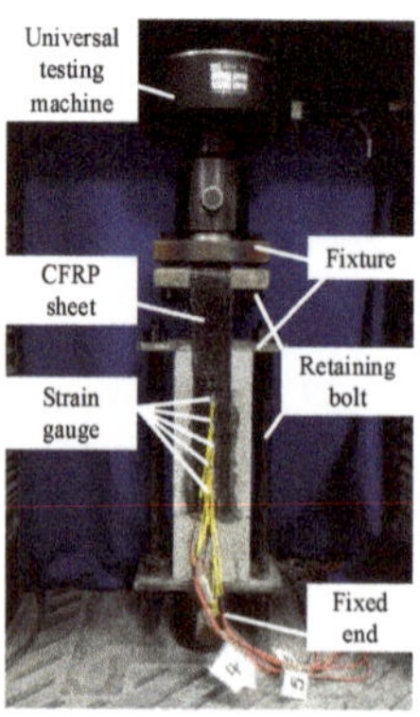

Figure 1. Diagram of test setup.

Strain measurement points are arranged along the central axis of the CFRP sheet's surface. The strain gauges were applied symmetrically to both sides of the CFRP sheets surrounding the test block, as illustrated in Figure 2. Considering the size of the strain gauges, an equal spacing method was used to apply the gauges with a spacing of 30 mm. The strain gauges were initially positioned 5 mm from the edge line of the unbonded area. Throughout the test, data were gathered automatically via the load sensor and the dynamic strain acquisition instrument.

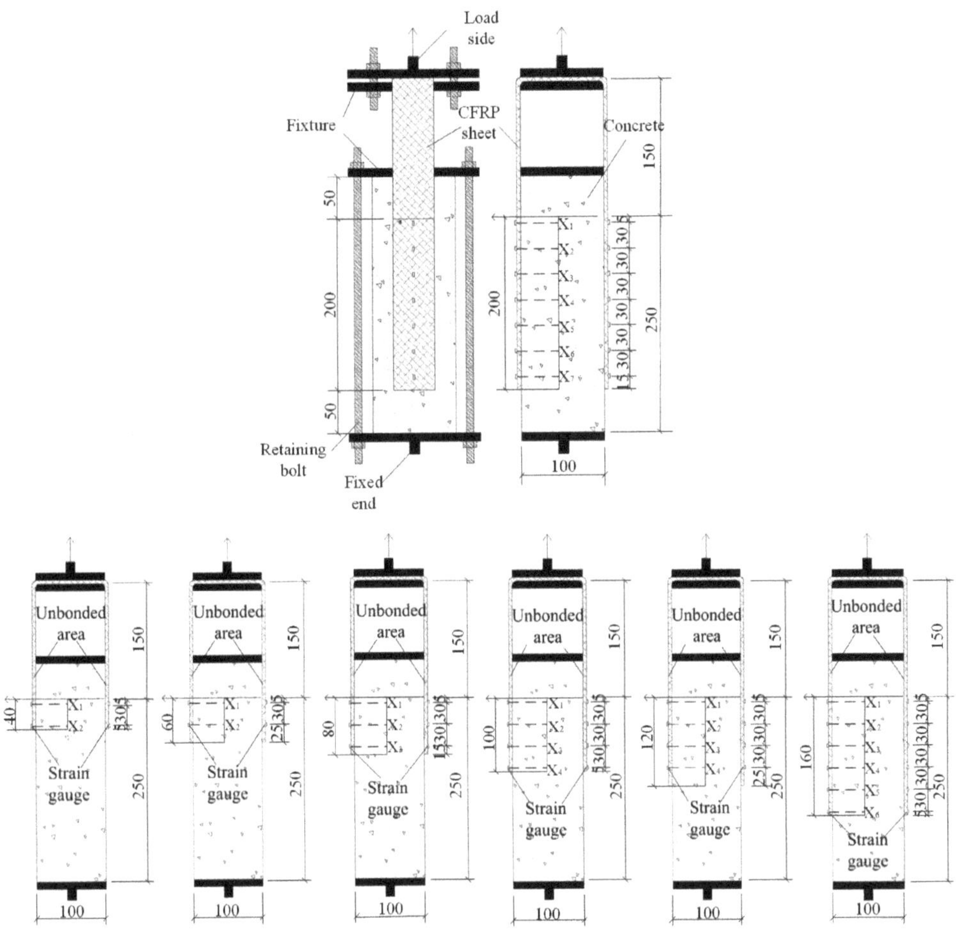

Figure 2. Specifications of test specimen and arrangement of strain gauges (unit: mm).

2.2. Specimen Preparation

The concrete was poured into the molds and cured for 28 days. In order to improve the bonding quality between the CFRP sheet and concrete, it is necessary to roughen the concrete surface. Firstly, the area for bonding the CFRP sheet is outlined on the concrete surface using a marker pen. As illustrated in Figure 3a,b, the area is chiseled out with a chisel machine to reveal the coarse aggregate of the concrete. After chiseling, transparent tape was applied to the non-bonding areas of the concrete to maintain their cleanliness and prevent the epoxy adhesive from adhering to these regions. Meanwhile, the CFRP sheets were trimmed to the necessary dimensions, and the concrete bonding area was initially coated with the epoxy adhesive. After even application, the CFRP sheets were affixed to the demarcated concrete area. Subsequently, another coat of the epoxy adhesive was spread over the CFRP sheets, and the surface was smoothed using a roller to ensure flatness, as depicted in Figure 3c. Finally, the affixed test specimens were allowed to solidify at room temperature for two weeks, with precautions taken to protect them from rain. It is important to mention that the distance from the loading end to the bond zone is the same for all conditions, and only the CFRP sheet bonded length is changed. In the preparation stage before the test, the CFRP sheet was secured with wooden supports, and a coating of the epoxy resin adhesive was applied to the loaded end. This adhesive was allowed to dry and harden before the commencement of testing, as shown in Figure 3d.

Figure 3. The specimen processing diagram: (**a**) delineate the area; (**b**) chisel the surface; (**c**) stick the CFRP sheet; (**d**) strengthen the CFRP sheet.

2.3. *Experimental Material*

2.3.1. Concrete

The concrete materials employed included a coarse aggregate in the form of gravel ranging in size from 5 to 20 mm, a fine aggregate in the form of medium sand, and cement possessing a strength rating of 42.5 MPa. According to the requirements of JGJ55-2011 [38], the mixing ratio of concrete was designed. The concrete mix proportion was water/cement/fine aggregate/coarse aggregate at ratios of 1:1.67:3.35:5.69. The concrete exhibited a compressive strength of 43.3 MPa and an axial tensile strength of 3.63 MPa.

2.3.2. CFRP Sheet

The fiber-reinforced composite used in this test was the CFRP sheet, which was produced by Carbon Technology Group Co., Ltd. (Tianjin, China). The nominal thickness of the CFRP sheet was 0.167 mm. It underwent tensile testing in accordance with GB/T 3354-2014 standards [39], and the static properties measured included a tensile strength of 3861.9 MPa, a modulus of elasticity of 240 GPa, and an ultimate strain of 1.7%.

2.3.3. Adhesive

The adhesive is an important material for bonding FRP to concrete, and adhesives with different material properties will have different test performances. This is due to the different transfer efficiency of interfacial shear stresses by different adhesives. An effective adhesive facilitates the transfer of shear stress across interfaces, enhancing the performance of FRP with concrete. The tensile strength of the adhesive used in this test was 60.9 MPa. The elastic modulus is 3700 MPa. The ultimate tensile strain is 2.0%.

2.4. *Experimental Protocol*

This test primarily investigates the impact of varying loading rates on the effective bond length at the CFRP–concrete interface, with parameters including the loading rate and

CFRP bond length. Details of the test program and ultimate bearing capacities are presented in Table 1. Loading rates were divided into three levels: S for quasi-static load, and DI and DII for dynamic loads, corresponding to 0.1 mm/min, 10 mm/min, and 100 mm/min (strain rates of 10^{-5} s^{-1}, 10^{-3} s^{-1}, and 10^{-2} s^{-1}), respectively. Considering the discreteness of the test results, three specimens were prepared for each working condition. The specimen number in turn indicates the loading level, bond length, loading rate, and serial number.

Table 1. Specifications of the specimens and main findings from the tests.

Specimen Number	L (mm)	v (mm/min)	Strain Rate (s^{-1})	P_u (kN)	
				Individual	Ave.
S-40-0.1-1	40	0.1	10^{-5}	11.03	11.31
S-40-0.1-2				11.64	
S-40-0.1-3				11.26	
DI-40-10-1		10	10^{-3}	12.31	12.34
DI-40-10-2				12.54	
DI-40-10-3				12.17	
DII-40-100-1		100	10^{-2}	13.43	13.09
DII-40-100-2				12.74	
DII-40-100-3				13.10	
S-60-0.1-1	60	0.1	10^{-5}	12.15	12.36
S-60-0.1-2				12.39	
S-60-0.1-3				12.54	
DI-60-10-1		10	10^{-3}	14.66	13.93
DI-60-10-2				13.15	
DI-60-10-3				13.98	
DII-60-100-1		100	10^{-2}	15.37	15.10
DII-60-100-2				15.11	
DII-60-100-3				14.82	
S-80-0.1-1	80	0.1	10^{-5}	13.11	12.79
S-80-0.1-2				12.84	
S-80-0.1-3				12.42	
DI-80-10-1		10	10^{-3}	15.19	15.73
DI-80-10-2				16.23	
DI-80-10-3				15.77	
DII-80-100-1		100	10^{-2}	16.66	16.75
DII-80-100-2				16.76	
DII-80-100-3				16.23	
S-100-0.1-1	100	0.1	10^{-5}	13.59	13.86
S-100-0.1-2				13.88	
S-100-0.1-3				14.10	
DI-100-10-1		10	10^{-3}	16.22	16.01
DI-100-10-2				15.89	
DI-100-10-3				15.92	
DII-100-100-1		100	10^{-2}	16.73	17.10
DII-100-100-2				17.65	
DII-100-100-3				16.92	
S-120-0.1-1	120	0.1	10^{-5}	13.74	13.98
S-120-0.1-2				14.24	
S-120-0.1-3				13.96	
DI-120-10-1		10	10^{-3}	16.28	16.16
DI-120-10-2				16.34	
DI-120-10-3				15.86	
DII-120-100-1		100	10^{-2}	17.65	17.28
DII-120-100-2				16.79	
DII-120-100-3				17.40	

Table 1. *Cont.*

Specimen Number	L (mm)	v (mm/min)	Strain Rate (s^{-1})	P_u (kN)	
				Individual	Ave.
S-160-0.1-1				14.17	
S-160-0.1-2		0.1	10^{-5}	14.30	14.07
S-160-0.1-3				13.74	
DI-160-10-1				15.67	
DI-160-10-2	160	10	10^{-3}	15.98	16.02
DI-160-10-3				16.41	
DII-160-100-1				16.73	
DII-160-100-2		100	10^{-2}	17.26	17.19
DII-160-100-3				17.58	
S-200-0.1-1				14.43	
S-200-0.1-2		0.1	10^{-5}	13.78	14.12
S-200-0.1-1				14.15	
DI-200-10-1				16.42	
DI-200-10-2	200	10	10^{-3}	16.19	16.18
DI-200-10-3				15.93	
DII-200-100-1				17.38	
DII-200-100-2		100	10^{-2}	17.09	17.15
DII-200-100-3				16.98	

Note: L = bond length; v = loading rate; P_u = ultimate bearing capacity.

3. Experimental Results and Discussion

3.1. Damage Pattern

The primary damage patterns observed in this test are depicted in Figure 4. Most specimens exhibited stripping of the concrete surface layer beneath the CFRP sheet, as shown in Figure 4a. This indicates that under external loading, the CFRP sheet effectively transfers the interfacial bond shear stress to the concrete surface via the adhesive layer, resulting in "ideally reliable" interfacial stripping damage. The surface stripping of concrete starts from the side near the loading end and progressively shifts towards the side away from the loading end. Under static loading, the specimen does not exhibit any noticeable changes initially. However, when the load reaches approximately 40% of its ultimate capacity, a slight stripping sound is audible at the interface. As the load nears the ultimate capacity, the specimen emits a distinct "crackling" sound before suddenly failing. Under dynamic loading, the CFRP sheet peels rapidly, accompanied by a loud sound. The failure process under static loading and dynamic loading is basically consistent with the results observed by Fang et al. [22]. As shown in Figure 4b, some specimens exhibited partial stripping and fracture of the CFRP sheet, potentially due to uneven stress from manual errors during bonding. Additionally, some CFRP sheets fractured, primarily under high strain rates and when using longer bond lengths, as shown in Figure 4c. This may be due to the uneven stress of some specimens during tension, and the interface bearing capacity is large under a high strain rate and long bond length, resulting in the ultimate strain of the CFRP sheet without debonding.

3.2. Load–Displacement Curve and Bearing Capacity

Figure 5 displays the load–displacement curves and bearing capacity changes in specimens with different bond lengths under different loading rates. The displacement data originate from the movements of the testing machine's beam, incorporating both interfacial slips and elastic distortions of the unbound CFRP sheet, as well as deformations of the loading equipment. Hence, these curves solely reflect the qualitative load–displacement behavior. Observing Figure 5a, it is evident that at an identical loading rate, specimens with varying bond lengths show similar slopes in the loading phase of the load–displacement curve. The load decreases rapidly to 0 upon reaching the ultimate bearing capacity, and the specimen reaches destruction. The load exhibited a gradual increase with the increase in bond length. The maximum load tends to stabilize when the bond length reaches a specific value, which can be considered the effective bond length [40]. When comparing the load–displacement curves across various loading rates, it is observed that the maximum loads consistently increase with higher loading rates. This observation aligns with the research findings reported by Shi et al. [32] and Huo et al. [41], suggesting the presence of a strain rate effect at the interface between CFRP sheets and concrete. Figure 5b shows the trend of bearing capacity under different conditions. It can be seen that when the loading rates are 0.1 mm/min, 10 mm/min, and 100 mm/min, the bearing capacity increases by 24.8%, 31.1%, and 32.0%, respectively. This shows that as the loading rate rises, the bonding length exerts a more pronounced influence on the bearing capacity. Combined with Figure 5a,b, it is evident that as the loading rate rises, the effective bond length range shows a gradual reduction. The effective bond length is reduced from approximately 100 mm to approximately 80 mm, a decrease of 20%. This finding is in alignment with the research outcomes presented by Shen et al. [33].

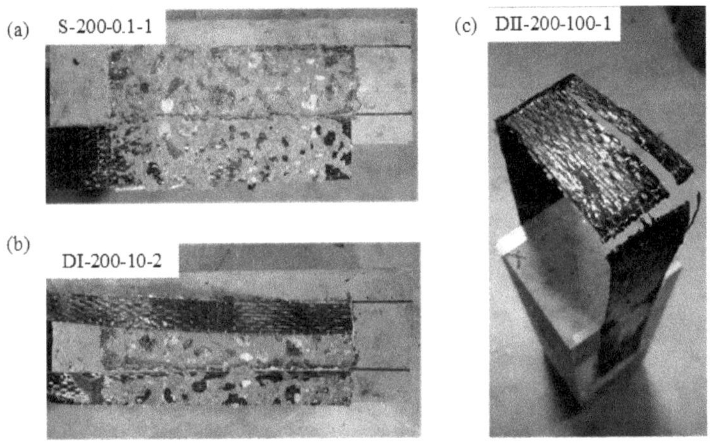

Figure 4. Primary damage patterns: (**a**) debonding in concrete; (**b**) concrete debonding and partial CFRP sheet fractured; (**c**) CFRP sheet fractured.

3.3. CFRP Strain Distribution

Figure 6 illustrates the development of CFRP strain distribution across specimens with varying loading rates and bond lengths. The horizontal axis represents the distance from each strain gauge measurement point to the loading end, while the vertical axis represents the strain data obtained at each point. It was observed that the strain distribution in the CFRP exhibits a consistent pattern across different strain rates as the load progressively increases. At the beginning of loading, the CFRP sheet in the bonded area shows only minor strain values in proximity to the loading end. As the loading progresses, the strain value of the CFRP sheet gradually increases and gradually reaches a maximum value, and further transfers away from the loading end. However, the strain development in each specimen is not evenly distributed throughout the bonding region, but stops at a certain point. It is predominantly located within 100 mm from the loading end, which indicates the presence of an "effective bond length" for the CFRP sheets. At a constant bond length, the strain transfer distance of CFRP decreases as the load rate increases, aligning with the findings reported by Shen et al. [32]. This indicates that as the loading rate increases, the effective bond length decreases. The maximum strain in the CFRP sheets increases with higher loading rates. For example, in the specimen with a 160 mm bond length illustrated in Figure 6f, the maximum CFRP strain increased from 5202 $\mu\varepsilon$ at a loading rate of 0.1 mm/min to 8963 $\mu\varepsilon$ at a loading rate of 100 mm/min. This result is in line with the trend of increasing bearing capacity at its interface. Figure 6d–g demonstrate that as the loading rate increases, the CFRP strain diminishes more swiftly from its peak to its minimum value.

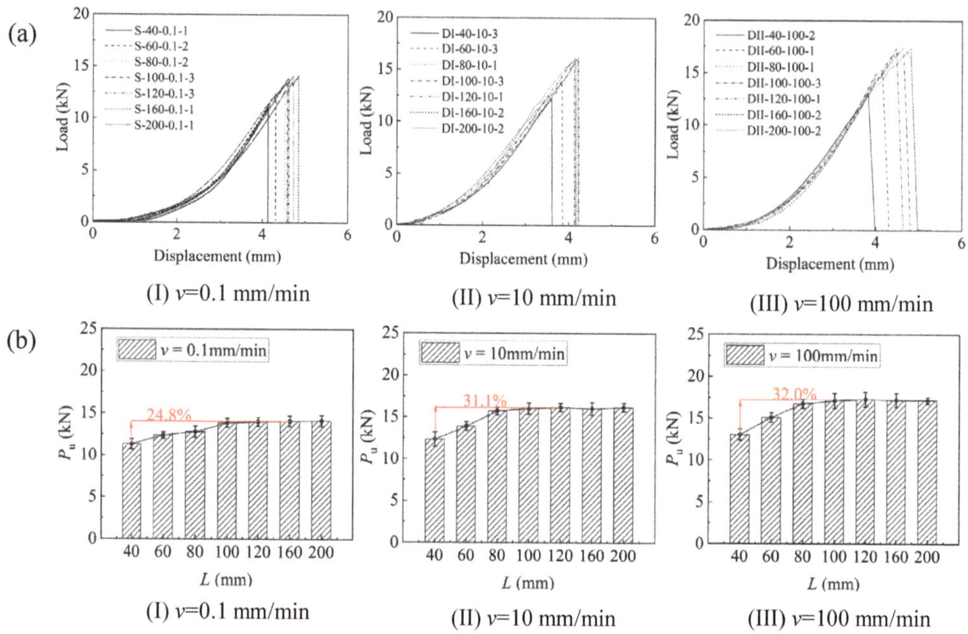

Figure 5. Load conditions of specimens with different bond lengths at different loading rates: (**a**) load–displacement curves; (**b**) development trend of bearing capacity.

Due to the rapid stripping of the CFRP sheet during the short period of time between the specimen reaching its ultimate bearing capacity and the occurrence of damage, there is a transfer of the maximum value of CFRP strain. This phenomenon is more obvious when the bond length exceeds 100 mm. In specimens with a bond length of less than 100 mm, the debonding process is more sudden, resulting in no further strain development curve after reaching the maximum bearing capacity. Figure 7 illustrates the strain development in the CFRP sheet of the specimen of Figure 6d–g as time progresses after P reaches P_u. The labels t_1, t_2, etc., in the graph indicate the change in CFRP strain with time in a short period of time after P reaches P_u. As illustrated in Figure 7, upon reaching the maximum load capacity of the interface, the maximum strain at the loaded end of the CFRP sheet exhibits minimal variation. As the strain progresses from the loaded end towards the free end, the strain at the free end increases progressively. The region where the maximum strain value is reached is increasing. It indicates that the stripping from the CFRP sheet is transmitted from the loaded end to the free end. Moreover, at lower loading rates, a greater range of a strain increase occurs. This aligns with the changes observed in the effective bond length depicted in Figure 5.

3.4. CFRP Interface Shear Stress and Local Bond Stress–Slip Relationships

To calculate the shear stress distribution at a given point, the following assumptions are made: (1) the strain change between two neighboring strain measurement points is linear, and (2) the strain is uniformly distributed across the thickness of the FRP sheet, meaning the upper bending moment on the CFRP sheet is considered to be zero.

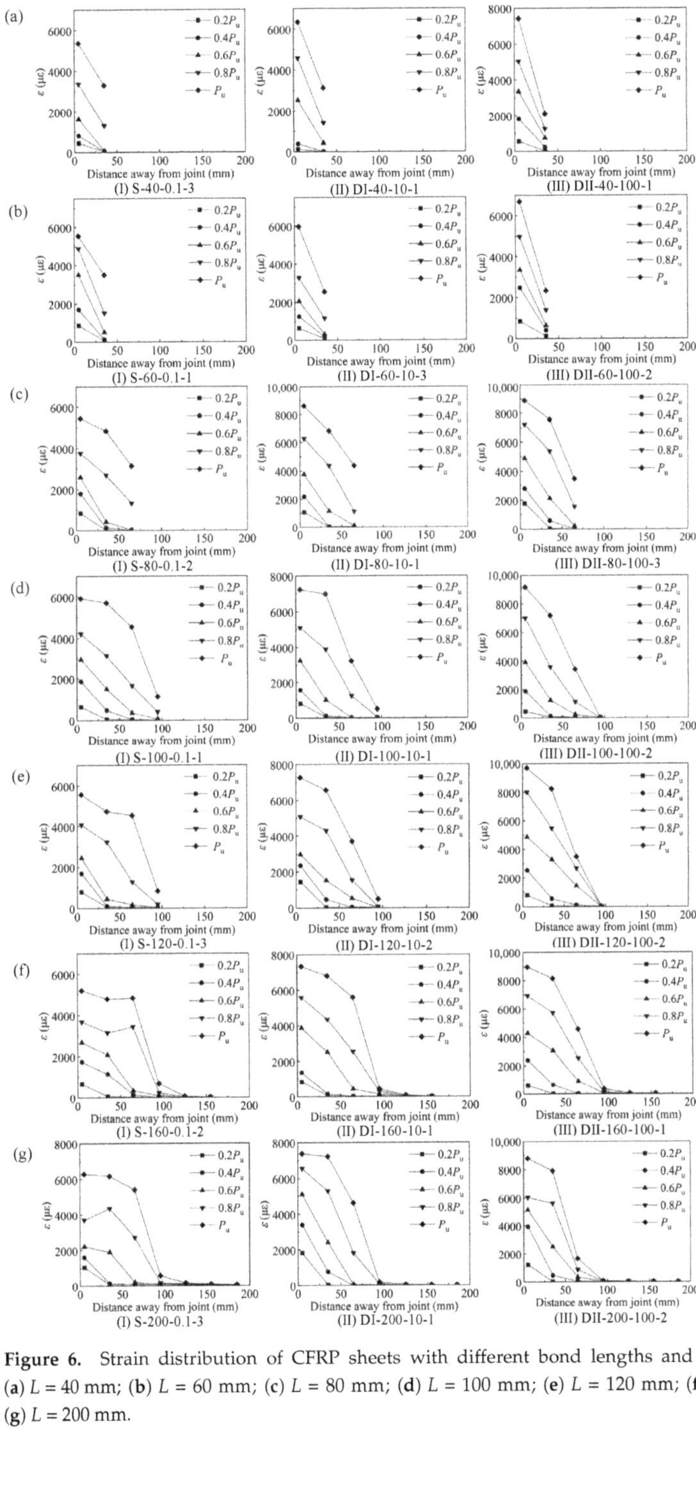

Figure 6. Strain distribution of CFRP sheets with different bond lengths and loading rates: (**a**) $L = 40$ mm; (**b**) $L = 60$ mm; (**c**) $L = 80$ mm; (**d**) $L = 100$ mm; (**e**) $L = 120$ mm; (**f**) $L = 160$ mm; (**g**) $L = 200$ mm.

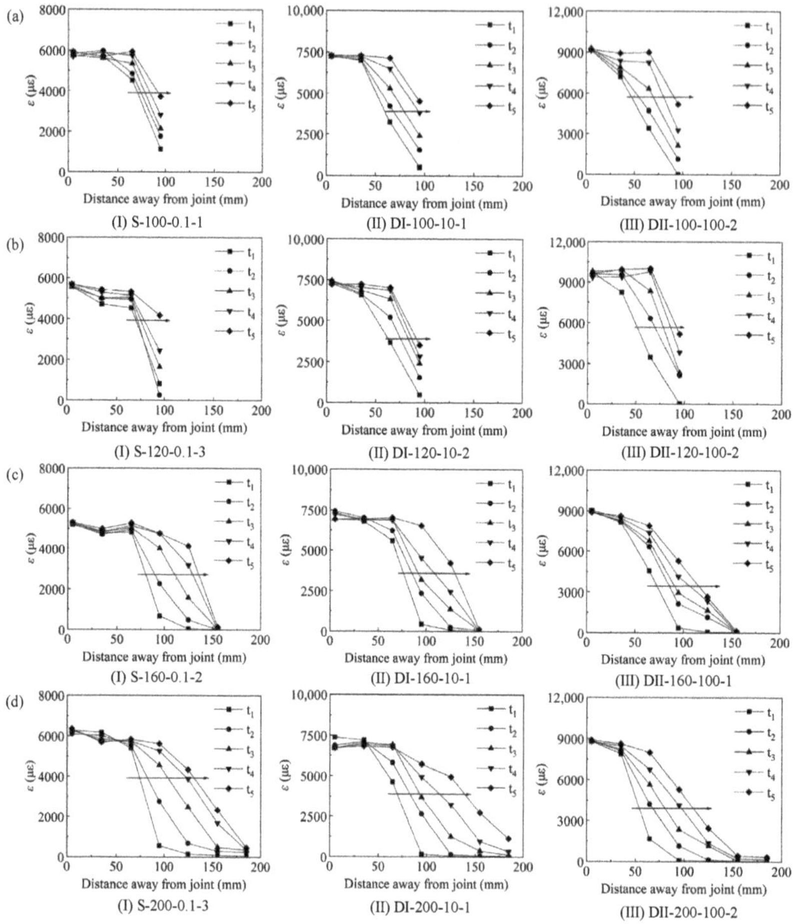

Figure 7. Strain distribution of CFRP sheets with different bond lengths and loading rates: (**a**) L = 100 mm; (**b**) L = 120 mm; (**c**) L = 160 mm; (**d**) L = 200 mm.

The formula for calculating the shear stress at a point is obtained through the force equilibrium relationship as shown below [42]:

$$\tau_i = \frac{\varepsilon_i - \varepsilon_{i+1}}{\Delta d} E_f t_f \tag{1}$$

where ε_i means the CFRP strain at Section i; ε_{i+1} means the CFRP strain at Section $i + 1$; t_f means the CFRP thickness; E_f means the modulus of elasticity of CFRP; and Δd means the interval between gauges (30 mm).

When the bonding length is short, there are few CFRP strain data points, which cannot fully reflect the development law of interfacial shear stress. Figure 8 shows the distribution of interfacial shear stress corresponding to different loads at different loading rates for specimens with bond lengths of 100 mm, 120 mm, 160 mm, and 200 mm, respectively.

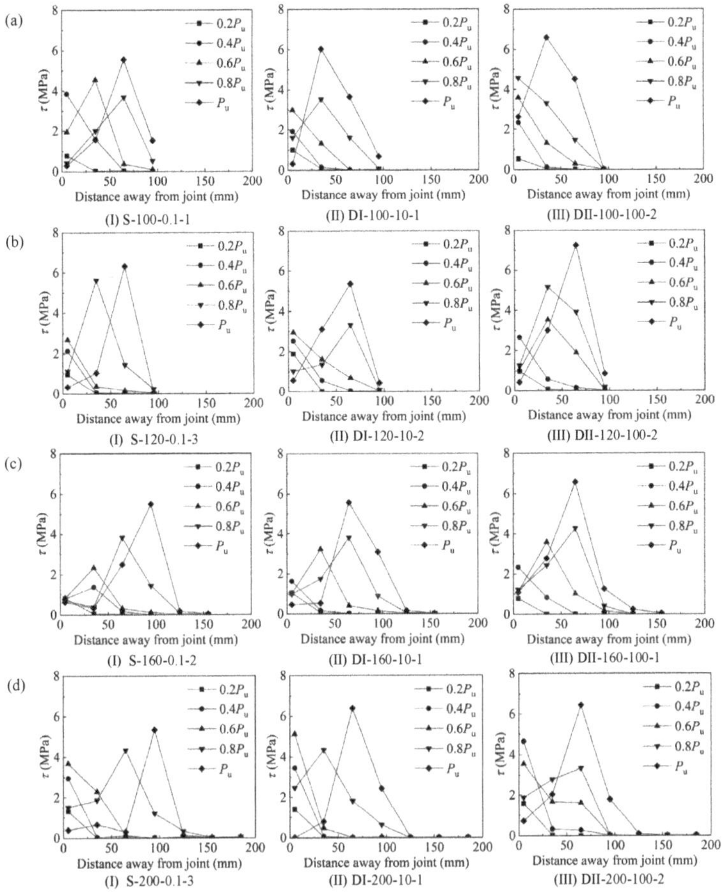

Figure 8. Distribution of interfacial shear stress in specimens featuring bond lengths of (**a**) L = 100 mm, (**b**) L = 120 mm, (**c**) L = 160 mm, and (**d**) L = 200 mm.

Although the peak shear stress on the CFRP sheet varied with the loading rates, the general pattern of the shear stress distribution curves remained consistent under various conditions. During the initial stage of load application to the specimen, shear stresses were primarily concentrated at the loaded end. With an increase in load, the shear stresses rapidly increased as well. However, in the region distant from the loading end, the shear stress was very small and did not change significantly. At this time, the shear stresses at the interface between CFRP and concrete were negligible away from the loaded end (i.e., free end). As the load is applied, the shear stress gradually attains its peak value at the loading end. At the same time, the peak shear stress at the interface shifts towards the free end, while the shear stress at the loading end gradually decreases. As the load approaches the interfacial damage threshold, shear stress at the loaded end is minimized, culminating in stripping damage at the interface between the CFRP sheet and the concrete. The force applied by the CFRP sheet to the concrete is primarily transferred through the bonding interface located close to the loading end, whereas the bonding interface near the free end does not contribute to load transmission. By comparing the distribution of interfacial shear stress across various loading rates, it is observed that the maximum interfacial shear stress intensifies as the loading rate increases. This aligns with the findings reported by Shi et al. [32], indicating that there is also a strain rate effect on interfacial shear stress.

Additionally, the extent of the maximum interfacial shear stress decreases, indicating a shortened effective bond length due to a reduced load transfer distance.

The bond stress–slip relationship between the CFRP and concrete interface reflects the bonding performance of the CFRP and concrete interface. It can visualize the stripping process of CFRP bonded to the concrete surface. The bond stress–slip relationship is characterized by two parameters: shear stress (τ) and slip (s). The interfacial shear stress of CFRP–concrete is described by Equation (1), and the calculation of slip is based on the following assumptions [43–46]: (1) the impact of concrete deformation on slip is negligible given the smaller strain in concrete relative to FRP; (2) strain values between two adjacent strain measurement points are assumed to be linearly distributed; (3) no slip occurs at the free end. The slip is determined using the following equation [42]:

$$s_i = \frac{\varepsilon_i + \varepsilon_{i+1}}{2}\Delta d + \varepsilon_{i+1}\Delta d + s_{i+1}\ (i = 1 \sim n) \tag{2}$$

where s_i and s_{i+1} represent the slip of strain measurement points i and i + 1, and ε_i and ε_{i+1} represent the strain values of FRP of strain measurement points i and i + 1, respectively. When $i = n$, s_{i+1} and ε_{i+1} are the slip and strain at the free end, both of which are zero.

Through the aforementioned procedure, the local bond stress–slip (τ-s) relationships for each gauge interval of all specimens can be obtained from the experimental data. Considering that the local bond stress–slip curve cannot be completely displayed when the bond length is short, Figure 9 depicts the local bond stress–slip curves for specimens featuring bond lengths of 100 mm, 120 mm, 160 mm, and 200 mm. The shapes of the bond–slip curves under various conditions are consistent, featuring distinct rising and falling sections and pronounced nonlinear characteristics. It is evident that the bond stress–slip curves include a smooth section following the descending section, which enlarges as the bond length increases. This shows that increasing the bond length of CFRP can effectively increase the ductility of the CFRP–concrete interface. Among specimens with identical bond lengths, it is observable that the peak bond stress at the CFRP–concrete interface rises with an increase in the loading rate. Additionally, the slip corresponding to this peak bond stress also increases, and the descending section of the bond stress–slip curve is gentler.

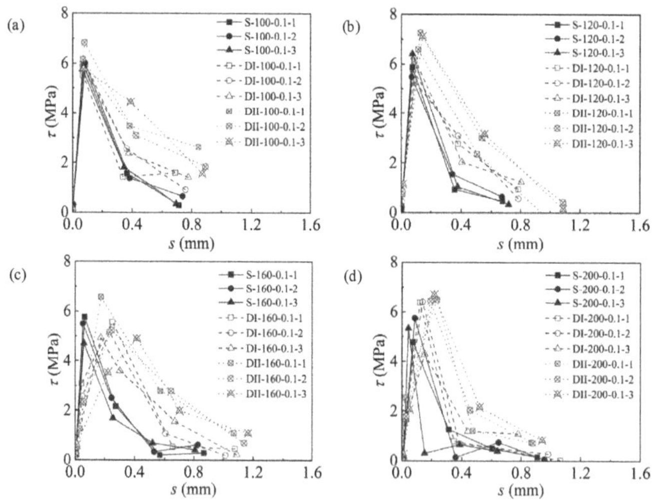

Figure 9. The local bond stress–slip (τ-s) curves for all specimens with different bond lengths: (**a**) L = 100 mm; (**b**) L = 120 mm; (**c**) L = 160 mm; (**d**) L = 200 mm.

The analysis of the test results reveals a notable strain rate effect on the CFRP–concrete interface. As the strain rate increases, both the ultimate bearing capacity and the peak shear stress of the interface exhibit an upward trend. In the dynamic strain rate range of this test, the interface damage pattern is basically the same as that under the static state, and it is still dominated by the shallow stripping failure of concrete. As the strain rate increases, the development range of CFRP strain at the interface gradually decreases, and the CFRP strain develops more rapidly. This makes the range of maximum interfacial shear stress decrease. This further demonstrates that the effective bond length of the CFRP–concrete interface diminishes as the strain rate rises. Identifying the effective bond length is crucial in investigating the bonding performance of the CFRP–concrete interface, and it holds significant importance for the practical application of CFRP sheets in engineering. A formula has been derived, based on experimental results and analyses, for computing the dynamic effective bond length while accounting for the impact of the strain rate.

4. Formula for Dynamic Effective Bond Length of CFRP Sheet

4.1. The Dynamic Effective Bond Length Formula

The effective bond length is primarily influenced by two factors: the stiffness of the FRP, $E_f t_f$, and the strength of the concrete, f_c [47]. The effective bond length is determined through the analysis of strain distributions (see Figure 6). A bond is deemed inactive if the strain gauge records a strain that is less than 5% of the peak strain. The effective bond length is calculated by measuring the distance from the loading end to the point where the strain reaches 5% of the peak strain [48]. If needed, interpolation techniques can be employed to facilitate this calculation. When the bond length is relatively short, it is possible that the effective bond length will not be reached. Consequently, the effective bond length cannot be accurately determined from the FRP strain distribution when the bond length is short. It can be judged from Figure 5 that the longest effective bond length in the test is about 100 mm. In this paper, the working conditions with the bond length of 100 mm and above were selected to determine the effective bond length. Table 2 shows the test values of effective bond lengths at different strain rates for bond lengths of 100~200 mm, denoted by $L_{e,t}$. The effective bond length mean values were determined to be 99.4 mm, 93.8 mm, and 86.1 mm for loading rates of 0.1 mm/min, 10 mm/min, and 100 mm/min, respectively.

Based on empirical data and fracture mechanics theories, several theoretical models have been developed to forecast the effective bond length at the FRP–concrete interface [49]. These models are commonly known as static effective bond length models. By comparing test values with theoretical values, seven analytical models for evaluating the static effective bond length were analyzed, as shown in Table 3. As we can see, all existing models belong to the category of static effective bond length models, which mainly consider the material properties of FRP and concrete. It should be noted that none of these formulas take into account the effect of the strain rate effect.

Table 2. Comparison of test and predicted values of effective bond length.

$\dot{\varepsilon}$ (s^{-1})	L (mm)	$L_{e,t}$ (mm)		$L_{e,m}$ (mm)	α	k
		Individual	Ave.			
10^{-5}	100	100	99.4	82.53	0.997	1.21
		98				1.19
		99				1.20
	120	98				1.19
		101				1.22
		99				1.20
	160	102				1.24
		100				1.21
		98				1.19
	200	99				1.20
		101				1.22
		98				1.19

Table 2. *Cont.*

$\dot{\varepsilon}$ (s^{-1})	L (mm)	$L_{e,t}$ (mm)		$L_{e,m}$ (mm)	α	k
		Individual	Ave.			
		94				1.21
	100	93				1.19
		92				1.18
		95				1.22
	120	93				1.19
10^{-3}		95	93.8	77.90	0.941	1.22
		94				1.21
	160	93				1.19
		95				1.22
		94				1.21
	200	92				1.18
		95				1.22
		85				1.18
	100	86				1.19
		87				1.21
		85				1.18
	120	86				1.19
10^{-2}		87	86.1	72.10	0.871	1.21
		88				1.22
	160	87				1.21
		84				1.17
		85				1.18
	200	86				1.19
		87				1.21
		Ave.				1.20

Table 3. Performance of static effective bond length models.

References	Equation	Consideration Factors	$L_{e,m}$ (mm)	IAE (%)
Wu et al. [50]	$L_e = 0.395\frac{(E_f t_f)^{0.54}}{f_{co}^{0.09}}$	E_f, t_f, f_{co}	87.92	10.36
Neubauer and Rostasy [51]	$L_e = \sqrt{\frac{E_f t_f}{2f_t}}$	E_f, t_f, f_t	71.68	26.92
Khalifa et al. [52]	$L_e = e^{6.13-0.580\ln(E_f t_f)}$	E_f, t_f	54.02	44.92
Chen and Teng [53]	$L_e = \sqrt{\frac{E_f t_f}{\sqrt{f_c'}}}$	E_f, t_f, f_c'	82.78	15.60
Lu et al. [24]	$L_e = 1.33\frac{\sqrt{E_f t_f}}{f_t}$	E_f, t_f, f_t	68.27	30.40
Yang et al. [54]	L_e = 100 mm	-	100	2.12
ACI 440.2R-08 [55]	$L_e = \frac{23{,}300}{(nE_f t_f)^{0.58}}$	E_f, t_f, n	49.85	49.18

Note: L_e = effective bond length; t_f = the thickness of FRP; E_f = the elastic modulus of FRP; f_c' or f_{co} = the cylinder axial compressive strength of concrete; f_t = the axial tensile strength of concrete; n = the number of FRP layers; $L_{e,m}$ = predicted value of effective bond length.

The corresponding values of effective bond lengths, denoted by $L_{e,m}$, for the static conditions of the different formulas were calculated using the formulas in Table 3. The accuracy of the different formulas was assessed by calculating the discrepancy between this predicted value and the value under quasi-static tests. In this paper, Equation (3) is used to calculate the absolute error rate:

$$\text{IAE} = \sum \frac{\left[(\text{Expe.} - \text{Theo.})^2\right]^{1/2}}{\sum|\text{Expe.}|} \tag{3}$$

where Expe. and Theo. denote test results and theoretical results, respectively. The results are shown in Table 3.

Among the formulas used to predict the effective bond length under static loading, the deviations of Khalifa [52] and ACI 440.2R-08 [55] are larger. Notably, in both of these models, the effective bond length decreases as the FRP stiffness increases. This can be attributed to the limited test observations used to calibrate the American ACI Code [55], which did not take into account the cohesive properties of the concrete substrates [56]. In the model developed by Yang et al. [54], the effective bond length is not affected by material stiffness, contrary to the results of many studies [57]. The effective bond length in other equations is directly proportional to the stiffness of FRP. As a result, existing models for calculating effective bond lengths use inconsistent forms. The predicted values of the formulas of Wu et al. [50] and Chen and Teng [53] are in good alignment with the test values. The formula of Wu et al. [50] is somewhat complicated in form and inconvenient to use. However, these models do not take into consideration the impact of the strain rate and are unable to accurately reflect the variations in effective bond length that occur under different strain rates.

The formula of Chen and Teng [53] is based on extensive test results and experience. It simplifies relevant assumptions and offers ease of use compared to other formulas. It is applicable to specific ranges of stress levels and material properties, thereby ensuring high reliability and wide recognition in the academic community. In this paper, based on the Chen and Teng formula [53], the calculation formula of dynamic effective bond length is proposed. It is specified as follows:

$$L_e = \alpha k \sqrt{\frac{E_f t_f}{\sqrt{f_c'}}} \tag{4}$$

where α is the strain rate effect coefficient; it expresses the relationship between the effective bond length under static and dynamic conditions. The coefficient α is defined as the ratio of the effective bond lengths at various strain rates in the test to that under static conditions in the test. It is important to note that the static effective bond length uses the average value of the static effective bond lengths obtained from the quasi-static tests. The relationship between the strain rate and the strain rate effect coefficient α is described by a quadratic function:

$$\alpha = A(\lg\dot{\varepsilon})^2 + B(\lg\dot{\varepsilon}) + C \tag{5}$$

where A, B, and C are the coefficients to be obtained by data fitting. The relationship between the strain rate effect coefficients and $\lg\dot{\varepsilon}$ is shown in Figure 10. In this paper, it is suggested that A be taken as -0.014, B as -0.14, and C as 0.647 for the working condition.

Considering that the Chen and Teng formula [53] exhibits a discrepancy between the calculated and test values of the effective bond length under static conditions, this paper introduces a scaling factor, k, to optimize the formula. The strain rate effect coefficients α at different strain rates based on Equation (5) are substituted into Equation (4) to obtain the scaling factor k at different strain rates, as shown in Table 2. $L_{e,m}$ is the resultant value of multiplying the calculated value of the formula of Chen and Teng [53] with the value of the corresponding strain rate. Based on different working conditions, k is calculated as the arithmetic mean value of 1.20.

4.2. Verification of the Dynamic Effective Bond Length Formula

Based on the above process, the dynamic effective bond length formula is proposed in this paper. In order to verify the reasonableness of the formula calculation, the theoretical value of the dynamic effective bond length formula by Equation (4) was calculated and compared with the results of this test as well as the existing test values [32,41,58], as shown in Figure 11. In the figure, $L_{e,t}$ is the test value of the effective bond length, and $L_{e,p}$ is the theoretical value of the effective bond length calculated by Equation (4), which gives an error line of 20%. The close match between the theoretical values calculated by the formula and the test values confirms the accuracy and reasonableness of the established dynamic effective bond length formula.

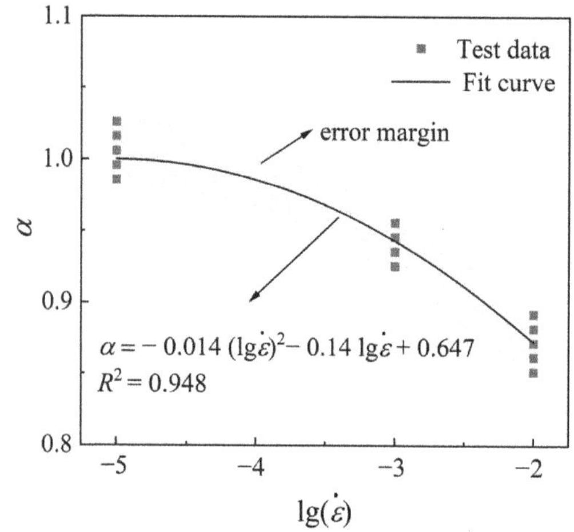

Figure 10. Strain rate effect coefficient fitting curve.

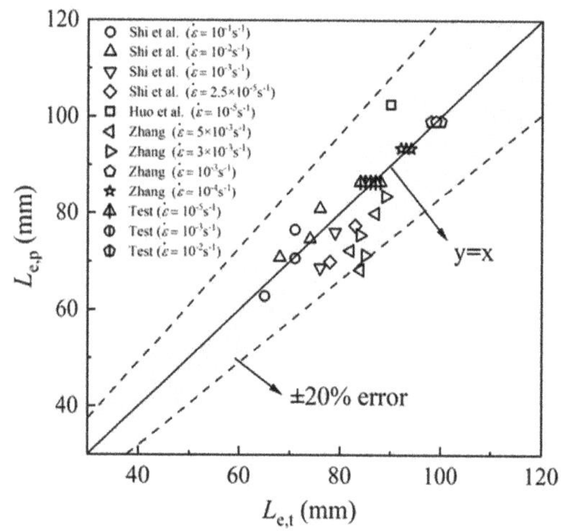

Figure 11. Comparison of theoretical and test values [32,41,58].

5. Conclusions

This paper delves into the bond performance characteristics of the CFRP–concrete interface through double-sided shear tests. It examines the impact of various loading rates on the damage pattern at the CFRP–concrete interface, the strain development of the CFRP sheet, and the effective bond length. Based on the test results, a formula for calculating the effective bond length under dynamic conditions was established. The main conclusions drawn are

(1) The damage pattern observed at the CFRP–concrete interface under dynamic loading is the same as that under static loading, which is mainly the shallow stripping damage of concrete under the CFRP sheet.

(2) At higher loading rates, the CFRP bond length has a more pronounced effect on bearing capacity. Specifically, as the bond length increases from 40 mm to 200 mm, the bearing capacity rises by 24.8% at a loading rate of 0.1 mm/min and by 32.0% at a loading rate of 100 mm/min.
(3) The distribution trend of CFRP strain is consistent under different loading rates. With an increase in the strain rate, the maximum strain value of CFRP increases by about 70%, but the strain transfer range decreases.
(4) At different loading rates, the profiles of the shear stress distribution curves on the CFRP sheet are similar. As the loading rate increases, the transfer range of the maximum shear stress decreases.
(5) When the strain rate increases from 10^{-5} s^{-1} to 10^{-2} s^{-1}, the effective bond length decreases from about 100 mm to about 80 mm, a decrease of 20%.
(6) The effective bond length formula proposed in this paper accounts for the strain rate effect, enhancing its predictive capability for effective bond length under both static and dynamic conditions.

It should be noted that this study only discussed the influence of the loading rate and CFRP sheet bond length on the bond properties of the CFRP sheet–concrete interface, while the influence of the CFRP bond width, concrete strength grade, and CFRP sheet type on CFRP sheet–concrete interfacial bond properties will be studied in the follow-up work. In addition, due to the limitations of the instruments selected in this paper, the range of the strain rate that can be explored is very limited. The strain rate range considered in this test is small, and more high-strain-rate tests are needed to verify the rationality and accuracy of the proposed prediction dynamic effective bond length formula. Subsequently, we will carry out more experimental and numerical simulation work to extend the results to a wider range of strain rates.

Author Contributions: Conceptualization, D.L. and L.J.; methodology, D.L.; software, X.W.; validation, D.L., X.W. and J.Z.; formal analysis, D.L., X.W. and J.Z.; investigation, J.Z.; resources, D.L.; data curation, X.W.; writing—original draft preparation, X.W.; writing—review and editing, D.L. and J.Z.; visualization, X.W.; supervision, X.D.; project administration, X.D.; funding acquisition, D.L. and L.J. All authors have read and agreed to the published version of the manuscript.

Funding: This research was funded by the Beijing Nova Program, grant number 20220484047; the National Natural Science Foundation of China, grant number 51978022; and the Natural Science Foundation of Beijing Municipality, grant number JQ22025.

Data Availability Statement: The data underlying this article are available in the article.

Conflicts of Interest: The authors declare no conflicts of interest.

References

1. Zhou, J.; Bi, F.; Wang, Z.; Zhang, J. Experimental investigation of size effect on mechanical properties of carbon fiber reinforced polymer (CFRP) confined concrete circular specimens. *Constr. Build. Mater.* **2016**, *127*, 643–652. [CrossRef]
2. Mugahed Amran, Y.H.; Alyousef, R.; Rashid, R.S.M.; Alabduljabbar, H.; Hung, C. Properties and applications of FRP in strengthening RC structures: A review. *Structures* **2018**, *16*, 208–238. [CrossRef]
3. Liu, H.B.; Zhao, X.L.; Al-Mahaidi, R. Effect of fatigue loading on bond strength between CFRP sheets and steel plates. *Int. J. Struct. Stab. Dy.* **2010**, *10*, 1–20. [CrossRef]
4. Wu, Y.; Cao, Y.G. Energy balance method for modeling ultimate strain of confined concrete. *ACI Struct. J.* **2017**, *114*, 373–381. [CrossRef]
5. Kalfat, R.; Al-Mahaidi, R. Improvement of FRP-to-concrete bond performance using bidirectional fiber patch anchors combined with FRP spike anchors. *Compos. Struct.* **2016**, *155*, 89–98. [CrossRef]
6. Singh, A.; Del Rey Castillo, E.; Ingham, J. FRP-to-FRP bond characterization and force-based bond length model. *Compos. Struct.* **2019**, *210*, 724–734. [CrossRef]
7. Teng, J.G.; Cao, S.Y.; Lam, L. Behaviour of GFRP-strengthened RC cantilever slabs. *Constr. Build. Mater.* **2001**, *15*, 339–349. [CrossRef]
8. Woo, S.; Lee, Y. Experimental study on interfacial behavior of CFRP-bonded concrete. *KSCE J. Civ. Eng.* **2010**, *14*, 385–393. [CrossRef]

9. Li, P.; Zhao, Y.; Wu, Y.; Lin, J. Effect of defects in adhesive layer on the interfacial bond behaviors of externally bonded CFRP-to-concrete joints. *Eng. Struct.* **2023**, *278*, 115495. [CrossRef]
10. Lee, J.; Lopez, M.M. Characterization of FRP Uwrap Anchors for Externally Bonded FRP-Reinforced Concrete Elements: An Experimental Study. *J. Compos. Constr.* **2016**, *20*, 4016012. [CrossRef]
11. Subramaniam, K.V.; Carloni, C.; Nobile, L. An understanding of the width effect in FRP-concrete debonding. *Strain* **2011**, *47*, 127–137. [CrossRef]
12. Ali-Ahmad, M.; Subramaniam, K.; Ghosn, M. Experimental Investigation and Fracture Analysis of Debonding between Concrete and FRP Sheets. *J. Eng. Mech.* **2006**, *132*, 914–923. [CrossRef]
13. Zheng, X.H.; Huang, P.Y.; Chen, G.M.; Tan, X.M. Fatigue behavior of FRP–concrete bond under hygrothermal environment. *Constr. Build. Mater.* **2015**, *95*, 898–909. [CrossRef]
14. Raoof, S.M.; Bournas, D.A. Bond between TRM versus FRP composites and concrete at high temperatures. *Compos. Part B Eng.* **2017**, *127*, 150–165. [CrossRef]
15. Pellegrino, C.; Tinazzi, D.; Modena, C. Experimental Study on Bond Behavior between Concrete and FRP Reinforcement. *J. Compos. Constr.* **2008**, *12*, 180–189. [CrossRef]
16. Wang, Z. Reformation for bonding stress of section of FRP strengthening RC girder. *Shanxi Archit.* **2005**, *21*, 79–80. (In Chinese)
17. De, L.L.; Miller, B.; Nanni, A. Bond of fiber-reinforced polymer laminates to concrete. *ACI Mater. J.* **2001**, *98*, 256–264.
18. Hassan, T.; Rizkalla, S. Bond Mechanism of NSM FRP bars for flexural strengthening of concrete structures. *Aci Struct. J.* **2004**, *101*, 830–839.
19. Mukhtar, F.M.; Faysal, R.M. A review of test methods for studying the FRP-concrete interfacial bond behavior. *Constr. Build. Mater.* **2018**, *169*, 877–887. [CrossRef]
20. Xue, W.; Zeng, L.; Tan, Y. Experimental studies on bond behaviour of high strength CFRP plates. *Compos. Part B Eng.* **2008**, *39*, 592–603. [CrossRef]
21. McIsaac, A.; Mak, K.; Fam, A. Influence of Resin Biocontent and Type on Bond Strength between FRP Wet Layup and Concrete. *J. Compos. Constr.* **2019**, *23*, 4019029. [CrossRef]
22. Fang, N.Q.; Liu, G.F.; Zhang, L.S. Experimental research on the bond performance of CFRP-concrete interface. *J. Build. Mater.* **2007**, *10*, 32–36. (In Chinese)
23. Yuan, J.; Yuan, W.; Hou, X.J.; Lin, J. Model analysis on effective bonding length of interfacial shear stress between AFRP and concrete. *Yangtze River* **2020**, *51*, 182–186+214. (In Chinese)
24. Lu, X.Z.; Teng, J.G.; Ye, L.P.; Jiang, J.J. Bond–slip models for FRP sheets/plates bonded to concrete. *Eng. Struct.* **2005**, *27*, 920–937. [CrossRef]
25. Pan, Y.; Liu, Q.; Ren, Y.; Guo, R. Experimental study on the bond behavior of CFRP strand sheet-concrete with different types of bond agents. *China Civil Eng. J.* **2021**, *54*, 26–37. (In Chinese)
26. Li, D.; Du, X.; Jin, L. Theoretical prediction method for nominal tensile strength of ordinary concrete considering strain rate effect and size effect. *Build. Struct.* **2021**, *51*, 122–128. (In Chinese)
27. Xiao, S.; Zhang, J. Compressive damage experiment of concrete at different strain rates. *China Civil Eng. J.* **2010**, *43*, 40–45. (In Chinese)
28. Buchan, P.A.; Chen, J.F. Blast resistance of FRP composites and polymer strengthened concrete and masonry structures—A state-of-the-art review. *Compos. Part B Eng.* **2007**, *38*, 509–522. [CrossRef]
29. Wu, C.; Oehlers, D.J.; Rebentrost, M.; Leach, J.; Whittaker, A.S. Blast testing of ultra-high performance fibre and FRP-retrofitted concrete slabs. *Eng. Struct.* **2009**, *31*, 2060–2069. [CrossRef]
30. Nam, J.; Kim, H.; Kim, S.; Yi, N.; Kim, J.J. Numerical evaluation of the retrofit effectiveness for GFRP retrofitted concrete slab subjected to blast pressure. *Compos. Struct.* **2010**, *92*, 1212–1222. [CrossRef]
31. Ha, J.; Yi, N.; Choi, J.; Kim, J.J. Experimental study on hybrid CFRP-PU strengthening effect on RC panels under blast loading. *Compos. Struct.* **2011**, *93*, 2070–2082. [CrossRef]
32. Shi, J.; Zhu, H.; Wu, Z.; Wu, G. Experimental study of the strain rate effect of FRP sheet-concrete interface. *China Civil Eng. J.* **2012**, *45*, 99–107. (In Chinese)
33. Shen, D.; Shi, H.; Ji, Y.; Yin, F. Strain rate effect on effective bond length of basalt FRP sheet bonded to concrete. *Constr. Build. Mater.* **2015**, *82*, 206–218. [CrossRef]
34. Li, Z.; Zhang, X.; Shi, Y. Experimental study on the dynamic bond behavior between CFRP and concrete under different slip rates. *Eng. Struct.* **2020**, *216*, 110788. [CrossRef]
35. Li, X.; Chen, J.; Lu, Y. A FRP-concrete bond slip model under dynamic load. In Proceedings of the 7th National Construction Engineering Application of FRP Academic Exchange, Chongqing, China, 15 May 2015; pp. 179–182. (In Chinese).
36. *GB 50608-2020*; Technical Standard for Fiber Reinforced Polymer (FRP) in Construction. Beijing China Planning Publishing House: Beijing, China, 2020. (In Chinese)
37. *T/CECS 146-2022*; Technical Specification for Strengthening Concrete Structures with Carbon Fiber Reinforced Polymer. China Architecture & Building Press: Beijing, China, 2022. (In Chinese)
38. *JGJ55-2011*; Specification for Mix Proportion Design of Ordinary Concrete. China Architecture & Building Press: Beijing, China, 2011. (In Chinese)

39. *GB/T 3354-2014*; Test Method for Tensile Properties of Orientation Fiber Reinforced Polymer Matrix Composite Materials. Standards Press of China: Beijing, China, 2014. (In Chinese)
40. Diab, H.M.; Farghal, O.A. Bond strength and effective bond length of FRP sheets/plates bonded to concrete considering the type of adhesive layer. *Compos. Part B Eng.* **2014**, *58*, 618–624. [CrossRef]
41. Huo, J.; Liu, J.; Dai, X.; Yang, J.; Lu, Y.; Xiao, Y.; Monti, G. Experimental Study on Dynamic Behavior of CFRP-to-Concrete Interface. *J. Compos. Constr.* **2016**, *20*, 4016026. [CrossRef]
42. Zhang, P.; Gao, D.Y.; Zhu, H. Numerical simulation and experimental study on the performance of wet-bonding interface between FRP plate and concrete. *China Civil Eng. J.* **2013**, *46*, 108–114. (In Chinese)
43. Mukhopadhyaya, P.; Swamy, R.N.; Lynsdale, C.J. Influence of aggressive exposure conditions on the behaviour of adhesive bonded concrete–GFRP joints. *Constr. Build. Mater.* **1998**, *12*, 427–446. [CrossRef]
44. Biscaia, H.C.; Chastre, C.; Silva, M.A.G. Nonlinear numerical analysis of the debonding failure process of FRP-to-concrete interfaces. *Compos. Part B Eng.* **2013**, *50*, 210–223. [CrossRef]
45. Toutanji, H.; Ueno, S.; Vuddandam, R. Prediction of the interfacial shear stress of externally bonded FRP to concrete substrate using critical stress state criterion. *Compos. Struct.* **2013**, *95*, 375–380. [CrossRef]
46. Mohammadi, T.; Wan, B.; Harries, K.A. Bond–slip behavior of fiber-reinforced polymer/concrete interface in single shear pull-out and beam tests. *J. Reinf. Plast. Compos.* **2015**, *35*, 375–386. [CrossRef]
47. Wu, Y.; Jiang, C. Quantification of Bond-Slip Relationship for Externally Bonded FRP-to-Concrete Joints. *J. Compos. Constr.* **2013**, *17*, 673–686. [CrossRef]
48. Bizindavyi, L.; Neale, K.W. Transfer Lengths and Bond Strengths for Composites Bonded to Concrete. *J. Compos. Constr.* **1999**, *3*, 153–160. [CrossRef]
49. Kang, T.H.K.; Howell, J.; Kim, S.; Lee, D.J. A State-of-the-Art Review on Debonding Failures of FRP Laminates Externally Adhered to Concrete. *Int. J. Concr. Struct. Mater.* **2012**, *6*, 123–134. [CrossRef]
50. Wu, Z.; Islam, S.M.; Said, H. A Three-Parameter Bond Strength Model for FRP—Concrete Interface. *J. Reinf. Plast. Compos.* **2008**, *28*, 2309–2323. [CrossRef]
51. Neubauer, U.; Rosta'sy, F.S. Design aspects of concrete structures strengthened with externally bonded CFRP plates. In Proceedings of the Seventh International Conference on Structural Faults and Repairs, Edinburgh, UK, 8 July 1997; ECS Publications: Edinburgh, UK, 1997; Volume 2, pp. 109–118.
52. Khalifa, A.; Gold, W.J.; Nanni, A.; MI, A.A. Contribution of Externally Bonded FRP to Shear Capacity of RC Flexural Members. *J. Compos. Constr.* **1998**, *2*, 195–202. [CrossRef]
53. Chen, J.F.; Teng, J.G. Anchorage Strength Models for FRP and Steel Plates Bonded to Concrete. *J. Struct. Eng.* **2001**, *127*, 784–791. [CrossRef]
54. Yang, Y.X.; Yue, Q.R.; Hu, Y.C. Experimental study on bond performance between carbon fiber sheets and concrete. *J. Build. Struct.* **2001**, *22*, 36–42.
55. *ACI 440.2R-08*; Guide for the Design and Construction of Externally Bonded FRP Systems for Strengthening Concrete Structures. ACI Committee 440: Farmington Hills, MI, USA, 2008.
56. D Antino, T.; Pellegrino, C. Bond between FRP composites and concrete: Assessment of design procedures and analytical models. *Compos. Part B Eng.* **2014**, *60*, 440–456. [CrossRef]
57. Yao, J.; Teng, J.G.; Chen, J.F. Experimental study on FRP-to-concrete bonded joints. *Compos. Part B Eng.* **2005**, *36*, 99–113. [CrossRef]
58. Zhang, Y.H. *Study on the Mechanism of FRP-Concrete Interfacial Debonding under Impact*; Ningbo University: Ningbo, China, 2017. (In Chinese)

buildings

MDPI

Article

Bond Behavior of CFRP–Concrete Bonded Joints with Additional GFRP Layer: Effect of Bonding Sequence

Hao Zhou [1], Jiahao Zhao [1], Yan Yang [1], Fengling Tan [2], Ya Ou [3], Yi Wang [1] and Chao Li [1,*]

[1] School of Civil Engineering, Central South University, Changsha 410075, China; hao.zhou@csu.edu.cn (H.Z.)
[2] Institute of Research and Design, Sinohydro Engineering Bureau 8 Co., Ltd., Changsha 410004, China
[3] School of Civil Engineering, Central South University of Forestry and Technology, Changsha 410004, China
* Correspondence: lichaocsu@csu.edu.cn

Abstract: Existing studies have shown that bonding a ±45° biaxial GFRP under CFRP laminate can significantly improve the load-carrying capacity and ultimate deformation of CFRP–concrete bonded joints. In such a bonding configuration, the GFRP interlayer is wider than the CFRP laminate so that the interfacial stress can be redistributed to achieve a higher fracture work; however, the effect of the bonding sequence,—specifically, the position of the GFRP layer—on the bond behavior is not yet clear. In this study, considering the same CFRP and GFRP usage, three types of CFRP–concrete bonded joints with CFRP bonded under, above, and between GFRP layers were prepared and tested under single-shear loading. Digital image correlation (DIC) was used to measure the deformation of the bonded joints during the test. Afterward, the failure mode, load–displacement behavior, and principal strain distribution were analyzed. The experimental results show that the dominant failure mode is the combined cohesion failure mode within the concrete and GFRP delamination, which is not affected by the bonding sequence. Compared to conventional CFRP–concrete bonded joints, bonding the CFRP laminate above, under, and between the GFRP layers achieved a 157.6%, 175.0%, and 177.2% increase in load-carrying capacity, respectively. Accordingly, the ultimate deformation also recorded an 83.0%, 103.6%, and 86.3% increase. However, the bonding sequence showed a slight influence on the initial stiffness of the load–displacement curve with a maximum difference of 16.1%, taking the minimum as a reference, which could be attributed to the differences in the strength and stiffness between the CFRP–concrete and CFRP–GFRP–concrete interfaces.

Keywords: bond behavior; CFRP–concrete; GFRP layer; bonding sequence

1. Introduction

Externally bonded (EB) carbon-fiber-reinforced polymers (CFRP) have been increasingly used in strengthening reinforced concrete (RC) structures due to their advantages of being lightweight, high in strength, and with excellent durability over conventional strengthening materials such as steel [1–5]. For EB–CFRP-strengthened RC members subjected to flexural loading, the stress is transferred between the CFRP and concrete through the adhesively bonded interface. As such, bond behavior plays an important role in assessing the strengthening performance. Existing studies have shown that two major failure modes, i.e., plate-end debonding and intermediate crack (IC)-induced debonding, often determine the load-carrying capacity. For plate-end debonding, additional anchors should be used to prevent premature debonding caused by the high interfacial peeling stress, whereas IC-induced debonding is caused by the high interfacial shear stress [6]. Therefore, it is of great significance to understand the bond behavior of CFRP–concrete bonded joints.

While investigating the bond behavior of CFRP–concrete bonded joints, among the existing test methods—such as the single-shear pull-off test [7], double-shear pull-off test [8], and flexural beam test [9]—the single-shear pull-off test is often used [10–12] due to its simplicity and acceptable accuracy, wherein a CFRP laminate is bonded to a concrete block before applying shear loading. Results of such experiments have clearly shown that the dominant failure mode is the cohesion failure within the concrete. Therefore, for a given CFRP laminate, the fracture toughness of the concrete mainly determines the load-carrying capacity of the bonded joints [13,14]. It is also found that when the bonding length is greater than the effective bonding length, the load-carrying capacity of the bonded joints cannot be further increased. As a result, the stress of the CFRP is merely 10–50% of the tensile strength of the CFRP laminate when the bonded joints reach the load-carrying capacity, resulting in a low material utilization rate of the CFRP [15,16]. This low utilization rate significantly hinders the application of CFRP in the strengthening of RC structures.

To improve the material efficiency of CFRP, numerous anchoring devices including steel–FRP plate anchors [17–19], shooting nail anchors [20], FRP anchors [21,22], and U-shaped anchors [23] have been proposed to prevent premature debonding. Taking advantage of the interlocking effect, dowel action, and the enhanced friction achieved by the anchorage devices, existing results have shown that the above approach can significantly improve the bonding performance. However, due to the stress concentration at the anchoring position, installing anchors through drilling the CFRP laminate may result in a brittle fracture of the CFRP [3], while U-jacketing requires access to the side of the RC members. In addition, the installation of anchors usually involves drilling and alignment procedures, during which the internal reinforcement may be damaged. In addition to anchoring the CFRP, changing the geometric or mechanical properties of the bonding surface on the concrete substrate is also proposed. These techniques include, but are not limited to, plastering high-performance concrete [24], grooving or drilling holes [25–28], or using flexible adhesives [29]; however, these measures are either labor-intensive or show limited improvement on the bonding performance and so their widespread application remains challenging.

Most recently, the Author's group proposed using a GFRP interlayer to enhance the bonding performance [30,31]. In this approach, a $\pm 45^\circ$ biaxial GFRP sheet was bonded between the CFRP and concrete substrate. Compared to the existing approach of using the CFRP patch beneath the CFRP laminate as an anchor [32], the GFRP interlayer has a lower stiffness that can effectively transfer the interfacial stress and enhance the deformation capability. In addition, GFRP sheets are much cheaper than CFRP, and thus, a more economical solution can be achieved. Experimental results showed that the width of the GFRP interlayer and the width ratio between the GFRP interlayer and CFRP laminate can significantly affect the bond behavior in terms of the load–displacement curve, the load-carrying capacity, and the ultimate deformation. Specifically, the force plateau diminished as the width ratio increased and the load-carrying capacity achieved, at most, a 203.0% increase compared to the conventional CFRP–concrete bonded joints. A simplified model was proposed to explain the bond behavior. It was shown that sequential activation of the GFRP–concrete interface in both the longitudinal and transverse directions determines the bond behavior. In addition, the existence of effective bonding length in such bonded joints depends on the width of the GFRP interlayer and vice versa. Detailed discussion can be found in [31].

Considering the high load-carrying capacity and ultimate deformation achieved by the CFRP–concrete bonded joints with a GFRP interlayer, bonding the GFRP fabric beneath the CFRP laminate in the intermediate region of the RC beam or slab is a promising application to improve the interfacial performance; for the plate-end debonding, it might be preferable to bond the GFRP on top of the CFRP laminate or sandwich the CFRP between two layers of GFRP sheet such that the peeling performance can be improved at the same time. However, compared to bonding the CFRP to the GFRP interlayer and then the concrete substrate, bonding the CFRP directly to the concrete substrate could affect the stress transfer path since part of the interfacial stress will be directly resisted by the concrete substrate before being redistributed by the GFRP interlayer. As a result, the fracture process and the activation of the GFRP interlayer in resisting external force could be affected.

Against this background, this paper conducts a series of single-shear pull-off tests on CFRP-to-concrete bonded joints with GFRP layers. Three bonding sequences are considered, i.e., bonding GFRP beneath the CFRP laminate, bonding GFRP on top of CFRP laminate, and bonding CFRP laminate between two GFRP sheets. The failure mode, load-carrying capacity, ultimate deformation, and principal strain development are presented and discussed to investigate the effects of bonding sequence on the stress transfer of CFRP–GFRP–concrete bonding joints.

2. Test Program

To investigate the effect of the bonding sequence, a total of 12 single-shear pull-off test specimens were prepared and tested. Among these specimens, three CFRP-to-concrete bonded joints without a GFRP layer were tested as reference specimens, and the other nine CFRP–GFRP–concrete bonded joints were prepared with three bonding sequences. Detailed descriptions of the specimen design, test set-up, and instrumentation are presented in the subsequent sections.

2.1. Materials

Table 1 summarizes the mechanical properties of the adhesive, CFRP, and GFRP used in this study. The structural epoxy adhesive supplied by Carbon was used to bond the CFRP and GFRP to the concrete substrate. The tensile strength was 60.9 MPa and the elastic modulus was 3.7 GPa, respectively, according to the manufacturer's data sheet. Unidirectional CFRP at 300 g/mm^2, used in all specimens, was supplied by Shanghai Niugu Construction Technology Company(Shanghai, China). The nominal thickness of the CFRP laminate after impregnation was 0.56 mm and the tensile strength and elastic modulus were 617.1 MPa and 75.6 GPa, respectively. Two types of ±45° biaxial GFRP, i.e., DB400-1270 (400 g/m^2) and DB800-1270 (800 g/m^2), were used in preparing the specimens. Both types of GFRP were provided by Changzhou Hualike New Material Company(Changzhou, China). The nominal thickness of the GFRP laminate after impregnation was 1.68 mm and 3.36 mm, respectively. In addition, the tensile strength of the GFRP along the fiber direction was 341.7 MPa and the elastic modulus was 22.9 GPa. The concrete blocks were cast with commercially available concrete. The cubic concrete compressive strength was tested to be 39.84 MPa according to GB/T 50081-2019 [26].

Table 1. Material properties.

Material	Mass per Unit Area [g/m^2]	Nominal Thickness [mm]	Tensile Strength [MPa]	Elastic Modulus [GPa]
Epoxy adhesive	--	--	60.9	3.7
CFRP	300	0.56	617.1	75.6
Low-density GFRP	400	1.68	341.7	22.9
High-density GFRP	800	3.36	341.7	22.9

2.2. Specimen Design

Figure 1 presents the schematics of single-shear pull-off test specimens investigated in this study. The dimensions of the concrete blocks are 400 mm × 200 mm × 150 mm. For CFRP–concrete bonded joints with a GFRP layer (Figure 1a), the CFRP sheet (50 mm × 700 mm) was bonded to the concrete block directly through the wet lay-up process (4 layers), whereas for CFRP–GFRP–concrete bonded joints, three bonding sequences were employed (Figure 1b–d). In the first bonding sequence, named the G/C group—in which 'C' and 'G', respectively, represent the CFRP and GFRP sheets, and the order of appearance of 'G' and 'C' represents the sequence in which CFRP and GFRP are bonded to the concrete block surface—a GFRP interlayer (300 mm × 150 mm) consisting of 1 layer of DB800-1270 sheet was bonded to the concrete prior to the bonding of CFRP sheet. For the second bonding sequence (named the C/G group), the CFRP sheet was bonded to the concrete prior to the bonding of the DB800-1270 GFRP sheet. Finally, for the third bonding sequence (named the G/C/G group), the CFRP sheet was bonded between two layers of DB400-1270 GFRP sheets. For all three bonding sequences, the planar dimension of the GFRP layer was identical and the fibers were positioned at ±45° to the longitudinal direction of the bonded joints. In addition, the free end and loaded end of the bonded joints were 50 mm away from the edges of the concrete block to avoid wedge failure. Table 2 summarizes the specimens prepared and tested in this study. Three repeated tests were conducted for each specimen group to reach a reliable test result.

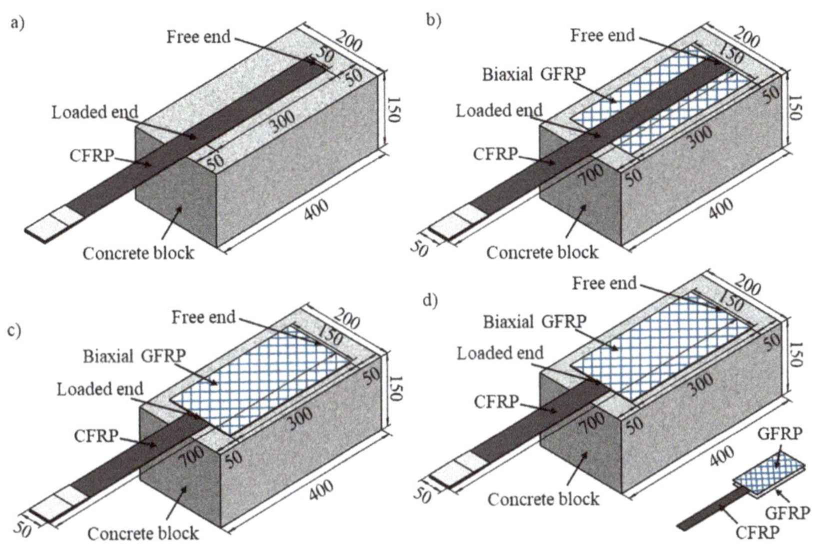

Figure 1. Schematics of (**a**) conventional CFRP–concrete bonded joints; (**b**) CFRP–GFRP bonded joints where a biaxial GFRP was placed in the interlayer between CFRP and concrete; (**c**) CFRP–GFRP bonded joints where CFRP was placed between GFRP and concrete; (**d**) CFRP–GFRP bonded joints where CFRP was placed between GFRP sheets.

Table 2. Parameters of specimens.

Specimen Group	Bonding Length [mm]	CFRP Width [mm]	GFRP Width [mm]	CFRP Layers	GFRP Layers	GFRP Density [g/mm^2]	Number of Specimens
C	300	50	--	4	--	--	3
G/C	300	50	150	4	1	800	3
C/G	300	50	150	4	1	800	3
G/C/G	300	50	150	4	2	400	3

2.3. *Specimen Preparation and Instrumentation*

Before bonding the CFRP and GFRP, the bonding area of the concrete surface was abraded with an electric hammer with a 32-millimeter-working-diameter point groove gouge (Figure 2a). Then, the dust and concrete debris were removed through a portable blower. To achieve clean bonding between the FRPs and the concrete substrate, the bonding areas were marked with masking tape (Figure 2b) to prevent excessive epoxy adhesive application outside these areas. After that, the GFRP and CFRP were successively bonded to the specified area according to the designated sequence (Figure 2c). During the wet lay-up process, a defoaming roller was used to squeeze out the trapped air bubbles. It should be noted that the screw roller should be moved in the fiber direction to avoid dispersing the fibers (Figure 2d).

Once all the FRP layers were applied, the excessive epoxy adhesive around the bonding area was carefully removed before curing under ambient temperature (Figure 2e). All prepared specimens were cured for more than 4 weeks before testing.

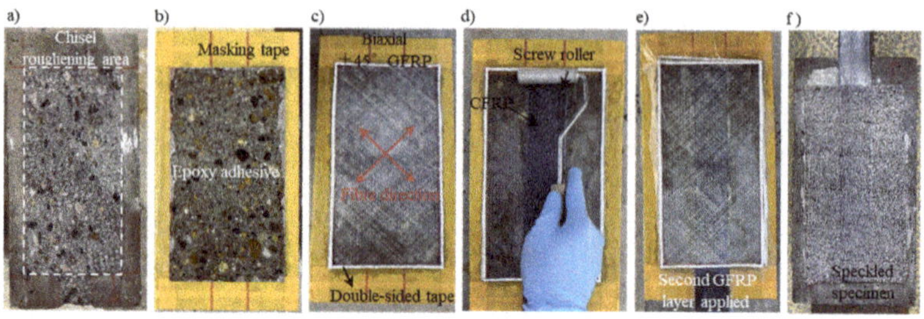

Figure 2. Procedures of specimens' preparation: (**a**) abrading the bonding area of concrete; (**b**) pasting masking tape around the bonding area; (**c**) applying the biaxial glass fabric saturated with epoxy adhesive; (**d**) squeezing out air bubbles; (**e**) curing specimen under ambient temperature; (**f**) a speckled specimen.

Figure 3 illustrates the set-up and configuration of the single-shear pull-off test. During the test, the specimen was fixed in a steel frame (Figure 3b), which was mounted to a 300 kN universal testing machine (Lishi MDC-350, shown in Figure 3a). During the test, the external load was applied at a loading rate of 0.05mm/min. To measure the deformation of the bonded interface during the test, a 3D digital image correlation (3D-DIC) system from Correlated Solutions (VIC-3D) was used (Figure 3a). More details about the concept and technical knowledge of DIC for calculating the deformation can be found in [33].

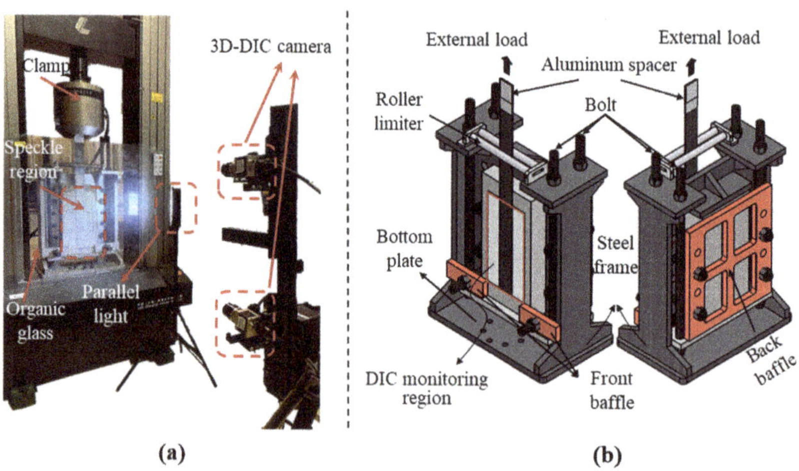

Figure 3. Test instrumentation: (**a**) specimens were fixed within the steel frame that was installed in the 300 kN universal testing machine and the 3D-DIC cameras were placed in front of the specimens to measure the deformation; (**b**) the specific steel frame fixture used to fix specimens.

3. Test Results

3.1. Failure Mode

The photographs of the failed specimens are presented in Figure 4. It can be seen that the typical failure mode of the CFRP–concrete bonded joints is cohesion failure within the concrete (Figure 4a), which is consistent with existing research [7,10,34]. For CFRP–concrete bonded joints with a GFRP layer, regardless of the bonding sequence, the main failure mode is the combined cohesion failure within the concrete and the GFRP delamination. Specifically, the former failure mode mainly occurred in regions close to the CFRP laminate, while the latter was observed in areas close to the edge of the bonding area (Figure 4b,c). Such behavior could be caused by the debonding propagation toward the far end that the interfacial stress near the free end was no longer purely shear stress. This complex interfacial stress state led to GFRP rupture and delamination failures. Such a failure mode has also been observed in CFRP–concrete- and CFRP–steel-bonded joint tests [23,24].

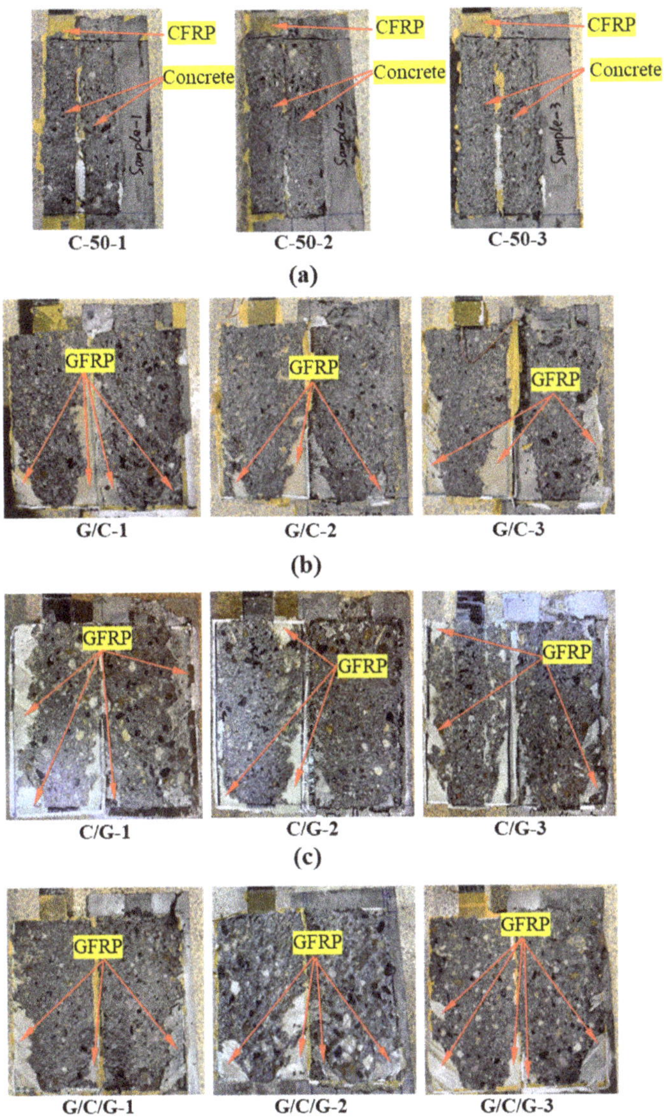

Figure 4. Failure modes include failure within concrete, GFRP delamination, and GFRP rupture: (**a**) C-50-1,C-50-2 and C-50-3; (**b**) G/C-1, G/C-2 and G/C-3; (**c**) C/G-1, C/G-2 and C/G-3; (**d**) G/C/G-1, G/C/G-2 and G/C/G-3.

A detailed inspection of the failure surfaces for specimens C-3, G/G-1, C/G-1, and G/C/G-2 is presented in Figure 5. In the conventional CFRP–concrete bonded joints, there were several deep cracks perpendicular to the loading direction (Figure 5a). Regardless of the bonding sequence, a similar crack pattern was observed in the overlapping area of CFRP and GFRP in the CFRP–GFRP–concrete specimens (zone III in Figure 5b–d). It is worth noting that deep cracks were also observed in the zones on either side of zone III where only the GFRP was present (zone I and II in Figure 5b). These cracks were at an angle of approximately 45° to the loading direction. However, in contrast to CFRP–concrete bonded joint specimens, a greater concrete chunk near the free end within the bonding area was pulled out in the CFRP–GFRP–concrete bonded joint specimens.

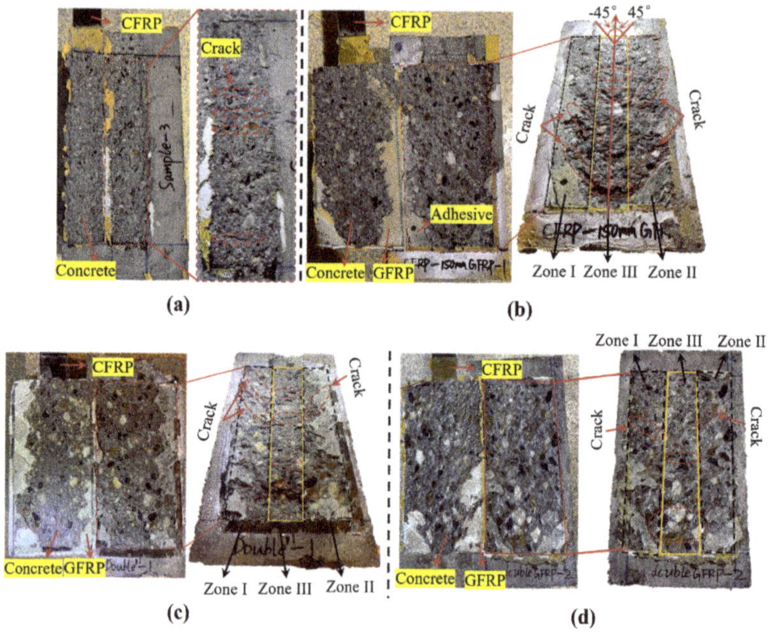

Figure 5. Failure surface of (**a**) C-3; (**b**) G/C-1; (**c**) C/G-1; (**d**) G/C/G-2.

3.2. Load–Displacement Behavior

The load–displacement curves at the loaded end of all specimens are shown in Figure 6. The displacement at the loaded end was measured by the VIC-3D system [33] and the corresponding force values were exported from the universal testing machine. As can be observed in Figure 6, for CFRP–concrete bonded joints, the force initially increases with the displacement at the loaded end before reaching point D1. After that, a general force plateau occurred. Such behavior has been commonly observed in the existing literature and the force plateau is caused by the debonding initiation and propagation of the bonded joints [11,14]. For CFRP–concrete bonded joints with GFRP layers, regardless of the bonding sequence, the force–displacement curves are significantly different to the CFRP–concrete bonded joints without GFRP layers. As indicated by the force and displacement values of A2, A3, and A4 in Figure 6b–d, the initial slope of the load–displacement curves at the loaded end of the bonded joints with the GFRP layer is generally greater than that of the CFRP–concrete bonded joints (calculated by C1 in Figure 6a). In addition, compared to the evident force plateau observed in the CFRP–concrete bonded joints, the force increases with the displacement before reaching the ultimate load-carrying capacity (F2, E3, and F4). After that, softening behavior in the load–displacement curve can be observed in specimens G/C-1, C/G-1 and 3, and G/C/G-2 and 3 as well as brittle failure in specimens

G/C-2 and 3, C/G-2, and G/C/G-1. Moreover, compared to specimens in the G/C and G/C/G groups, the load–displacement curves of the C/G specimens show a more scattered behavior after point C. Such difference might be attributed to the non-homogeneity of the concrete substrate that may affect the fracture path and toughness when debonding propagates. More advanced non-destructive testing methods, such as 3D-CT [35], should be employed in the future to investigate the fracture process of the concrete during the test. Except for this difference, the load–displacement curves of the bonded joints with GFRP layers are similar to each other, regardless of the bonding sequence.

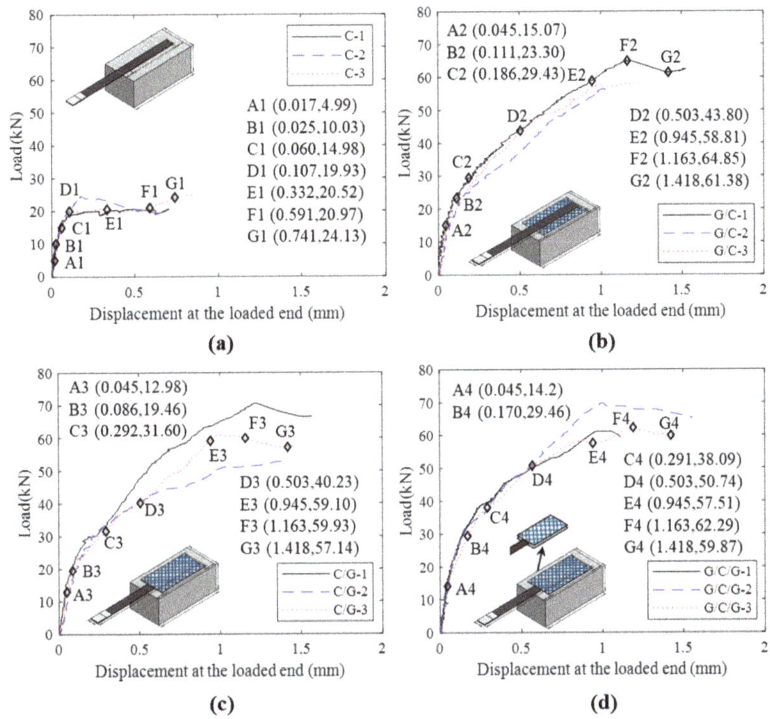

Figure 6. Load–displacement curves of (**a**) C specimens; (**b**) G/C specimens; (**c**) C-G specimens; (**d**) G-C-G specimens.

Table 3 summarizes the load-carrying capacity and ultimate displacement at the loaded end of each specimen. Compared to the CFRP–concrete bonded joints, the bonded joints using GFRP layers exhibited a significantly higher load-carrying capacity and ultimate displacement. Specifically, for the load-carrying capacity, G/C specimens show a 157.6% increase, C/G specimens record a 175.0% increase, and G/C/G specimens show a 177.2% increase on average. Regarding the ultimate displacement at the loaded end, the C/G specimens exhibited the maximum increase at 103.6%, while the G/C and G/C/G specimens recorded a similar increase of 83.0% and 86.3%, respectively. From the above experimental results, it can be briefly concluded that among the three bonding sequences considered in this study, bonding the CFRP between or under the GFRP layer provided a higher load-carrying capacity and ultimate displacement compared to bonding CFRP on top of the GFRP layer. Nevertheless, all three bonding sequences can effectively improve the bonding performance.

Table 3. Load-carrying capacity and ultimate displacement of tested bonded joints.

Specimen Group	No.	Load		Loaded-End Displacement	
		Ultimate Load [kN]	Average Ultimate Load [kN] Increase %	Ultimate Displacement [mm]	Average Ultimate Displacement [mm] Increase
C	1 2 3	20.51 24.45 24.81	23.26	0.712 0.700 0.861	0.758
G/C	1 2 3	64.88 56.37 58.47	59.91 (157.6%)	1.532 1.302 1.326	1.387 (83.0%)
C/G	1 2 3	70.75 60.02 61.11	63.9 (175.0%)	1.595 1.570 1.465	1.543 (103.6%)
G/C/G	1 2 3	61.19 69.89 62.37	64.48 (177.2%)	1.136 1.639 1.460	1.412 (86.3%)

3.3. *Strain Distributions on CFRP and GFRP Laminates*

To further investigate the bond behavior of CFRP–concrete bonded joints with GFRP layers, the principal strain distribution on the GFRP and CFRP of the specimens at different displacement levels (corresponding to the marked points on the load–displacement curve in Figure 6) are presented in Figure 7. As the principal distribution of the different specimens from the same group shows a very similar pattern, only the principal strain distribution of one representative of the specimens from each group is presented. In the figures, the black arrows represent the direction of the principal strains. However, as the regions with purple colors in the principal strain distribution contour were not activated while loading, the strain measurement was interpreted as measurement noise and is not discussed in the subsequent sections.

From the principal strain distribution contour of specimen C-3, it can be observed that when the displacement at the loaded end is less than 0.107 mm (i.e., point D1 in Figure 6a), only a small region (with greenish color) close to the loaded end was activated to resist the external force. In addition, the principal strain direction in most areas on the CFRP was consistent with the loading direction, only a few areas at the edge of the CFRP exhibited an angle with the external load direction. When the displacement at the loaded end was 0.332 mm (point E1 in Figure 6a), more bonding regions were activated to resist the external force. In the meantime, the regions where the direction of the principal strain of CFRP was diagonal to the loading direction expanded (marked with ellipses in Figure 7a). Such behavior might be attributed to the non-uniform debonding within CFRP–concrete in the width direction. As the displacement at the loaded end increases (i.e., at 0.741 mm—point G1 in Figure 6a), more bonding areas have debonded as a general uniform strain can be observed in the rectangular region. The above experimental observation is consistent with the existing literature, thus validating the test set-up and data acquisition.

The principal strain development of specimens G/C-1, C/G-3, and G/C/G-3 are presented in Figure 7b–d. For all three types of specimens, when the slip at the loaded end is 0.045 mm, only a small area of the CFRP and GFRP laminate showed visible strain distribution, which indicated the activation of these bonding areas in resisting the external force. Such behavior also explains the slightly higher initial slope of the load–displacement curves at the loaded end compared to that of the C-3. However, G/C-1 exhibited the highest force among the three types of specimens. As all specimens were in the elastic stage (Figure 6), G/C-1 can provide the greatest initial stiffness. As the slip at the loaded end increased to 0.503 mm, the activated bonding area expanded significantly in both longitudinal and transverse directions. However, it is evident that the G/C-1 specimen expanded wider in the longitudinal direction and C/G-3 showed further expansion in the transverse direction, while specimen C/C/G-3 exhibited a more balanced activation in both directions. Meanwhile, specimen G/C/G-3 recorded a much higher external force at this point than the other two specimens. Such activation pattern of the bonding area continued when the bond slip increased to 0.945 mm when the principal strain distribution of specimens C/G-3 and G/C/G-3 reached 132 mm and 130 mm in the transverse direction, and 176 mm and 205 mm in the longitudinal direction, respectively. For the above two specimens, the activated bonding area reached the edge when the displacement at the loaded end reached 1.163 mm (indicated by the alignment of the principal strain and loading direction). As such, no more bonding area in the width direction can be activated to provide additional interfacial resistance to the external force. Such behavior explained the softening in the load–displacement curves in specimens C/G-1 and 3 and G/C/G-2 and 3. As for the brittle failure of the other specimens in both specimen groups, this might be attributed to the non-homogeneity and dynamic effect when the complete debonding is about to happen; however, more experimental and numerical research should be conducted in the future. As a comparison, the principal strain distribution of specimen G/C-1 reached 138 mm and 217 mm in the transverse and longitudinal directions, respectively, and it is evident that the activated bonding area reached the edge when the displacement at the loaded end reached 1.418 mm.

It should be noted that regardless of the bonding sequence, the direction of principal strain on the GFRP layer is approximately 45° to the loading direction, which aligns with the fiber direction of the GFRP layer. Such behavior can be attributed to the fact that the GFRP layer is subjected to in-plane shear loading. As a result, the principal stress within the GFRP layer is approximately 45° to the loading direction. Such principal stress will further cause cracks within the concrete substrate in zones I and II, as observed in the failed specimens (Figure 5). Additionally, the activated bonding area is significantly greater in the bonded joints with GFRP layers compared to their CFRP–concrete bonded joints counterparts. Such phenomena can explain the remarkable increase in the load-carrying capacity in the bonded joints with GFRP layers.

To further investigate the effect of the GFRP layers on the bond behavior of the CFRP–concrete bonded joints, the strain distribution curves on CFRP and GFRP at different loading levels are presented in Figure 8. For specimen C-3, the strain distribution on the CFRP sheet is presented, while for specimen G/C-1, the principal strain distribution along the CFRP sheet and five lines along the fiber direction of the GFRP layer are displayed. As the CFRP sheet is covered by the GFRP layer in specimens C/G-3 and G/C/G-3, only the strain in the loading direction along the midline of the bonded joints (i.e., axis-x) as well as the principal strain along the fiber direction of GFRP on one side of the CFRP sheet are presented. However, considering the deformation continuity of the bonded CFRP and GFRP, the strain on the centerline of the GFRP can be taken as that of the CFRP laminate.

The longitudinal strain distribution curve of the C-3 specimen is shown in Figure 8a-1. A clear strain plateau can be observed when the displacement at the loaded end reaches 0.332 mm. This indicates that the CFRP–concrete bonded joints started debonding. As the external force increases, the increase in the strain plateau indicates the debonding propagation. In comparison, the evident strain gradient of the CFRP in the loading direction of specimen G/C-1 throughout the test can be observed. The maximum strain of CFRP reached as high as 15,000 $\mu\varepsilon$. However, for specimens C/G-3 and G/C/G-3, a similar strain plateau with specimen C-3 can be observed when the displacement at the loaded end reached 0.945 mm and 0.503 mm, respectively. Such strain plateaus meant the interfacial stress transfer was minimal in this region. Additionally, the maximum strain that both specimens can reach are approximately 12,000 $\mu\varepsilon$ and 10,000 $\mu\varepsilon$, which were less than that in G/C-1. Such differences indicate a distinct interfacial stress transfer in the G/C-1 specimen.

Regarding the principal strain distribution along the fiber direction of the GFRP layer, a similar strain distribution along lines L1 to L5 can be observed in the bonded joints with GFRP layers. For the principal strain along a specific line (i.e., L1 to L5), a clear increase in the strain values close to the loaded side can be observed. In addition, the activated bonding area with the significant strain propagated toward the free side as the external load increased. When the displacement at the loaded end increased to 0.945 mm, evident strain plateaus occurred along L1 of specimens G/C-1 and G/C/C-3. Such plateaus can be also observed along L3 of specimen G/C-1 and L4 of all three discussed specimens as the displacement further increased to 1.418 mm. From the above observations, it can be seen that the GFRP layer is sequentially activated in both transverse and longitudinal directions to resist the external load.

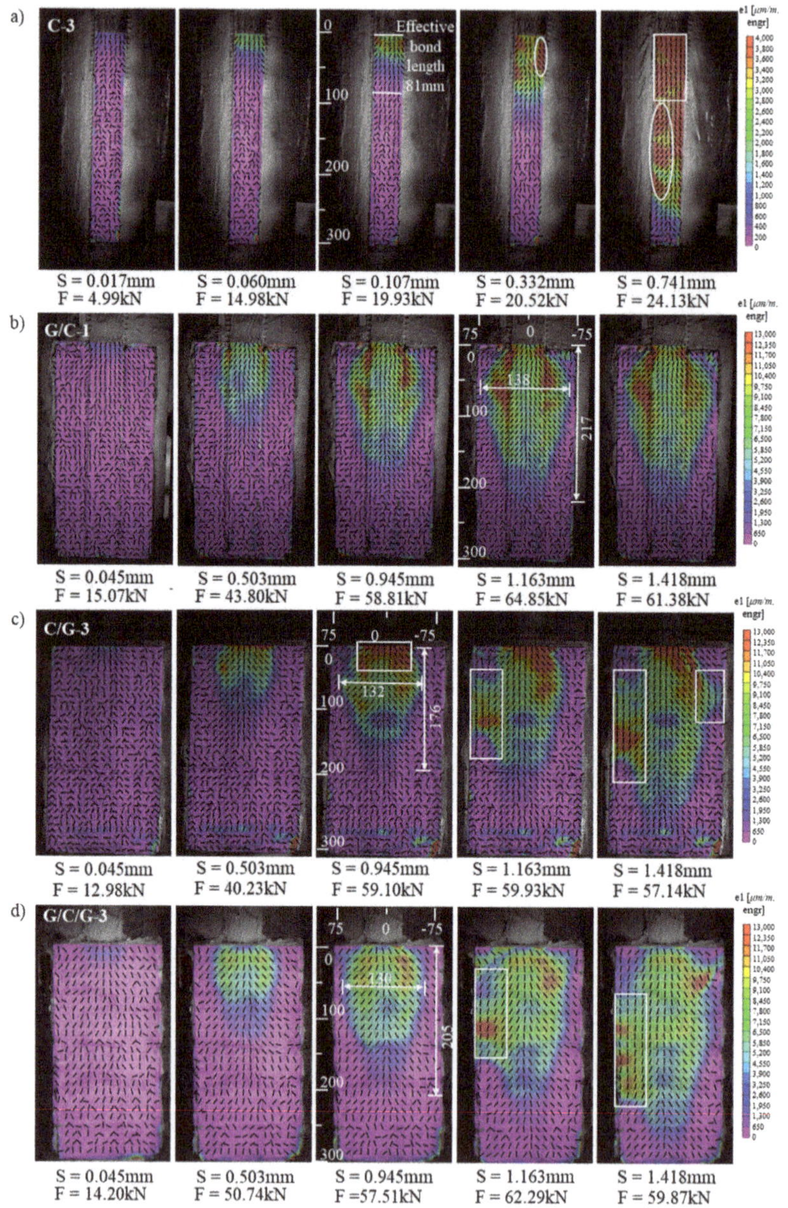

Figure 7. Principal strain distribution on the CFRP laminate and GFRP interlayer: (**a**) C-3; (**b**) G/C-1; (**c**) C/G-3; and (**d**) G/C/G-3.

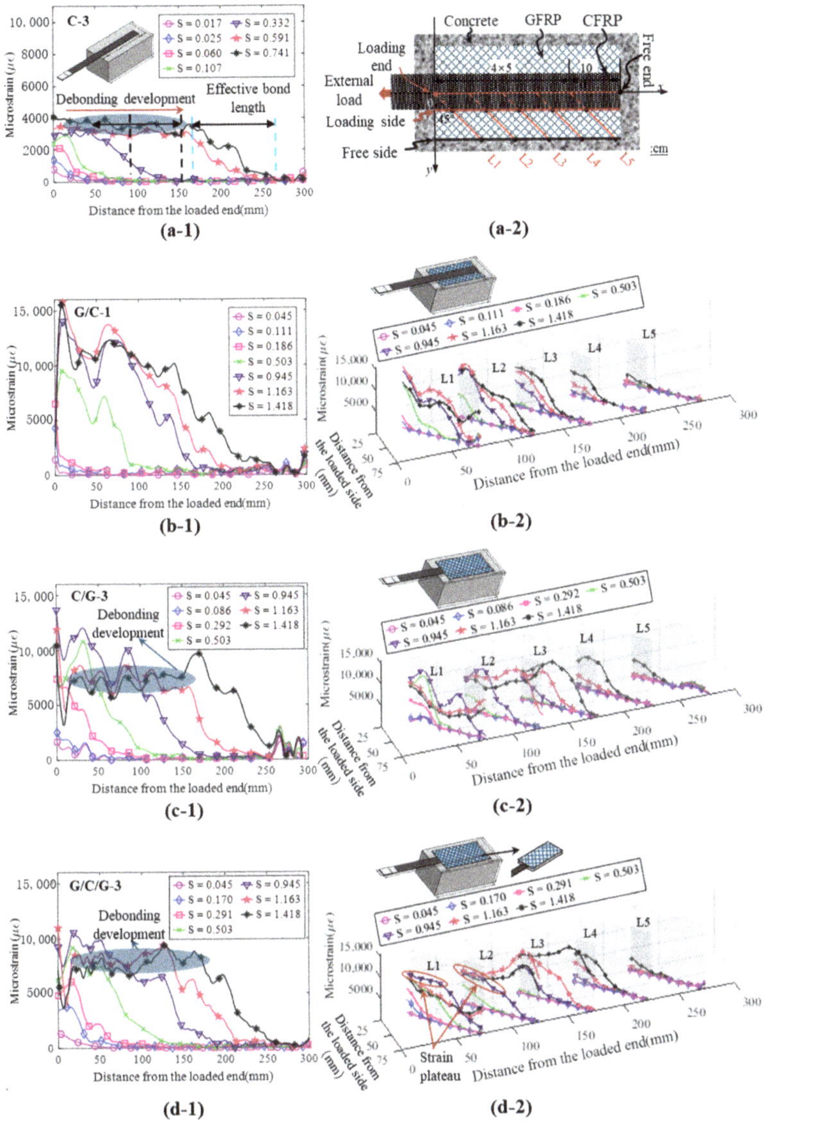

Figure 8. The principal strain distribution along the CFRP laminate: (**a-1**) specimen C-3; (**a-2**) schematic of the principal strain extraction. The principal strain distribution in the GFRP interlayer: (**b-1**,**b-2**) specimen G/C-1; (**c-1**,**c-2**) specimen C/G-3; (**d-1**,**d-2**) specimen G/C/G-3.

4. Discussion

The above results clearly indicate that inclusion of a GFRP layer can significantly affect the bond behavior of CFRP–concrete bonded joints regarding the load-carrying capacity, ultimate deformation, and strain distribution. As discussed in Zhou et al. [31], when the GFRP layer is positioned under the CFRP laminate as an interlayer (i.e., the G/C specimen group), the axial stress within the CFRP is firstly transferred to the GFRP interlayer and then distributed to the concrete substrate. The sequential activation of the bonding area in the transverse and longitudinal directions can provide a greater active bonding area when subjected to external force. As the load-carrying capacity of the bonded joints is mainly determined by the fracture work, a greater bonding area leads to a higher total fracture work, thus, a significant increase in the load-carrying capacity can be observed. Detailed discussions can be found in Zhou et al. [31].

Regardless of the bonding sequence, the failure modes of the CFRP–concrete bonded joints with a GFRP layer include the combined cohesion failure within the concrete and the GFRP delamination at the edge of the bonding area. Except for the approximately 45° cracking to the load direction, the above failure mode is consistent with the conventional CFRP–concrete bonded joints. Such consistency indicates that with proper concrete surface treatment, the inclusion of a GFRP layer would not introduce a weak interface to the conventional CFRP–concrete bonded joints. Additionally, as the interfacial stress is redistributed by the GFRP layer in the principal direction (i.e., approximately 45° to the loading direction), a similar crack pattern in the concrete substrate can be expected.

When bonding the CFRP laminate under or between the GFRP laminate, though the bond behavior of the GFRP–concrete interface on both sides of the CFRP laminate is identical as they possess the same thickness and fiber volume ratio, the slight change in the stress transfer path will affect the initial elastic stiffness of the load–displacement curve (maximum difference of 16.1% taking the minimum as a reference). Specifically for the C/G specimens, the axial stress within the CFRP laminate is transferred to the GFRP laminate and concrete substrate simultaneously, then, the GFRP laminate on both sides of the CFRP will distribute the stress to the concrete substrate. However, for the G/C specimen, the axial stress within the CFRP laminate is transferred to the GFRP interlayer first and then redistributed to the concrete substrate. Therefore, compared to the C/G specimens, more concrete is activated to resist the external force in the G/C specimen, which leads to a higher initial slope in the load–displacement curve. Meanwhile, for the G/C/G specimens, the axial stress is transferred to the single layer under and above the CFRP laminate; however, since only one layer of GFRP laminate was bonded between the CFRP and concrete, the overall stiffness will be greater than that of the C/G specimen but less than the G/C specimen.

Except for affecting the initial stiffness, the bonding sequence also exhibited influence on the principal strain distribution and propagation. Compared to the G/C specimens, the CFRP laminate was bonded to the concrete directly in the C/G specimens. As the CFRP–concrete interface tended to debond earlier than the CFRP–GFRP (two layers)–concrete interface due to the deformation and stress distribution of the GFRP interlayer, the displacement of the CRRP laminate will be greater for a given load. Considering the displacement continuity between the CFRP and bonded GFRP, more of the GFRP–concrete interface will be activated to resist the force. As such, the principal strain would propagate more in the transverse direction and failure in the width direction can be first observed. In comparison to the G/C specimens, the principal strain tends to develop in the longitudinal direction, while for the G/C/G specimens, the stiffness of the CFRP–GFRP (one layer)–concrete interface lies between the C/G and G/C specimens in that the principal strain distribution shows development in both transverse and longitudinal directions when subjected to shear loading.

Nevertheless, the bonding sequence showed less of an effect on the overall activated bonding area when the maximum load was attained, thus, the load-carrying capacity was similar for different specimens. As a result, the bonding scheme can be customized to achieve the optimal strengthening performance in practice. For example, when the GFRP layer is employed for strengthening a reinforced concrete aqueduct structure with CFRP laminate, the GFRP can be bonded on top of the CFRP at the inner side of the aqueduct so that a smooth surface can be achieved to reduce the water blocking. The water permeability can be also improved in such as bonding configuration. For the RC beam or slab strengthening, the GFRP layer can also be bonded on top of the CFRP laminate at the plate end to improve the shear and peeling performance simultaneously. However, if a large bonding area is expected, e.g., bonding the GFRP layer in the intermediate region of the RC beam or slab, a bonding GFRP layer beneath the CFRP sheet is preferred to avoid the possible gaps at the edge of the CFRP laminate, especially when a thick CFRP laminate is used. Additionally, the C-G-C bonding configuration could be used in producing a commercial hybrid carbon and glass fiber fabric so that only one-time bonding is required in practice. However, more research should be conducted to investigate the water-resistance performance, Mode-I bond behavior, and production process for the above-mentioned applications of different bonding schemes.

5. Conclusions

This paper investigates the influence of the bonding sequence on the behavior of CFRP–concrete bonded joints with a GFRP layer. Three types of specimens, i.e., G/C, C/G, and G/C/G, were prepared and tested through a series of single-shear pull-off tests. The failure modes, load-carrying capacity, ultimate displacement at the loaded end, and strain distribution of the specimens are presented and discussed. The following conclusions can be drawn:

1. The bonding sequence of the CFRP and GFRP showed negligible influence on the failure mode of tested specimens. The dominant failure mode was combined cohesion failure within the concrete and GFRP delamination. As such, the proper bonding position of the GFRP layer can be selected in different scenarios so that the benefits provided by the improvements in the Mode-I bond behavior and water permeability by the GFRP layer can be fully utilized;
2. Compared to the CFRP–concrete bonded joints, the inclusion of a GFRP layer can significantly improve the bonding performance. A minimum of 157.6% increase in the load-carrying capacity and 83.0% in the ultimate deformation can be expected in G/C specimens. However, the effect of the bonding sequence on the bonding performance is insignificant. Bonding the CFRP laminate between two GFRP layers recorded the highest increase in load-carrying capacity (177.2%); the results for the other two types of specimens were 157.6% and 175.0%, respectively. The C/G specimen exhibited the maximum increase in ultimate displacement (103.6%), although 83.0% and 86.3% increases were achieved by the G/C and G/C/G specimens;
3. The bonding sequence of the CFRP and GFRP sheets showed an influence on the initial stiffness of the load–displacement curves and the principal strain distribution propagation direction. Such influences could be attributed to the bonding stiffness and strength of the CFRP–concrete and CFRP–GFRP–concrete interfaces. However, the overall activated bonding area is insignificantly influenced by the bonding sequence.

Author Contributions: Conceptualization, H.Z.; Data curation, J.Z.; Formal analysis, J.Z., Y.Y., F.T., Y.O., Y.W. and C.L.; Funding acquisition, H.Z.; Investigation, H.Z., J.Z. and Y.Y.; Methodology, H.Z. and Y.Y.; Project administration, H.Z.; Resources, H.Z., F.T. and C.L.; Software, J.Z. and Y.Y.; Supervision, H.Z.; Validation, H.Z.; Writing—original draft, J.Z.; Writing—review and editing, H.Z., Y.Y., Y.O. and C.L. All authors have read and agreed to the published version of the manuscript.

Funding: This research was funded by the National Natural Science Foundation of China, Grant No. 52208325, and the Scientific Research Fund of Hunan Provincial Education Department (Grant No. 21B0250).

Data Availability Statement: The raw/processed data required to reproduce these findings cannot be shared at this time as the data also form part of an ongoing study.

Conflicts of Interest: The authors declare no potential conflicts of interest with respect to the research, authorship, and publication of this article.

References

1. Al-Ghrery, K.; Al-Mahaidi, R.; Kalfat, R.; Oukaili, N.; Al-Mosawe, A. Externally Bonded CFRP for Flexural Strengthening of RC Beams with Different Levels of Soffit Curvature. *J. Compos. Constr.* **2022**, *26*, 04021062. [CrossRef]
2. Milev, S.; Tatar, J. Durability Assessment of the First Externally Bonded FRP Repair of a Publicly Owned Bridge in the United States after 26 Years of Service. *J. Compos. Constr.* **2023**, *27*, 04023049. [CrossRef]

3. Zhou, Y.W.; Sui, L.L.; Huang, X.X.; Guo, M.H.; Luo, M.S.; Hu, B.; Chen, C. Enhancing the EB-FRP strengthening effectiveness by incorporating a cracking-control layer of ECC with different thicknesses. *Constr. Build. Mater.* **2021**, *286*, 122975. [CrossRef]
4. de Waal, L.; Fernando, D.; Nguyen, V.T.; Cork, R.; Foote, J. FRP strengthening of 60 year old pre-stressed concrete bridge deck units. *Eng. Struct.* **2017**, *143*, 346–357. [CrossRef]
5. Zhou, C.Y.; Wang, L.; Wang, Y.; Fang, Z. Experimental study on the flexural strengthening of one-way RC slabs with end-buckled and/or externally bonded CFRP sheets. *Eng. Struct.* **2023**, *282*, 115832. [CrossRef]
6. Smith, S.T.; Teng, J.G. FRP-strengthened RC beams. II: Assessment of debonding strength models. *Eng. Struct.* **2002**, *24*, 397–417. [CrossRef]
7. Mazzotti, C.; Savoia, M.; Ferracuti, B. A new single-shear set-up for stable debonding of FRP-concrete joints. *Constr. Build. Mater.* **2009**, *23*, 1529–1537. [CrossRef]
8. Ko, H.; Matthys, S.; Palmieri, A.; Sato, Y. Development of a simplified bond stress-slip model for bonded FRP-concrete interfaces. *Constr. Build. Mater.* **2014**, *68*, 142–157. [CrossRef]
9. Tatar, J.; Hamilton, H.R. Comparison of laboratory and field environmental conditioning on FRP-concrete bond durability. *Constr. Build. Mater.* **2016**, *122*, 525–536. [CrossRef]
10. Yao, J.; Teng, J.G.; Chen, J.F. Experimental study on FRP-to-concrete bonded joints. *Compos. Part B Eng.* **2005**, *36*, 99–113. [CrossRef]
11. Zhou, H.; Fernando, D.; Chen, G.M.; Kitipornchai, S. The quasi-static cyclic behaviour of CFRP-to-concrete bonded joints: An experimental study and a damage plasticity model. *Eng. Struct.* **2017**, *153*, 43–56. [CrossRef]
12. Li, P.; Zhao, Y.; Wu, Y.-F.; Lin, J.-P. Interfacial Behavior of Carbon FRP-to-Granite Joints with Mechanical Fastening. *J. Compos. Constr.* **2022**, *26*, 04021066. [CrossRef]
13. Lu, X.Z.; Teng, J.G.; Ye, L.P.; Jiang, J.J. Bond–slip models for FRP sheets/plates bonded to concrete. *Eng. Struct.* **2005**, *27*, 920–937. [CrossRef]
14. Yuan, H.; Teng, J.G.; Seracino, R.; Wu, Z.S.; Yao, J. Full-range behavior of FRP-to-concrete bonded joints. *Eng. Struct.* **2004**, *26*, 553–565. [CrossRef]
15. Kalfat, R.; Al-Mahaidi, R.; Smith, S.T. Anchorage Devices Used to Improve the Performance of Reinforced Concrete Beams Retrofitted with FRP Composites: State-of-the-Art Review. *J. Compos. Constr.* **2013**, *17*, 14–33. [CrossRef]
16. Lu, T.; Li, P.; Cui, C.; Wu, J.; Fu, B. Shear transferring mechanism of the FPR-to-concrete bonded joint with end U-jacketing: A theoretical study. *Structures* **2023**, *56*, 104991. [CrossRef]
17. Razaqpur, A.G.; Cameron, R.; Mostafa, A.A.B. Strengthening of RC beams with externally bonded and anchored thick CFRP laminate. *Compos. Struct.* **2020**, *233*, 111574. [CrossRef]
18. Zhou, Y.W.; Wang, X.W.; Sui, L.L.; Xing, F.; Huang, Z.Y.; Chen, C.; Li, P.D.; Mei, L. Effect of mechanical fastening pressure on the bond behaviors of hybrid-bonded FRP to concrete interface. *Compos. Struct.* **2018**, *204*, 731–744. [CrossRef]
19. Wang, Y.; Surahman, R.; Zhou, C.; He, X.; Chen, Y. Analytical study on flexural strengthening of narrow RC beams with hybrid anchored CFRP sheets. *J. Build. Eng.* **2024**, *88*, 109095. [CrossRef]
20. McCoy, B.C.; Bourara, Z.; Lucier, G.W.; Seracino, R.; Liu, M.; Lin, S.H. Prestressed MF-FRP: Experimental study of rapid retrofit solution for deteriorated prestressed C-channel beams. *J. Perform. Constr. Facil.* **2021**, *35*, 04020124. [CrossRef]
21. Smith, S.T.; Hu, S.; Kim, S.J.; Seracino, R. FRP-strengthened RC slabs anchored with FRP anchors. *Eng. Struct.* **2011**, *33*, 1075–1087. [CrossRef]
22. Al-Atta, B.; Kalfat, R.; Al-Mahaidi, R.; Al-Mosawe, A. Mitigation of IC debonding in FRP-plated concrete slabs using patch anchors. *Eng. Struct.* **2020**, *214*, 110626. [CrossRef]
23. Zhang, S.S.; Ke, Y.; Smith, S.T.; Zhu, H.P.; Wang, Z.L. Effect of FRP U-jackets on the behaviour of RC beams strengthened in flexure with NSM CFRP strips. *Compos. Struct.* **2020**, *256*, 113095. [CrossRef]
24. Zhang, P.; Shang, J.Q.; Fan, J.J.; Chen, Q.Z.; Zhu, H.; Gao, D.Y.; Sheikh, S.A. Experimental study on the bond behavior of the CFRP plate-ECC-concrete composite interface under freeze–thaw cycles. *Constr. Build. Mater.* **2022**, *316*, 125822. [CrossRef]
25. Yuan, C.; Chen, W.S.; Pham, T.M.; Hao, H.; Chen, L.; Zhang, M. New epoxy anchor for better bonding between FRP sheets and concrete. *Constr. Build. Mater.* **2020**, *248*, 118628. [CrossRef]
26. Moghaddas, A.; Mostofinejad, D.; Saljoughian, A.; Ilia, E. An empirical FRP-concrete bond-slip model for externally-bonded reinforcement on grooves. *Constr. Build. Mater.* **2021**, *281*, 122575. [CrossRef]
27. Jiang, C.; Wan, B.L.; Wu, Y.F.; Omboko, J. Epoxy interlocking: A novel approach to enhance FRP-to-concrete bond behavior. *Constr. Build. Mater.* **2018**, *193*, 643–653. [CrossRef]
28. Liu, Y.; Zhou, C.; Chen, L.; Wang, Y. Experimental study on bond behavior between CFRP and concrete with longitudinal grooves. *Structures* **2024**, *65*, 106654. [CrossRef]
29. Dai, J.G.; Ueda, T.; Sato, Y.; Ito, T. Flexural strengthening of RC beams using externally bonded FRP sheets through flexible adhesive bonding. In *International Symposium on Bond Behaviour of FRP in Structures (BBFS 2005)*; International Institute for FRP in Construction: Kingston, ON, Canada, 2005.
30. Etemadi, E.; Zhou, H.; Dao, V.; Fernando, D. Improving the load-carrying capacity of FRP-to-concrete bonded joints using a biaxial GFRP interlayer. *J. Compos. Constr.* **2024**, *28*, 04024027. [CrossRef]
31. Zhou, H.; Yang, Y.; Fernando, D.; Tianli, H.; Ou, Y. Bond behaviour of CFRP-to-concrete bonded joints with a GFRP interlayer: An experimental study. *Constr. Build. Mater.* **2024**, *438*, 137042. [CrossRef]

32. Kalfat, R.; Al-Mahaidi, R. Investigation into bond behaviour of a new CFRP anchorage system for concrete utilising a mechanically strengthened substrate. *Compos. Struct.* **2010**, *92*, 2738–2746. [CrossRef]
33. Correlated, S. *Vic-3D-2010 Manual*; Correlated Solutions: Irmo, SC, USA, 2010.
34. Zhou, H.; Fernando, D.; Thuan Nguyen, V.; Dai, J.-G. The bond behaviour of CFRP-to-concrete bonded joints under fatigue cyclic loading: An experimental study. *Constr. Build. Mater.* **2021**, *273*, 121674. [CrossRef]
35. Huang, Y.J.; Natarajan, S.; Zhang, H.; Guo, F.Q.; Xu, S.L.; Zeng, C.; Zheng, Z.S. A CT image-driven computational framework for investigating complex 3D fracture in mesoscale concrete. *Cem. Concr. Compos.* **2023**, *143*, 105270. [CrossRef]

buildings

MDPI

Article

Evaluation of the Strengthening Effects on Prestressed Carbon-Fiber-Reinforced-Polymer-Strengthened Steel Beam Bridges Using Macro-Strain Influence Lines

Bitao Wu [1,*], Qingquan Xia [1], Yan Gong [1], Sicheng Fu [1], Haitao Wang [2,*] and Zhongzhao Guo [1]

[1] State Key Laboratory of Performance Monitoring Protecting of Rail Transit Infrastructure, East China Jiaotong University, Nanchang 330013, China; xiaqingquan@ecjtu.edu.cn (Q.X.)
[2] College of Civil and Transportation Engineering, Hohai University, Nanjing 210098, China
* Correspondence: wubitao@ecjtu.edu.cn (B.W.); cewht@hhu.edu.cn (H.W.)

Abstract: Effectively evaluating the effectiveness of bridge strengthening is a necessary means to ensure the normal operation of existing strengthened bridges, especially when evaluating the effectiveness of bridge strengthening without interrupting normal traffic. Based on a distributed long-gauge Fiber Bragg Grating (FBG) sensor, this paper derived the macro-strain influence line (MSIL) formula for a simply supported beam bridge under a moving vehicle load, studied the changes in the MSIL at the bottom of the beam under the vehicle load before and after the prestressed CFRP plate strengthening, and proposed a rapid evaluation method for the strengthening effect based on the amplitude of the MSIL as the evaluation index for the strengthening effect. Finally, the prestressed CFRP-strengthened steel beam was tested under the moving vehicle load. The theoretical analysis and the experimental results confirm that under the load of moving vehicles, the macro-strain–time history amplitude of the strengthened steel beams under different prestressed tensioning conditions is different. The amplitude of the macro-strain time history of the strengthened bridge is reduced compared to before strengthening, and the local strengthening effect of the bridge can be monitored by the amplitude change in a single sensor. The change in global stiffness can be evaluated by monitoring the MSIL obtained from multiple long-gauge strain sensors.

Keywords: prestressed CFRP plate; long-gauge sensor; bridge strengthening; strengthening effect; bridge influence line

1. Introduction

Due to the design load being lower than the load actually subjected, existing bridges in service gradually enter a stage of aging. Simultaneously, factors such as natural wear and tear, vehicle overloading, and structural deterioration have significantly reduced their load-bearing capacity and durability, resulting in significant safety hazards. Consequently, the strengthening of these bridges has become a prominent research focus in the industry [1–5]. To enhance the mechanical properties of these bridge structures and prolong their service life, strengthening measures are necessary [6–10]. Particularly, using externally bonded fiber-reinforced polymer (FRP) to strengthen concrete structures has become a common practice. Over the past three decades, the use of carbon-fiber-reinforced polymer (CFRP) to strengthen existing structures has garnered considerable attention, primarily due to its high strength, light weight, excellent corrosion resistance, and ease of installation [11–17]. As a result, utilizing CFRP to strengthen a structural component is an economical and environmentally friendly option [18–20].

For the commonly used externally bonded CFRP plate strengthening technology, the effectiveness of the CFRP plates depends on the strength of the bonding interface. However, due to the low interfacial bonding interface strength, the role of CFRP plates is often difficult to fully exert. When structures strengthened with CFRP plates fail, the strength of the

CFRP plates is only about 1/5 of their ultimate tensile strength [21,22]. To address this issue, an anchorage technique can be utilized. Various types of anchorage systems have been considered by a number of researchers [23–27]. One promising technique is the use of a mechanical anchorage, which has proven to be an effective approach to keeping the CFRP engaged with the concrete beams beyond premature debonding [28,29].

In practical engineering, to overcome the problems of the easy peeling and low strength utilization of CFRP plates caused by direct bonding strengthening, scholars have studied and proposed a prestressed CFRP strengthening method [30–34]. Research has shown that prestressed CFRP strengthening technology can effectively improve the strengthening effect and strength utilization rate of CFRP, avoiding end delamination of the CFRP [25,29,31]. However, prestressed CFRP strengthening of bridges inevitably leads to prestressed loss, which affects the strengthening effect. Many scholars have conducted extensive monitoring experiments and obtained results such as those on the mechanism of prestressed loss and the approximate range of loss [35,36]. In the existing studies on prestressed CFRP-strengthened beams, the main focus was on conducting ultimate load tests on the prestressed CFRP-strengthened beams to assess the effectiveness of strengthening methods. Overall, research on the effectiveness of prestressed CFRP plate strengthening has predominantly centered on laboratory beams, with the evaluations typically involving static load tests. However, conducting ultimate load tests to evaluate the strengthening effectiveness in actual bridges is impractical.

Static load tests on bridges are commonly used to assess bridges' load-bearing capacity, simulating vehicle loads equivalent to the design specifications. These tests evaluate whether the stress conditions and deformations under test loading meet the regulatory requirements for bridge load capacity. Nevertheless, load tests necessitate traffic disruption and entail significant time and economic costs. Furthermore, to assess the effectiveness of bridge strengthening, comparing the results before and after strengthening through two rounds of load testing would incur even greater expenses. To expedite the assessment of existing bridge conditions, researchers propose employing influence lines to evaluate bridge stiffness under moving vehicle loads [37–40]. Most of the monitoring methods involve attaching resistance strain gauges to the surface of CFRP plates. Traditional "point" strain gauges are prone to being affected by minor surface cracks on CFRP plates, especially under high-stress conditions, making it difficult to obtain effective monitoring data, which is unsuitable for long-term stress monitoring of CFRP plates in practical engineering.

To evaluate the effect of prestressed CFRP-plate-strengthened bridges without interrupting traffic and to overcome the limitations of the measurement range of resistance strain gauges that cannot achieve distributed monitoring, this paper studied the mechanical relationship for the strain influence line of strengthened beams under moving vehicle loads. Then, a bridge strengthening effect evaluation index was proposed based on the distributed MSIL. In order to verify the feasibility of the proposed evaluation method, a verification test was conducted on a model bridge strengthened with prestressed CFRP plates under moving vehicle loads. In the verification test, a self-developed long-gauge strain sensor was used to monitor the distributed macro-strain, and the test results confirmed the feasibility of the proposed method.

2. Derivation of the MSIL for a Prestressed CFRP-Plate-Strengthened Bridge

Due to the limitations of the measurement range of "point type" measuring sensors, the strain influence line (SIL) has mainly been used for load identification [41], and the principle of a bridge's SIL is shown in Figure 1. For a simply supported beam AB, the span of the beam is L, the height of the beam is h, the height of the neutral axis in the beam section is y, and the bending stiffness of the beam section is EI. Assuming a moving load P moves from the A end to the B end of the simply supported beam, the equation for the bending moment influence line at any point C on the beam can be obtained. By using the material mechanics calculation formula $\sigma = \frac{My}{EI}$ and Hooke's law $\varepsilon = \frac{\sigma}{E}$, the bending moment influence line equation can be transformed into the SIL equation:

$$\varepsilon(x) = \begin{cases} \frac{(L-x_i)yx}{EIL} & 0 \le x \le x_i \\ \frac{(L-x)yx_i}{EIL} & x_i \le x \le L \end{cases} \quad (1)$$

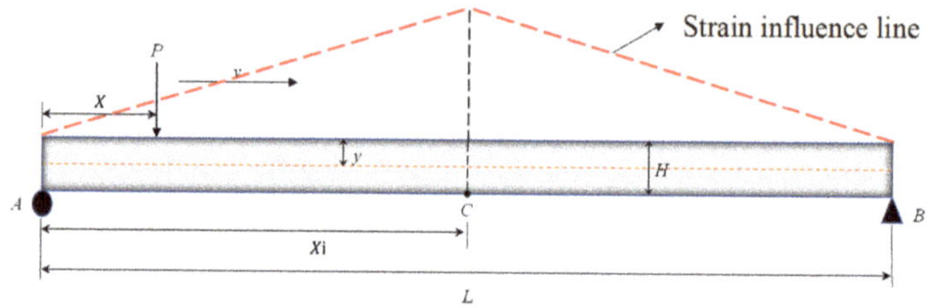

Figure 1. SIL of simply supported beam under moving load.

In practical bridge monitoring, bridges often have large spans, making it difficult for traditional point strain gauges to achieve distributed monitoring of an entire bridge. With their distributed measurement capability, long-gauge strain sensors enable monitoring of the entire bridge using only a small number of sensors. FBG sensors can be conveniently self-made into strain sensors with a longer gauge through secondary packaging. A schematic diagram of the long-gauge sensor developed in this paper is shown in Figure 2. The innermost layer of the sensor is an FBG sensor, which is protected by a 3 mm plastic hose for the sensing section to ensure that the bare fiber can move freely in the hose. The two ends of the sensor are anchor sections, generally with a length of 5 cm to ensure that the fiber does not slip, and are dipped in modified epoxy resin using a 5 mm plastic hose. The outermost layer of the sensor is a 5 mm diameter stainless steel tube for protection, and the anchor section is fixed to the structure using stainless steel gasket plates. It is worth noting that if the FBG is pre-strained during the fabrication process, this can ensure that the long-gauge FBG strain sensor has the ability to measure compressive strains. Additionally, by using long-gauge FBG strain sensors with different wavelengths, it is possible to achieve the purpose of full-bridge coverage monitoring by serially connecting multiple sensors.

Due to the fact that the strain measured by the long-gauge strain sensor is the average strain within its segment, i.e., the macro-strain, the MSIL is different from the traditional point strain influence line (SIL). When using long-gauge strain sensors to monitor and obtain the MSIL for bridge evaluation, it is necessary to study the calculation formula for the MSIL and clarify the mechanical relationship between macro-strain and the bridge's longitudinal stiffness. As shown in Figure 3, for a simply supported beam covered by n long-gauge FBG strain sensors, to obtain the MSIL at the cross-section corresponding to the m-th sensor, it is necessary to integrate the area covered by the m-th sensor and take the average value:

$$\overline{\varepsilon}_m(x) = \begin{cases} \frac{1}{l_e}\int_{(m-1)l_e}^{ml_e} \frac{(L-x_i)yx}{(\overline{EI})_m L} dx_i & 0 \le x \le (m-1)l_e \\ \frac{1}{l_e}\int_{(m-1)l_e}^{x} \frac{(L-x)yx_i}{(\overline{EI})_m L} dx_i + \frac{1}{l_e}\int_{x}^{ml_e} \frac{(L-x_i)yx}{(\overline{EI})_m L} dx_i & (m-1)l_e \le x \le ml_e \\ \frac{1}{l_e}\int_{(m-1)l_e}^{ml_e} \frac{(L-x)yx_i}{(\overline{EI})_m L} dx_i & ml_e \le x \le L \end{cases} \quad (2)$$

in which $\overline{\varepsilon}_m(x)$ represents the MSIL equation for the m-th segment; $(\overline{EI})_m$ represents the average bending stiffness within the section; and l_e represents the gauge length of a long-gauge FBG strain sensor. We solve the above integral equation and derive the expression for the MSIL of the long-gauge FBG strain sensor:

$$\overline{\varepsilon_m}(x) = \begin{cases} \frac{[2L-(2m-1)l_e]yx}{2(\overline{EI})_m L} & 0 \le x \le (m-1)l_e \\ \frac{y}{2(\overline{EI})_m L}\left\{-\frac{L}{l_e}x^2 + [2mL-(2m-1)l_e]x - (m-1)^2 l_e L\right\} & (m-1)l_e \le x \le ml_e \\ \frac{(2m-1)(L-x)yl_e}{2(\overline{EI})_m L} & ml_e \le x \le L \end{cases} \tag{3}$$

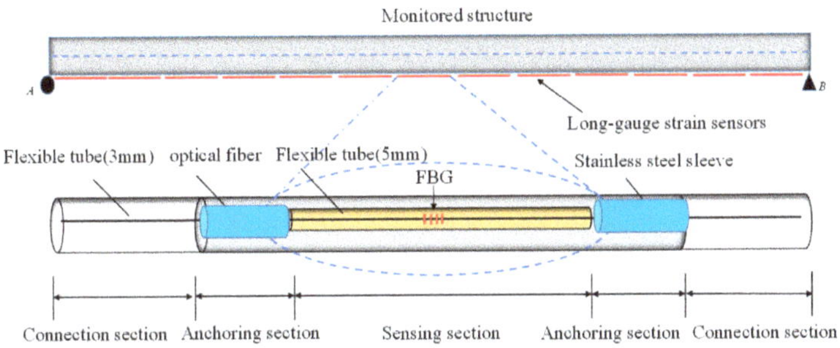

(**a**)

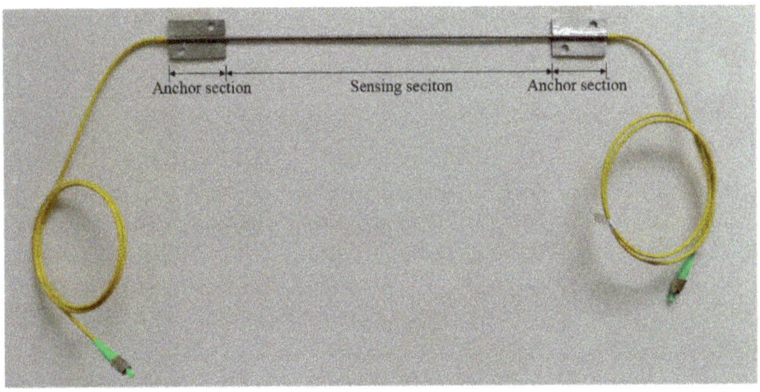

(**b**)

Figure 2. Self-developed long-gauge FBG strain sensor: (**a**) the packaging of a long-gauge FBG strain sensor; (**b**) a physical picture of a long-gauge FBG strain sensor.

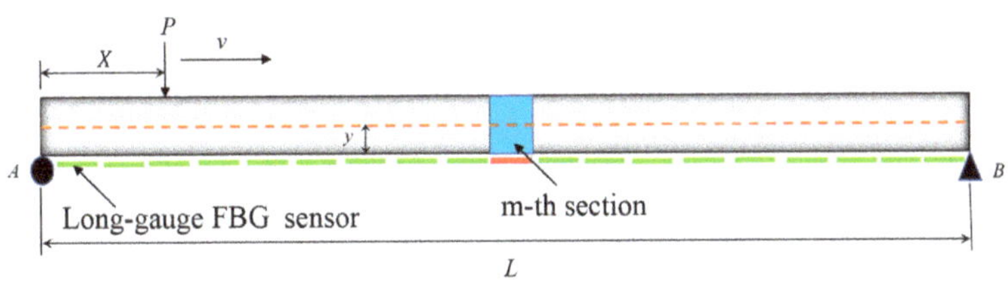

Figure 3. Beam monitored by long-gauge FBG strain sensors.

According to Equations (3), when the moving load P does not move to the range of the m-th section, the shape of the MSIL remains a straight line; but when the load P is within its range, the shape of the MSIL is a quadratic curve. The shape of the MSIL at the m-th section of the final section is composed of a segmented curve consisting of a straight segment, a curved segment, and a straight segment, as shown in Figure 4.

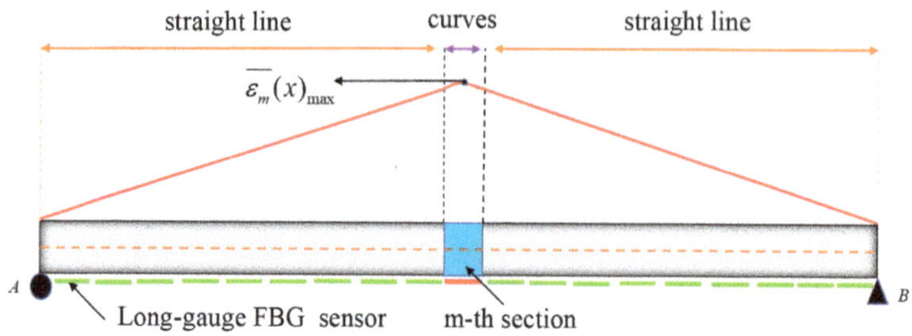

Figure 4. MSIL of the m-th section in the bridge.

Obviously, for the m-th section, the maximum amplitude of the MSIL occurs in the area covered by the m-th long-gauge sensor. Solving and integrating the expression of the MSIL can obtain the macro-strain amplitude value. For m-th section m, when x takes the value shown in the equation, its MSIL reaches its maximum amplitude:

$$x = \frac{[2mL - (2m-1)l_e]l_e}{2L} \tag{4}$$

The expression for the maximum macro-strain amplitude value of the m-th section at this time is as follows:

$$\bar{\varepsilon}_{m0,\max} = \frac{2y}{(\overline{EI})_m L}\left[\left(l_e^3 - 2Ll_e^2\right)m^2 + \left(Ll_e^2 - l_e^3 + 2L^2 l_e\right)m - L^2 l_e\right] \tag{5}$$

According to analysis of Equations (5), the height of the neutral axis y, the bending stiffness of the section $(\overline{EI})_m$, the bridge span L, and the gauge length of FBG strain sensor l_e are constant. The distribution curve of the MSIL measured by long-gauge FBG strain sensors under a single load is a quadratic curve. In practical situations, the vehicle load acting on the bridge is usually multiple loads, with different force sizes and positions. The MSIL under multiple loads can be solved based on the principle of the superposition of influence lines. Assuming that there are n moving loads acting on a simply supported beam AB, each of which is $P_1, P_2, P_3 \ldots P_n$, the distance between each load and P_1 can be expressed as $d_1, d_2, d_3 \ldots d_{n-1}$, as shown in Figure 5.

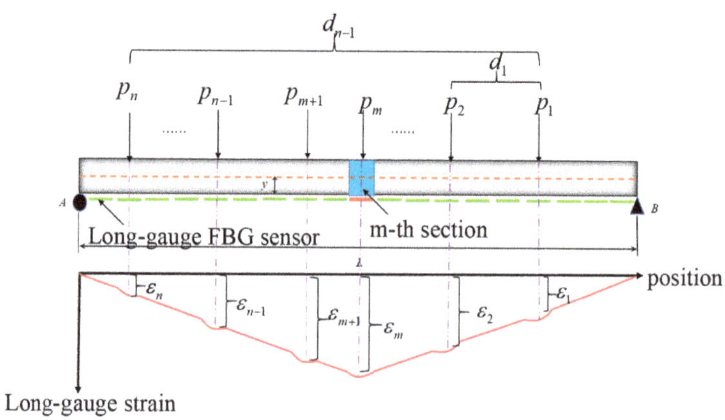

Figure 5. MSIL of segment m under vehicle loads.

According to the principle of linear superposition, the macro-strain response of the m-th section under multiple loads $\bar{\varepsilon}_{mp}(x)$ can be expressed as follows:

$$\bar{\varepsilon}_{mp}(x) = \sum_{k=1}^{n} P_k \bar{\varepsilon}(x - d_{k-1}) \tag{6}$$

The area value $A(m)$ enclosed by the time history curve can be obtained by integrating the macro-strain time history.

$$A(m) = \begin{cases} \frac{l_e{}^2 y(2L - l_e \times (2m-1)) \times (m-1)^2}{4\left(\overline{EI}\right)_m L} & 0 \le x \le (m-1)l_e \\ \frac{-\left(yl_e{}^2\left(8L + 3l_e - 12mL - 12ml_e + 12l_e m^2\right)\right)}{12\left(\overline{EI}\right)_m L} & (m-1)l_e \le x \le ml_e \\ \frac{yl_e(2m-1)(L - ml_e)^2}{4\left(\overline{EI}\right)_m L} & ml_e \le x \le L \end{cases} \tag{7}$$

From Equation (7), we can see that the MSIL of the m-th section is composed of three parts: the first part is the initial straight section, the second part is the middle-curved section, and the third part is the segmented curve of the end straight section. After the beam is strengthened with prestressed CFRP plates, the beam will experience arching and initial strain. When the load P travels from point A to point B of the beam, the macro-strain–time history curve of the strengthened beam shows a decrease in amplitude compared to the macro-strain–time history curve of the unreinforced beam. This phenomenon can be used for rapid evaluation and monitoring of the effectiveness of prestressed CFRP plate strengthening on bridges. Furthermore, the change in the area enclosed by the strain–time history curve can be used to assess the strengthening effect:

$$\Delta A = \frac{Amw1 - Amw2}{Amw1} \times 100\% \tag{8}$$

in which $Amw1 = P\int_0^L \bar{\varepsilon}_m(x)dx$ represents the integral value of the macro-strain influence line of load P along the time axis in the m-th section and $Amw2$ represents the integral value of the macro-strain influence line along the time axis of load P in the m-th section after strengthening the bridge with prestressed CFRP plates.

3. Evaluation of the Effect of Strengthening on Bridges Based on the MSIL

After strengthening the bridge with prestressed CFRP plates, the CFRP plates and the bridge are considered a whole. The bending stiffness of the bridge section is slightly increased, the neutral axis height of the section decreases, and there will be an initial value in the macro-strain–time history curve measured in the m-th section. The magnitude of this initial value is determined by the prestressing force of the CFRP plate, and the macro-strain amplitude of the m-th section after strengthening is as follows:

$$\bar{\varepsilon}_{m1,max} = \frac{2y_{com}}{\left(\overline{EI}\right)_{com} L}\left[\left(l_e^3 - 2Ll_e^2\right)m^2 + \left(Ll_e^2 - l_e^3 + 2L^2 l_e\right)m - L^2 l_e\right] - \bar{\varepsilon}_p \tag{9}$$

in which $\bar{\varepsilon}_p$ is related to the prestress P_p applied to the CFRP plate. Extract the constant term from the formula, and under the condition that factors such as the beam span L and the sensor gauge length *le* remain unchanged, for the monitoring segment m, Δ_{cons} is a constant.

$$\Delta_{\text{cons}} = \frac{2}{L}\left[\left(l_e^3 - 2Ll_e^2\right)m^2 + \left(Ll_e^2 - l_e^3 + 2L^2 l_e\right)m - L^2 l_e\right] \tag{10}$$

The above equation can be simplified as follows:

$$\bar{\varepsilon}_{m1,max} = \frac{y_{com}}{(\overline{EI})_{com}}\Delta_{\text{cons}} - \bar{\varepsilon}_p \tag{11}$$

From the above equation, it can be seen that the macro-strain amplitude $\bar{\varepsilon}_{m1,max}$ of the monitoring m-th section after it is strengthened is inversely proportional to the bending stiffness $(\overline{EI})_{com}$ of the strengthened beam, directly proportional to the height y_{com} of the neutral axis, and closely related to the strain $\bar{\varepsilon}_p$ generated by the prestressing applied to the CFRP plate. Therefore, the changes in the macro-strain influence lines after strengthening are shown in Figure 6.

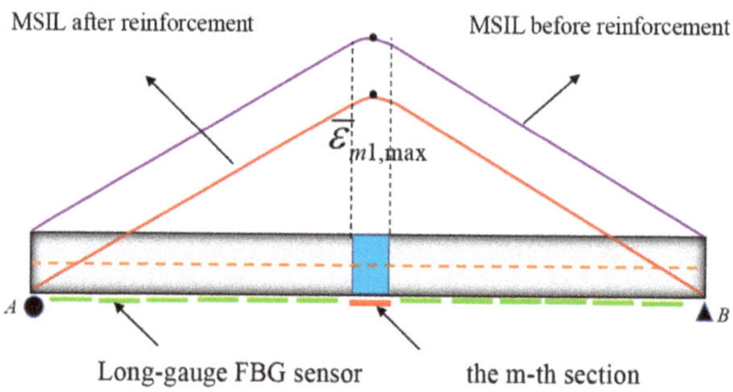

Figure 6. MSIL of prestressed CFRP plates before and after strengthening.

Therefore, it can be concluded that after strengthening the components with prestressed CFRP plates, the strengthening effect is mainly divided into two parts:

Part 1: Under the action of prestressed CFRP plates, the beam is transformed from an initial single-material section into a composite section that works together with CFRP plates. The bending stiffness of the section is improved, which reduces the deformation of the beam under external loads after strengthening the beam bridge with prestressed CFRP plates; Part 2: Due to the presence of prestress, the strengthened beam will experience a reverse arch, reducing the height of the neutral axis of the beam. At the same time, the prestress causes an initial strain in the beam, improving the safety reserve of the bridge. Usually, this part plays a major role. The strengthening effect on the MSIL is manifested as an overall decrease in the value of the macro-strain measured by the long-gauge FBG strain sensor after strengthening compared to the value before strengthening. Therefore, the effect of prestressed CFRP plate strengthening on beam bridges can be quickly evaluated by the amplitude change in the MSIL before and after strengthening using moving vehicles.

4. Evaluation Test of Prestressed CFRP Plate Strengthening Based on the MSIL

To study the evaluation method for the strengthening effect of prestressed CFRP-plate-strengthened beam bridges based on the bridges' MSIL, verification tests were conducted on prestressed CFRP-plate-strengthened steel beams. Self-developed long-gauge FBG strain sensors were installed at the bottom of the steel beam. In order to identify different tensile stress conditions during the experimental process, a non-bonded strengthening method was adopted by adjusting the magnitude of the tensile force. Different strengthening conditions with various pre-tensioning forces were designed, and after each condition was tensioned, a vehicle load test was performed while simultaneously collecting the dynamic macro-strain of the steel beam. The MSILs were extracted for different pre-tensioning forces to monitor and evaluate the effectiveness of the prestressed CFRP plate strengthening on the steel beam. The mechanical properties of the steel beams and CFRP plates are shown in Table 1.

Table 1. Mechanical properties.

Materials	Tensile Strength (MPa)	Elastic Modulus (GPa)
Steel beam	420	202
CFRP plate	2531	165

The beam parameters for this experiment are as follows: the beam height is 200 mm, and the width of the upper and lower flanges is both 200 mm. The flange thickness is 12 mm, and the web thickness is 8 mm. The bottom of the steel beam is strengthened with CFRP plates, and the width and thickness of the CFRP plates used in the experiment are 50 mm and 3 mm, respectively. Three sets of strengthening cases (JG1–JG3) and one initial condition without strengthening (CS-1) were designed, as shown in Table 2. The tension control force is the maximum tensile force applied to the prestressed CFRP plate, and a hydraulic jack with a digital display is used for control in the experiment. To achieve different strengthening effects, the tension levels are divided into 30%, 70%, and 100%, with the maximum tension force controlled at 20% of the ultimate tensile stress of the CFRP plate.

Table 2. CFRP plate tensioning cases.

Cases	Tensioning Level (%)	Tensioning Control Force (kN)	CFRP Plate Strain (με)
CS-1	/	/	/
JG-1	30	25.8	1015
JG-2	70	60.1	2369
JG-3	100	85.8	3384

The beam ends are simply supported, and the total length of the steel beam is 4000 mm, with a clear span of 3600 mm. There are stiffeners at the bearing, the mid-span, and the mid-point between the bearing and the mid-span, with a thickness of 10 mm. The specific cross-sectional parameters of the specimen are shown in Figure 7 (the shaded area represents the stiffeners), and the overall structure and dimensions can be referred to in Figure 8.

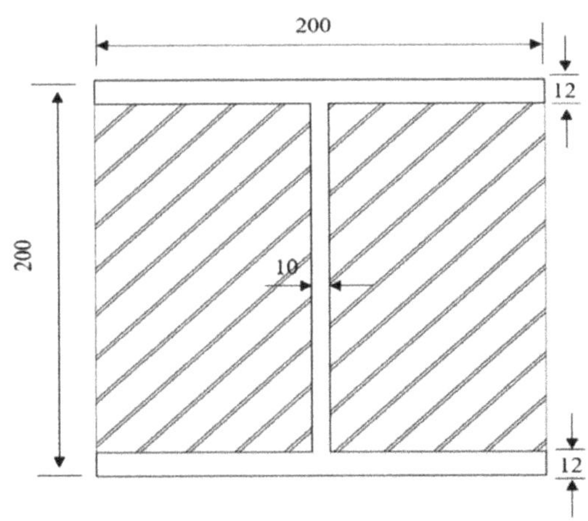

Figure 7. A cross-section of the main beam.

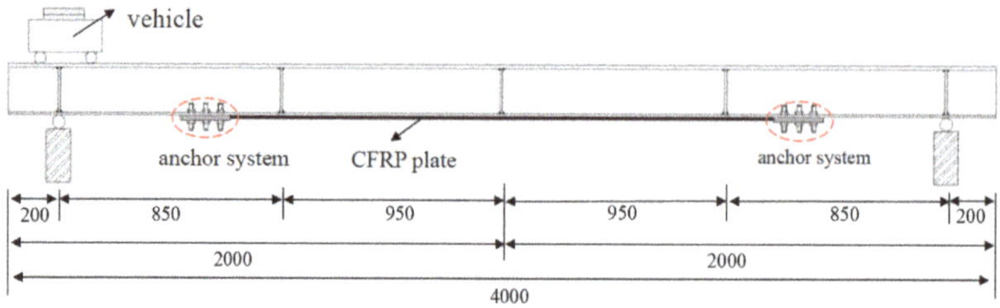

Figure 8. Experimental verification of the strengthening effect.

The CFRP plate is positioned at the mid-span of the steel beam, with a distance of 700 mm from each end of the beam. Seven long-gauge FBG strain sensors are arranged at 300 mm intervals along the bottom of the steel beam. The experiment is mainly aimed at conducting traffic tests after strengthening the steel beams with prestressed CFRP plates. The test model vehicle is a two-axle car, with a longitudinal distance of 0.28 m between the two axles and a lateral distance of 0.18 m between the wheelsets. The total mass of the model car is 80 kg. After each strengthening condition is completed, the load is sustained for 10 min, followed by conducting traffic tests and collecting strain data for the long-gauge distance, as shown in Figure 9. Figure 10 shows a schematic diagram of the evaluation test under a moving vehicle load. In order to precisely control the model vehicle's speed to keep it constant, a numerical control motor is used to pull the model vehicle. The speed of the model vehicle can be adjusted by changing the rotation speed of the numerical control motor.

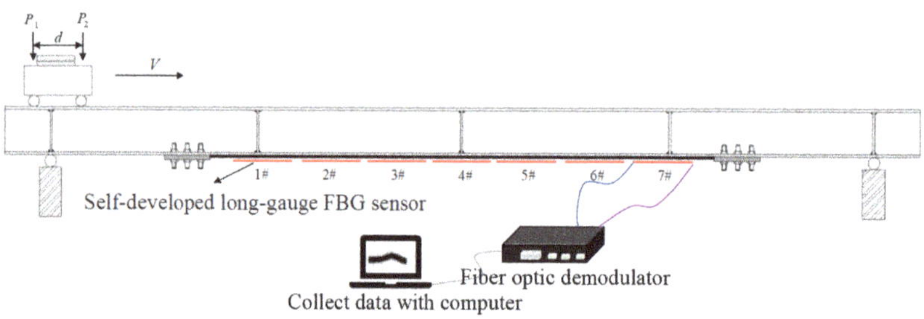

Figure 9. Layout of long-gauge FBG sensors at the bottom of the beam.

Figure 10. Evaluation test under moving vehicle load.

When the vehicle passes over the bridge, the vibration signal generated by the bridge is a complex signal composed of multiple-frequency components. Therefore, it is necessary to filter and denoise the signal to extract the quasi-static macro-strain signals generated by the moving vehicle load. The threshold decomposition method in wavelet analysis is used to denoise the original signal and extract the quasi-static long-gauge distance strain under moving vehicle loads to evaluate the strengthening effects.

After each moving vehicle load test under different strengthening cases was completed, an analysis was conducted on the time history–strain curves of each long-gauge FBG strain sensor at the bottom of the steel beam. Dynamic signal denoising is performed using wavelet denoising, with the selection of the db10 wavelet function and a decomposition level of 6 layers. Taking the initial condition CS-1 of the steel beam without strengthening as an example, a comparison between the original signals measured by the FBG strain sensors at positions 1# to 7# and the static long-gauge strain processed by the improved morphological wavelet threshold decomposition method is shown in Figure 11.

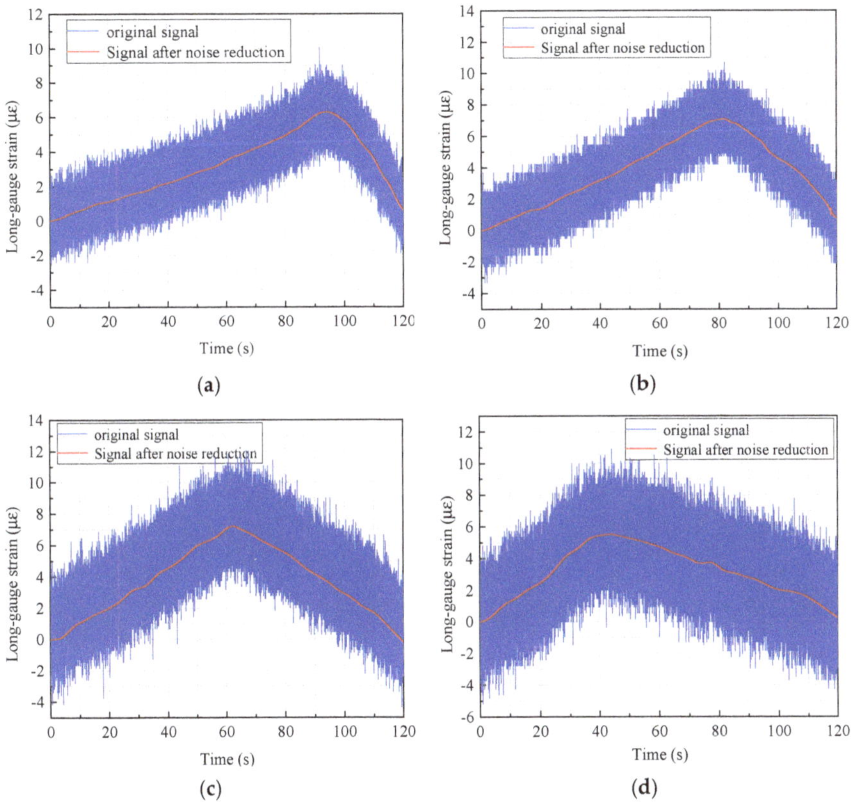

Figure 11. Comparison of original signal and denoised signals of long-gauge sensor: (**a**) 1# sensor; (**b**) 3# sensor; (**c**) 5# sensor; (**d**) 7# sensor.

By comparing the original signals with the denoised signals in Figure 11, it can be observed that the noise signals have effectively been removed from the static long-gauge strain–time history curves obtained after wavelet threshold decomposition and appear relatively smooth.

The quasi-static macro-strain–time history curves after the denoising of sensors 1# to 7# under test condition CS-1 are shown in Figure 12. The maximum values of the quasi-static long-gauge strain–time history curves for each sensor are plotted as macro-strain envelope curves. Figure 12a shows the macro-strain envelope curve for the steel beam in its initial state before strengthening. From Figure 12b, it can be observed that the measured maximum macro-strain value is 7.51 με, occurring at the position of sensor 4# at the mid-span of the test steel beam. The macro-strain values measured by the other sensors decrease successively along the mid-span positions towards the ends. The macro-strain envelope curves for each sensor approximately form a parabolic shape, which is consistent with the theoretical distribution law of macro-strain values in simply supported beams. From Figure 12b, it can be seen that there is a 1 με difference in the monitoring data between sensor 1# and sensor 7#. One possible reason for this is that the weight of the model vehicle is too small relative to the load-bearing capacity of the steel beam, resulting in approximately 6 με. The monitoring accuracy of the thermal sensors is ±1 με, which is in the same order of magnitude as the monitoring data, leading to some variance in the results.

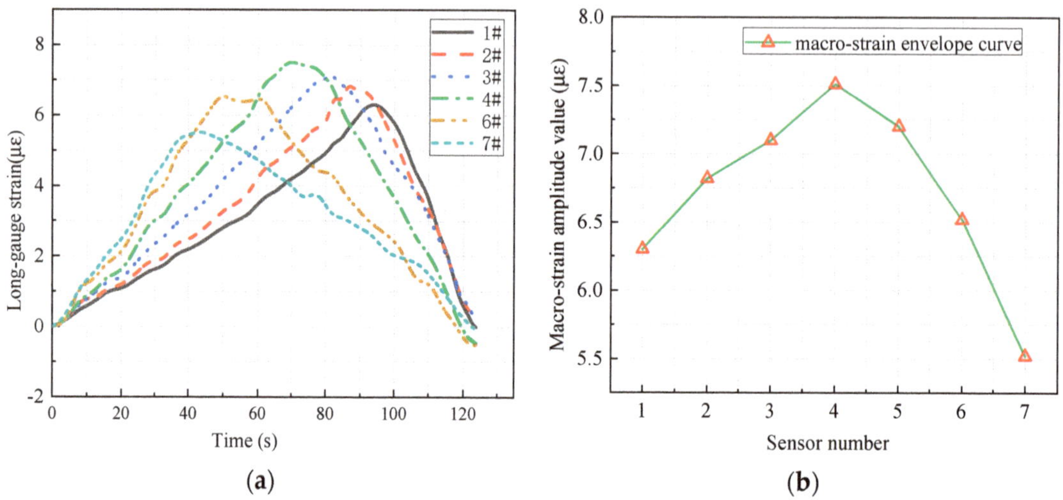

Figure 12. Macro-strain-time history and influence envelope curve: (**a**) long-gauge strain time history; (**b**) macro-strain amplitude envelope curve.

The previous analysis focused on the trends in the macro-strain–time history curves measured by the long-gauge strain sensors at the bottom of the test beam before and after prestressed CFRP plate strengthening. The results indicate that the mid-span macro-strain amplitude of the steel beam under moving loads decreases with an increase in the level of prestress of the CFRP plate. The distributed macro-strain distribution under moving vehicle load tests and under different strengthening conditions is shown in Figure 13. From Figure 13a, it can be seen that as the prestress of the CFRP plate increases, the stress monitored by the long-gauge strain sensors decreases, clearly illustrating the strengthening effect of the prestressed CFRP plate. This is because the prestress applied to the model bridge is greater than the strain generated by the self-weight of the model vehicle load. In practical bridge strengthening, since the self-weight of the bridge is significant, the strain generated by the bridge's self-weight is much larger than the longitudinal strain generated by the tension applied during strengthening, making it difficult to assess the strengthening effect by directly monitoring the internal forces of the bridge. Therefore, an evaluation of the strengthening effect can be carried out using the MSIL composed of the area encompassed by the macro-strain time history, as shown in Figure 13b.

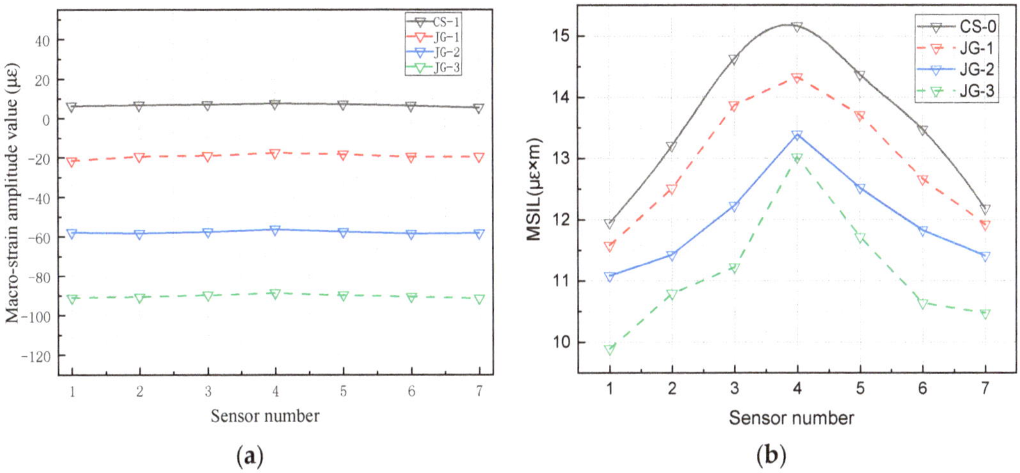

Figure 13. Macro-strain and MSIL under different strengthening cases: (**a**) macro-strain of different cases; (**b**) MSIL of different cases.

Based on the MSIL under moving vehicle loads, we calculate the difference between the macro-strain amplitude within the sections of the long-gauge strain sensors under various strengthening conditions and the initial macro-strain amplitude. We substitute these differences into Equation (7) to determine the strengthening effect of the prestressed CFRP plate on the test beam. The strengthening effects of the steel beam are shown in Figure 14, where the MSIL amplitudes of sensors 3#, 4#, and 5# are selected to evaluate the strengthening effect of the steel beam. It can be observed that there are some differences in the evaluation results of the strengthening effect between sensors at different locations. The evaluation accuracy can be improved by taking multiple measurements and averaging the results. Compared to the pre-strengthening state, the JG-1 condition increased the strengthening effect by about 6%, the JG-2 condition increased the bearing capacity by around 15%, and the JG-3 condition increased the bearing capacity by approximately 30%.

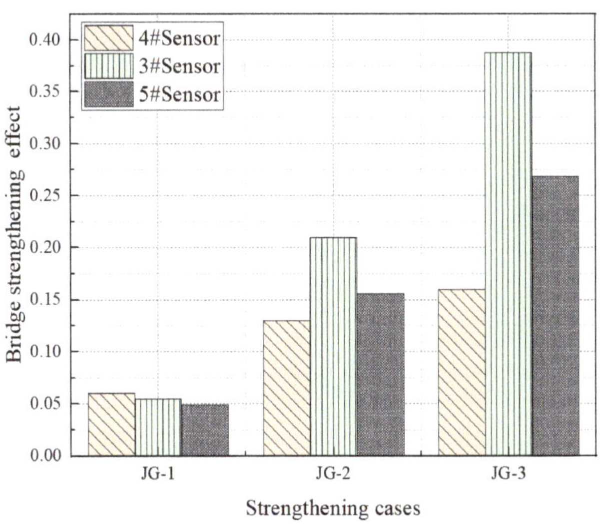

Figure 14. Evaluation of strengthening effect under different strengthening cases.

5. Conclusions

This study proposes an evaluation method for the strengthening effect of prestressed CFRP plates on bridges under moving vehicle loads using distributed long-gauge FBG strain sensors. The main conclusions drawn are as follows:

(1) From the mechanical relationship between the macro-strain impact envelope curve under moving vehicle loads and the stiffness of the bridge, it is known that an increase in the longitudinal internal forces in the bridge will lead to a decrease in the macro-strain amplitude of the bridge. Consequently, the area enclosed by the macro-strain–time history curve also decreases. By integrating to obtain the area enclosed by the macro-strain–time history curve, the pattern of variation in the longitudinal internal forces in the bridge can be more clearly revealed. This phenomenon can be utilized to evaluate the strengthening effect of the bridge.

(2) The results of the accelerated test on the prestressed CFRP-plate-strengthened beam show that compared to the unstrengthened beam, the peak value of the strain at the mid-span of the beam significantly decreases after the strengthening with the prestressed CFRP plates. The degree of the reduction is closely related to the level of prestress in the CFRP plates. Additionally, the area under the strain curve along the time axis also decreases. By substituting the change in area under the strain curve along the time axis into the evaluation index for the strengthening effect of the bridge based on the distributed sensing technology proposed in this article, the strengthening effect of the steel beam can be determined.

(3) A wavelet threshold decomposition method with a db10 wavelet function was employed, resulting in a good denoising effect. The extracted envelope of the macro-strain amplitude before strengthening roughly follows a parabolic distribution, while the envelope of the macro-strain amplitude after strengthening shows an overall decreasing trend with the increase in the prestress of the CFRP plates. It should be noted that the proposed method evaluates the changes in the distributed macro-strain before and after the strengthening of beams under moving vehicle loads, and it is also suitable for assessing the strengthening effects of concrete bridges during their operational period.

Author Contributions: Conceptualization, B.W. and H.W.; methodology, B.W. and H.W.; investigation, Q.X., Y.G. and S.F.; writing—original draft preparation, B.W.; writing—review and editing, B.W., H.W. and Z.G. All authors have read and agreed to the published version of the manuscript.

Funding: This work was supported by the National Natural Science Foundation of China (Grant No. 52368042, 52108141), the Jiangxi Provincial Cultivation Program for Academic and Technical Leaders of Major Subjects (Grant No. 20225BCJ23025), and the Natural Science Foundation of Jiangxi Province (Grant No. 20242BAB231008; Grant No. 20224BAB214066).At the same time, the authors would like to express our gratitude to the Jiangxi Provincial Institute of Transportation Science for providing the experimental conditions (Grant No. 2022C0007).

Data Availability Statement: The data presented in this study are available in the article.

Conflicts of Interest: The authors declare no conflicts of interest.

References

1. Havez, A.A.; Al-Mayah, A. Flexural strengthening of concrete structures using externally bonded and unbonded prestressed CFRP laminates: A Literature review. *J. Compos. Constr.* **2023**, *27*, 03123001. [CrossRef]
2. Gong, X.; Song, X.; Cai, C.S.; Li, G.Q.; Xiong, W. A temperature-driven approach for quantitative assessment of strengthening effect of continuous bridges using structural health monitoring data. *Struct. Health Monit.* **2024**, *23*, 1053–1070. [CrossRef]
3. Dardar, S.; Eamon, C.D.; Parra-Montesinos, G. Flexural Reliability of CFRP-Strengthened Bridge Girders. *J. Perform. Constr. Facil.* **2024**, *38*, 04023065. [CrossRef]
4. Chen, L.; Qiang, X.; Sun, P.; Zhang, S.; Jiang, X. Experimental and theoretical study on flexural behavior of prestressed concrete beams strengthened by CFRP tendons with external unbonded retrofit system. *J. Build. Eng.* **2024**, *89*, 109301. [CrossRef]
5. Nasera, M.Z.; Hawileh, R.A.; Abdall, J.A. Fiber-reinforced polymer composites in strengthening reinforced concrete structures: A critical review. *Eng. Struct.* **2019**, *198*, 109542. [CrossRef]

6. Tahsiri, H.; Belarbi, A.; Gencturk, B. Thermal effects on transfer length and prestress losses in CFRP prestressed prisms. *Constr. Build. Mater.* **2024**, *416*, 135160. [CrossRef]
7. Wang, H.T.; Wu, Q.; Li, L.; Wei, F.F.; Zhu, D.C.; Yu, X.P. Mechanical behaviors of steel CFRP plate bonded joints after freezing-thawing cycles. *Constr. Build. Mater.* **2024**, *412*, 134876. [CrossRef]
8. Wang, H.T.; Wu, Q.; Zhu, C.Y.; Mao, Y.H.; Guo, C.C. An innovative prestressing system of prestressed carbon fiber sheets for strengthening RC beams under flexure. *Constr. Build. Mater.* **2024**, *411*, 134409. [CrossRef]
9. Shi, J.W.; Wu, Q.Q.; Li, B.; Liu, Y.; Cao, W.H.; Wang, H.T. Fatigue bond behavior of FRP-to-concrete joints with various bonding adhesives. *Eng. Struct.* **2024**, *301*, 117311. [CrossRef]
10. Wang, H.T.; Tang, C.; Chen, M.S.; Shi, J.; Cao, X.Y. Experimental study on the flexural performance of prestressed RC beams with post-tensioned CFRP strands. *Eng. Struct.* **2024**, *309*, 118118. [CrossRef]
11. Azevedo, A.S.; Firmo, J.P.; Correia, J.R.; Chastre, C.; Biscaia, H.; Franco, N. Fire behaviour of CFRP-strengthened RC slabs using different techniques-EBR. *Compos. Part B Eng.* **2022**, *230*, 109471. [CrossRef]
12. Pang, Y.; Wu, G.; Wang, H.; Gao, D.; Zhang, P. Experimental study on the bond behavior of the CFRP-steel interface under rapid loading. *Thin. Wall. Struct.* **2021**, *159*, 107233. [CrossRef]
13. Eslami, A.; Shayegh, H.R.; Moghavem, A.; Ronagh, H.R. Experimental and analytical investigations of a novel end anchorage for CFRP flexural retrofits. *Compos. Part B Eng.* **2019**, *176*, 107309. [CrossRef]
14. Bodzak, P. Prestressing and preloading effect in RC slabs strengthened with normal and pretensioned CFRP EBR strips. *Compos. Part B Eng.* **2019**, *169*, 19–28. [CrossRef]
15. Rezazadeh, M.; Costa, I.; Barros, J. Influence of prestress level on NSM CFRP laminates for the flexural strengthening of RC beams. *Compos. Part B Eng.* **2014**, *116*, 489–500. [CrossRef]
16. Pang, Y.Y.; Wu, G.; Wang, H.T.; Su, Z.L.; He, X.Y. Experimental study on the bond behavior of the CFRP-steel interface under the freeze–thaw cycles. *J. Compos. Mater.* **2020**, *54*, 13–29. [CrossRef]
17. Xian, G.J.; Zhou, P.; Bai, Y.B.; Wang, J.Q.; Li, C.G.; Dong, S.C.; Guo, R.; Li, J.H.; Du, H.Q.; Zhong, J. Design, preparation and mechanical properties of novel glass fiber reinforced polypropylene bending bars. *Constr. Build. Mater.* **2024**, *429*, 136455. [CrossRef]
18. Rasheed, H.A.; Zaki, M.A.; Foerster, A.S. Efficient bidirectional U-wrap system to anchor CFRP sheets bonded to reinforced concrete T-girders. *Structures* **2022**, *38*, 226–236. [CrossRef]
19. Alhoubi, Y.; Mahaini, Z.; Abed, F. The flexural performance of BFRP-reinforced UHPC beams compared to steel and GFRP-reinforced beams. *Sustainability* **2022**, *14*, 15139. [CrossRef]
20. Guo, R.; Xian, G.; Li, C.; Hong, B. Effect of fiber hybrid mode on the tension–tension fatigue performance for the pultruded carbon/glass fiber reinforced polymer composite rod. *Eng. Fract. Mech.* **2022**, *260*, 108208. [CrossRef]
21. ACI Committee 440.2R. *Guide for the Design and Construction of Externally Bonded Frp Systems for Strengthening Concrete Structures (ACI 440.2 R-22)*; American Concrete Institute: Farmington Hills, MI, USA, 2022. [CrossRef]
22. Aram, M.R.; Czaderski, C.; Motavalli, M. Debonding failure modes of flexural FRPstrengthened RC beams. *Compos. Part B Eng.* **2008**, *39*, 826–841. [CrossRef]
23. Wang, H.T.; Bian, Z.N.; Wu, Q.; Hu, L.; Chen, M.S.; Tang, C. Development of a novel anchorage and tensioning system for strengthening steel beams with unbonded prestressed CFRP plates. *Eng. Struct.* **2024**, *312*, 118234. [CrossRef]
24. Smith, S.T.; Hu, S.; Kim, S.J.; Seracino, R. FRP-strengthened RC slabs anchored with FRP anchors. *Eng. Struct.* **2011**, *33*, 1075–1087. [CrossRef]
25. Wang, H.T.; Bian, Z.N.; Xu, G.W.; Chen, M.S.; Xiong, H.; Liu, S.S. Flexural strengthening of large-scale RC beams with nonprestressed and prestressed CFRP using different anchorages. *Polymers* **2022**, *14*, 5498. [CrossRef] [PubMed]
26. Zaki, M.A.; Rasheed, H.A. Behavior of reinforced concrete beams strengthened using CFRP sheets with innovative anchorage devices. *Eng. Struct.* **2020**, *215*, 110689. [CrossRef]
27. Muciaccia, G.; Khorasani, M.; Mostofinejad, D. Effect of different parameters on the performance of FRP anchors in combination with EBR-FRP strengthening systems: A review. *Constr. Build. Mater.* **2022**, *354*, 129181. [CrossRef]
28. Zaki, M.A.; Rasheed, H.A. Impact of efficiency and practicality of CFRP anchor installation techniques on the performance of RC beams strengthened with CFRP sheets. *Can. J. Civ. Eng.* **2019**, *46*, 796–809. [CrossRef]
29. Wang, H.T.; Liu, S.S.; Zhu, C.Y.; Xiong, H.; Xu, G.W. Experimental study on the flexural behavior of large-scale reinforced concrete beams strengthened with prestressed CFRP plates. *J. Compos. Constr.* **2022**, *26*, 04022076. [CrossRef]
30. Puurula, A.M.; Enochsson, O.; Sas, G.; Blanksvärd, T.; Ohlsson, U.; Bernspång, L.; Täljsten, B.; Carolin, A.; Paulsson, B.; Elfgren, L. Assessment of the strengthening of an RC railway bridge with CFRP utilizing a full-scale failure test and finite-element analysis. *J. Struct. Eng.* **2015**, *141*, D4014008. [CrossRef]
31. Wang, H.T.; Bian, Z.N.; Chen, M.S.; Hu, L.; Wu, Q. Flexural strengthening of damaged steel beams with prestressed CFRP plates using a novel prestressing system. *Eng. Struct.* **2023**, *284*, 115953. [CrossRef]
32. Ghafoori, E.; Hosseini, A.; Al-Mahaidi, R.; Zhao, X.L.; Motavalli, M. Prestressed CFRP-strengthening and long-term wireless monitoring of an old roadway metallic bridge. *Eng. Struct.* **2018**, *176*, 585–605. [CrossRef]
33. Grace, N.; Jensen, E.; Matsagar, V.; Penjendra, P. Performance of an AASHTO beam bridge prestressed with CFRP tendons. *J. Bridge Eng.* **2013**, *18*, 110–121. [CrossRef]

34. Wu, B.; Zhou, Y.; Yin, X.; Lin, Z.; Zhou, Z. The anti-arch inhibition effect of multispan continuous girder bridge strengthened with prestressed CFRP plates. *Structures* **2022**, *35*, 845–855. [CrossRef]
35. Wang, H.T.; Liu, S.S.; Shi, J.W.; Xu, G.W.; Chen, M.S.; Bian, Z.N. Bonded and unbonded strengthening of RC beams with the prestressed CFRP plate system: An experimental study. *Constr. Build. Mater.* **2024**, *438*, 137118. [CrossRef]
36. Kianmofrad, F.; Ghafoori, E.; Elyasi, M.M.; Motavalli, M.; Rahimian, M. Strengthening of metallic beams with different types of pre-stressed un-bonded retrofit systems. *Compos. Struct.* **2017**, *159*, 81–95. [CrossRef]
37. He, W.Y.; Ren, W.X.; Zhu, S.Y. Baseline-free damage localization method for statically determinate beam structures using dual-type response induced by quasi-static moving load. *J. Sound Vib.* **2017**, *400*, 58–70. [CrossRef]
38. Zheng, X.; Yi, T.H.; Yang, D.H.; Li, H.N. Stiffness estimation of girder bridges using influence lines identified from vehicle-induced structural responses. *J. Eng. Mech.* **2021**, *147*, 04021042. [CrossRef]
39. Zhang, G.W.; Cheng, Y.Y.; Xia, Q.; Zhang, J. Curvature envelope area based rapid identification method of bridge distributional element stiffness using microwave interference radar. *Mech. Syst. Signal Process.* **2023**, *197*, 110390. [CrossRef]
40. Hester, D.; Brownjohn, J.; Huseynov, F.; Obrien, E.; Gonzalez, A.; Casero, M. Identifying damage in a bridge by analysing rotation response to a moving load. *Struct. Infrastruct. Eng.* **2020**, *16*, 1050–1065. [CrossRef]
41. Ojio, T.; Yamada, K. Bridge Weigh-In-Motion Systems Using Stringers of Plate Girder Bridges. Iowa State University: Ames, IA, USA, 2002.

buildings

MDPI

Article

The Effect of Volcanic Stone and Metakaolin on the Compressive Properties of Ultrahigh-Performance Concrete Cubes

Yushi Yin [1,2,*] and Zeyu Ma [3]

[1] College of Architectural Engineering, Yangzhou Polytechnic Institute, Yangzhou 225127, China
[2] Department of Civil Engineering, Tongji University, Shanghai 200092, China
[3] School of Architectural and Engineering, Yuncheng Vocational and Technical University, Yuncheng 044000, China; mazy@ypi.edu.cn
* Correspondence: yinys@alumni.tongji.edu.cn

Abstract: Over the past few decades, ultrahigh-performance concrete (UHPC) has been widely studied and applied because of its outstanding mechanical properties, such as its high strength and notable durability. However, because of its high cost and easy shrinkage cracking during early pouring in mass concrete construction, to reduce the cost of UHPC and reduce the cracks caused by early pouring, volcanic stone was used as a new type of UHPC coarse aggregate, while metakaolin (MK) was added to the system at the same time, and then two parameters, namely the volcanic rock particle size group and the MK dispersion ratio, were set. Scanning electron microscopy (SEM), X-ray diffraction (XRD) and thermogravimetric (TG) microanalysis methods were used to reveal the influence of changes in the material microstructure, phase composition, material composition and crystallinity of the mineral composition on the compressive properties of the UHPC cubes. The results show that the mechanical "lock-in effect" of the structure formed by the volcanic rock holes and mortar can effectively improve the shear resistance of the UHPC–volcanic rock interface, and the compressive strength of the UHPC cubes increases with the volcanic stone's particle size. When the MK dispersion ratio is less than 4%, the cube compressive strength of the UHPC and the contents of $CaCO_3$ crystals, C-S-H gel and travertine in the UHPC increase with an increasing MK dispersion ratio. At an age of 28 days, compared with the addition of 1% MK, the addition of 4% MK increases the production of C-S-H gel and travertine in the UHPC matrix by 24.82%. When the MK dispersion ratio is 4%, the crystallinity values of the C-S-H gel, travertine and limestone in the UHPC are greater. Adding MK at a 4% dispersion ratio can promote the crystallization of limestone into a large amount of calcite, which can increase the strength of UHPC. On the one hand, the addition of volcanic coarse aggregate results in the retention of more free water and bound water; on the other hand, it also makes it difficult to crystallize $CaCO_3$. The combined action of MK at a 4% dispersion ratio and volcanic rock significantly inhibits $CaCO_3$ crystallization.

Keywords: ultrahigh-performance concrete; coarse aggregate; metakaolin; volcanic stone; compressive strength; microstructure

1. Introduction

Ultrahigh-performance concrete (UHPC) is a durable and high-performance concrete material with excellent properties, such as high strength and durability [1,2]. Since its emergence more than 20 years ago, UHPC has attracted the attention of civil engineering circles at home and abroad. While UHPC achieves ultrahigh performance, the reality of its high production cost cannot be ignored. The cost of its raw materials is generally above 4000 RMB/m^3, with the highest reaching 10,000 RMB/m^3 [3]. In addition, due to its low water–binder ratio and large amount of self-shrinkage, UHPC is prone to shrinkage cracking during the early casting process [4]. Early cracking does not close, which not only reduces the compressive strength of UHPC in the later stage but also affects the overall quality of the project. High cost and high shrinkage seriously restrict the promotion and applications of UHPC in mass concrete engineering [5–8].

In response to the above problems, scholars at home and abroad have proposed the formation of coarse aggregate–UHPC (CA-UHPC) containing coarse aggregate by adding natural coarse aggregate to UHPC to replace part of the active powder without significantly reducing the excellent mechanical properties of UHPC. CA-UPC is generally made by replacing UHPC components with basalt gravel of a certain particle size. For example, Shi et al. reported that the compressive strength of CA-UHPC is not much different from that of UHPC without coarse aggregates [9–12], Wille et al. reported that coarse aggregates with larger particle sizes can improve the compressive strength of UHPC to a certain extent [2] and the compressive strength of CA-UHPC tends to first increase and then decrease with an increasing coarse aggregate-to-binder ratio [13,14].

Studies have found that metakaolin (MK) not only improves the early strength of concrete but also enhances its durability and resistance to chemical attack. Metakaolin reacts pozzolanically to produce additional calcium silicate hydrate (C-S-H), filling the pores in the concrete, thus increasing its density and crack resistance [15,16]. Additionally, metakaolin exhibits water-reducing properties, which can lower the water demand in concrete and improve its workability [17]. The combined use of volcanic ash and metakaolin in concrete demonstrates significant synergistic effects. Research indicates that the combination of these two materials can further enhance the compressive strength, tensile strength and durability of concrete [18,19]. Studies have also shown that the specific surface area of MK has a significant effect on cement-based materials [20], and the filling effect and active reaction of MK can effectively improve the early mechanical properties of concrete [21,22], improve the pore structure of cement and improve its self-shrinking ability; at the same time, the addition of MK can fully improve the strength and toughening ability of steel fibers [12]. Microscopically, the amount of chemical bonding water in the gel in the UHPC matrix increases with an increasing MK content, while the amount of $Ca(OH)_2$ decreases continuously [23,24]. When MK is used to replace 10~15% of the cement base, the compressive performance of UHPC can reach the best level [25–27].

Scholars have effectively improved the early mechanical properties of concrete by carrying out structural improvement and material modification of cement-based materials, especially for the placement of coarse aggregates in UHPC, which has led to some beneficial research results. However, research on the effect of MK on the properties of cement-based materials has been concentrated on non-high-strength concrete materials, and research on its impact on UHPC's mechanical properties is still relatively rare; moreover, related research on the impact of MK on the mechanical properties of CA-UHPC is rare. In addition, basalt gravel is generally used as a coarse aggregate in UHPC, while other coarse aggregate substitutes have rarely been studied. Thus, the limitations of the related research are obvious.

In this paper, by using volcanic rocks as a new type of coarse aggregate for UHPC, the goal was not only to maintain the compressive strength of UHPC but also to significantly reduce the construction cost of UHPC projects, resulting in the development of a new type of high-strength engineering material. The conclusions of this paper provide reliable data support for the UHPC theoretical framework and its application in large-scale green CA-UHPC engineering projects.

2. Experimental Program

2.1. Material Selection

As shown in Figure 1, the cement is ordinary Portland cement with a strength grade of P·O42.5, produced by the Liaoning Daying Cement Plant in China, and its loss on ignition (LOI) is 3%.The silica fume was produced by Lingshou Yunda Mineral Products Co., Ltd. (Huainan, China), with a particle size of 2.5 μm, and its LOI is 0.5%. MK was produced by Inner Mongolia Baotou Steel Hefa Rare Earth Co., Ltd. (Baotou, China), with a particle size of 3.75 μm. River sand with a particle size of 0.075~2.36 mm from certain tailing impoundments in Liaoning Province was adopted. Volcanic stone with a particle size of 9.5~31.5 mm was obtained from Lingshou Weifeng Mineral Products Co., Ltd. (Huainan, China), which has a crushing value of 24.36%. The steel fibers used were steel fibers produced by Tengzhou Sida Shimisi Metal Products Co., Ltd. (80% end-hook-type and 20% straight-wire-type, length of 13 mm, length–diameter ratio of 65, tensile strength of 2850 MPa, Tengzhou, China), and the water-reducing agent used was a polyhydroxy acid superplasticizer mother liquor produced by Weihe Technology Admixture Factory (water-reducing rate of 40%, Lianyungang, China). All the raw materials were sourced from China. The chemical compositions of the cement and kaolin are shown in Table 1.

Figure 1. Experimental raw materials; (**a**) cement; (**b**) silica fume; (**c**) metakaolin; (**d**) steel fiber; (**e**) volcanic stone; (**f**) river sand; (**g**) water reducer.

Table 1. Chemical composition of cement and MK (w/%).

Raw Material	SiO_2	Al_2O_3	Fe_2O_3	CaO	MgO	Na_2O	SO_3	K_2O	TiO_2	The Others
Cement	23.22	4.51	2.33	60.47	1.16	0.87	4.22	1.23	-	1.99
MK	51.92	44.52	0.42	-	-	-	-	-	1.23	1.91

2.2. Mix Proportion Design

The dispersion ratio of MK refers to the proportion of MK in cementitious materials. Fly ash is not used in this study because it is also a cementitious material and would affect the analysis of MK's impact on cement performance. Therefore, the MK proportion is expressed as a percentage total mass of cement and silica fume (1–4%). The factor level of the experimental design is shown in Table 2. The MK dispersion ratio and the volcanic stone particle size group were used as two variable factors, A and B. The A factor level was 4, and the B factor level was 3. At the same time, a UHPC mix ratio design table was established, with a total of 12 groups, as shown in Table 3.

Table 2. Parameters and levels.

Levels	Parameters	
	A: MK Dispersion Ratio/%	B: Volcanic Stone Granular Formation/mm
1	1	9.5–16
2	2	16–19
3	3	19–31.5
4	4	-

Table 3. Mix proportion of UHPC.

No.	Mass Fraction/%							Water–Binder Ratio	Particle Size of Volcanic Stone/mm
	Cement	Silica Fume	MK	Volcanic Stone	River Sand	Steel Fiber	Water-Reducing Agent		
U	35.46	8.00	0	0	40.0	6.28	0.87	0.18	0
G1	35.46	8.00	0.43	0	40.0	6.28	0.87	0.18	0
G2	35.46	8.00	0.86	0	40.0	6.28	0.87	0.18	0
G3	35.46	8.00	1.30	0	40.0	6.28	0.87	0.18	0
G4	35.46	8.00	1.73	0	40.0	6.28	0.87	0.18	0
H1	29.79	6.72	0	16	33.6	5.28	0.73	0.18	9.5–16
H2	29.79	6.72	0	16	33.6	5.28	0.73	0.18	16–19
H3	29.79	6.72	0	16	33.6	5.28	0.73	0.18	19–31.5
GH1	29.79	6.72	0.36	16	33.6	5.28	0.73	0.18	9.5–16
GH2	29.79	6.72	0.73	16	33.6	5.28	0.73	0.18	9.5–16
GH3	29.79	6.72	1.09	16	33.6	5.28	0.73	0.18	9.5–16
GH4	29.79	6.72	1.46	16	33.6	5.28	0.73	0.18	9.5–16

Notes: U stands for UHPC; G represents UHPC doped with MK (GA, a number represents the horizontal factor of factor A in Table 2); H represents UHPC mixed with volcanic stone (Hb, B represents the horizontal factor of factor B in Table 2); GH represents UHPC mixed with MK and volcanic stone (GHC, C represents the horizontal factor of factor A in Table 2).

2.3. Test Specimens

With 6 specimens as a group, a total of 72 cube compressive standard specimens were used to carry out the experimental research. By calculating the weights of the required materials, the MK and water were mixed according to the mixing ratio and stirred evenly to make the MK slurry, and the volcanic stone was soaked in the MK slurry for 12 h and then removed. Then, the weighed cementing material (cement, silica fume), river sand, volcanic stone, steel fiber, water-reducing agent and MK slurry were poured into the mixer, wetted by the cement slurry in turn. Each time a material was poured, it was stirred for 30 s until it was uniform. After all the raw materials were added, the mixture was stirred again for 2 to 3 min and then quickly poured into a 150 mm × 150 mm × 150 mm trial mold. After pouring, the test specimen was placed in a normal-humidity environment of 20 ± 1 °C for 48 h and then demolded. After demolding, the test specimens were put into a standard curing box with a temperature of 20 ± 1 °C and a humidity greater than 95% and cured for 3 days, 7 days, 14 days or 28 days. The compressive strength of the cubes was tested at the corresponding curing age, and XRD and TG microscopic tests were carried out at the corresponding curing age.

2.4. Microscopic Test Method

To observe the microscopic morphology of the materials and the specimens, according to the scanning electron microscopy (SEM) sampling requirements, samples of 10 mm × 10 mm × 10 mm specimens were collected after destruction. The samples were fixed, rinsed, dehydrated, replaced, dried and coated. The processed samples were scanned by SEM point scanning using a Shimadzu scanning electron microscope (SS-550, Kyoto, Japan), and high-definition SEM images of the samples were obtained.

To investigate the phase composition and crystallinity of the mineral components of each group of specimens, according to the X-ray diffraction (XRD) sampling requirements, multiple powder samples above 20 mg with a distance of approximately 25 mm from the surface of the specimen were collected. The samples were ground to within an order of magnitude of 40 μm, and a SmartLab SE instrument (Rigaku, Tokyo, Japan) was used to perform XRD. The scanning angle range of the instrument was set to 5~65°, the step size was set to 0.02° and the scanning speed was set to 4°/min to obtain the diffraction pattern of the samples. Then, JADE9.0 software was used for comparative analysis.

To understand the material composition of the test specimen and conduct quantitative analysis, according to the thermogravimetric (TG) analysis and sampling requirements, 10 mm × 10 mm × 10 mm specimens were collected after destruction, and a German NETZSCH STA-449C synchronous thermal analyzer (Thauern, Bavaria, Germany) was used to perform thermogravimetric treatment on the samples, which were preheated for 3 min after starting. The target temperature was set at 1000 °C, and the heating rate was 20 °C/min. The samples were placed in a ceramic crucible to obtain TG and DTG data.

3. Test Results and Analysis

3.1. Experimental Phenomenon

To investigate the 28-day cubic compressive strength values of the test specimens, all the specimens were tested on a yE-200A 2000 kN hydraulic pressure testing machine with displacement control and a loading rate of 0.02 mm/s. During the tests, cracks began to appear when the specimens were loaded to approximately 80% of their ultimate load, accompanied by a clear "ding-ding" sound from the peeling of the steel fibers. When a specimen was completely destroyed, multiple vertical cracks appeared, and the pressure curve dropped sharply. The fracture load threshold of the test specimen was 1100 kN, which was specifically expressed at the moment of complete failure. If the instantaneous load of the test specimen did not exceed 1100 kN, the specimen would have obvious cracks, and there were no obvious sounds. When the instantaneous load of the specimen exceeded 1100 kN, the failure of the test specimen was in a "semiburst" style, which was manifested as the specimen splitting into three or four pieces of gravel, accompanied by loud noise, but because of the existence of steel fibers, the gravel pieces were connected to each other and did not scatter around. The failure forms of each specimen in the compressive strength test at 28 days of age are shown in Table 4.

Table 4. Failure forms of different specimens at the age of 28 d.

No.	U	G4	H3	GH4
Description of failure mode	There are many penetrating vertical cracks on the front and side.	There are two thick vertical through-cracks on the front, which divide the specimen into three pieces, and many crushed stones and steel fibers collapse around.	The test piece is broken into several pieces, and the whole piece has been seriously deformed. The internal volcanic stone is seriously broken, and there is a volcanic stone peeling phenomenon.	Similar to H3, but without H3, the overall deformation is serious, and the degree of peeling for the volcanic rock coarse aggregate is weaker than that for the H3 specimen.
Damage pattern picture				

3.2. Test Results

In general, the cubic compressive strength of a specimen increased rapidly in the early stage and gradually decreased in the later stage, as shown in Figure 2. It is worth noting that the cubic compressive strength in the later stage for UHPC basically tended to be stable, but the cubic compressive strength in the later stage for UHPC with added MK still continuously improved.

At an age of 3 days, the cubic compressive strength values of the test specimens decreased in the order of G4 > G2 > G3 > U > G1. At an age of 7 days, the growth rate of the cubic compressive strength of G4 was significantly lower than that of the other groups, for which the cubic compressive strength decreased to the lowest level. The reason for this phenomenon is mainly the extremely low water–cement ratio of UHPC and relatively little free water. When cement and silica fume undergo a hydration reaction with free water at the same time, the MK activity of the high specific surface area is larger, and more free water is required for the hydration reaction, which affects its strength development. At 14 days of age, the growth rate of U's cubic compressive strength was significantly greater than that of the other groups, and its cubic compressive strength was highest, which indicates that there was more free water in U at this age, the cement hydration rate was the fastest and the growth rate of the cubic compressive strength was the highest. At 28 days of age, the growth rate of the cubic compressive strength of G4 was significantly greater than that of the other groups, the cubic compressive strength was the highest, reaching 136.78 MPa, and the cubic compressive strength of U was 123.14 MPa. The cubic compressive strength of G4 was 11.077% greater than that of U, and the cubic compressive strength of the specimens was in the order of G4 > G3 > U > G2 > G1, showing that the cubic compressive strength of the 28-day-old UHPC increased with an increasing MK dispersion ratio when the MK dispersion ratio was in the range of 4%.

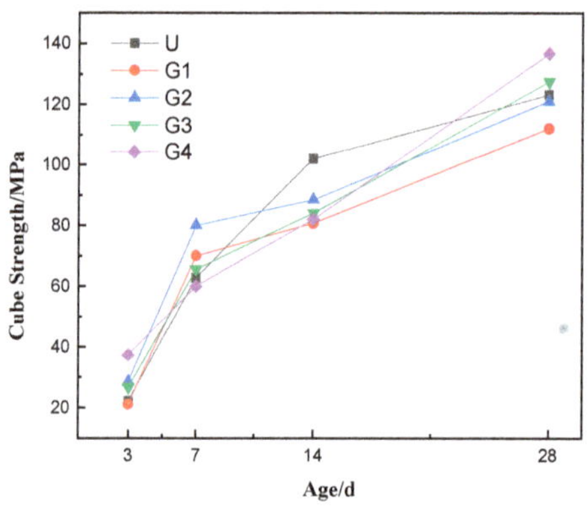

Figure 2. Cube compressive strength change rule for MK-UHPC.

By observing the cubic compressive strength curves of UHPC at ages of 3–28 days, it could be seen that the growth rate of the cubic compressive strength of G4 was the smallest during the entire process, and the cubic compressive strength curve was almost a straight line. The strength values at 3 and 28 days were the highest, which is obviously different from those of the other groups; the growth rates of the cubic compressive strength values in the early stage for all of the UHPC with MK added were lower than those of UHPC and the growth rate values of the cubic compressive strength in the later stage were higher than those of UHPC. This was because the MK particles absorbed part of the free water in the early stage of hydration, resulting in relatively slow hydration of the cement in the early stage of hydration, and the MK that absorbed free water in the later stage of hydration gradually underwent an active reaction, which will be further analyzed and explained in Section 2.2.

As shown in Figure 3a, at 28 days, the cubic compressive strength values of the test specimens were U > H3 > H2 > H1 in descending order, indicating that when the volcanic stone particle size was in the range of 9.5~31.5 mm, the cubic compressive strength of CA-UHPC increased with an increasing particle size. This was because the larger particle size of the volcanic stone aggregates led to a coarser network matrix composed of UHPC slurry, which caused the overall strength of the UHPC specimen to be higher. Moreover, volcanic stone has a higher crushing value (as much as 24.36% higher) than road basalt gravel. Therefore, although the compressive strength of CA-UHPC based on volcanic stone was lower than that of UHPC without coarse aggregates, the price of volcanic stone itself is very low, the compressive strength of volcanic stone CA-UHPC is still within the range of high-performance concrete (UPC) and it is still of high use value in the field of construction, while reducing the cost of building materials.

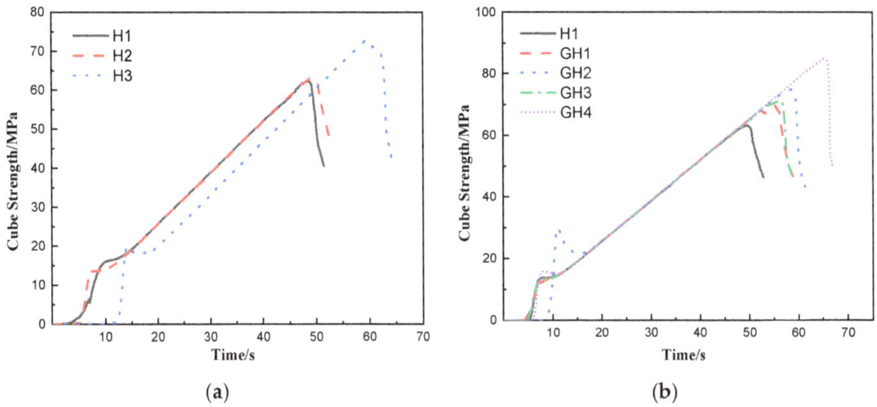

Figure 3. Cube compressive strength curves of different specimens at the age of 28 d: (**a**) specimens of H1, H2 and H3; (**b**) specimens of H1, GH1, GH2, GH3 and GH4.

As shown in Figure 3b, at 28 days, the compressive strength of U was also the highest. The compressive strength of GH4 in CA-UHPC with MK was the highest, reaching 84.77 MPa, which is consistent with the test results for UHPC supplemented with MK, indicating that when the MK dispersion ratio was in the range of 4%, the cubic compressive strength of UHPC decreased first and then increased with an increasing MK dispersion ratio. The cubic compressive strength of CA-UHPC increased with an increasing MK dispersion ratio. The cubic compressive strength of GH4 was 34.27% greater than that of H1, and the improvement was more obvious than that of UHPC.

The stress curves for the CA-UHPC are similar; that is, there is a "weak segment" in the early stage, and then the stress curve gradually increases until failure. The reason for the "weak segment" is that there were occasional volcanic stone aggregates on the stress surface of a specimen. This part of the volcanic stone aggregate having first contact with the pressure testing machine resulted in a long pressure time history, and the curve thus produced had a "weak" effect.

4. Micro Analysis

4.1. SEM Image of the "Interlock Effect"

When the interface is magnified 200–300 times, Figure 4 shows that the interface between the volcanic stone and the concrete matrix was relatively dense, and the surface of the volcanic stone and the concrete matrix were well combined. Figure 5 clearly shows that the pores of the volcanic stone were filled with mortar, indicating that the pores of the volcanic stone better formed mechanical occlusive "interlocking" with the mortar matrix, which effectively enhanced the shear resistance performance of the volcanic stone–UHPC interface.

In addition, due to the porous structure of volcanic stone, the interface of the pore structure was the most unfavorable interface in the compressive process of the CA-UHPC, which was one of the main reasons for the decrease in the compressive strength of the CA-UHPC.

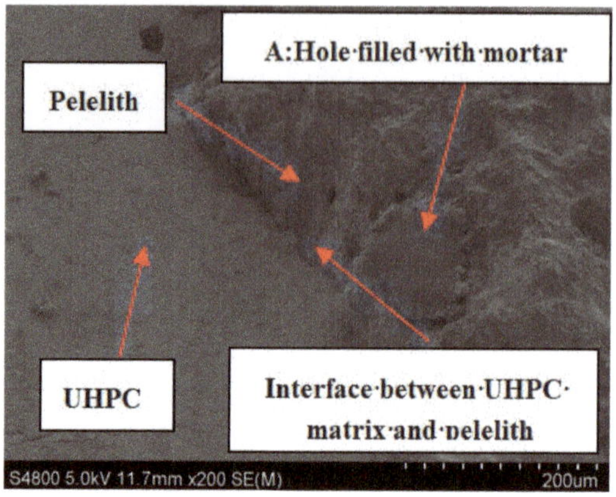

Figure 4. Interface between concrete and volcanic stone.

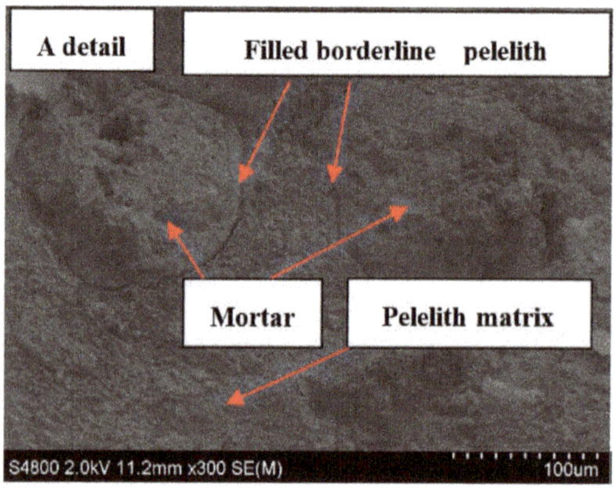

Figure 5. Mechanical lock between volcanic stone and mortar.

4.2. XRD Comparative Analysis

Figure 6 shows a comparison of the XRD patterns of the UHPC and CA-UHPC sample groups with added MK. Figure 6 shows that the reaction was mainly concentrated in the 25~35° section (2θ), forming an amorphous "wave crest" centered at 27~29° (2θ), in which the "wave crest" was the limestone phase. A small sharp peak was found at approximately 27°~30° (2θ), which is usually related to C-(A)-S-H gel or calcite formed by carbonatation. According to the standard diffraction pattern, it was determined that a C-S-H gel product was generated by the reaction.

Figure 6a–d shows that at different ages, the orders of the C-S-H diffraction peaks of each sample were basically the same as the orders of the compressive strength values. The diffraction peak of C-S-H represents the crystallinity of C-S-H, and the diffraction peak was proportional to the crystallinity, indicating that at the age of 28 days, the crystallinity of C-S-H was greater when the dispersion ratio of the MK was 4%, and the growth of the crystal plane was more orderly. A high dispersion ratio of MK promotes the crystallization of C-S-H, which can increase the strength of the material. The above analysis results verify the change rule of the cubic compression test described above.

Figure 6e shows that at 28 days, the C-S-H diffraction peak of the GH1 sample was significantly greater than those of the other dispersion ratios, indicating that the CA-UHPC with added MK at a dispersion ratio of 4% had a higher C-S-H crystallinity, which is consistent with the results shown in Figure 6d and also verifies the results of the previous cubic compression tests. In Figure 6, hydrargillite was found near 50.7° (2θ), and wustite was found near 60.7° (2θ). The intensities of the diffraction peaks at these two angles did not change significantly but decreased slightly in comparison in UHPC with added MK.

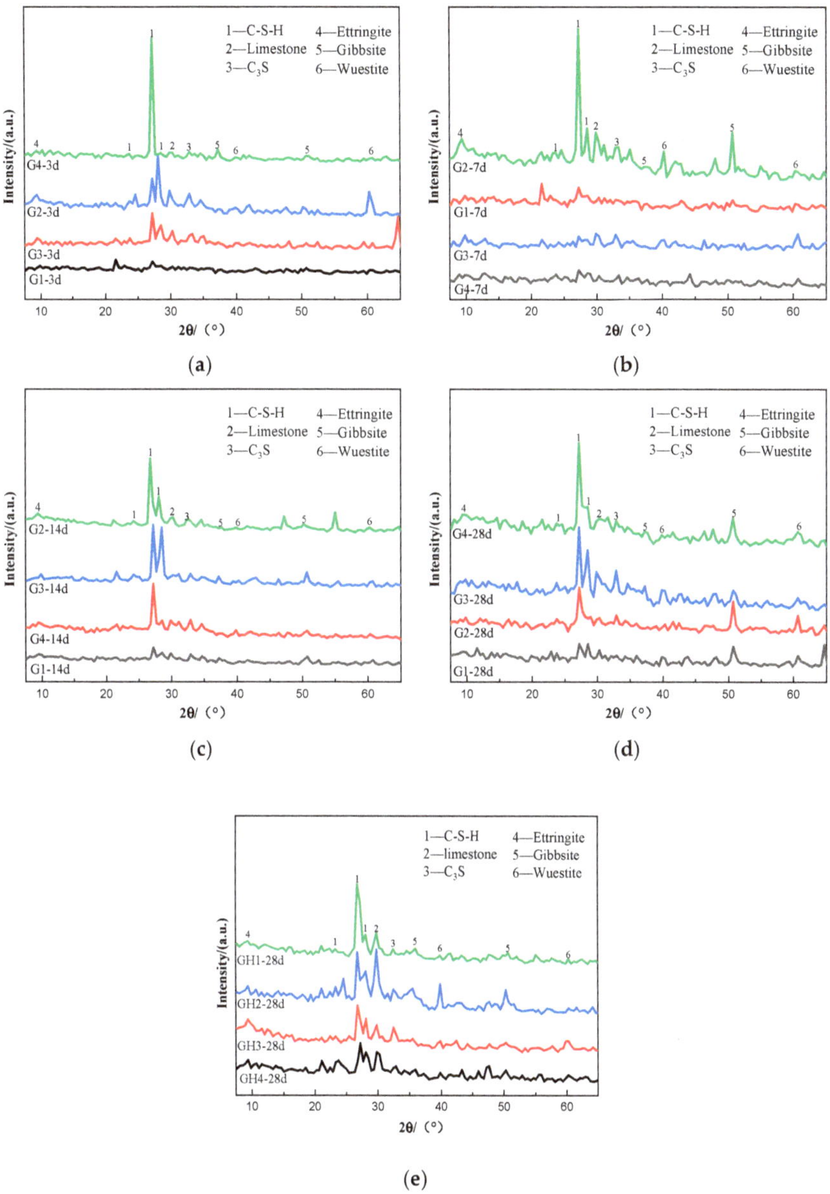

Figure 6. XRD diffraction patterns of different specimens: (**a**) specimens of MK-UHPC at the age of 3 d; (**b**) specimens of MK-UHPC at the age of 7 d; (**c**) specimens of MK-UHPC at the age of 14 d; (**d**) specimens of MK-UHPC at the age of 28 d; (**e**) sspecimens of MK-CA-UHPC at the age of 28 d.

4.3. TG/DTG Analysis

Figure 7a–e are comparison diagrams of the TG and DTG curves, respectively, for the UHPC and CA-UHPC sample groups with added MK. The MK-added UHPC and CA-UHPC sample groups had obvious thermal decomposition reactions in the temperature ranges of 50~150 °C and 550~700 °C. The mass loss at 50~150 °C and 550~700 °C was caused by the evaporation of free water and bound water with an increasing temperature and the decomposition of some poorly crystallized $CaCO_3$ crystals to generate CO_2 with increasing temperature.

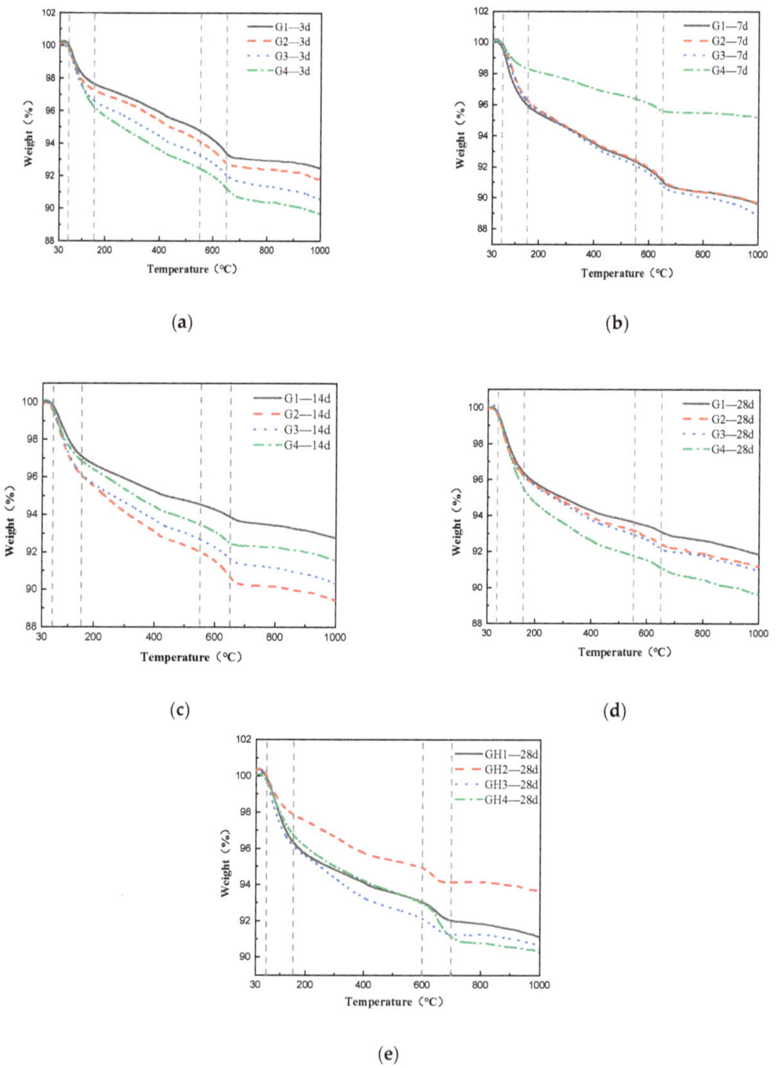

Figure 7. TG curves of different specimens: (**a**) specimens of MK-UHPC at the age of 3 d (**b**) specimens of MK-UHPC at the age of 7 d; (**c**) specimens of MK-UHPC at the age of 14 d; (**d**) specimens of MK-UHPC at the age of 28 d; (**e**) specimens of MK-CA-UHPC at the age of 28 d.

Comparing Figures 7a–d and 7e, it can be seen that the mass loss rates of the CA-UHPC sample group with MK added were faster in the temperature ranges of 50~150 °C and 600~700 °C, which indicates that, on the one hand, the incorporation of volcanic stone coarse aggregate retains more free water and bound water, and on the other hand, it increases the crystallization of $CaCO_3$, and the decomposition temperature of $CaCO_3$ increased by 50 °C compared with that at other ages, indicating that the $CaCO_3$ crystals of the samples at 28 days of age were more stable as the crystal planes became more ordered. Figure 7e shows that the mass loss of GH4 in the temperature range of 600–700 °C was significantly greater than those of the other specimens, indicating that the amount of $CaCO_3$ crystals in GH4 was the greatest, and MK with a high dispersion ratio helped to generate more $CaCO_3$ crystals and was able to provide a higher strength. The results of the cubic compression tests described above also verified that MK with a high dispersion ratio had the most significant strength improvement for the CA-UHPC system.

Figure 7d shows that the decomposition temperature of ettringite is 70 °C, the weight loss at 70~150 °C was basically the weight loss of the bound water in the C-S-H gel and the ettringite and the weight loss ratios were 2.97%, 2.99%, 3.16% and 3.71%, respectively. At 28 days of age, the masses of C-S-H gel and ettringite in G4 were approximately 1.24 times greater than those in G1, and the cubic compressive strength of G4 was approximately 1.22 times greater than that of G1, which are basically equal ratios. The TGA results further verified the results of the cubic compressive strength test described above.

The decomposition temperatures of $Ca(OH)_2$ and $CaCO_3$ crystals with poor crystallization and $CaCO_3$ crystals with better crystallization are 430–550 °C, 550–750 °C and 750–950 °C, respectively. Figure 8 shows that there was a relatively obvious "concave section" in the temperature ranges of 50–150 °C and 600–700 °C in the UHPC and CA-UHPC sample groups with MK added, which is consistent with Figure 7, and there was a smaller "concave section" at approximately 430 °C. After analysis, it was found that the "concave section" in the temperature range of 50~150 °C was due to the acceleration of the weight loss caused by the evaporation and the loss of bound water in the C-S-H gel ($CaO{\cdot}2SiO_2{\cdot}3H_2O$) and ettringite ($3CaO{\cdot}Al_2O_3{\cdot}3CaSiO_4{\cdot}32H_2O$). The "concave section" in the temperature range of 600~700 °C was due to the acceleration of weight loss caused by the decomposition of some poorly crystallized $CaCO_3$ crystals with an increasing temperature to produce CO_2, which is consistent with the analysis of the TG curve. The "concave section" near 430 °C was caused by the conversion of $Ca(OH)_2$ into CaO and free water with increasing temperature and the loss of free water with high-temperature evaporation.

As shown in Figure 8a–d, it can be observed that G4 was at the lowest point in the "concave section" in the temperature range of 50~150 °C, indicating that the contents of C-S-H and ettringite in the G4 specimen at the age of 28 days were relatively greater than those in the other UHPC specimens with added MK. Figure 8e shows that the position of GH4 was significantly lower those that of the other specimens in the concave section in the temperature range of 600–700 °C, indicating that the $CaCO_3$ content in the GH4 specimen at the age of 28 days was significantly greater than those in the other CA-UHPC specimens with added MK and different dispersion ratios. High contents of C-S-H, ettringite and $CaCO_3$ led to the higher compressive strength values for the G4 and GH4 specimens, which was verified by the results of the cubic compressive test described above.

Comparing the TG curves of the MK-added UHPC sample groups at different ages, it can be seen that the "concave section" of the UHPC samples with a 4% MK dispersion ratio added at 3 days and 28 days of age was the most distinctive in the temperature range of 600–700 °C, indicating that the $CaCO_3$ contents of UHPC with a 4% MK dispersion ratio added were greater at 3 and 28 days of age, which is consistent with the results on the cubic compressive strength.

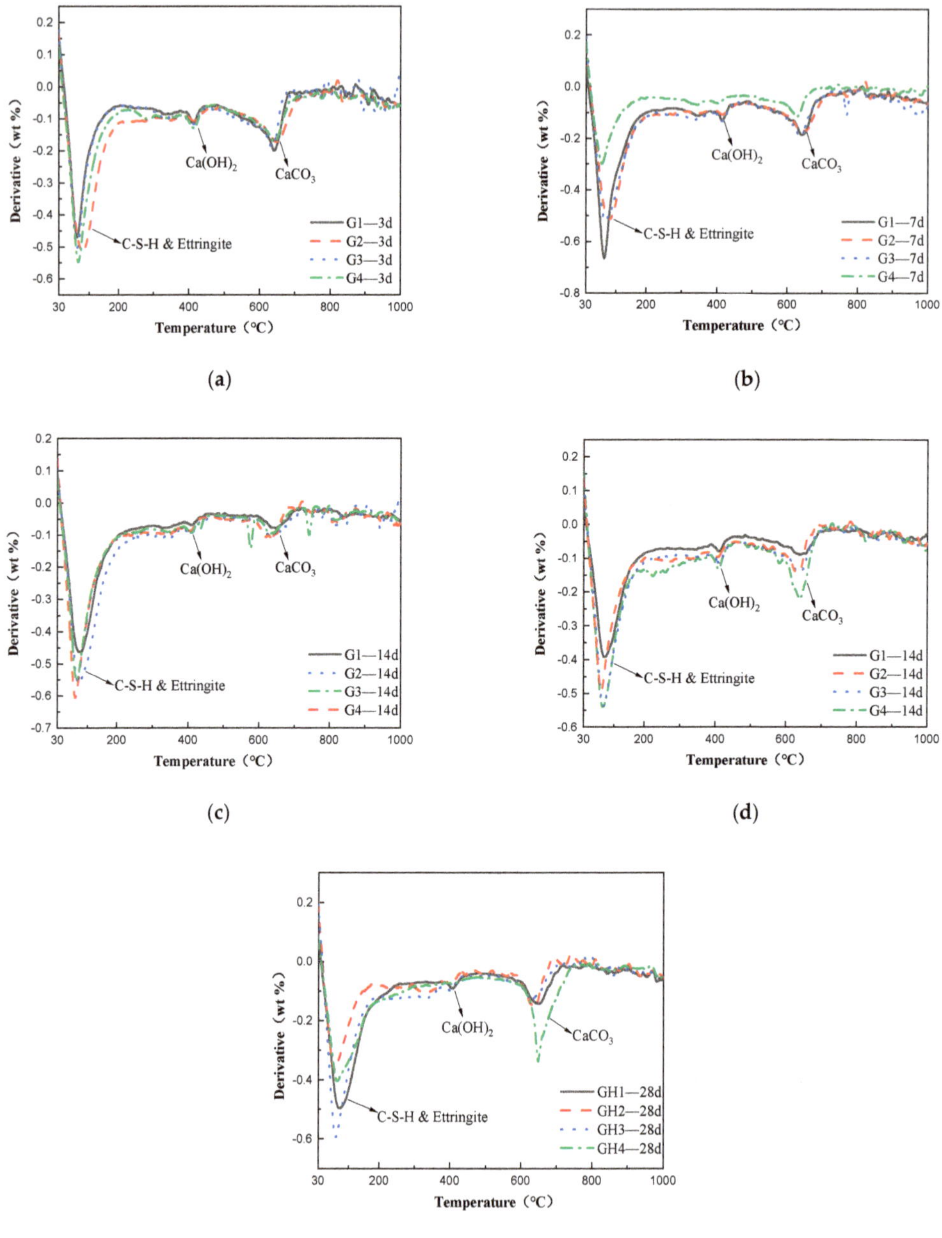

Figure 8. DTG curves of different specimens: (**a**) specimens of MK-UHPC at the age of 3 d; (**b**) specimens of MK-UHPC at the age of 7 d; (**c**) specimens of MK-UHPC at the age of 14 d; (**d**) specimens of MK-UHPC at the age of 28 d; (**e**) specimens of MK-CA-UHPC at the age of 28 d.

5. Conclusions

(1) Volcanic rocks as a new type of coarse aggregate for UHPC were used not only to maintain the compressive strength of UHPC but also to significantly reduce the

construction cost of UHPC projects, resulting in the development of a new type of high-strength engineering material. The conclusions of this research provide reliable data support for the UHPC theoretical framework.

(2) The volcanic stone pore structure and the mortar matrix form a mechanical "interlocking effect" structure, which can effectively improve the interfacial shear resistance of CA-UHPC. At the same time, the filling interface of volcanic stone pores was one of the main reasons for the decrease in the compressive strength of CA-UHPC, and the combined effect of the two was that the latter dominated.

(3) The crystallinity of C-S-H was greater for both UHPC and CA-UHPC with MK added at a dispersion ratio of 4%. MK at a content of 4% promoted the crystallization of C-S-H, which increased the strength of UHPC. On the one hand, the incorporation of volcanic stone coarse aggregate retained more free water and bound water, and on the other hand, it also increased the crystallization of $CaCO_3$. The combined effect of MK with a high dispersion ratio and volcanic stone promoted the crystallization of $CaCO_3$.

(4) When the MK dispersion ratio was less than 4%, the contents of $CaCO_3$ crystals, C-S-H gel and calcium vanadium increased with an increasing MK dispersion ratio. At the age of 28 days, an MK dispersion ratio of 4% improved the quality of the C-S-H gel and ettringite by 24.82% compared with an MK dispersion ratio of 1%. A high content of C-S-H, ettringite and $CaCO_3$ caused the cubic compressive strength of the GH4 specimen to be higher.

Author Contributions: Conceptualization, Y.Y.; methodology, Y.Y.; software, Z.M.; validation, Z.M.; formal analysis Z.M.; investigation Z.M.; resources, Y.Y.; data curation, Y.Y.; writing—original draft preparation Z.M.; writing—review and editing, Z.M. and Y.Y.; All authors have read and agreed to the published version of the manuscript.

Funding: This research was sponsored by the National Natural Science Foundation of China (Nos. 51978501, 51774163), and also funded by the Basic Science (Natural Science) Research Project of Higher Education Institutions in Jiangsu Province (No. 23KJA560008).

Data Availability Statement: The datasets in the current study are available from the corresponding author upon reasonable request.

Conflicts of Interest: The authors declare no potential conflicts of interest with respect to the research, authorship and publication of this article.

References

1. Zhang, L.F.; Wang, H.C. Study on the influence of metakaolin and silica fume on the strength and microstructure of cement concrete. *Transp. Technol.* **2017**, *31*, 133–137.
2. Wille, K.; Naaman, A.E.; Parra-Montesinos, G.J. Ultra-High Performance Concrete with Compressive Strength Exceeding 150 MPa. *ACI Mater. J.* **2011**, *108*, 189–203.
3. Chen, B.C.; Ji, T.; Huang, Q.W. Review of research on ultra high performance concrete. *J. Build. Sci. Eng.* **2014**, *3*, 11–24.
4. Pyo, S.; Kim, H.K.; Lee, B.Y. Effects of coarser fine aggregate on tensile properties of ultra high performance concrete. *Cem. Concr. Compos.* **2017**, *84*, 28–35. [CrossRef]
5. Meng, W.; Valipour, M. Optimazation and performance of cost-effective ultra-high performance concrete. *Mater. Struct.* **2017**, *50*, 29–41. [CrossRef]
6. Yu, R.; Spiesz, P.; Brouwers, H. Mix design and propeties assessment of ultra-high performance fibre reinforced concrete. *Cem. Concr. Res.* **2014**, *56*, 29–39. [CrossRef]
7. Xiao, J.F. Preparation and Performance Research of Ultra High Performance Concrete under Conventional Technology. Ph.D. Thesis, Hunan University, Changsha, China, 2013.
8. Wang, Y.; Shao, X.D.; Cao, J.H. Experimental study on basic performances of reinforced UHPC bridge deck with coarse aggregates. *J. Bridge Eng.* **2019**, *24*, 04019119. [CrossRef]
9. Shi, Z.C.; Su, Q.T.; Shao, C.Y.; Chen, L. Evaluation method of basic mechanical properties and bending toughness of coarse aggregate UHPC. *J. Civ. Eng.* **2020**, *53*, 86–97.
10. Orgass, M.; Dehn, F. Comparative investigations on ultra-high performance concrete with and without coarse aggregates. Proceedings of International Symposium on Ultra High Performance Concrete (UHPC), Kassel, Germany, 13–15 September 2004; Kassel University Press: Kassel, Germany, 2004; pp. 205–212.

11. Cheng, J.; Liu, J.P.; Liu, J.Z.; Zhang, Q.Q.; Zhang, L.H.; Lin, W.; Han, F.Y. Study on mechanical properties and mechanism analysis of ultra-high performance concrete containing coarse aggregate. *Mater. Guide* **2017**, *31*, 115–119+131.
12. Zhang, X.B.; Han, F.Y.; Liu, J.Z. Influence of metakaolin on rheology and fiber distribution of ultra high performance concrete. *J. Silic.* **2021**, *49*, 2375–2383.
13. Shen, C.Q.; Li, B.X. Influence of coarse aggregate on mechanical properties of ultra-high performance concrete and its fitting analysis. *J. Mater. Sci. Eng.* **2021**, *39*, 35–40.
14. Huang, Z.Y.; Li, S.G. Study on mechanical properties of ultra high performance concrete containing coarse aggregate. *J. Hunan Univ. (Nat. Sci. Ed.)* **2018**, *45*, 47–54.
15. Li, N.; Mo, L.; Unluer, C. Emerging CO_2 utilization technologies for construction materials: A review. *J. CO2 Util.* **2022**, *65*, 102237. [CrossRef]
16. Siddique, R. Properties of concrete made with volcanic ash. *Resour. Conserv. Recycl.* **2012**, *66*, 40–44. [CrossRef]
17. Deveshan; Pillay, L.; Oladimeji; Olalusi, B.; Moses; Kiliswa, W.; Awoyera, P.O.; Kolawole, J.T.; Babafemi, A.J. Engineering performance of metakaolin based concrete. *Clean. Eng. Technol.* **2022**, *6*, 100383.
18. Massazza, F. Pozzolana and Pozzolanic Cements. In *Lea's Chemistry of Cement and Concrete*, 4th ed.; Hewlett, P.C., Ed.; Elsevier: Amsterdam, The Netherlands, 1998; pp. 471–635.
19. Dilaria, S.; Secco, M.; Bonetto, J.; Ricci, G.; Artioli, G. Making Ancient Mortars Hydraulic. How to Parametrize Type and Crystallinity of Reaction Products in Different Recipes. In *Conservation and Restoration of Historic Mortars and Masonry Structures*; Bokan Bosiljkov, V., Padovnik, A., Turk, T., Eds.; Springer: Berlin/Heidelberg, Germany, 2023; p. 42. [CrossRef]
20. Biljana, I.; Vlastimir, R.; Miodrag, Z. Effects of mechanical and thermal activation on pozzolanic activity of kaolin containing mica. *Appl. Clay Sci.* **2016**, *19*, 135–147.
21. Dinakar, P.; Pradosh, K.S. Effect of Metakaolin Content on the Properties of High Strength Concrete. *Int. J. Concr. Struct. Mater.* **2013**, *7*, 135–146. [CrossRef]
22. Mo, Z.Y.; Bai, L.G.; Wang, D.G. Study on the influence of metakaolin on the performance of ultra-high performance concrete. *Nonmet. Ore* **2018**, *41*, 30–42.
23. Mo, Z.Y.; Liu, Y.L.; Wang, D.G. Research Progress on mechanical properties of metakaolin cement-based materials. *Silic. Bull.* **2018**, *37*, 911–917.
24. Mo, Z.Y. Strength Development and Rehydration Characteristics of UHPC Matrix Doped with Metakaolin. Ph.D. Thesis, Harbin Institute of Technology, Harbin, China, 2021.
25. Ahmed, T.; Gilles, E.; Soltane, L. Metakaolin in the formulation of UHPC. *Constr. Build. Mater.* **2008**, *23*, 1440–1454.
26. Fang, Y.H.; Zheng, B. Metakaolin and its application in high performance concrete. *J. Silic.* **2003**, *8*, 801–805.
27. Jiang, G.W.; Lv, G.Y. Study on the mechanical properties and durability of metakaolin on concrete. *ZheJiang Archit.* **2019**, *36*, 56–59+63.

buildings

MDPI

Article

Tensile Mechanical and Stress-Strain Behavior of Recycling Polypropylene Fiber Recycled Coarse Aggregate Concrete

Jianchao Wang [1], Jiahe Liang [2], Yucheng Li [2,*] and Wei Hou [3]

[1] School of Science, Shenyang Jianzhu University, Shenyang 110168, China; wangjianchaook@sjzu.edu.cn
[2] School of Civil Engineering, Shenyang Jianzhu University, Shenyang 110168, China; liangjiahe_2021@163.com
[3] School of Marxism, Shenyang Jianzhu University, Shenyang 110168, China; houwei@sjzu.edu.cn
* Correspondence: yucheng_li1@163.com

Abstract: To effectively recycle waste petroleum products and construction waste, recycling polypropylene fiber (RPF) and recycled aggregate can be mixed into concrete to make RPF recycled coarse aggregate (RCA) concrete. In this study, the RPF recycled from a polypropylene (PP) packaging belt was used as the test material and manually cut into the shape required for the experiment. The effects of RCA and RPF on the tensile mechanical behavior of concrete are researched. The failure modes and constitutive relationship of the specimens under axial tension and splitting tension are further investigated. The results show that the axial tensile strength of RPF RCA concrete first increased and then decreased with the increase in fiber volume content, and was the largest when the fiber volume content was 1.5%, and its strength increased by 21.14% compared with that of recycled concrete. Its lifting rate relative to recycled concrete is between 13.14–21.41%. The change trend of axial tensile strength with the substitution rate of RCA is that it decreases with the increase in substitution rate, and the substitution rate decreases by 9.64% when the substitution rate is 100% compared with 0%.The peak strain first increased and then decreased with the increase in fiber volume content, and the maximum fiber volume content was 1.5%, which increased by 28.19% compared with that of recycled concrete. The peak strain first increased and then decreased with the increase in fiber length-diameter ratio, and the maximum length-diameter ratio was 47.85, which increased by 18.22% compared with that of recycled concrete. The peak strain increased with the increase in the replacement rate of RCA, and the peak strain at 30%, 60% and 100% was 96.22%, 102.45% and 118.09% when the replacement rate was 0%, respectively.

Keywords: recycled coarse aggregate concrete; polypropylene fiber; tensile mechanical properties; stress-strain constitutive model

1. Introduction

With the acceleration of global urban renewal, nearly 10 billion tons of building rubbish will be produced every year in the world due to the demolition and renovation of old buildings [1]. After classifying, screening, crushing and grinding construction waste, it can be prepared into RCA (with a diameter of 2–3 mm) and used to partially replace natural fine aggregate to prepare RCA Concrete [2,3]. However, during the preparation of RCA, there will be several micro-cracks and initial defects, which will increase the porosity and crushing index of RCA [4,5], thus causing the basic dynamical performances of RCA concrete to generally decline with the comparison of natural coarse aggregate concrete [6,7]. This is also an important reason why it is difficult to popularize recycled concrete for application in engineering. A study by Bárbara Oliveira Paiva and José Maria Franco de Carvalho [8] evaluates the properties of fresh and hardened lightweight concretes, aiming for an optimal mixture design exclusively with lightweight aggregates and mostly plastic waste aggregates. The study shows that, with an optimum dosage, reusing plastic waste in concrete is a viable alternative, contributing to environmental sustainability.

Fiber Recycled Concrete (FRC) is made by mixing discontinuous small randomly distributed fibers in RCA concrete. When fiber is added, the compressive tensile strength [9], creep property, flexural strength [10], tensile strength, toughness [11] and other mechanical properties of fiber recycled concrete will be improved accordingly. It can be seen that adding a certain amount of fiber to RCA concrete is a significant method to enhance the mechanical performance of RCA concrete.

Scholars around the world have drawn important conclusions on the mechanical performances of RCA concrete. Bagherzadeh [12] and Ahmed [13] indicated that when the adjunction of PP fiber with a diameter of 22 μm and length of 19 mm is 0.1–0.3% in RCA concrete, the microscopic structure of RCA concrete can be enhanced and the formation and growth of concrete micro-cracks can be inhibited. In addition, when the RCA concrete is defective, the stress in the concrete can be transferred to the PP fiber due to the bridge connection efficiency of PP fiber, thus enhancing the tensile performance of RCA concrete [14–17]. However, the tensile strength of RCA concrete will be adversely affected by the excessive content of waste PP fiber [16]. Das C S et al. believe that this is due to the high stress generated at the end of the fiber and shorter fibers in the cleavage process, resulting in a decreased RCA concrete tensile strength [17–20]. In addition, the optimal fiber addition rate of RCA concrete will also change with the replacement rate of RCA. Nam J [18] found that adding polyvinyl alcohol and nylon fiber into RCA concrete is helpful to improve the interface bonding ability between cement matrix and RCA, thus improving the dynamical performances of RCA concrete. Thwe Thwe Win [21] found that introducing FM fibers, particularly an FM with a 5 mm diameter and 10 mm length, in the mortar increased both the tensile and flexural strengths. Among the various combinations of FMs studied, a mixture containing 0.15% FMs exhibited the best performance. The findings of this research reveal that FMs can be reused as fibers to enhance the tensile and flexural strengths of cement mortar.

PP packing belts are one of the most widely used packaging strapping materials, but the recycling rate is low and they are usually disposed of in landfills, which causes considerable environmental pollution due to the difficult degradation characteristics. With people paying more and more attention to the environment, new thinking has been carried out on the reuse value of waste, including waste in the form of fiber as a reinforcing material for concrete. This is considered a good waste treatment method to save resources and protect the environment, and has broad development prospects. PP material has the characteristics of non-water absorption [22,23], light weight [24,25] and good acid and alkali resistance [26,27], which makes it possible for it to be used as a concrete reinforcement material. In this regard, a new type of fiber recycled concrete was proposed: recycled PP packing belts were prepared into fibers, and RPF recycled concrete was prepared according to the construction process of fiber reinforced concrete, so as to improve the strength and toughness of recycled concrete. At present, there are few articles that analyze the influence of various parameters of RPF on the mechanical properties of concrete in detail, so this article has research significance. Based on the fact that RPF can enhance the tensile strength, crack resistance and toughness of concrete, this article aims to enhance the tensile performance of concrete by collecting waste materials, such as discarded packing belts in daily life that contain RPF, in order to achieve efficient waste utilization, environmental protection and sustainable development. The RPF used in the experiment was from the non-standard hot melt machine with a transparent packaging belt, with a density of 0.748 g/cm^3, which is manually cut into the desired shape. Using the RPF recycled by the PP packing belt as the test material, several groups of RPF RCA concrete were designed and fabricated. The length-diameter ratio, volume admixture and RCA substitution rate of RPF were taken as variables. The impact of RPF on the tensile performances of RCA concrete was investigated in depth, which provided a theoretical basis for the recycling of industrial recycled PP packing belts as a strengthening and toughening material of recycled concrete.

2. Experimental Scheme

2.1. Experimental Material

Cement: grade P.O 42.5. Table 1 shows the mineral and chemical component of P.O 42.5 cement.

Table 1. Mineral and chemical component of P.O 42.5 cement.

Mineral Component of P.O 42.5 (%)					
C_3S	C_2S	C_3A	C_4AF	$CaSO_4$	Other
58.72	18.54	6.91	9.03	3.16	3.64
Chemical component of P.O 42.5 (%)					
SiO_2	Al_2O_3	Fe_2O_3	CaO	MgO	SO_3
21.76	5.15	2.93	66.59	1.21	2.36

Aggregate: natural gravel and RCA are mixed in a certain proportion as coarse aggregate. The coarse aggregate has a particle size of 5–20 mm. Fine aggregate adopts machine-made sand with fineness modulus of 2.6 and particle size range of 0–4.75 mm. The RCA comes from the abandoned test wall, and its compressive strength is 27.6 MPa, as measured by the rebound meter. After crushing, screening, cleaning and preparation, RCA with a similar size of three axes is selected as far as possible, and needle-shaped aggregate is avoided. Table 2 lists the performance indicators of RCA.

Table 2. Mechanical performance of RCA.

Material	Size (mm)	Performance Density ($kg{\cdot}m^{-3}$)	Water Absorption Rate (%)	Crush Index (%)	Stacking Density ($kg{\cdot}m^{-3}$)	Mud Content (%)
RCA	5–20	2353	2.68–3.05	12.7	1457	3.55

Fiber: discarded PP fibers with a density of 0.748 g/cm^3 are used, which are derived from discarded packing belts in daily life. They are manually cut into the required shape. The mechanical performance indicators can be seen in Table 3.

Table 3. Performance of PP belt.

Material	Peak Load (N)	Peak Strain	Peak Stress (MPa)	Elasticity Modulus (GPa)
RPF	881	0.066	164.59	2.49

RPF size: the size of RPF is measured by digital vernier caliper and steel tape measure with an accuracy of 0.01 mm. The average width is 1.72 mm, the thickness is between 0.45–0.55 mm, the average value is 0.5 mm, the equivalent diameter of the cross-section is 1.046 mm, and the average length of each group of fibers is 30.02 mm, 39.77 mm, 50.07 mm and 58.94 mm, respectively. The length-diameter ratios are 28.69, 38.01 and 47.

2.2. Proportioning of RCA Concrete

The proportioning of RCA concrete was designed according to the Chinese standard JGJ/T443 [28] and 55-2011 [29]. The incorporation of RPF will produce a large number of fiber mortar clumps during mixing, and the fluidity will decrease, resulting in uneven distribution of concrete fiber and aggregate [30–32]. Therefore, in the process of material configuration, cement, sand, RCA and RPF are dry mixed, and water is added in two batches for secondary mixing after mixing evenly.

As there are many initial imperfections in the RCA, it will absorb water, resulting in a decrease in hydration water content and the water-binder ratio of RCA concrete [33].

To reduce the RCA water absorption to prevent RPF RCA concrete water cement ratio changes [34–36], the mixing water of recycled concrete is hydration water and additional water; the additional water is determined by the water absorption rate. The number and mix ratio of all specimens in this project can be seen in Table 4.

Table 4. Proportion of group.

Group	Proportioning of RCA Concrete (kg/m^3)							
	Cement	Sand	Water	RCA	Crushed Stones	RPF	Additional Water	Slump (mm)
RC-100-0	590	704.48	295	810.53	0	0	23.26	48
FRC-100-5-50	590	704.48	295	810.53	0	3.74	23.26	47
FRC-100-10-50	590	704.48	295	810.53	0	7.48	23.26	45
FRC-100-15-50	590	704.48	295	810.53	0	11.22	23.26	44
FRC-100-20-50	590	704.48	295	810.53	0	14.96	23.26	43
FRC-100-15-30	590	704.48	295	810.53	0	11.22	23.26	44
FRC-100-15-40	590	704.48	295	810.53	0	11.22	23.26	44
FRC-100-15-60	590	704.48	295	810.53	0	11.22	23.26	44
FRC-60-15-50	590	704.48	295	486.32	324.21	11.22	13.96	46
FRC-30-15-50	590	704.48	295	243.16	567.37	11.22	6.98	47
FNC-0-15-50	590	704.48	295	0	810.53	11.22	0	48

RC: Recycled Concrete, FRC: Fiber Recycled Concrete, FNC: Fiber Reinforced Concrete.

2.3. Specimen Preparation

A total of 11 experimental groups are shown in Table 4. There are six specimens in each group and each group has three test blocks for cleavage tensile strength, stress-strain curve and axial tensile strength test. Specimen size of 150 × 150 × 150 mm is for cleavage tensile strength specimen and 150 × 100 × 460 mm is for axial tensile strength of the fusiform specimen. In addition, the tensile elastic modulus, axial tensile peak and ultimate strain of RPF RCA concrete can be measured synchronously in the axial tensile $\sigma - \varepsilon$ test.

To ensure the same environmental temperature and other factors during the production of specimens, the same batch of concrete specimens was completed at the same time in accordance with standard curing requirements, at constant temperature and humidity for 28 days.

2.4. Test Methods

2.4.1. Cleavage Tensile Strength Test

The experimental methods refer to GB/T50080 [37], GB/T50081-2019 [38], BS EN 12350 [39] and BS EN 12390 [40–43]. After curing the test block for 28 d, check that the dimension deviation of the test block does not exceed 1 mm. The test block is installed on the cleavage tensile test fixture. The pad, pad strip and specimen are placed on the loading platform of 60 t universal testing machine. The test apparatus and data collection and analysis system are arranged as can be seen in Figure 1. First, start the loading machine at the loading rate of 0.2 mm/min. When the testing machine pressure increase is slowed down, keep the loading machine at a constant rate until the specimen is thoroughly destroyed.

Figure 1. The device used for loading.

2.4.2. Axial Tensile Strength Test

After the test block reaches the curing stage, install and place the test block and the test device on the 60 t universal test machine and set the test parameters. Start the testing machine and load it to 0.5 MPa. Check whether the strain difference on the 4 screws is less than 10%. If so, adjust the strain value of each screw through the knob nut until the strain difference between the 4 screws is less than 10%. Before the failure of the specimen, the loading has been carried out at the displacement rate of 0.04 mm/min. The test steel frame is composed of 48.9-level M20 screws and 2 25 mm thick high-strength steel plates, as shown in Figure 2.

Figure 2. Test setups.

3. Results and Discussion

3.1. Cleavage Tensile Strength

Through Figure 3, it is shown that the cleavage tensile failure mode of RPF RCA concrete is different from that of non-fiber recycled concrete. Its failure cracks are not completely perpendicular to the bearing surface and are more irregular, and the concrete on the failure surface is less spalling. By checking the cross-section, it can be found that the fiber failure in concrete is both fiber tensile fracture and fiber bonding failure.

(a) (b)

Figure 3. Failure mode. (**a**) Failure mode of split tensile; (**b**) Section of splitting tensile.

The cleavage tensile experimental results of specimens in each group are listed in Table 5. A conclusion can be drawn that the cleavage tensile performances of RCA concrete have made a big difference due to the incorporation of RPF.

Table 5. Each performance index of each test block.

Group	Cleavage Tensile Strength	Axial Tensile			
	(MPa)	Peak Strength (MPa)	Initial Crack Strength (MPa)	Initial Crack Strain (με)	Peak Strain (με)
RC-100-0	2.46	2.14	1.14	87.65	152.98
FRC-100-5-50	2.55	2.15	1.17	91.15	158.97
FRC-100-10-50	2.69	2.34	1.22	96.10	178.93
FRC-100-15-50	2.98	2.53	1.30	107.65	196.10
FRC-100-20-50	2.61	2.26	1.17	94.88	166.52
FRC-100-15-30	2.87	2.49	1.34	109.02	178.85
FRC-100-15-40	2.91	2.49	1.31	108.87	178.20
FRC-100-15-60	2.79	2.43	1.25	101.11	182.22
FRC-60-15-50	2.99	2.56	1.35	108.65	170.12
FRC-30-15-50	3.11	2.62	1.42	109.52	166.78
FNC-0-15-50	3.23	2.80	1.50	111.86	166.05

The Influence of RCA Replacement Rate

Figure 4c shows that, with the rise in the replacement ratio of RCA, the cleavage tensile strength of RPF RCA concrete gradually goes down. This is because there are several initial cracks in the RCA concrete. Due to the rise in RCA replacement rate in concrete, the initial cracks in concrete also increase, thus producing a negative effect on RPF RCA concrete strength. From the standard deviation distribution in the three figures, it can be seen that the standard deviation value is the largest under the aspect ratio condition, and the standard deviation value is close to that under the fiber volume parameter and aggregate substitution rate condition.

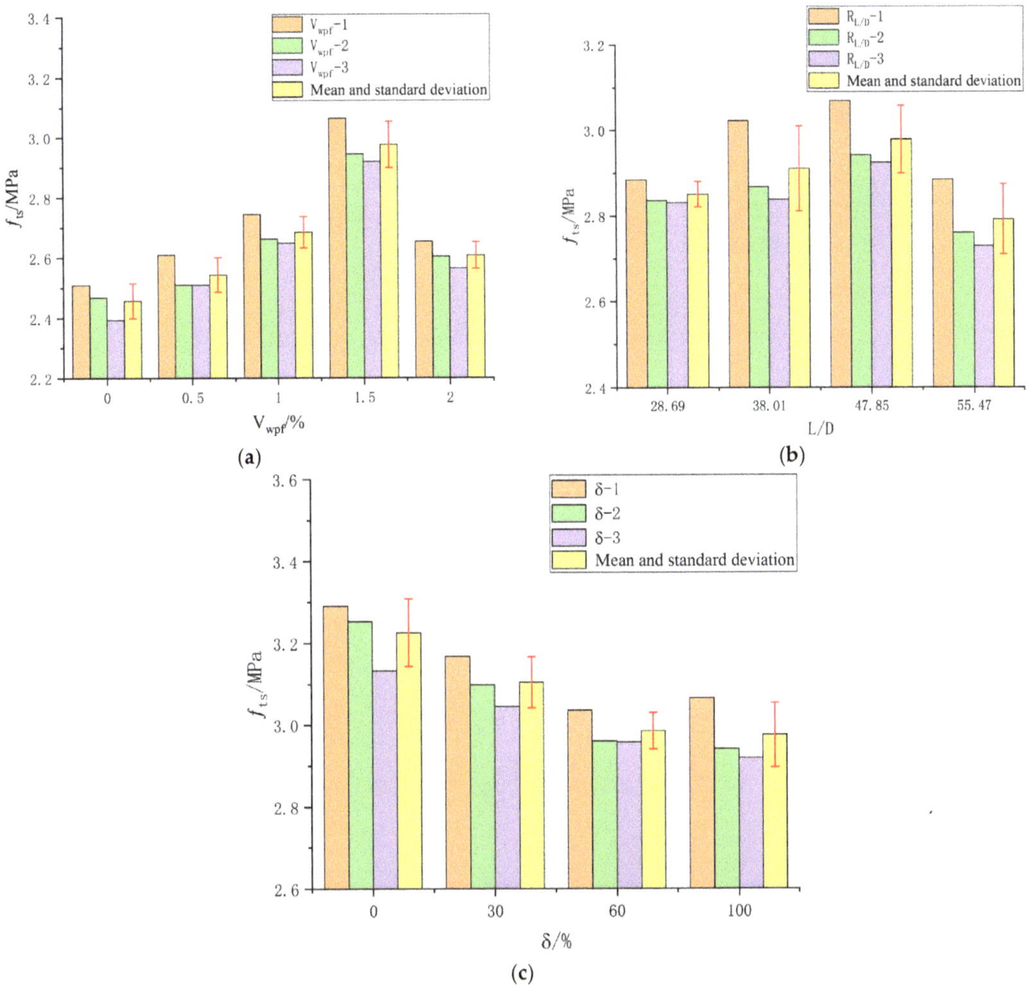

Figure 4. Influence curves of various parameters on cleavage tensile strength. (**a**) Influence curve of fiber volume content group; (**b**) Influence curve of fiber length-diameter ratio group; (**c**) Influence curve of replacement rate of an RCA group.

3.2. Axial Tensile Test Strength

Figure 5 shows the tensile failure form of recycled concrete tends to be irregular following the addition of RPF. The section is basically the minimum section, but the development direction is irregular, and there are several secondary cracks near the main section; the failure shows certain plastic characteristics.

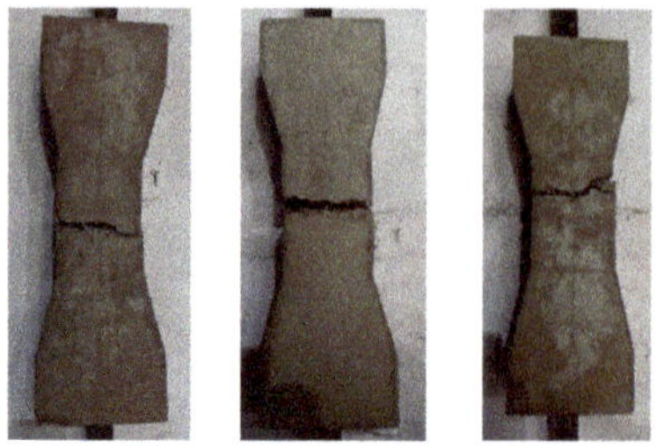

Figure 5. Failure mode of axial tensile test.

As indicated in Table 3, the incorporation of RPF improves the axial tensile strength of RCA concrete.

Figure 6 shows the incidence of various impact parameters on the initial-crack (I-C) strength and axial tensile strength of RPF RCA concrete. Figure 6a indicates the impact of the fiber volume addition ratio on the axial tensile strength and I-C strength. It can be found that the axial tensile strength and I-C strength increase first with the rise in the fiber volume addition ratio. When the fiber volume addition ratio reaches 1.5%, the axial tensile strength and I-C strength of concrete reach the maximum. Subsequently, when the fiber volume addition ratio exceeds 1.5%, the axial tensile strength and initial crack strength decrease with the rise in the RPF volume addition ratio. The influence principle of fiber volume addition ratio on axial tensile strength is the same as that of cleavage tensile.

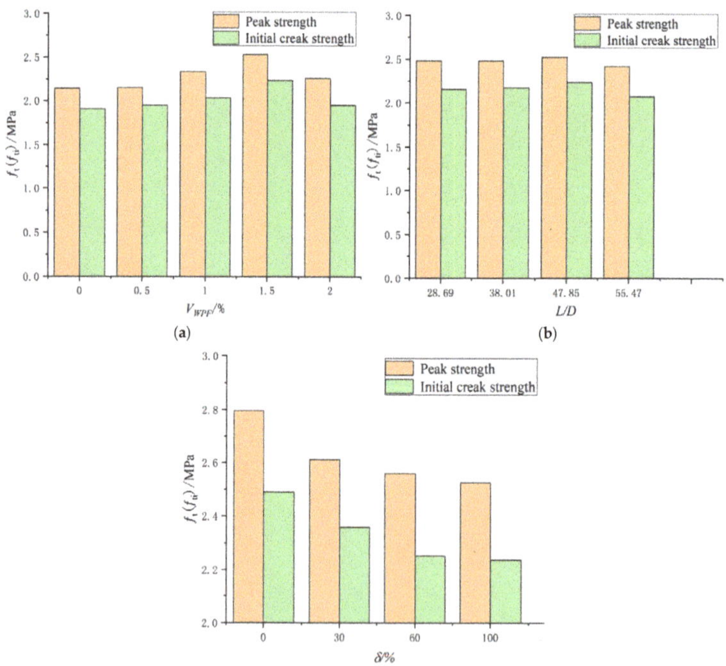

Figure 6. Initial crack strength and peak strength of RPF RCA concrete. (**a**) Influence curve of fiber volume content; (**b**) Influence curve of fiber length-diameter ratio; (**c**) Influence curve of replacement rate of RCA.

It can be found from Figure 6b, the axial tensile strength and initial crack strength of RPF RCA concrete increase at first and then drop with the rise in the fiber length-diameter ratio. When the aspect ratio of RPF reaches 47.85, the axial tensile strength and initial crack strength reach the peak.

It can be shown through Figure 6c that, with the increase in RCA replacement ratio, both the axial tensile strength and initial cracking strength are decreased.

3.3. Axial Tensile Initial Crack Strain and Peak Strain

The impact of RPF volume additional ratio on the initial crack (I-C) strain and peak strain of RPF concrete can be seen in Figure 7a. The conclusion can be drawn that both I-C strain and peak tensile strain increased first and decreased later, due to the increase in the RPF volume additional ratio.

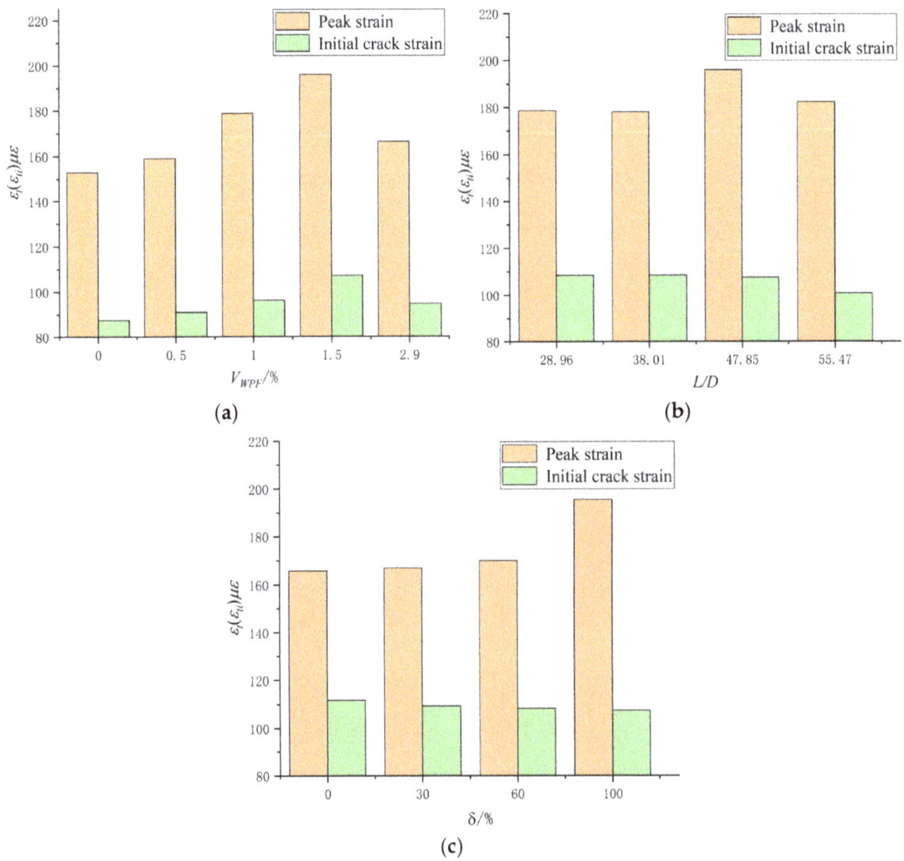

Figure 7. Peak strain and initial cracking strain of RPF RCA concrete. (**a**) Influence curve of fiber volume content; (**b**) Influence curve of fiber length-diameter ratio; (**c**) Deformation influence curve of replacement rate of RCA.

Figure 7b analyzes the influence of the RPF length-diameter (L-D) ratio on the initial crack strain and peak strain of RPF RCA concrete. Through Figure 7b, the fiber aspect ratio has a slight effect on the I-C strain, and the overall initial crack strain decreases with the increase in the L-D ratio. Due to the increase in the L-D ratio, the peak strain of RCA concrete increases at first and then decreases last.

Figure 7c shows the influence of the replacement ratio of RCA on the deformation. It can be found that the incipient strain goes down with the increase in the replacement rate, while the peak strain raises with the rise in the replacement rate of RCA.

3.4. Analysis of the Stress-Strain Relationship

As shown in Figure 8, the impacts of RCA and RPF on the $\sigma - \varepsilon$ behavior of concrete are significant. The $\sigma - \varepsilon$ curve of RPF RCA concrete is similar to that of FNC. Compared with FNC, RCA reduces the peak strength of RCA concrete. It is precisely because the addition of RPF can not only enhance the axial tensile strength of RCA concrete, but also increase the peak tensile strain of RCA concrete. The influence of different factors on the axial tensile $\sigma - \varepsilon$ behavior of FRC can be seen in Figure 9.

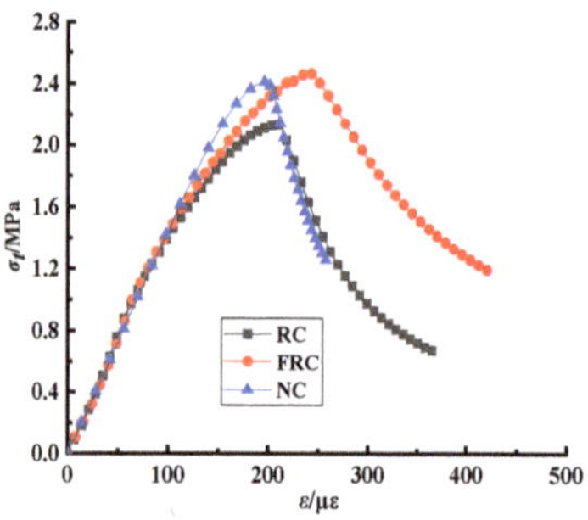

Figure 8. The axial tensile stress-strain behavior of concrete.

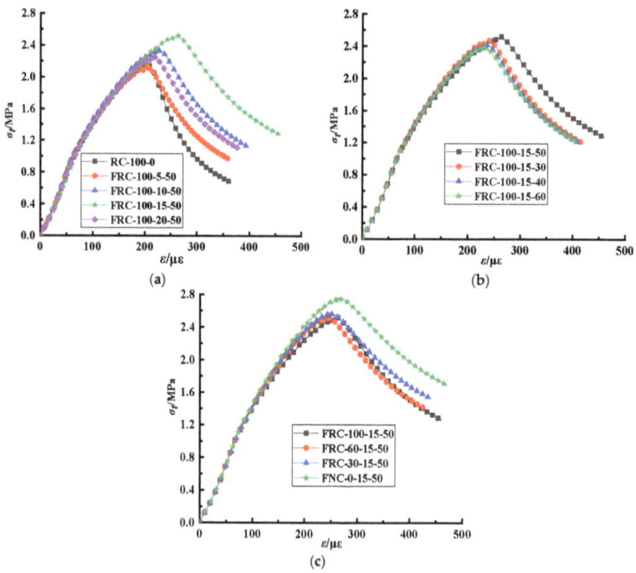

Figure 9. The axial tensile stress-strain behavior of FRC. (**a**) Influence of fiber volume; (**b**) Influence of fiber length-diameter ratio; (**c**) Influence of replacement rate of RCA.

3.5. Stress-Strain Constitutive Models

In order to deeply and comprehensively understand the tensile properties of RCA concrete materials, it is necessary to study the tensile stress-strain relationship.

3.5.1. Axial Tensile Stress-Strain Constitutive Model of NC

Recently, researchers have proposed many NC axial tensile stress-strain constitutive models. Among them, the segmented model was proposed by Guo Zhenhai [44,45] in 1993. Due to its simple form and clear meaning, it has been broadly used in the world. The mathematical equation of the $\sigma-\varepsilon$ model is as follows in Equation (1):

$$\sigma/f_t=\begin{cases}1.2(\varepsilon/\varepsilon_p)-0.2(\varepsilon/\varepsilon_p)^6 & (\varepsilon/\varepsilon_p)\leq 1\\ \frac{(\varepsilon/\varepsilon_p)}{\alpha_t[(\varepsilon/\varepsilon_p)-1]^{1.7}+(\varepsilon/\varepsilon_p)} & (\varepsilon/\varepsilon_p)>1\end{cases} \quad (1)$$

$$\alpha_t=0.312f_t^2 \quad (2)$$

The CDP model is a widely used concrete constitutive model in the world [46]. It has become an important reference for concrete structure design [47]. The mathematical expression of the constitutive model is as follows (3):

$$\sigma=(1-d_t)E_c\varepsilon \quad (3)$$

$$d_t=\begin{cases}1-\rho_t[1.2-0.2x^5] & x\leq 1\\ 1-\frac{\rho_t}{\alpha_t(x-1)^{1.7}+x} & x>1\end{cases} \quad (4)$$

$$x=\varepsilon/\varepsilon_{t,r} \quad (5)$$

$$\rho_t=f_{t,r}/E_c\varepsilon_{t,r} \quad (6)$$

3.5.2. Axial Tensile Stress-Strain Constitutive Model of FRC

Equation (1) is simple in form and has good accuracy in fitting the $\sigma-\varepsilon$ curve of axial tensile performances. In this paper, this equation is used to fit the full curve of the axial tensile $\sigma-\varepsilon$ of RPF RCA concrete.

$$\sigma/f_t=\begin{cases}a(\varepsilon/\varepsilon_p)+(1.5-1.25a)(\varepsilon/\varepsilon_p)^2+(0.25a-0.5)(\varepsilon/\varepsilon_p)^6 & (\varepsilon/\varepsilon_p)\leq 1\\ \frac{(\varepsilon/\varepsilon_p)}{\alpha_t[(\varepsilon/\varepsilon_p)-1]^{1.45}+(\varepsilon/\varepsilon_p)} & (\varepsilon/\varepsilon_p)>1\end{cases} \quad (7)$$

Equation (7) is used to fit the measured $\sigma-\varepsilon$ curve, and the fitted curve is shown in Figure 10. The fitted parameter values are shown in Table 6.

Table 6. Tensile constitutive parameters.

Group	Parameters	
	a	α_t
RC-100-0	1.742	5.883
FRC-100-5-50	1.788	3.214
FRC-100-10-50	1.780	2.947
FRC-100-15-50	1.733	2.645
FRC-100-20-50	1.670	2.844
FRC-100-15-30	1.681	2.876
FRC-100-15-40	1.646	2.731
FRC-100-15-60	1.693	2.597
FRC-60-15-50	1.691	2.140
FRC-30-15-50	1.760	1.844
FNC-0-15-50	1.714	1.682

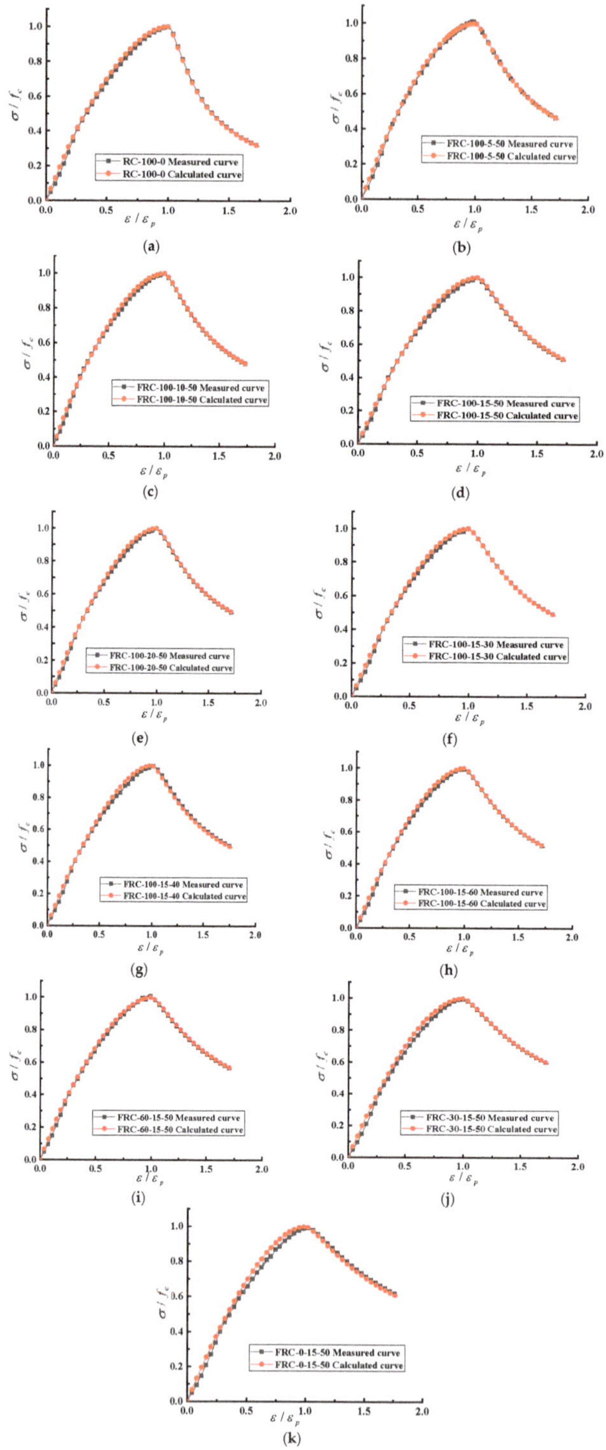

Figure 10. Tensile stress-strain fitting curve. (**a**) RC-100-0 tensile stress-strain fitting curve; (**b**) FRC-100-5-50

tensile stress-strain fitting curve; (**c**) FRC-100-10-50 tensile stress-strain fitting curve; (**d**) FRC-100-15-50 tensile stress-strain fitting curve; (**e**) FRC-100-20-50 tensile stress-strain fitting curve; (**f**) FRC-100-15-30 tensile stress-strain fitting curve; (**g**) FRC-100-15-40 tensile stress-strain fitting curve; (**h**) FRC-100-15-60 tensile stress-strain fitting curve; (**i**) FRC-60-15-50 tensile stress-strain fitting curve; (**j**) FRC-30-15-50 tensile stress-strain fitting curve; (**k**) FNC-0-15-50 tensile stress-strain fitting curve.

3.6. Tensile Toughness

The equivalent tensile toughness index of RPF recycled concrete can be calculated by the formula (8), as shown in Table 7 and Figure 11.

$$W_{tu} = f_{te} \times \frac{\delta}{l} = \frac{\Omega_u}{A\delta_u} \times \frac{\delta_u}{l} = \frac{\Omega_u}{Al} \tag{8}$$

Table 7. Equivalent tensile toughness index.

Group	$W_{tu}^{1.0}$ (Pa)	Gain Ratio
RC-100-0	207.6905	100.00%
FRC-100-5-50	217.9078	104.92%
FRC-100-10-50	266.6963	128.41%
FRC-100-15-50	309.6578	149.10%
FRC-100-20-50	234.8848	113.09%
FRC-100-15-30	234.8848	113.09%
FRC-100-15-40	274.8961	132.36%
FRC-100-15-60	276.7001	133.23%
FRC-60-15-50	272.0142	130.97%
FRC-30-15-50	276.1605	132.97%
FNC-0-15-50	291.8175	140.51%

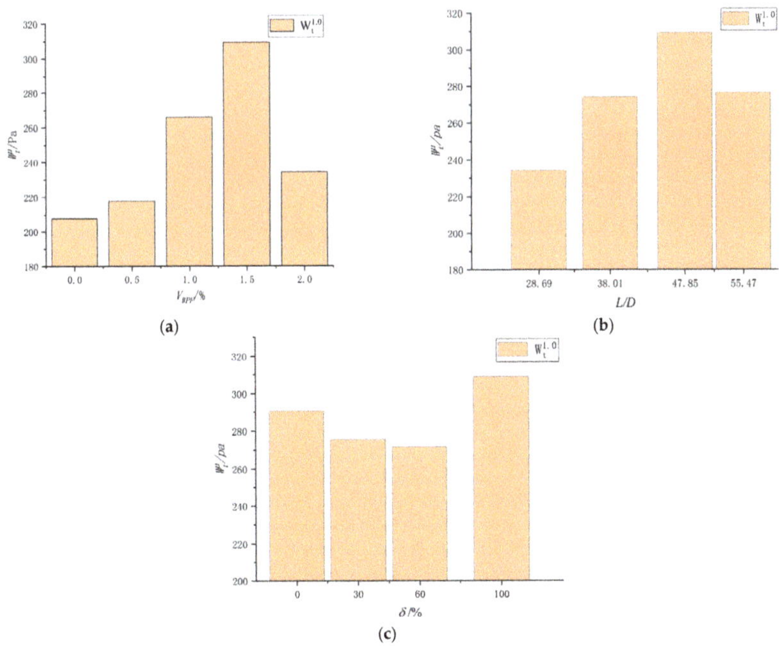

Figure 11. Equivalent tensile toughness index curve. (**a**) Influence of fiber volume; (**b**) Influence of fiber length-diameter ratio; (**c**) Influence of replacement rate of RCA.

4. Conclusions

In this chapter, the axial tensile strength, tensile stress-strain curve and peak tensile strain under different fiber volume content, fiber length-diameter ratio and RCA substitution rate are obtained, using 11 groups of RPF recycled concrete axial tensile test blocks, and the following conclusions are obtained:

(1) The axial tensile strength first increased and then decreased with the increase in fiber volume content, and was largest when the fiber volume content was 1.5%. Its strength increased by 18.22% compared with recycled concrete, and the change trend of axial tensile strength with fiber length-diameter ratio was roughly increased first and then decreased with the increase in length-diameter ratio, but the change range was not large, and its lifting rate relative to recycled concrete was between 5.61–18.22%. The change trend of axial tensile strength with the substitution rate of RCA is that it decreases with the increase in substitution rate, and the substitution rate decreases by 9.64% when the substitution rate is 100% compared with 0%.

(2) The peak strain first increased and then decreased with the increase in fiber volume content, and the fiber volume content was largest when the fiber volume content was 1.5%, which was 28.19% higher than that of recycled concrete. The peak strain increased first and then decreased with the increase in the fiber length-diameter ratio, and the maximum was 47.85. The peak strain increased with the increase in the substitution rate of RCA, and the peak strain at 30%, 60% and 100% was 100.44%, 102.45% and 118.09% at 0%, respectively.

(3) The equivalent tensile toughness index $W_{tu}^{1.0}$ of RPF recycled concrete first increased and then decreased with the increase in fiber volume content, and the volume content was the best when the volume content was 1.5%. With the increase in the fiber aspect ratio, the tensile toughness index $W_{tu}^{1.0}$ first decreased and then increased, and the substitution of natural aggregate by RCA may enhance the energy dissipation capacity of RPF recycled concrete.

(4) The constitutive model of the tensile rising section of the RPF recycled concrete is proposed, which is consistent with the actual rising section of the stress-strain curve.

Author Contributions: Conceptualization, J.L.; Methodology, J.L. and Y.L.; Software, J.L.; Validation, J.L.; Formal analysis, J.L., Y.L. and W.H.; Investigation, Y.L.; Resources, Y.L. and W.H.; Data curation, Y.L.; Writing – original draft, J.W.; Writing – review & editing, J.W.; Visualization, J.W.; Supervision, J.W.; Project administration, J.W. and W.H.; Funding acquisition, W.H. All authors have read and agreed to the published version of the manuscript.

Funding: This research received no external funding.

Data Availability Statement: The data presented in this study are available in the article.

Conflicts of Interest: The authors declare no conflict of interest.Notation

Notation

ε	Concrete tensile strain
σ	Concrete tensile stress
f_t	Peak tensile stress in concrete
ε_p	Peak tensile strain of concrete
d_t	evolution parameter of concrete uniaxial tensile damage
α_t	Parameter values for the falling section of the uniaxially tensile stressed concrete strain curve
$\varepsilon_{t,r}$	Peak tensile strains in concrete corresponding to representative values of uniaxial tensile strength $f_{t,r}$
W_{tu}	Equivalent tensile toughness index of concrete (Pa)
l	Vertical deformation scale of axial tensile test block (mm)
f_{te}	Equivalent tensile strength of concrete (MPa)
Ω_u	The area under the load-deformation curve when the vertical deformation is δ_u(N·mm)
δ_u	The longitudinal deformation when the load is reduced to a multiple of the ultimate load u (mm)
A	Pressure area of axial compression test block (mm^2)

References

1. Mejia-Ballesteros, J.E.; Rodier, L.; Filomeno, R.; Savastano, H., Jr.; Fiorelli, J.; Rojas, M.F. Influence of the fiber treatment and matrix modification on the durability of eucalyptus fiber reinforced composites. *Cem. Concr. Compos.* **2021**, *124*, 104280. [CrossRef]
2. Ferriz-Papi, J.A.; Weekes, E.; Whitehead, N.; Lee, A. A cost-effective recycled aggregates classification procedure for construction and demolition waste evaluation. *Constr. Build. Mater.* **2022**, *324*, 126642. [CrossRef]
3. Frazão, C.; Díaz, B.; Barros, J.; Bogas, J.A.; Toptan, F. An experimental study on the corrosion susceptibility of Recycled Steel Fiber Reinforced Concrete. *Cem. Concr. Compos.* **2019**, *96*, 138–153. [CrossRef]
4. Tran, D.L.; Mouret, M.; Cassagnabère, F.; Phung, Q.T. Effects of intrinsic granular porosity and mineral admixtures on durability and transport properties of recycled aggregate concretes. *Mater. Today Commun.* **2022**, *33*, 104709. [CrossRef]
5. Wu, H.; Wang, C.; Ma, Z. Drying shrinkage, mechanical and transport properties of sustainable mortar with both recycled aggregate and powder from concrete waste. *J. Build. Eng.* **2022**, *49*, 104048. [CrossRef]
6. Yang, Z.C.; Han, L.H.; Hou, C. Performance of recycled aggregate concrete-filled steel tubular columns under combined compression and shear load. *Eng. Struct.* **2022**, *253*, 113771. [CrossRef]
7. El-Hawary, M.; Al-Sulily, A. Internal curing of recycled aggregates concrete. *J. Clean. Prod.* **2020**, *275*, 122911. [CrossRef]
8. Resende, D.M.; de Carvalho, J.M.F.; Paiva, B.O.; Gonçalves, G.d.R.; Costa, L.C.B.; Peixoto, R.A.F. Sustainable Structural Lightweight Concrete with Recycled Polyethylene Terephthalate Waste Aggregate. *Buildings* **2024**, *14*, 609. [CrossRef]
9. Wang, J.; Li, Y.; Qiu, Z.; Zhang, Y. Experimental research on compressive properties of recycling polypropylene (PP) fiber recycled coarse aggregate concrete. *J. Build. Eng.* **2023**, *76*, 107403. [CrossRef]
10. Li, Z.; Shen, A.; Zeng, G.; Chen, Z.; Guo, Y. Research progress on properties of basalt fiber-reinforced cement concrete. *Mater. Today Commun.* **2022**, *33*, 104824. [CrossRef]
11. Aslani, F.; Hou, L.; Nejadi, S.; Sun, J.; Abbasi, S. Experimental analysis of fiber-reinforced recycled aggregate self-compacting concrete using waste recycled concrete aggregates, polypropylene, and steel fibers. *Struct. Concr.* **2019**, *20*, 1670–1683. [CrossRef]
12. Bagherzadeh, R.; Sadeghi, A.H.; Latifi, M. Utilizing polypropylene fibers to improve physical and mechanical properties of concrete. *Text. Res. J.* **2012**, *82*, 88–96. [CrossRef]
13. Kazmi, S.M.S.; Munir, M.J.; Wu, Y.F.; Patnaikuni, I.; Zhou, Y.; Xing, F. Axial stress-strain behavior of macro-synthetic fiber reinforced recycled aggregate concrete. *Cem. Concr. Compos.* **2019**, *97*, 341–356. [CrossRef]
14. Hanumesh, B.M.; Harish, B.A.; Venkata Ramana, N. Influence of polypropylene fibers on recycled aggregate concrete. *Mater. Today Proc.* **2018**, *5*, 1147–1155. [CrossRef]
15. Hunashikatti, G.M.; Pradhan, S.; Bara, S.V. Partially hydrated recycled aggregate concrete: A systematic approach towards sustainable development. *Constr. Build. Mater.* **2018**, *186*, 537–549. [CrossRef]
16. Ye, P.; Chen, Z.; Su, W. Mechanical properties of fully recycled coarse aggregate concrete with polypropylene fiber. *Case Stud. Constr. Mater.* **2022**, *17*, e01352. [CrossRef]
17. Das, S.; Sinah, P.; Negi, V.S. Strength and durability properties of polypropylene fiber reinforced recycled coarse aggregate concrete. *Indian Concr. J.* **2018**, *94*, 62–69.
18. Nam, J.; Kim, G.; Yoo, J.; Choe, G.; Kim, H.; Choi, H.; Kim, Y. Effectiveness of fiber reinforcement on the mechanical properties and shrinkage cracking of recycled fine aggregate concrete. *Materials* **2016**, *9*, 131. [CrossRef] [PubMed]
19. Mudadu, A.; Tiberti, G.; Germano, F.; Plizzari, G.A.; Morbi, A. The effect of fiber orientation on the post-cracking behavior of steel fiber reinforced concrete under bending and uniaxial tensile tests. *Cem. Concr. Compos.* **2018**, *93*, 274–288. [CrossRef]
20. Xiao, J.; Poon, C.S.; Zhao, Y.; Wang, Y.; Ye, T.; Duan, Z.; Peng, L. Fundamental behavior of recycled aggregate concrete—Overview II: Durability and enhancement. *Mag. Concr. Res.* **2022**, *74*, 1011–1026. [CrossRef]
21. Win, T.T.; Jongvivatsakul, P.; Jirawattanasomkul, T.; Prasittisopin, L.; Likitlersuang, S. Use of polypropylene fibers extracted from recycled surgical face masks in cement mortar. *Constr. Build. Mater.* **2023**, *391*, 131845. [CrossRef]
22. Al-Quraishi, H.; Abdulkhudhur, R.; Abdulazeez, A. Shear Strength Behavior of Fiber Reinforced Recycled Aggregate Concrete Beams. *Int. Rev. Civ. Eng.* **2021**, *12*, 4183. [CrossRef]
23. McGinnis, M.J.; Davis, M.; de la Rosa, A.; Weldon, B.D.; Kurama, Y.C. Quantified sustainability of recycled concrete aggregates. *Mag. Concr. Res.* **2017**, *69*, 1203–1211. [CrossRef]
24. Carneiro, J.A.; Lima, P.R.L.; Leite, M.B.; Toledo Filho, R.D. Compressive stress–strain behavior of steel fiber reinforced-recycled aggregate concrete. *Cem. Concr. Compos.* **2014**, *46*, 65–72. [CrossRef]
25. Dinh, N.H.; Park, S.H.; Choi, K.K. Effect of dispersed micro-fibers on tensile behavior of uncoated carbon textile-reinforced cementitious mortar after high-temperature exposure. *Cem. Concr. Compos.* **2021**, *118*, 103949. [CrossRef]
26. Feng, Z.; Shen, D.; Luo, Y.; Huang, Q.; Liu, Z.; Jiang, G. Effect of polypropylene fiber on early-age properties and stress relaxation of ultra-high-performance concrete under different degrees of restraint. *J. Build. Eng.* **2023**, *68*, 106035. [CrossRef]
27. Vijayan, V.; Jayakesh, K.; Anand, K.B. Mechanical properties of recycled aggregates concrete with sisal fiber and silica fume. *Mater. Today Proc.* **2022**, *65*, 1887–1894. [CrossRef]
28. *JGJ/T 443-2018*; Technical Standard for Recycled Concrete Structure. Ministry of Housing and Urban Rural Development of the People's Republic of China: Beijing, China, 2018.
29. *JGJ 55-2011*; Specification for Mix Proportion Design of Ordinary Concrete. Ministry of Housing and Urban Rural Development of the People's Republic of China: Beijing, China, 2011.

30. Magdalena, P.M.; Kaszynska, M. Mechanical performance and environmental assessment of sustainable concrete reinforced with recycled end-of-life tyre fibers. *Materials* **2021**, *14*, 1–20.
31. Ahmadi, M.; Farzin, S.; Hassani, A.; Motamedi, M. Mechanical properties of the concrete containing recycled fibers and aggregates. *Constr. Build. Mater.* **2017**, *144*, 392–398. [CrossRef]
32. Bui, N.K.; Satomia, T.; Takahashia, H. Recycling woven plastic sack waste and PET bottle waste as fiber in recycled aggregate concrete: An experimental study. *Waste Manag.* **2018**, *78*, 79–93. [CrossRef]
33. Ferrara, G.; Pepe, M.; Martinelli, E.; Tolêdo Filho, R.D. Tensile behavior of flax textile reinforced lime-mortar: Influence of reinforcement amount and textile impregnation. *Cem. Concr. Compos.* **2021**, *119*, 103984. [CrossRef]
34. Katkhuda, H.; Shatarat, N. Improving the mechanical properties of recycled concrete aggregate using chopped basalt fibers and acid treatment. *Constr. Build. Mater.* **2017**, *140*, 328–335. [CrossRef]
35. Tam, V.W.; Wattage, H.; Le, K.N.; Butera, A.; Soomro, M. Hybrid methods to improve microstructure of recycled concrete and brick aggregate for high-grade concrete production. *Mag. Concr. Res.* **2022**, *75*, 17–31. [CrossRef]
36. Li, L.; Ruan, S.; Zeng, L. Mechanical properties and constitutive equations of concrete containing a low volume of tire rubber particles. *Constr. Build. Mater.* **2014**, *70*, 291–308. [CrossRef]
37. *GB/T 50080-2016*; Standard for Test Method of Performance on Ordinary Fresh Concrete. Ministry of Housing and Urban Rural Development of the People's Republic of China: Beijing, China, 2016.
38. *GB/T 50081-2019*; Standard for Test Method of Mechanical Properties on Ordinary Concrete. Ministry of Housing and Urban Rural Development of the People's Republic of China: Beijing, China, 2019.
39. *BS EN 12350-2-2009*; Testing Fresh Concrete-Slump-Test. Comité Européen de Normalisation: Brussels, Belgium, 2009.
40. *BS EN 12390-3-2009*; Testing Hardened Concrete-Compressive Strength of Test Specimens. Comité Européen de Normalisation: Brussels, Belgium, 2009.
41. *BS EN 12390-5-2009*; Testing Hardened Concrete-Flexural Strength of Test Specimens. Comité Européen de Normalisation: Brussels, Belgium, 2009.
42. *BS EN 12390-6-2009*; Testing Hardened Concrete-Tensile Cleavage Strength of Test Specimens. Comité Européen de Normalisation: Brussels, Belgium, 2009.
43. *BS EN 12390-13-2013*; Testing Hardened Concrete-Determination of Secant Modulus of Elasticity in Compression. Comité Européen de Normalisation: Brussels, Belgium, 2013.
44. Guo, Z.; Zhang, X.; Zhang, D. Experimental study on the stress-strain curve of concrete. *J. Build. Struct.* **1982**, 1–12.
45. Guo, Z.; Nan, J.; Shi, X. Temperature stress coupled constitutive relationship of concrete. *J. Tsinghua Univ. (Nat. Sci. Ed.)* **1997**, 89–92. [CrossRef]
46. Han, R.; Zhao, S.; Qu, F. Experimental study on the tensile performance of steel fiber reinforced concrete. *J. Civ. Eng.* **2006**, 63–67.
47. Yang, M.; Huang, C. Experimental study on the axial tensile performance of steel fiber high-strength concrete. *J. Civ. Eng.* **2006**, 55–61.

buildings

MDPI

Article

Experimental Study on the Effect of Steel Reinforcement Ration on the Cracking Behaviour of FRP-Strengthened RC Elements

Andrea Armonico [1], Laurent Michel [1], Mohamed Saidi [2] and Emmanuel Ferrier [1,*]

[1] Laboratory of Composite Materials for Construction (LMC2), University Claude Lyon 1, 82 bd Niels Bohr, 69622 Villeurbanne, France; andrea.armonico@univ-lyon1.fr (A.A.); laurent.michel@univ-lyon1.fr (L.M.)

[2] LOCIE, Université Savoie Mont-Blanc, Boulevard du Lac, 73370 Le Bourget-du-Lac, France; mohamed.saidi@univ-smb.fr

* Correspondence: emmanuel.ferrier@univ-lyon1.fr

Abstract: This paper examines the cracking behaviour of reinforced concrete beams strengthened by externally bonded fiber-reinforced polymer. The crack opening of RC structures is a key parameter for the durability of concrete structures. It is of vital importance for designers to be able to make correct estimations of the crack opening values of strengthened structures. FRP strengthening affects the cracking behaviour of RC beams with different steel percentages. Beams have been tested under four-point bending mechanical tests until failure with three steel ratios and two layers of externally bonded wet carbon fibers (CFRP). In order to measure the crack opening during loading, Digital Image Correlation is used to obtain the crack opening along the beam during load functioning. The results allow for a comparison of the RC beams with and without FRP and enhance the effect of FRP on crack opening. The crack width was compared with the theoretical values obtained based on the relation proposed by Eurocode 2 (EC2). The comparison enhanced the need to propose a modified relation. Subsequently, an empirical model was established as a modification of EC2, considering the presence of a CFRP system. The corresponding results were compared and discussed to validate the model. For the same level of loads, the crack opening can be reduced by 20 to 50% depending on the level of steel ratio.

Keywords: CFRP; digital image correlation; cracking; RC beams; strengthening

1. Introduction

In the past century, concrete structures have been used worldwide. The aging process of concrete and the upgrade of loading conditions have enhanced the need to strengthen these structures [1–6]. Since the 1990s, many structures have been strengthened to face these problems using externally bonded material, such as steel cladding and fiber-reinforced polymer. In the past 20 years, there has been an important research development in fiber-reinforced polymer (FRP) systems that are bonded in different forms (i.e., rebars, plates, fabrics, and grids) and with different fibers (i.e., carbon, glass, and aramid). These materials have several advantages including high strength, low weight, and excellent resistance to external agents [7]. Carbon fibers are preferred because of their properties and advantages compared with other synthetic fiber composites, such as higher mechanical properties, more excellent fatigue resistance, corrosion resistance, and creep resistance [8]. The main advantage is that to increase the tension capacity of a beam, the FRP material acts like an additional tension reinforcement; this allows for a slight improvement in the bending stiffness and the ultimate load [9–13]. These systems improve the capacity of RC elements with an increase in ductility and a limit in crack openings. Based on 30 years of research, several design codes and testing procedures have been developed [14–17]. This research has shown that failures in strengthened RC beams are obtained after large deflections and steel rebars yielding with FRP debonding. This debond can occur in the concrete part of the beam through intermediate debonding or through peeling-off at the end of the plate [18].

In a beam strengthened by FRP, researchers have enhanced that crack behavior is modified, crack spacing is modified with more small cracks in between the large cracks, and crack opening is also modified. Based on this development, an improvement of the strengthening technic has been performed with prestress FRP [11]. In this purpose anchorage system, a prestress process has been developed with excellent results [11]. However, this technique is not commonly used in the field because the process is not easy to implement in reality.

For durability purposes, the crack opening reduction is important to quantify. These effects have not been studied deeply in the literature; thus, it is necessary to conduct intense research to provide data that propose a design model suitable for RC strengthened by FRP.

At the same time, Digital Image Correlation has been developed in the construction industry, in order to measure displacements or deflections, or to localize the presence of cracks in reinforced concrete.

The feasibility of using an optical full-field Digital Image Correlation (DIC) technique for the measurement of strain fields on FRP materials used in the civil engineering industry has been investigated and the level of error in the DIC method when using more traditional methods was determined by Del Rey Castillo et al. [19]. The main advantage of using DIC over more traditional methods is the capacity to measure full field strains instead of strains at local points, which has been enhanced by providing the measurements of various specimens of FRP materials [20]. The reported strain fields are examples of what were obtained during an experimental campaign to understand the behavior of strengthened beams using externally bonded CFRP [21–23].

The present research program enhances changes in the mechanical behaviour of RC beams strengthened by externally bonded FRP systems and focuses on crack opening using DIC methods. The FRP considered in this study is a wet lay-up applied carbon sheet of 200 g/m^2. Three sets of beams were prepared with steel ratios of 0.4%, 0.6%, and 0.8%. These three levels of steel ratio correspond to structures with very low, low, and normal steel ratio reinforcement in building structures. The CFRP is applied by the wait lay-up method. The cross-section of the CFRP is fixed to 16.9 mm^2, based on the Young modulus of Carbon fibers (245 GPa) and the literature review. The FRP strengthening ratio is high enough to obtain a real strengthening effect on RC beams whatever the level of the steel ratio. The effects of applying two layers of CFRP were evaluated and discussed in detail regarding the influence of the steel ratio using beams tested for four-point bending.

Herein, the first section discusses the experimental program with material characterization and specimen setup. Next, the results are reviewed and deep analysis is performed on the changes in cracking behaviour, strength improvement, and midspan deflection.

The last section proposes the model based on the Eurocode 2 [24] formula, which is appropriately fitted for unstrengthened RC elements. In addition, the experimental and calculated results are compared and discussed, and some perspectives are provided.

2. Experimental Program

Ten RC beams with rectangular cross-sections of 150 by 250 mm are strengthened using two layers of CFRP sheets. Two layers have been retained in this study since, normally, 1 to 6 layers of CFRP can be added to RC structures, based on previous researchers' experience. Two layers of CFRP corresponds to a normal strengthening ratio.

2.1. Materials

The concrete used in this study has a characteristic concrete strength of C30/37, based on European standard [24].

2.1.1. Reinforced Concrete

The beams were cast using concrete with a characteristic compressive strength (f_{ck}) of 35 MPa as measured during the compression tests. The concrete mixtures used for the beam casting are listed in Table 1.

Table 1. Concrete composition.

Component	Cement CEM I 52.5 R	Sand	Gravel	Water
[kg/m^3]	420	890	890	200

The compression tests on the concrete were performed according to the NF EN 12390-1 standard [25]. The standard specifies the size of the specimens, limiting conditions, test velocity, and test procedure. The results are shown in Table 2 below.

Table 2. Concrete strength.

Specimen	1	2	3	4	5	6	Average	Standard Deviation
σ [MPa] at 28 days	37.4	38.2	42.2	38.65	38.85	44.1	39.9→35	2.41

The steel reinforcement consists of high limiting S500B rebar, having a characteristic yield strength (f_{yk}) of 500 MPa.

2.1.2. CFRP Reinforcement Material

The elastic modulus (E_f) of carbon sheet was 245 GPa, whereas the composite single-layer thickness was 0.169 mm, as reported in the technical datasheets provided by the manufacturer (Fibre NET, Pavia di Udine (UD), Italy). The FRP was applied by wet lay-up method of CFRP fabrics bonded using a bicomponent epoxy resin. The beams were retrofitted using a carbon-fiber-reinforced polymer consisting of unidirectional carbon fabric Betontex FB-GV330U-HT produced by Fiber Net s.p.a. The layers had a nominal thickness of 0.169 mm and Young's modulus (E_f) of 245 GPa, as reported in the technical datasheets provided by the manufacturer. The fibers were glued using a bicomponent epoxy resin Betontex FB-RC02 with a declared Young's modulus (E_m) of 3000 MPa and an ultimate strain equal to 2.9%. To verify the material properties, tests were performed, and results are given in Table 3. The elastic modulus was evaluated following the NF EN ISO 527-5 standard [26]; test coupons were prepared with 2-layer of CFRP fabrics. Uniaxial traction tests were performed using a universal testing machine (UTM). The tensile strength and Young modulus are reported in Figure 1 and Table 3, respectively, based on a calculation with an equivalent cross-section of the carbon fiber (thickness equal to 0.338 mm).

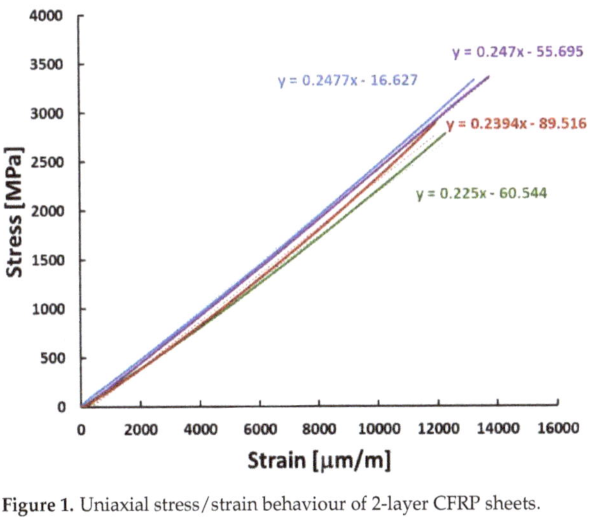

Figure 1. Uniaxial stress/strain behaviour of 2-layer CFRP sheets.

Table 3. Mechanical properties of 2-layer CFRP sheet.

Specimen	Young's Modulus—E_f [MPa]	Maximal Strain—ε_u [%]	Ultimate Stress—f_{tk} [MPa]
1	225,000	1.23%	2786.4
2	247,000	1.38%	3363.4
3	242,200	0.88%	2141.2
4	247,700	1.28%	3329.5
5	239,400	1.20%	2891.3
Average	240,260	1.19%	2902.4
Standard deviation	9194.45	0.0018	496.99
Coeff. of variation	4%	16%	17%

2.2. Specimen Preparation

The concrete mix was used to prepare ten RC beams with a rectangular cross-section of 150 mm × 250 mm, a 2.00-m span, and two reinforcing steel bars. The beams were prepared with different steel percentages: (i) 10 mm-diameter bars; (ii) 12 mm-diameter bars; and (iii) 14 mm-diameter bars, and hence steel ratios of 0.41%, 0.60%, and 0.82%. Details of the specimens are provided in Table 4.

Table 4. Specimen naming.

Specimen Name	R10-NR *	R12-NR	R14-NR	R10-RB **	R12-RB0	R12-RB1	R12-RB2	R12-RB3	R12-RB4	R14-RB
Diameter of reinforcing bars	10	12	14	10	12	12	12	12	12	14
Presence of CFRP	No	No	No	Yes	Yes	Yes	Yes	Yes	Yes	Yes

* RXNR: Rebars × mm Non-Reinforced. ** RXRBY Rebars × mm Reinforced before loading number Y.

Before bonding the composite layers, the beam's bottom surface was mechanically grouted and then a first layer of polymer was applied to improve CFRP adhesive performance onto the concrete. (Figure 2) The CFRP had a total final length equal to 1.9 m to avoid having CFRP on the bearing support during the test.

(a)

(b)

Figure 2. *Cont.*

(**c**)

Figure 2. Strengthening procedure. (**a**) Surface preparation. (**b**) First layer of polymer application. (**c**) CFRP sheet applied on support with a last layer of epoxy.

For each configuration, a beam was non-reinforced to be used as a reference. The test set up is reported in Figure 3.

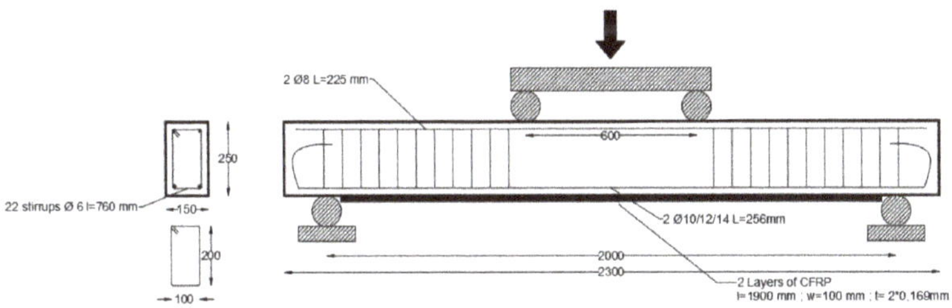

Figure 3. Beam setup.

Strain sensors were placed in the central area on longitudinal steel rebar and on the CFRP after the full polymerization of the CFRP.

The vertical displacement was monitored by two linear variable displacement transducers (LVDTs) placed in the mid part of the beam (Figure 4).

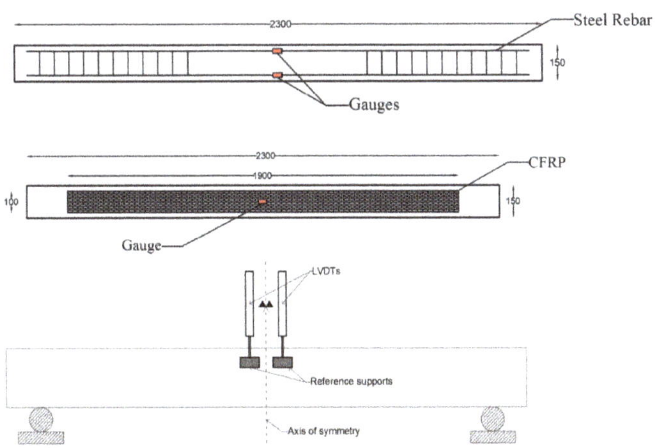

Figure 4. Strain gauge and LVDT location.

2.2.1. Two-Dimensional Digital Image Correlation (DIC 2D)

The crack was measured on the lateral surface of the beam in the middle where there is only normal stress without shear. The measurement techniques were based on the comparison of two images at two different levels of load. This noncontact optical technique was based on the theory of field displacement correlation to detect changes in pixel positions in a comparison of two stages (i.e., displacements or deformations) through consecutive image comparison. The surfaces of the specimens must be prepared with a white background and black speckles [27,28]. The image acquisition procedure was performed according to good practice for the DIC [29]. During the loading of beams, a set of data was obtained with several grayscale photos captured using a standard high resolution camera and a commercial correlation tools named GOM Correlate software 2015 was employed for post-processing [30]. The software algorithm requires the definition of a region of interest (ROI). There are several small facets called areas of interest (AOI) with well-defined pixel sizes [31]. The camera used was a Kramer electronics VP 211K with a resolution of 2560 × 2048 pixels^2 and a lens KOWA LM25HC:f = 25 mm; F 1.8–16. The focal was between f/6 and f/9. The size of measurement (ROI) was 600 mm × 250 mm. The density of the speckles was a fundamental factor in the accuracy of the results, which was related to the camera's proximity and resolution. Generally, as a guideline, A. D. Terani [29] suggests using 5-pixel size speckles covering 50% of the investigated surface. The tested specimens were inspected using the commercial software GOM Correlate®, which can process a set of 2D images by defining a scale factor, an ROI, and several AOIs. The size in pixels of the area of interest was equal to 30 px, with the distance of 20 px between their middles. The size of the AOI was directly proportional to the speckle dimension (large speckles correspond to large AOIs). Each AOI had a centre with a spatial location, which was used as a reference for the following steps. The axial strain (ε_x) or longitudinal displacement field map was obtained by comparing two images (see Figure 5). In Figure 5b, red values correspond to 2 mm and blue values to less than 0.02 mm.

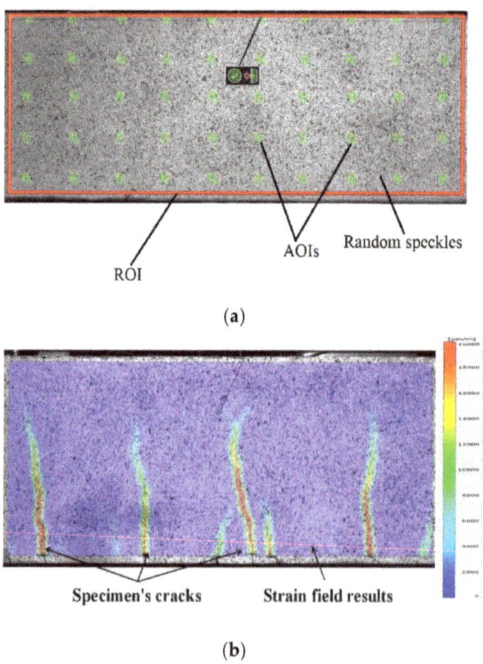

Figure 5. DIC procedure: (**a**) definition of ROI and AOIs; (**b**) strain map with crack locations.

2.2.2. Loading Device

The test set-up is based on a high stiffness steel frame equipped with a hydraulic jack which has a load capacity of 200 kN. The beams were simply supported, and their spans were 2 m. A four-point load was applied in the middle with a force distance of 600 mm [32–35]. Tests were performed using a displacement control system at a rate of 1 mm/min in isostatic conditions. The complete setup is illustrated in Figure 6.

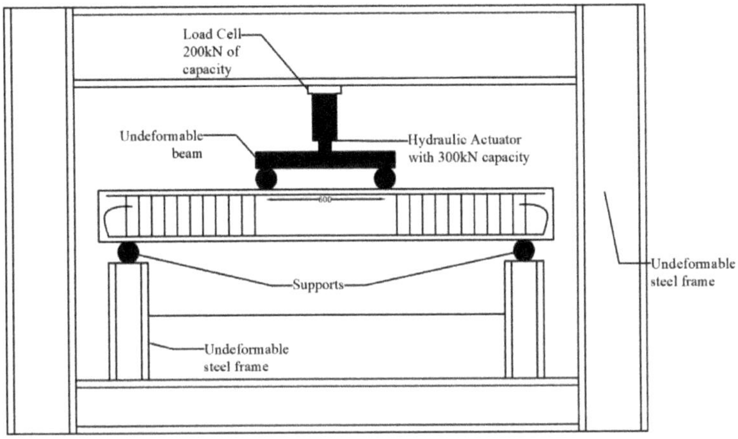

Figure 6. Bending test setup.

3. Results and Discussion

The data were analyzed based on the load-displacement curve and crack load–crack opening curve. These allow for both global and local behaviour.

The load displacement curves were analyzed to obtain a different global mechanical stage (cracking load, non-linear behavior, and failure). The different values for each stage are given in Table 5, with the load and midspan deflection corresponding to the first cracking, steel elastic limit strain, and failure.

Table 5. Representation of load and midspan deflection for key events during 4-point bending tests.

Name	First Crack		Steel Yielding		Failure	
	Load [kN]	Deflection [mm]	Load [kN]	Deflection [mm]	Load [kN]	Deflection [mm]
R10-NR	25	1.85	57.8	7.21	70.58	31.61
R12-NR	25	1.55	63.4	6.16	92.57	31.10
R14-NR	30	1.26	85.1	6.84	112.11	27.72
R10-RB	40	3.30	55.50	5.48	126.25	27.82
R12-RB0	35	2.49	69.12	6.78	114.20	20.81
R12-RB1	45	3.13	68.40	5.71	120.81	19.43
R12-RB2	50	3.77	67.53	5.82	110.97	17.80
R12-RB3	45	3.20	69.14	5.96	121.28	21.12
R12-RB4	50	3.80	70.55	6.16	109.65	17.18
R14-RB	55	4.10	67.84	5.27	148.24	24.15

3.1. Ultimate Load

A CFRP external strengthening system allows for the modification of flexural behaviour in an RC beam in accordance with the literature review [35–40]. The increase in bending stiffness and/or ultimate bending moment is more effective for beams with a low steel ratio (0.4%). In this case, the ultimate capacity was increased by 80%. When the steel ratio is equal to 0.6% or 0.80% of the concrete section, the increase is given to be equal to 30%. Figure 7 demonstrates that ultimate load improvement was evident in the R10 beam, reaching almost 80% of the corresponding reference. By contrast, in the R12 and R14 groups, the ultimate load was not significantly different; both achieved a 30% improvement. Comparable results were obtained in the R12 group. The R12 and R14 group results were very close because the steel ratios (0.6% and 0.8%, respectively) were very close and the more the beam is already strengthened by longitudinal steel rebars the less FRP is efficient. The load–displacement curve clearly enhanced the effect of FRP on the ultimate load, as mentioned before. These results are comparable to those obtained by previous researchers; for example, Chajes et al. [41] shows that CFRP is more effective for load capacity increases than for increasing the stiffness.

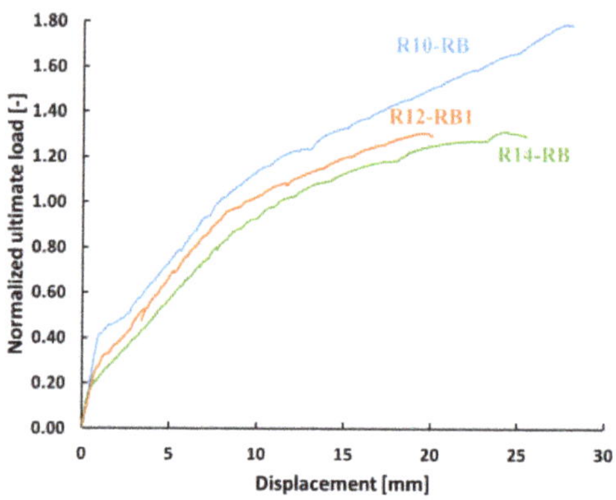

Figure 7. Improvement in ultimate load in the tested groups normalized to the reference one.

Since the design according to Eurocode 2 is based on behaviour governed by steel yielding and not on concrete crushing, the ultimate load of an RC beam is mostly dependent on the tensile capacity of the cross-section of the beam. The limit of the strengthening by externally bonded CFRP is the compressive strength of the top part of the beam to avoid concrete crush in compression zone.

3.2. Mid-Span Displacement

The strengthening using FRP modifies the bending stiffness of the beam because of externally bonded carbon fiber strengthening. Figure 8 enhances the load displacement curve, before the steel rebars yielding the bending stiffness is slightly increased by only 10%; at this stage, the bending stiffness value mainly depends on the concrete being under compression and the steel rebars being under tension. FRP contribution is low. Based on the cross-section on the beam, steel ratio, and FRP ratio reinforcement, the cracking inertia of the non-strengthened beam is around 93.10^6 mm^4 while it grows to 116.10^6 mm^4 when two layers of FRP are applied. When steel yields, its Young Modulus decreases to a low value, and, at this stage, all additional loss is undertaken by the CFRP and then the strengthening material is dependent on the highest second slope of the load displacement curve for the beams strengthened by CFRP. The R10-RB group exhibits a more significant improvement after strengthening. Therefore, the deflection is proportionally higher in the R10 beams. It is also important to outline that the more the beams are strengthened, the more the ultimate deflection is reduced with a loss in beam ductility. This observation is linked to the failure mode of the beam with FRP debonding.

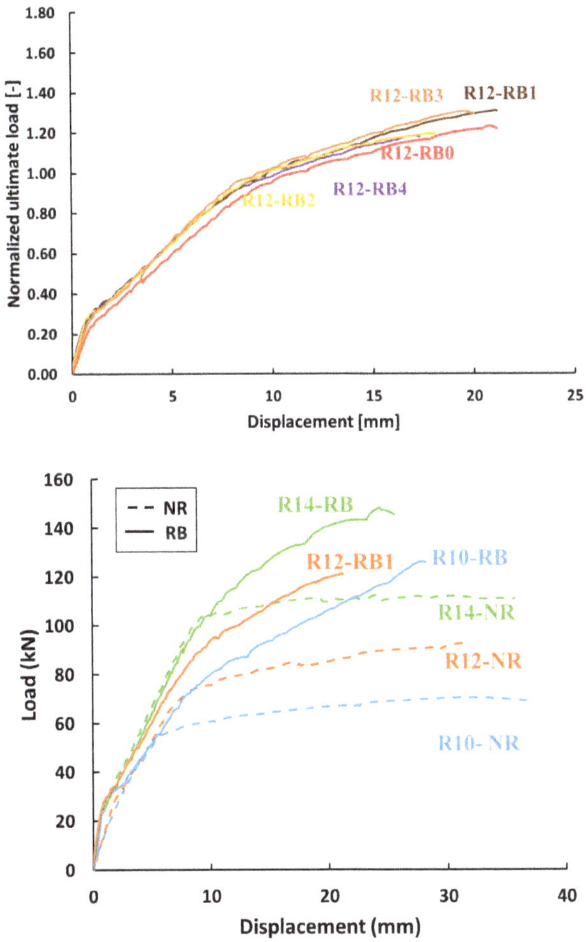

Figure 8. Reduction in deflection for the tested beams.

Composite Strain

In flexural strengthening, the task of the composite material is to increase the tensile strength of the section, which is partially guaranteed by the steel reinforcement. Knowledge of the stress level plays a crucial role in understanding the material's behaviour. For example, traction in the fiber produces shear stress within the resin used for bonding. An increase in the tangential stress is consequently the cause of the premature debonding of the reinforcement. Externally bonded fabric was applied on the concrete surfaces as presented in Section 2.2. The FRP is not prestressed before bonding and needs a crack to appear to develop tension [37].

In addition, due to the higher stiffness, when compared with the steel, the composite strain at a reference load is less than the corresponding steel strain. Similarly, as far as the influence of the steel ratio is concerned, due to what has been folded in the previous section, the composite is activated by reaching higher values in beams with less steel. Table 6 shows the difference in strain between steel and composite for several load levels for each test group. It can be seen that the composite exhibits more significant deformation than steel in the early stages of loading. This inconsistency may be due to the strain gauge's positioning, which is not aligned with the correspondent placed on the steel bar.

Table 6. Comparison of strain between steel and composite for several load levels.

Load [kN]	R10-RB		R12-RB		R14-RB	
	Steel Strain [μm/m]	Composite Strain [μm/m]	Steel Strain [μm/m]	Composite Strain [μm/m]	Steel Strain [μm/m]	Composite Strain [μm/m]
10	19	22	58.19	70.46	63	68
35	1299	1696	1083	1350	1071	651
60	2730	2837	2112	2378	2132	1296
80	5433	4491	3314	3296	3020	1968
100	9798	7591	8257	6055	4398	2850
120	17,836	10,186	-	-	8379	4593
140	-	-	-	-	12,931	7161

3.3. Failure Modes

For all the strengthened beams using externally bonded CFRP, failure occurs during the debonding of the concrete covering at the end of the FRP layers near the beam support. This mode of failure is commonly called peeling-off in the literature. The shear stress in this area is located between the FRP and is caused when the steel longitudinal rebar is too high in comparison with the concrete tensile-shear strength (Figure 9).

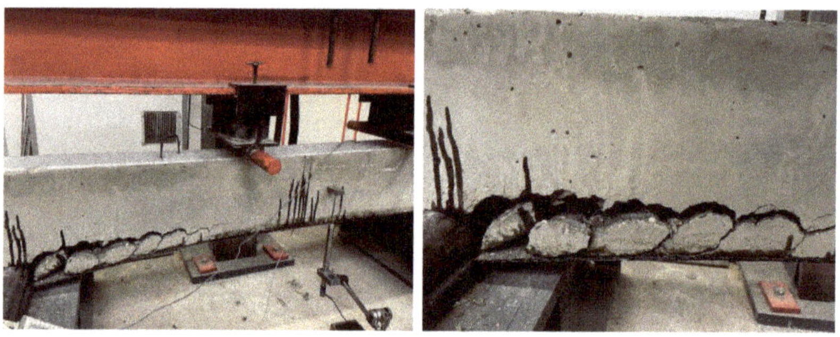

Figure 9. Failure modes of strengthened beams.

3.4. Cracking Behaviour

The crack width was calculated thanks to the DIC method by obtaining images from a virtual displacement sensor located at 25 mm from the bottom right of the cracking zone. The displacement length of the displacement sensor was 1 cm so that the change in value was assumed to be only due to crack opening and not tensile concrete stress. The results were then analysed based on load–crack opening curves. The presence of CFRP fabric was found to influence the cracking behaviour of the tested beams, with a significant impact on the crack formation phase (CFP) [42,43]. Typically, when a crack appears, the effect of FRP allows to bridge the crack and limit the opening. For the cracking load, the FRP strain increased and controlled the tension in the steel rebars [44,45]. The load was transferred along the FRP bond length and when the load increased a new crack was initiated; this explained that the crack spacing was reduced [46]. In non-strengthened beams, the crack is controlled around steel rebars in tension. Hognestad et al. [44] fixed the minimum value, a_{min}, given by Equation (1)

$$a_{min} = \frac{A_e f_t'}{u\Sigma o} \quad (1)$$

where A_e is the effective area of concrete in tension, f_t' is the concrete tensile strength, u is the average bond stress, and Σo represents the sum of the bar perimeters.

In the experimental tests, strengthening further reduced the crack width in the R10 group compared with the other groups. The group with the lowest steel ratio exhibited a more significant effect at low-load levels. In fact, in the R10 group, steel yielding occurred at a lower load level, and the crack behaviour exhibited a divergent trend from 40 kN. However, the R12 and R14 groups did not exhibit consistent improvement in the first crack load because of the better steel–concrete adherence owing to the higher stiffness of the beam. In addition, the crack behaviour diverged for higher load values because of the later yielding of the steel bars (see Figure 10).

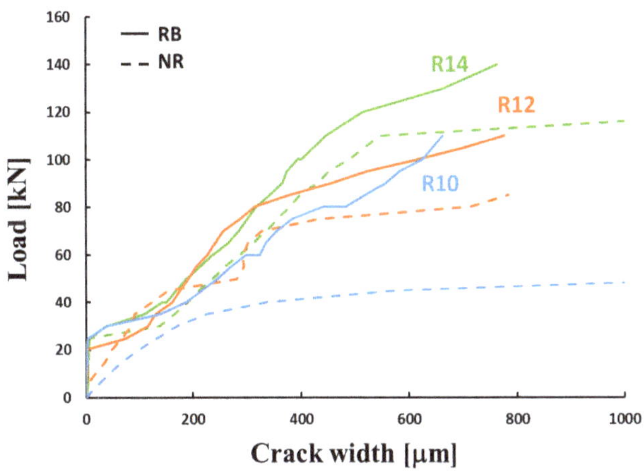

Figure 10. Crack–load curve.

A detailed overview of the crack width values for different load levels is provided in Table 7 and Figure 11.

Table 7. Values of crack opening values in function of loads.

Load [kN]	Crack Width [μm]					
	R10-NR	R12-NR	R14-NR	R10-RB	R12-RB1	R14-RB
25	133	124	137	3	72	7
30	169	145	166	7	115	37
40	334	190	210	33	161	142
50	1221	237	258	122	191	188
60	-	278	311	200	226	236
70	-	335	355	244	255	284
80	-	877	398	313	312	316
90	-	-	451	444	458	364
100	-	-	545	516	618	394
110	-	-	-	-	776	446
120	-	-	-	-	-	514

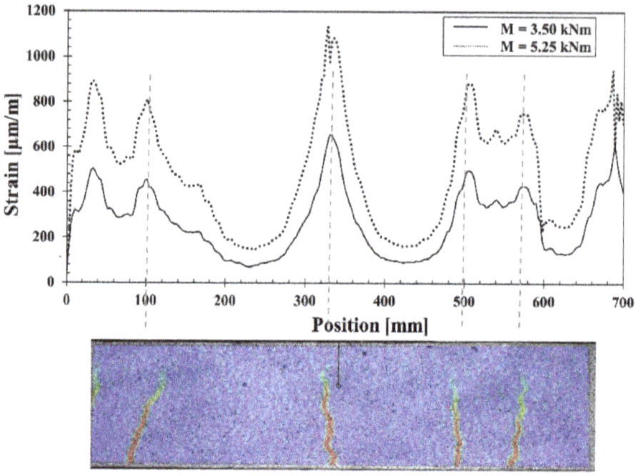

Figure 11. Evolution of strain and location of cracks for beams R10-RU80 for a bending moment of 3.5 and 5.25 kN·m based on DIC.

3.5. Model Proposal

Several codes have been proposed to predict bent RC crack spacing and the width of RC beams [24,47–49] under flexural loading. However, the presence of a strengthening system changes the usual crack behaviour, reducing the crack spacing and width [50,51]. The Eurocode 2 (EC2) model [24] is based on the principle of strain distribution between two adjacent cracks, between concrete and steel. The crack width is calculated as the product of the maximum spacing ($S_{r,max}$) and the steel–concrete mean strain difference ($\varepsilon_{sm} - \varepsilon_{cm}$), written as Equation (2).

$$w_k = S_{r,max} \cdot (\varepsilon_{sm} - \varepsilon_{cm}) \quad (2)$$

In this study, an EC2-modified model was proposed that considers the presence of a reinforcing system. The main hypothesis is that stress in steel rebar is reduced on the same level of load due to the hypothesis that the sum of the tensile force in a cross-section should be equal before and after strengthening under the same bending stiffness. Then the tensile strain in steel is also reduced. The simplicity of the model was maintained, and only a modification of the factor was presented to consider the effect of CFRP.

3.5.1. Eurocode Model

The EC2 formulation can predict the crack opening in an RC beam under a bending moment. The assumption is the relative difference between the steel strain and the concrete tensile strain before crack, and the difference is due to crack opening. The difference between the mean strains of the materials is based on a hypothesis of the perfect bonding of steel rebars with concrete. Then, a crack occurs when the steel strain is incompatible with the admitted concrete tensile strain [52]. EC2 provides a formula for calculating the mean strain difference between the RC constituent materials, depending on the geometrical and mechanical properties, given by Equation (3)

$$(\varepsilon_{sm} - \varepsilon_{cm}) = \left(\sigma_s - \frac{k_t \left(\frac{f_{ct,eff}}{\rho_{p,eff}} \right) \left(1 + n\rho_{p,eff} \right)}{E_s} \right) \geq \frac{0.6\sigma_s}{E_s} \quad (3)$$

where σ_s is the stress in the steel rebar, $f_{ct,eff}$ is the tensile concrete strength, n is the steel–concrete elastic modulus ratio, $\rho_{p,eff}$ is the steel–concrete geometrical ratio referred to as the effective concrete area around the steel rebar $A_{c,eff}$, E_s is the steel elastic modulus, and k_t is a coefficient that considers the load duration (0.6 and 0.4 for short and long duration, respectively). The value of $A_{c,eff}$ is also provided by EC2 with reference to the American Concrete Institute (ACI) [47] and is given by Equation (4)

$$A_{c,eff} = b_w Min\left\{2.5(h-d); \frac{h-x}{3}; 0.5h\right\} \tag{4}$$

Here, h is the height of the cross-section, d is the distance measured from the centroid of the steel reinforcement to the top concrete fiber, b_w is the minimum section width, and x is the neutral axis depth.

Meanwhile, the maximum crack spacing was calculated using Equation (5)

$$S_{r,max} = 3.4c + \frac{0.425k_1k_2k_4}{\rho_{p,eff}} \tag{5}$$

where c is the concrete cover, ϕ is the reinforcing steel diameter, and k_1, k_2, and k_4 are the factors given by EC2.

The proposed equation fits the experimental data for the non-strengthened beams but is too conservative for beams strengthened using FRP. In particular, as illustrated in Figure 12, the EC2 formula is too conservative in the case of group R10, whereas it deviates less from the overall behaviour in the case of group R14, where the steel ratio is the highest. However, this result is understandable if one refers to Section 3.2, noting that the influence of the composite material was less evident in the group with a higher reinforcement level.

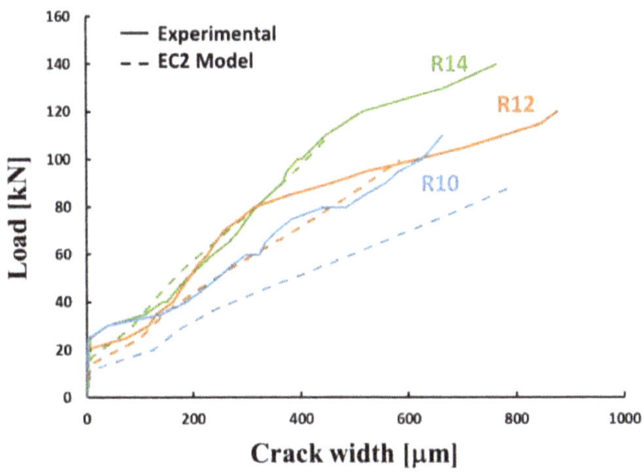

Figure 12. Comparison between experimental and predicted results using the existing EC2 formula.

3.5.2. Modified Model

The EC2 model accurately predicted the cracking evolution in an RC beam subjected to a bending moment. However, the increasing use of CFRP external strengthening of concrete structures highlights the need to adopt a new crack model for more extensive use by taking into consideration the strengthening effect. To consider the effect of strengthening by FRP and to input the geometrical and mechanical factors, a new relation was proposed that can better fit the model in this study. A new equation was proposed to modify the equation of steel strain (relation 6). The main objective was to introduce the effect of FRP on the changes in steel stress, to introduce the FRP cross-section in the equation, and to consider the modification of the neutral axis position due to FRP to obtain a better accuracy. The main objective was to introduce the geometrical properties of FRP in the model for design purposes.

The Young modulus of steel was modified to introduce FRP and a new model would consider the steel and composite moduli and material volume ratio. The homogenization factor n^* was the ratio between the equivalent and concrete elastic moduli, and the geometrical reinforcement ratio. The percentage of tensile reinforcement was also modified to consider the effect of FRP $\rho_{p,eff}$ (relation (8)).

$$(\varepsilon_{sm} - \varepsilon_{cm})^* = \left(\sigma_s^* - \frac{k_t \left(\frac{f_{ct,eff}}{\rho_{p,eff}^*} \right) \left(1 + n^* \rho_{p,eff}^* \right)}{E_r} \right) \geq \frac{0.6\sigma_s^*}{E_r} \tag{6}$$

$$E_r = \frac{E_s A_s + E_f A_f}{A_s + A_f} = \frac{E_s A_s + E_f A_f}{A_r} \tag{7}$$

$$\rho_{p,eff} = \frac{A_r}{A_{c,eff}} \tag{8}$$

Ceroni and Pecce [50] also modified the reinforcement ratio, which considers the use of CFRP written as Equation (9).

$$\rho_{p,eff} = \frac{A_s + \frac{A_f E_f}{E_s}}{A_{c,eff}} = \frac{A_s + n_f A_f}{A_{c,eff}} \tag{9}$$

The CFRP also modified the mechanical properties of the beam. Due to the addition of the tensile force on the bottom, the neutral axis position was changed to a lower position in comparison with non-strengthened beams. To calculate the new position of the neutral axis position, a numerical calculus process was proposed to obtain the neutral axis depth for each load level. At each step, compression and tensile forces were calculated to obtain the force cross-section equilibrium (10). The Bernoulli hypothesis and perfect bond between the steel and concrete was assumed. This allowed all strains to be written as functions of the steel strain (11). The material constitutive laws for concrete is linear-parabolic and the steel mechanical law is considered to be elastic-plastic, with an elastic behaviour for FRP. Therefore, three steps of calculation were undertaken: (i) all materials are elastic, (ii) concrete overpasses the pseudo-elastic range, and (iii) reinforcing steel yields.

The calculation process provided the assignment of steel strain. Then, starting from the geometrical centre y_g, a neutral axis depth was given, and the equilibrium was verified. A new neutral axis value was assigned if the section was not equilibrated. The flexural moment was calculated using the rotational equilibrium when the translational equilibrium was reached.

$$A_s \cdot E_s \cdot \varepsilon_s + A_f \cdot E_f \cdot \varepsilon_f = A_c \cdot E_c \cdot \varepsilon_c + A_s' \cdot E_s \cdot \varepsilon_s' \tag{10}$$

$$\frac{\varepsilon_s}{d - x} = \frac{\varepsilon_f}{H - x} = \frac{\varepsilon_c}{x} = \frac{\varepsilon_s'}{x - c} \tag{11}$$

The corrected neutral axis position changed the post-cracking behaviour of the beam cross-section because of the change in the damaged section inertia I_{II}, calculated as in (12). This value will change for each load level influencing the steel stress, calculated using Navier's Equation (13), where M_{ser} is the flexural moment obtained from the rotational equilibrium, I_I is the moment of inertia of the undamaged cross-section, and M_{cr} is the flexural moment corresponding to the first cracking calculated using Equation (14) including the concrete mean tensile strength f_{ctm}.

$$I_{II} = \frac{b_w h^3}{3} + n A_s (d - x)^2 + n_f A_f (h - x)^2 \tag{12}$$

$$\sigma_s^* = \frac{n^* M_{ser}}{I_{II}} \cdot (d - x) + \frac{n^* M_{cr}}{I_I} \cdot (d - y_g) \tag{13}$$

$$M_{cr} = \frac{f_{ctm}}{y_g} \cdot I_I \tag{14}$$

The crack spacing value was determined using the same formulation reported in (5), substituting the $\rho_{p,eff}$ coefficient with the modified one.

The results accurately represent the crack behaviour until the steel yielded; however, the curve trend was linear rather than a double slope. Thus, the formulation was again modified to consider non-linear behaviour. The weight of the reinforcement materials (i.e., steel and concrete) in the equivalent elastic modulus following steel yielding was considered and discussed. Equation (7) was used in the steel elastic range and replaced with (15) following its yielding.

$$E_r = \frac{\alpha \cdot \sqrt{\beta} \cdot A_s \cdot E_s + A_f E_f}{A_f + A_s} \tag{15}$$

The β factor considers potential damage in the beam after the first cracking and is the ratio between the first cracking moment M_{cr} and the effective load at the time of verification M_{ser}. Finally, α had an experimental value of 2.1. As illustrated in Figure 13, the theoretical model predicted an earlier failure than the observed behaviour. However, because the cracking phenomenon mainly affected the service limit states, it was considered that assessing the accuracy of the empirical model up to theoretical failure was sufficient to describe the phenomenon. For each load level, a pair of crack width values were analysed to estimate the coefficient of determination R^2 and root mean square error (RMSE). The results are listed in Table 8 for each specimen group. In Figure 14, a comparison between the experimental- and model-obtained values is presented for the load level of 60 kN using the average value of the R12 series.

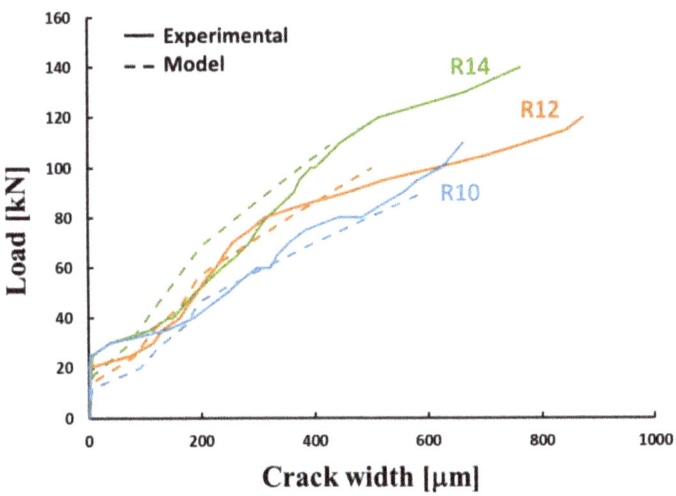

Figure 13. Comparison between the EC2-modified model and the experimental results.

Table 8. Coefficient of determination and root mean square error of the empirical model calculated for the tested groups.

Group	R10	R12	R14
R^2	0.85	0.96	0.98
$RMSE$ [μm]	76.31	39.11	37.63

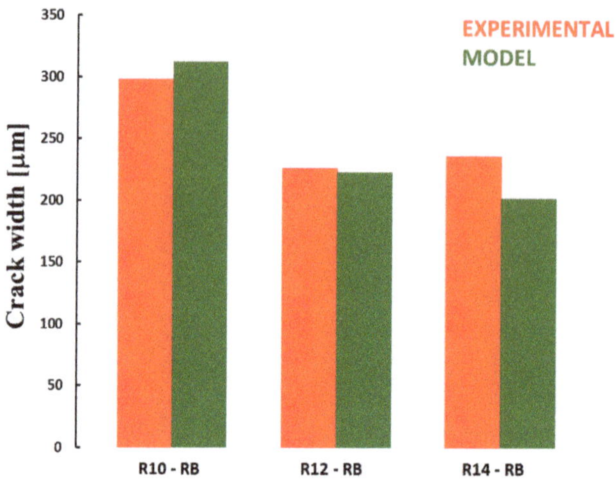

Figure 14. Comparison between the measured and calculated values of crack width corresponding to load level of 60 kN.

4. Conclusions

The present study aimed to evaluate the effect of externally bonded FRP on the cracking behaviour of reinforced concrete beams. Tests were undertaken on ten beams with three ratios of steel reinforcement and the FRP were reinforced by two layers of CFRP. The experimental results demonstrate the influence of the steel ratio on strength improvement, loss of ductility, and crack opening. The main results show that

- with a lower level of steel reinforcement, the contribution of the composite material was more significant.
- the beams with a low steel ratio (0.4%) suffered less loss in ductility and cracking.
- the crack opening calculation based on Eurocode 2 formula is too conservative compared with those measured.
- an empirical model was proposed based on the formulation of the Eurocode with appropriate modifications, taking into account the geometric and mechanical aspects of a strengthened structure. The calculated values were compared with the measured values and showed good agreement.

The presented model is proposed as an easy-to-use tool for design, allowing for the consideration of FRP for crack opening calculations. Future developments can be accomplished by comparing the model to others and comparing with other results from the literature review to validate this model and to introduce the effect of FRP on beams already damaged before strengthening. The model should also be extended to all kinds of FRP (laminate, higher or lower FRP ratio).

Author Contributions: Conceptualization, A.A. and L.M.; methodology, A.A.; software, A.A.; validation, M.S., L.M. and E.F.; resources, E.F.; writing—original draft preparation, E.F., A.A..; writing—review and editing, L.M.; supervision, L.M. and E.F. All authors have read and agreed to the published version of the manuscript.

Funding: This research received no external funding.

Data Availability Statement: The data presented in this study are available on request from the corresponding author.

Conflicts of Interest: The authors declare no conflict of interest.

References

1. Maruyama, I.; Lura, P. Properties of early-age concrete relevant to cracking in massive concrete. *Cem. Concr. Res.* **2019**, *123*, 105770. [CrossRef]
2. Adamo, N.; Al-Ansari, N.; Sissakian, V.; Laue, J.; Knutsson, S. Dam Safety: Technical Problems of Ageing Concrete Dams. *J. Earth Sci. Geotech. Eng.* **2020**, *10*, 241–279.
3. Ahn, I.-S. Creep and Shrinkage of Early-Age Concrete at Cold Temperature. *Adv. Civ. Eng. Mater.* **2021**, *10*, 300–312. [CrossRef]
4. Wan-Wendner, R. Aging concrete structures: A review of mechanics and concepts. *Die Bodenkult. J. Land Manag. Food Environ.* **2018**, *69*, 175–199. [CrossRef]
5. Binda, L.; Tiraboschi, C.; Folli, R.T. On-site and Laboratory Investigation of Materials and Structure of a Belltower in Monza/Untersuchungen an den Werkstoffen und am Tragwerk im Labor und am Objekt am Beispiel eines Glockenturms in Monza. *Restor. Build. Monum.* **2000**, *6*, 41–62. [CrossRef]
6. Issa, M.; Ismail, E.-S. Long-term deflections of FRP reinforced concrete beams. *HBRC J.* **2020**, *16*, 269–282. [CrossRef]
7. Di Tommaso, A.; Neubauer, U.; Pantuso, A.; Rostasy, F. Behavior of Adhesively Bonded Concrete-CFRP Joints at Low and High Temperatures. *Mech. Compos. Mater.* **2001**, *37*, 327–338. [CrossRef]
8. Hu, C.; Sang, L.; Jiang, K.; Xing, J.; Hou, W. Experimental and numerical characterization of flexural properties and failure behavior of CFRP/Al laminates. *Compos. Struct.* **2021**, *281*, 115036. [CrossRef]
9. Bakis, C.E.; Bank, L.C.; Brown, V.L.; Cosenza, E.; Davalos, J.F.; Lesko, J.J.; Machida, A.; Rizkalla, S.H.; Triantafillou, T.C. Fiber-reinforced polymer composites for construction—State-of-the-art review. *J. Compos. Constr.* **2002**, *6*, 73–87. [CrossRef]
10. Nayak, A.N.; Kumari, A.; Swain, R.B. Strengthening of RC Beams Using Externally Bonded Fibre Reinforced Polymer Composites. *Structures* **2018**, *14*, 137–152. [CrossRef]
11. Peng, H.; Zhang, J.; Shang, S.; Liu, Y.; Cai, C.S. Experimental study of flexural fatigue performance of reinforced concrete beams strengthened with prestressed CFRP plates. *Eng. Struct.* **2016**, *127*, 62–72. [CrossRef]
12. Delle Ricerche, C.N. *Guide for the Design and Construction of Fiber-Reinforced Concrete Structures*; SN: CNR-DT; National Research Council-Advisory Committee on Technical Recommendations for Construction: Rome, Italy, 2006; Volume 204, p. 2006.
13. Triantafillou, T.; International Federation for Structural Concrete (Eds.) *Externally Bonded FRP Reinforcement for RC Structures: Technical Report on the Design and Use of Externally Bonded Fibre Reinforced Polymer Reinforcement (FRP EBR) for Reinforced Concrete Structures*; International Federation for Structural Concrete: Lausanne, Switzerland, 2001.
14. AFGC. Réparation et Renforcement des Structures en Béton au Moyen des Matériaux Composites. 2011. Available online: https://www.afgc.asso.fr/publication/reparation-et-renforcement-des-structures-en-beton-au-moyen-des-materiaux-composites/ (accessed on 1 February 2024).
15. Darby, A.; Ibell, T.; Clarke, J. *TR55 Design Guidance for Strengthening Concrete Structures Using Fibre Composite Materials*; The Concrete Society: Sandhurst, United Kingdom, 2004.
16. *SIA166*; Klebebewehrungen (Externally Bonded Reinforcements. Schweizer I. Architektenverein: Zürich, Switzerland, 2004.
17. ACI Committee 440. *440.2R-17: Guide for the Design and Construction of Externally Bonded FRP Systems for Strengthening Concrete Structures*; American Concrete Institute: Farmington Hills, MI, USA, 2017. [CrossRef]
18. Gorbunov, I.; Kakusha, V. Behavior of concrete beams reinforced with FRP during bending. *E3S Web Conf.* **2021**, *263*, 02052. [CrossRef]
19. del Rey Castillo, E.; Allen, T.; Henry, R.; Griffith, M.; Ingham, J. Digital image correlation (DIC) for measurement of strains and displacements in coarse, low volume-fraction FRP composites used in civil infrastructure. *Compos. Struct.* **2019**, *212*, 43–57. [CrossRef]
20. Zhou, K.; Lei, D.; Chun, P.-j.; She, Z.; He, J.; Du, W.; Hong, M. Evaluation of BFRP strengthening and repairing effects on concrete beams using DIC and YOLO-v5 object detection algorithm. *Constr. Build. Mater.* **2024**, *411*, 134594. [CrossRef]
21. Armonico, A.; Saidi, M.; Michel, L.; Bel, S.; Ferrier, E. Inverse analysis based on DFOS technology for the study of the bonding behaviour of FRP to concrete beams. *Mater. Today Commun.* **2023**, *35*, 105655. [CrossRef]
22. Hoult, N.A.; Andy Take, W.; Lee, C.; Dutton, M. Experimental accuracy of two dimensional strain measurements using Digital Image Correlation. *Eng. Struct.* **2013**, *46*, 718–726. [CrossRef]
23. Mehdikhani, M.; Aravand, M.; Sabuncuoglu, B.; Callens, M.G.; Lomov, S.V.; Gorbatikh, L. Full-field strain measurements at the micro-scale in fiber-reinforced composites using digital image correlation. *Compos. Struct.* **2016**, *140*, 192–201. [CrossRef]
24. *Eurocode 2: Design of Concrete Structures—Part 1-1: General Rules and Rules for Buildings*; European Union: Brussels, Belgium, 2004; p. 230.
25. *BS EN 12390-3:2009*; Testing Hardened Concrete. Compressive Strength of Test Specimens. British Standards Institution: London, UK, 2009.

26. *ISO 527-5:2021*; Plastics—Determination of Tensile Properties—Part 5: Test Conditions for Unidirectional Fibre-Reinforced Plastic Composites. ISO: Geneva, Switzerland, 2021; Volume 5.
27. Sutton, M.A.; Orteu, J.J.; Schreier, H. *Image Correlation for Shape, Motion and Deformation Measurements: Basic Concepts, Theory and Applications*; Springer Science & Business Media: Berlin/Heidelberg, Germany, 2009.
28. Pan, B.; Qian, K.; Xie, H.; Asundi, A. Two-dimensional digital image correlation for in-plane displacement and strain measurement: A review. *Meas. Sci. Technol.* **2009**, *20*, 062001. [CrossRef]
29. International Digital Image Correlation Society. *A Good Practices Guide for Digital Image Correlation*; International Digital Image Correlation Society (iDICs): Boston, MA, USA, 2018.
30. Tehrani, A.D.; Kohankar Kouchesfehani, Z.; Najafi, M. Pipe profiling using digital image correlation. In *Pipelines 2020*; American Society of Civil Engineers Reston: Reston, VA, USA, 2020; pp. 36–45.
31. Saidi, M.; Gabor, A. Use of distributed optical fibre as a strain sensor in textile reinforced cementitious matrix composites. *Measurement* **2019**, *140*, 323–333. [CrossRef]
32. Bravo, M.; Sáenz, J.; Navas, M.; Lopez-Amo, M. Concrete Beam Bending Test Monitorization Using a High Strain Fiber Optic Sensor. *J. Light. Technol.* **2012**, *30*, 1085–1089. [CrossRef]
33. Ferrier, E.; Caggegi, C.; Michel, L. Influence of EB-CFRP on Cracks for Reinforced Concrete Beams Strengthening. *Spec. Publ.* **2018**, *237*, 44.1–44.16.
34. Yusra, A.; Opirina, L.; Triwulan, T.; Raka, G.; Safriani, M. Comparative Study of Steel Reinforced Concrete Beams with Bamboo Reinforced Concrete Beams. *Solid State Technol.* **2021**, *63*, 12857–12863.
35. Banjara, N.K.; Ramanjaneyulu, K. Investigations on behaviour of flexural deficient and CFRP strengthened reinforced concrete beams under static and fatigue loading. *Constr. Build. Mater.* **2019**, *201*, 746–762. [CrossRef]
36. Shamass, R.; Mahi, B.; Cashell, K.A.; Abarkan, I.; El-Khannoussi, F. Experimental investigation into the behaviour of continuous concrete beams reinforced with basalt FRP. *Eng. Struct.* **2022**, *255*, 113888. [CrossRef]
37. Al-Khafaji, A.; Salim, H.; El-Sisi, A. Behavior of RC beams strengthened with CFRP sheets under sustained loads. *Structures* **2021**, *33*, 4690–4700. [CrossRef]
38. Ali, H.; Assih, J.; Li, A. Flexural capacity of continuous reinforced concrete beams strengthened or repaired by CFRP/GFRP sheets. *Int. J. Adhes. Adhes.* **2020**, *104*, 102759. [CrossRef]
39. Bocciarelli, M.; di Feo, C.; Nisticò, N.; Pisani, M.A.; Poggi, C. Failure of RC beams strengthened in bending with unconventionally arranged CFRP laminates. *Compos. Part B Eng.* **2013**, *54*, 246–254. [CrossRef]
40. Pisani, M. Evaluation of Bending Strength of RC Beams Strengthened with FRP Sheets. *J. Compos. Constr.* **2006**, *10*, 313–320. [CrossRef]
41. Chajes, M.J.; Thomson, T.A., Jr.; Januszka, T.F.; Finch, W.W., Jr. Flexural strengthening of concrete beams using externally bonded composite materials. *Constr. Build. Mater.* **1994**, *8*, 191–201. [CrossRef]
42. Euro-International Committee for Concrete; Favre, R. *CEB Design Manual on Cracking and Deformations*; École Polytechnique Fédérale de Lausanne: Lausanne, Switzerland, 1985.
43. Ng, P.L.; Kwan, A. A rigorous analytical model for shrinkage cracking of reinforced concrete. *Mag. Concr. Res.* **2017**, *69*, 120–133. [CrossRef]
44. Hognestad, E. High-strength bars as concrete reinforcement, Part 2: Control of flexural cracking. *J. PCA Res. Dev. Lab.* **1962**, 46–53.
45. Watstein, D.; Parsons, D.E. Width and spacing of tensile cracks in axially reinforced concrete cylinders. *J. Res. Natl. Bur. Stand.* **1943**, *31*, 1–24. [CrossRef]
46. Borosnyói, A.; Balázs, G.L. Models for flexural cracking in concrete: The state of the art. *Struct. Concr.* **2005**, *6*, 53–62. [CrossRef]
47. *ACI 318R-08*; Building Code Requirements for Structural Concrete (ACI 318-08) and Commentary: An ACI Standard. American Concrete Institute (ACI): Farmington Hills, MI, USA, 2008.
48. *ECP 203-2007*; Egyptian Code for Design and Construction of Reinforced Concrete Structures. Ministry of Housing, Utilities and Urban Communities: El Sayeda Zeinab, Egypt, 2007.
49. *BS 8110-2:1985*; Structural Use of Concrete. Part 2: Code of Practice for Special Circumstances. British Standards Institution: Milton Keynes, UK, 1985.
50. Ceroni, F.; Pecce, M. Design provisions for crack spacing and width in RC elements externally bonded with FRP. *Compos. Part B Eng.* **2009**, *40*, 17–28. [CrossRef]
51. Barris, C.; Torres, L.; Vilanova, I.; Miàs, C.; Llorens, M. Experimental study on crack width and crack spacing for Glass-FRP reinforced concrete beams. *Eng. Struct.* **2017**, *131*, 231–242. [CrossRef]
52. Leonhardt, F. *Crack Control in Concrete Structures*; PCI Journal: Chicago, IL, USA, 1977.

buildings

MDPI

Article

Effect of Textile Layers and Hydroxypropyl Methylcellulose on Flexural Behavior of TRLC Thin Plates

Jiyang Wang [1,2,*], Dan Yu [1], Chen Zeng [1], Xiaohua Ji [1], Lingpeng Ye [2,3], Pinghuai Zhou [4] and Senlin Zhao [5]

1 College of Civil Engineering and Architecture, Zhejiang University, Hangzhou 310058, China; 22112269@zju.edu.cn (D.Y.); 22012216@zju.edu.cn (C.Z.); jixiaohua@zju.edu.cn (X.J.)
2 Center for Balance Architecture, Zhejiang University, Hangzhou 310007, China; ylpccea@zju.edu.cn
3 The Architectural Design & Research Institute of Zhejiang University Co., Ltd., Hangzhou 310007, China
4 Zhejiang Province Institute of Architectural Design and Research, Hangzhou 310006, China; anji18@126.com
5 Zhejiang Building—Tech Energy Dissipation Technology Co., Ltd., Shaoxing 312030, China; zsl@zjjkjz.com
* Correspondence: kyotowang@zju.edu.cn

Abstract: To examine the flexural toughness characteristics of textile-reinforced lightweight aggregate concrete (TRLC), a four-point bending test was conducted to assess the impact of varying numbers of textile layers and the inclusion of hydroxypropyl methylcellulose on the ultimate load-bearing capacity and deformation capacity of TRLC thin plates. Six groups of specimens were prepared for the experiment, and the bending capacity of the thin plates in each group was evaluated. The flexural toughness index was utilized to quantify the bending performance of TRLC thin plates. The findings revealed that increasing the number of textile layers improved the initial cracking load, initial cracking deflection, ultimate load, ductility, and flexural toughness of the thin plates. For the specimens without HPMC, the initial cracking load was increased by up to 36.1%, the ultimate load by up to 40.9%, and the flexural toughness index by up to 292% as the number of textile layers was increased. For specimens doped with HPMC, the initial cracking load was increased by up to 61.7%, the ultimate load by up to 246.7%, and the flexural toughness index by up to 65%. The TRLC thin plate containing hydroxypropyl methylcellulose exhibited a reduced initial cracking load yet displayed a stronger matrix consistency and good flexural toughness. Moreover, the enhancement in the ultimate load of TRLC thin plates with hydroxypropyl methylcellulose was more pronounced with an increased number of textile layers, resulting in a significantly higher number of cracks compared to TRLC without hydroxypropyl methylcellulose and an 11.40-fold increase in the flexural toughness index.

Keywords: textile-reinforced concrete; lightweight aggregate; flexural toughness index; thin plate; hydroxypropyl methylcellulose

1. Introduction

Lightweight aggregate concrete refers to concrete formulated with light coarse aggregate, light sand (or ordinary sand), cementitious materials, admixtures, water, etc., and its dry apparent density is not greater than 1950 kg/m^3. Light aggregate concrete is widely used in engineering practice due to its characteristics, such as light deadweight, but it needs to be optimized due to its disadvantages, such as brittleness and poor mechanical properties [1]. Research on textile-reinforced concrete (TRC) started in the late 1980s. Gardiner et al. [2] proved it is an effective reinforcement method by incorporating a fabric net with fixed intervals in the warp and weft yarns in cementitious materials. It also has been shown that the mechanical properties of concrete can be improved by incorporating textiles into the concrete [3–5]. The bonding strength between the textile and the concrete matrix influences the synergetic performance of the TRC [6]. The bonding performance of the TRC interface depends mainly on the mechanical properties and geometrical characteristics of the textile itself as well as on the strength of the concrete matrix [7,8]. To enhance the interface bonding performance, Xu and Yin [9–11] proved that epoxy impregnation and sand bonding, mixing of short-cut fibers, and applying pre-stressing can all enhance the bonding performance through a large number of experimental analyses, numerical simulations, and mechanism studies.

However, in practical structural engineering retrofitting and maintenance, the total weight of the structure cannot be significantly increased, as this may alter the inherent characteristics of the structure, such as the natural frequency, thereby adversely affecting the seismic behavior of the structure. Building on the previously discussed enhancement effect of textiles on the mechanical properties of concrete matrix, researchers have incorporated them into lightweight aggregate concrete to optimize its mechanical properties. Vast [12] designed a coffee table using textile-reinforced lightweight aggregate concrete (TRLC). Cibulka [13] et al. produced several thin-walled slabs reinforced with two-dimensional carbon and 3-dimensional glass textiles. The tests confirmed a significant increase in the strength of all reinforced specimens. Rozycki et al. [14] employed "Stikloporas" lightweight aggregate, which has a porosity of 41%, in the preparation of TRLC. They conducted experimental investigations on the flexural and compressive mechanical characteristics. The findings indicated that thinner specimens exhibited improved toughness and excellent interfacial bonding between glass fibers and the lightweight aggregate cement mortar matrix. Additionally, they utilized this TRLC material to design a canoe named "PKanoe", showcasing the versatility and strength of the material. Naseri et al. [15] fabricated 18 TRLC thin plates using various textiles and subjected them to three-point bending tests. The findings revealed that the TRLC thin plates using nylon textiles exhibited good ductility but had a lower flexural load-bearing capacity. In contrast, the TRLC thin plates using steel fiber net displayed a high flexural load-bearing capacity and adequate energy absorption capabilities despite their limited ductility. Both thin-plate types demonstrated ample wind resistance, making them suitable for use in the construction of walls and roofs for prefabricated houses. Lightweight aggregate concrete is commonly employed in industrial and civil buildings, as well as other projects, to reduce structural weight and conserve materials. Incorporating textiles into lightweight aggregate concrete produces structures that are lightweight yet highly resilient. TRLC technology can be utilized in the production of walls, partition walls, heat insulation, roofs, insulation panels, floor slabs, staircases, pipes, bridges, and various other building components.

While the aforementioned studies have highlighted the potential of textiles as reinforcement in lightweight aggregate concrete, the precise bonding dynamics and synergistic performance between the lightweight aggregate concrete matrix and textiles remain ambiguous. The bonding properties between the textile and the concrete matrix significantly influence the structure's mechanical properties. Lightweight aggregate concrete, characterized by numerous internal pores, high water content, and brittleness, presents unclear bonding properties with textiles. While hydroxypropyl methylcellulose exhibits a notable thickening effect and enhances the tensile bond strength of cement mortar, its impact on TRLC remains uncertain. Moreover, the effects of critical factors, such as the number of textile layers, on the mechanical properties of TRLC are not yet fully elucidated. Therefore, further investigation is necessary to explore the influence of varying numbers of textile layers and the external incorporation of hydroxypropyl methylcellulose into the matrix on the flexural behavior of TRLC thin plates. In this study, we conducted four-point bending tests on six sets of 36 TRLC thin plates to examine the impact of varying numbers of textile layers and external doping of the matrix with hydroxypropyl methylcellulose on the ultimate load-bearing capacity and deformation capacity of TRLC thin plates. Additionally, we utilized the bending toughness index to quantitatively assess the bending performance of TRLC thin plates. The results provide valuable insights into the bending capacity of TRLC thin plates and the potential effects of different material compositions on their mechanical properties.

2. Experimental Program

2.1. Specimen Design

In this study, the effect of the number of textile layers and the presence of hydroxypropyl methylcellulose in the cement mortar matrix on the flexural properties of a TRLC thin plate was examined. HPMC enhances the water retention of cement mortar and acts as a thickening agent in cement slurry. Studies have demonstrated that a 0.5% HPMC content can substantially improve the plastic viscosity and water retention of cement mortar [16]. Based on the proportion of TRC cement mortar matrix obtained from previous studies [17,18], the light aggregate concrete matrix proportion and experimental groups for this experiment are detailed in Tables 1 and 2, respectively. Six specimens were prepared for each experimental group.

Table 1. The proportion of light aggregate concrete matrix (kg/m^3).

Cement	Fly Ash	Silica Fume	Shale Ceramic Sand	Mixing Water	Pre-Wetted Water	Superplasticizer
445	160	32	835	158	167	3.6

Table 2. Specimen index and main information.

Specimen	Number of Textile Layers	Hydroxypropyl Methylcellulose Dosage
Y1	1	0%
MY1	1	0.5%
Y2	2	0%
MY2	2	0.5%
Y3	3	0%
MY3	3	0.5%

Note: Y—epoxy-impregnated textile specimen; 1/2/3—number of textile layers; M—externally doped with 0.5% hydroxypropyl methylcellulose by mass ratio to cement.

Peled A [19] found that a 3 mm cement mortar protective layer was adequate to safeguard the textile in TRC members up to 30 mm thickness, effectively preventing cracks in the protective layer. To ensure the durability of TRC members, a 5 mm cement mortar protective layer is recommended. Consequently, all specimen dimensions were standardized to 500 mm × 100 mm × 16 mm, with the thickness of TRLC varying depending on the number of textile layers, as illustrated in Figure 1.

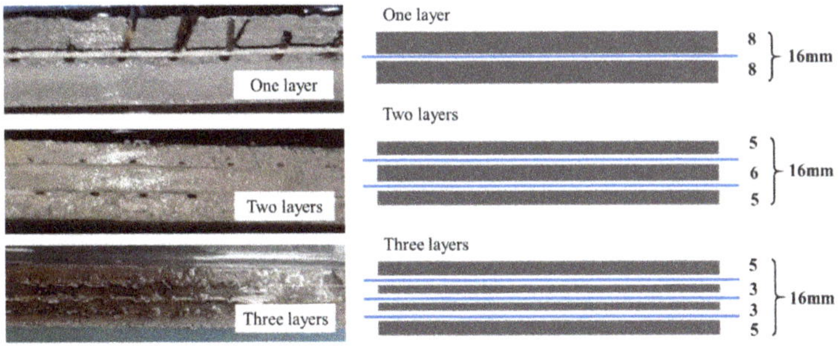

Figure 1. TRLC thin plate thickness and number of textile layers.

2.2. Raw Materials

In the preparation of TRLC thin plate specimens, PII 52.5 grade silicate cement and Class II fly ash were employed in the light aggregate concrete matrix, and the relevant parameters are shown in Tables 3 and 4, respectively. The ceramic sand utilized was the 700-grade crushed shale ceramic sand manufactured by Hubei Yichang Guangda Ceramic

Sand Co., Ltd. (Yichang, China), featuring a particle size ranging from 0 mm to 1.2 mm; a fineness modulus of 1.7, placing it in the category of fine sand in Zone II; an apparent density of 1740 kg/m^3; and a water absorption rate of 20%. The superplasticizer used was a BASF polycarboxylic acid high-efficiency superplasticizer with a water-reducing rate of 27%. The thickening agent applied was hydroxypropyl methylcellulose (HPMC) from Fuqiang Chemical Industry (Jinzhou, China), characterized by a viscosity of 200,000 MPa.s. The light aggregate concrete matrix exhibited a 28-day compressive strength of 55.59 MPa and a 28-day flexural strength of 8.26 MPa.

Table 3. Properties of fly ash.

Fineness (>45 μm, %)	Water Demand (%)	Loss on Ignition (%)	Moisture Content (%)	SO_3 (%)	Free CaO (%)	28d Activity Index (%)
9.7	93	1.8	0.02	0.7	0.05	79

Table 4. Properties of silica fume.

Fineness (>45 μm, %)	SiO_2 (%)	Loss on Ignition (%)	Moisture Content (%)	Moisture Content (kg/m^3)
0.5	94	0.5	0.4	0.282

The textile selected was crafted from basalt–glass fibers, with its overall morphology and fiber structure under electron microscope scrutiny depicted in Figure 2. This textile was a two-dimensional weave comprising basalt fiber bundles and glass fiber bundles, featuring a mesh size of 10 mm × 10 mm, a fiber bundle width of 1 mm, a fiber density of 1550 g/m^2, a fiber linear density of 0.80 g/m, a fiber mesh thickness of 0.2 mm, and individual fiber monofilaments with a diameter of approximately 13 μm, as discerned via SEM microscopy. We conducted five sets of mechanical property tests on basalt fiber bundles, and the results obtained are presented in Table 5. As determined through testing, the ultimate tensile strength of the basalt fiber bundles ranged from 427 to 586 MPa, with an average value of 490.8 MPa and a standard deviation of 56 MPa. In addition, Young's modulus was 27.8 GPa, and the ultimate tensile strain was 1.77%. The glass fibers, considered unstressed fibers, exhibited a measured monofilament tensile strength of 3100 MPa, Young's modulus of 72 GPa, and an elongation at break of 4.5%. The test results showed some variation, which may be due to the difficulty of achieving complete consistency during material processing, transport, and storage. Additionally, as noted by Dai et al. [20], epoxy impregnation can notably bolster the tensile strength of fiber bundles. For instance, the basalt fiber bundles experienced a 170% surge in ultimate tensile strength post-impregnation compared to pre-impregnation levels. However, the extent of epoxy impregnation and its uniformity could not be precisely regulated, potentially contributing to the variability observed in the test results.

Table 5. Mechanical properties of basalt fiber bundles of textile.

	Sample 1	Sample 2	Sample 3	Sample 4	Sample 5	Average
Tensile strength (MPa)	512.7	427.3	482.2	445.6	586.0	490.8
Young's Modulus (GPa)	29.5	26.7	26.9	28.6	27.2	27.8
Ultimate strain (%)	1.74	1.60	1.79	1.59	2.15	1.77

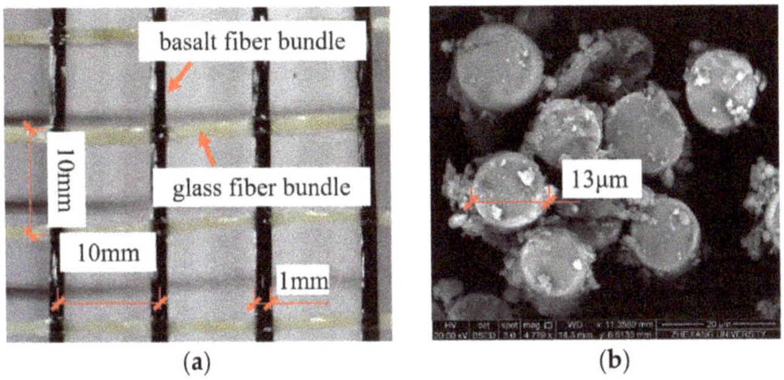

Figure 2. (**a**) Mesh size and (**b**) micro-morphology of basalt–glass textile.

2.3. Specimens and Preparation

To streamline the mass production of TRLC thin plate specimens, wooden mold frames measuring 650 mm × 500 mm were utilized. As illustrated in Figure 3, wooden molds of varying thicknesses—3 mm, 5 mm, 6 mm, and 8 mm—were crafted, with each frame combination capable of producing six specimens. The specimen fabrication process unfolds as follows:

(1) Pre-treatment: Affix the textile onto the mold using adhesive, ensuring that the radial basalt fibers align parallel to the 500 mm side during textile placement to preserve their utility as tensile fibers post-cutting into thin plates later. In addition, the subsequent cutting position was calculated to avoid cutting into the basalt fiber bundles. A mixture of epoxy resin and hardener in a 1:1 ratio was prepared, adding 1/4 to 1/5 xylene as a thinner based on the external temperature conditions. The textile was cured for 24 h before the subsequent pouring of the cement mortar matrix began.

(2) Specimen production: First, we fixed the wooden frame on the base plate and mixed the light aggregate concrete, in which the hydroxypropyl methylcellulose was mixed with the dry powder and stirred evenly. The protective layer of light aggregate concrete was poured first, then the mold with the textile was fixed, and finally, the light aggregate concrete was poured. After 24 h of initial setting, the mold was removed, and the specimens were cured. The specimen preparation process is shown in Figure 3.

(3) Specimen processing: After the molded specimens were cured for 28 d, the specimens were cut into standard bending specimens of 500 mm × 100 mm using a cutting machine, as shown in Figure 4.

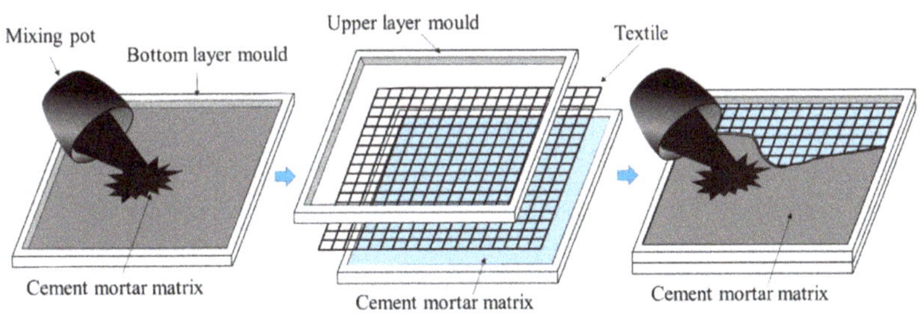

Figure 3. Preparation schematic of TRLC thin plates.

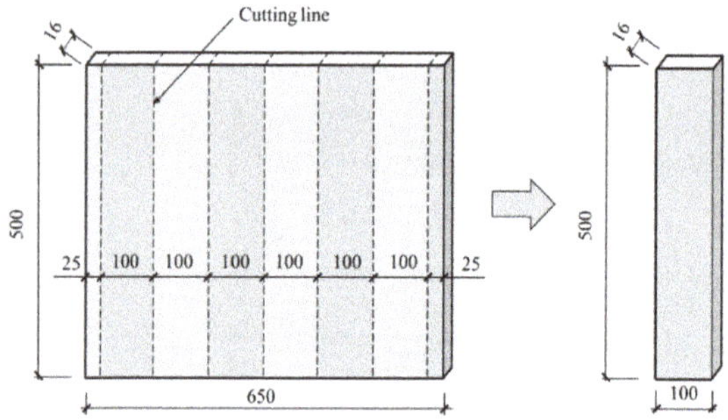

Figure 4. Schematic of specimen cutting from an entire thin plate. (Note: All units in mm).

2.4. Testing Method

The four-point bending test was conducted by the "Standard for test methods of concrete physical and mechanical properties" (GB/T 50081-2019) [21] and the "Standard test methods for fiber reinforced concrete" (CECS 13:2009) [22]. The INSTRON universal testing machine was employed for the loading tests. The length of the thin plate measured 500 mm and a span of 300 mm to preserve the length of the pure moment section as much as possible and the shear span of 100 mm on each side. Displacement loading control was adopted, with a loading speed set at 0.5 mm/min. When the textile was pulled off or pulled out, or when the specimen was severely damaged, the loading was stopped. Linear variable displacement transducers (LVDT) were positioned at the span center and span end. The schematic diagram of the loading device is shown in Figure 5.

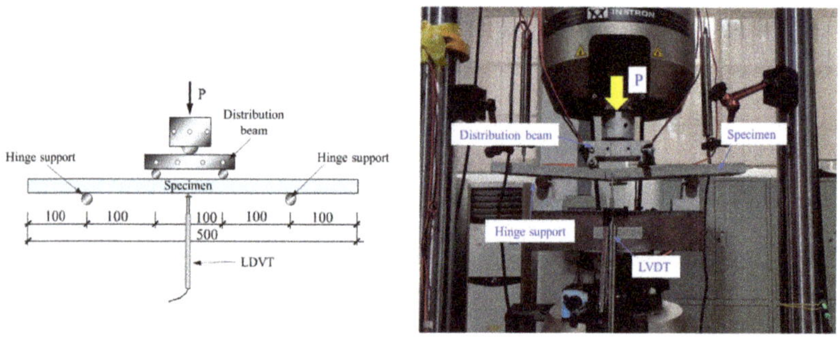

(a) (b)

Figure 5. Four-point bending test device and specimen. (**a**) Dimensions and loading device arrangement; (**b**) a photo of the loading test. Note: the unit for the numbers is mm.

3. Test Results and Analysis

3.1. Characterization of Damage at Various Stages of Loading

The load-deflection relationship curves for the six groups of tests are shown in Figure 6. As shown in the figures, the trends of the load-deflection curves are relatively consistent, but the results still differ between specimens. In this study, the fiber bundles were impregnated with epoxy resin to improve the bonding properties of the fiber bundles. However, the degree of epoxy resin impregnation and its uniformity could not be precisely controlled in advance, resulting in a relatively discrete tensile strength of the fiber bundles [20]. This may be one of the reasons for the differences in the test results. In order to make a more scientific comparison, an average curve was obtained for each test group for subsequent analysis.

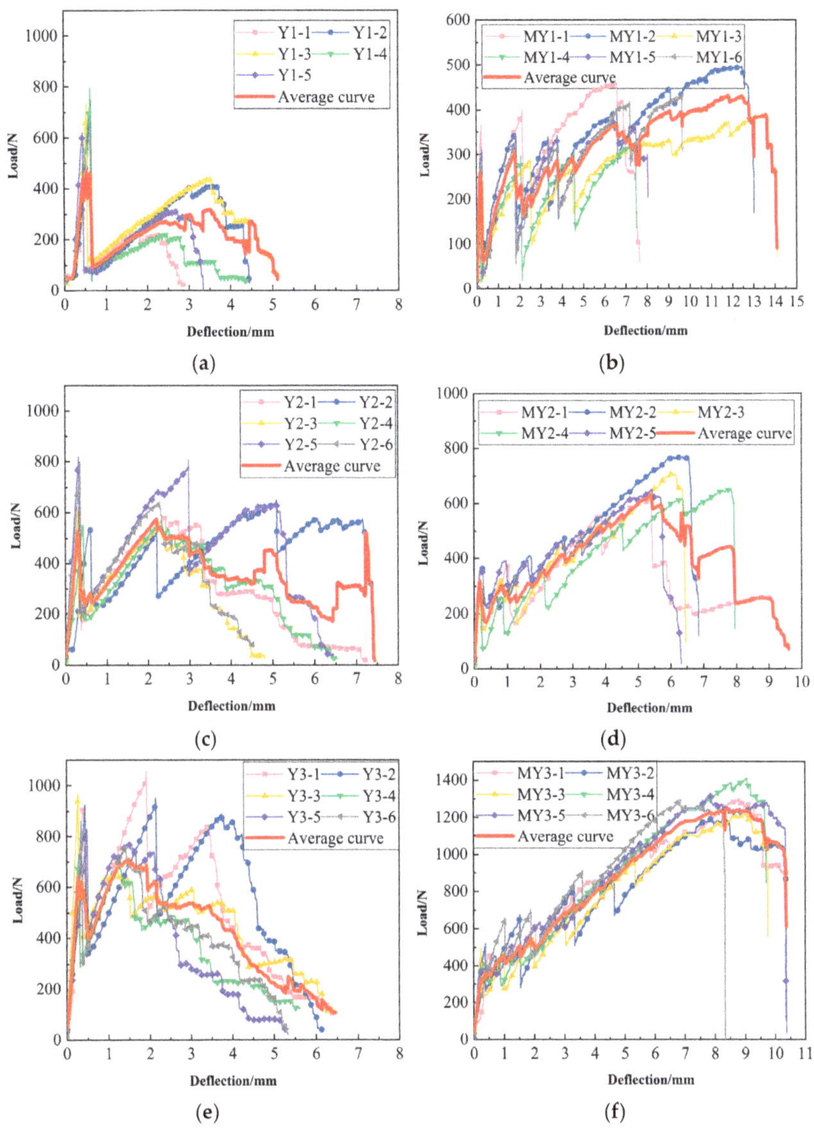

Figure 6. Load-deflection relation curves: (**a**) Y1; (**b**) MY1; (**c**) Y2; (**d**) MY2; (**e**) Y3; and (**f**) MY3.

The initial cracking point is the first spike of the load on the curve. After the cracking of the specimen, there was a sudden drop in the load, and due to the bonding and bridging effect between the textile and the concrete matrix, the specimen was not completely damaged despite the sudden drop in the load. The textile began to be tensed, and the load continued to increase. With the increase in the number of textile layers and the incorporation of hydroxypropyl methylcellulose, the initial cracking load and ultimate load of the specimen increased. The damage of the specimen showed a tendency for multi-seam cracking. The damage mode of the TRLC thin plate can be divided into four stages. The first is the concrete matrix cracking; thereafter, due to the bridging effect of the textile, the TRLC thin plate is synergistically stressed, forming a state of multi-seam cracking. As the load continues to increase, a certain main crack appears to bear the main stress. As the deflection increases, it continues to expand, and finally, the final damage occurs along the main crack. The damage form of the TRLC thin plate is mainly realized in the form of textile pulling off, and thus the overall fracture. The specific load-deflection behavior and crack development state are shown in Figure 7.

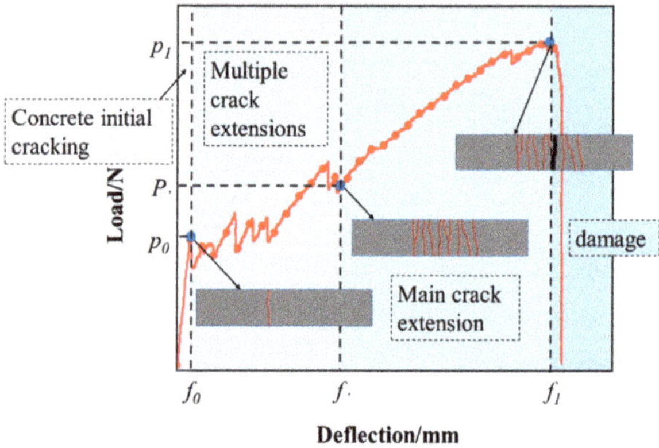

Figure 7. Bending damage stages of TRLC thin plate.

The initial cracking load, initial cracking deflection, peak load, and ultimate deflection for each test group are shown in Table 6. All data in Table 5 are averages of the results from each group. To quantify the bending stress, the equivalent bending stress σ was introduced and calculated as follows:

$$\sigma = \frac{3PL}{2bh^2} \tag{1}$$

where P is the peak load, L is the distance between the hinge supports in Figure 5, b is the width of the thin plate, and h is the thickness of the TRLC thin plate.

Table 6. Test results of each specimen group.

Specimen	Initial Cracking Load (N)	Initial Cracking Stress (MPa)	Initial cracking Deflection (mm)	
Y1	670.24 (71.75)	11.78 (1.26)	0.40 (0.13)	
Y2	693.22 (117.78)	12.19 (6.49)	0.41 (0.13)	
Y3	912.16 (50.7)	16.03 (0.89)	0.37 (0.09)	
MY1	283.11 (23.43)	4.98 (2.74)	0.20 (0.03)	
MY2	317.40 (10.58)	5.58 (0.18)	0.17 (0.01)	
MY3	457.79 (36.62)	8.05 (0.65)	0.34 (0.09)	
Specimen	**Peak Load (N)**	**Peak Stress (MPa)**	**Ultimate Deflection (mm)**	**Number of Cracks (Average Value)**
Y1	670.24 (71.75)	11.78 (1.26)	3.37 (0.77)	1.00
Y2	719.08 (81.04)	12.64 (1.59)	6.42 (1.32)	1.50
Y3	944.38 (78.3)	16.60 (1.38)	6.17 (0.62)	1.33
MY1	378.51 (55.93)	6.65 (0.99)	8.43 (1.12)	3.33
MY2	665.32 (46.08)	11.7 (0.81)	8.00 (1.57)	4.40
MY3	1312.33 (92.16)	23.07 (1.62)	9.62 (0.39)	6.33

Note: Each data point is taken as the average of the test results of the six specimens in the same experimental group, with standard deviations in parentheses. Peak load and peak stress correspond to the maximum load during loading, and the ultimate deflection is the deflection corresponding to the last peak point on the load-deflection curve.

The ideal state of cracking under the bending load is multi-seam cracking in TRLC thin plates. The cracking of thin plates in the different experimental groups is given in

Figure 8, and the average values of the crack number statistics for the pure bending section are given in Table 6. The results show that the enhancement of the number of textile layers increases the likelihood of the occurrence of the number of multi-cracks, and its effect is more pronounced on the HPMC-doped TRLC thin plate. The number of cracks also tends to increase due to the rise in bond strength between the textile and the concrete matrix.

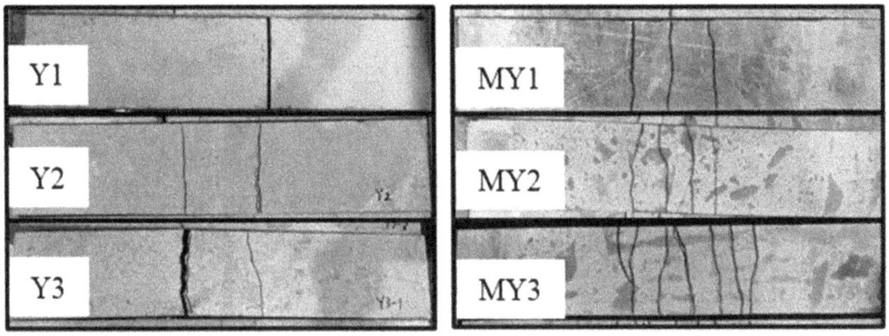

Figure 8. TRLC thin plate cracking conditions.

3.2. *Effect of the Number of Textile Layers on the Bending Properties of TRLC Thin Plate*

Figure 9 illustrates the load-deflection curves for TRLC thin plates with varying numbers of textile layers, and the average curves of each group were taken for comparison.

Due to the bonding and bridging interaction between the textile and the concrete matrix, the concrete retained the capability to distribute the external loads with the textile even after cracking occurred. As the number of textile layers increased, the residual stress following the initial concrete cracking also rose, leading to a higher incidence of cracks, improved ultimate load-carrying capacity and ductility in the TRLC thin plate, and a more favorable damage pattern. The transition from one to two textile layers resulted in a more substantial increase in the initial cracking load and damage load. Upon reaching three textile layers, multiple cracks emerged in the thin plate post-initial cracking, allowing for a greater energy absorption capacity. Relative to the Y1 specimens, the initial cracking loads of Y2 and Y3 were elevated by 3.4% and 36.1%, the ultimate loads by 7.3% and 40.9%, and the ultimate deflections by 90.5% and 83.1%, respectively. The number of textile layers had little effect on the initial cracking deflection of the specimens in group Y. Relative to the MY1 specimens, the initial cracking loads of MY2 and MY3 were elevated by 12.1% and 61.7%, and the ultimate loads by 75.8% and 246.7%, respectively. The ultimate deflection of the MY2 specimens is slightly lower than that of the MY1 specimens. The reason for this may be that the addition of HPMC has significantly increased the ultimate deflection of the specimens, and the increase in the number of textile layers from one to two has a very limited effect on the ultimate deflection at this point. When the number of layers was increased to three, the ultimate deflection of the group's MY3 specimens increased by 14% compared to the group's MY1 specimens.

In the case of the HPMC-doped TRLC thin plate (Groups MY), the enhanced bonding between the concrete matrix and textile resulted in a stronger bridging effect post-initial cracking. This effect hindered crack propagation and promoted the development of multiple cracks in the TRLC thin plate. All specimens within this group exhibited damage characterized by multiple cracks, a preferred form of structural damage. The addition of more textile layers notably improved the flexural properties, leading to an increase in crack formation, as well as enhancements in the initial cracking load and ultimate load.

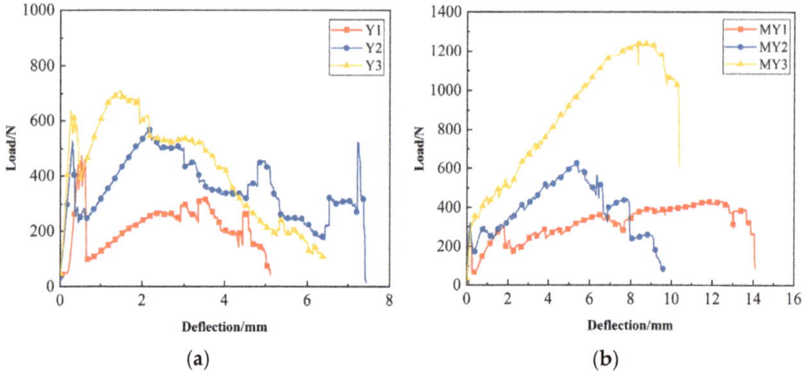

Figure 9. Effect of the number of textile layers on the bending properties of TRLC thin plate. (**a**) Groups Y; (**b**) Groups MY.

Figure 10 illustrates the linear relationship between the number of textile layers and the ultimate load of groups Y and groups MY. In examining the correlation between the ultimate load and the number of textile layers, it is observed that for a standard lightweight aggregate concrete matrix, an increase in the number of textile layers can indeed boost the ultimate load of the TRLC thin plate. A positive correlation is noted between the number of textile layers and the ultimate load within the test group. Specifically, a higher number of textile layers resulted in a considerable increase in the ultimate load within the test group.

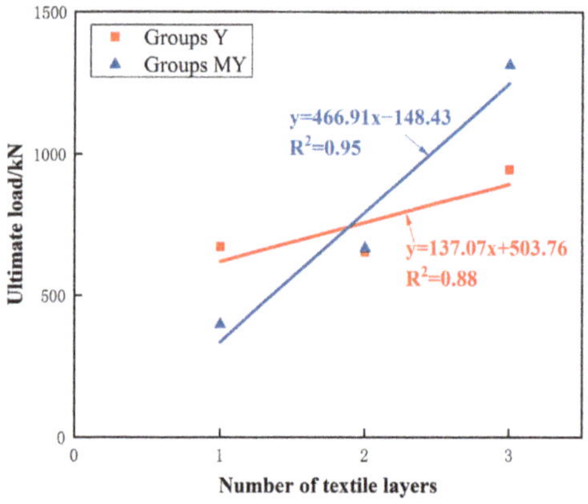

Figure 10. Linear relationship between the number of textile layers and ultimate load of groups Y and groups MY.

On the whole, augmenting the number of textile layers proves to be a beneficial strategy for enhancing the flexural properties of TRLC thin plates. This enhancement is primarily evident in several aspects: (1) enhancement of the initial cracking load and initial cracking deflection; (2) improvement in residual stress post-initial cracking to prevent premature damage; (3) promotion of multi-crack formation following initial cracking, thereby improving the ductility of the thin plates; and (4) delaying textile pull-out or pull-off, consequently enhancing the ultimate load capacity of the thin plates.

3.3. Effect of HPMC Doping on Flexural Properties of TRLC Thin Plate

Two types of concrete matrix were selected for this research: a standard lightweight aggregate concrete matrix and a lightweight aggregate concrete matrix with HPMC. Figure 11 illustrates the load-deflection curves of TRLC thin plates with varying concrete matrices, and the average curves of each group were taken for comparison. The graph indicated a significant decrease in the initial cracking load and stress of the HPMC-doped thin plate under bending stress. This decline can be attributed to the presence of voids within the HPMC-doped thin plate itself, resulting in lower stiffness and an earlier onset of initial cracking. Moreover, the bond strength between the HPMC-doped matrix and the textile was notably robust. Following the initial cracking, the textile exhibited enhanced bridging capabilities, impeding crack propagation and leading to the formation of multiple cracks in the TRLC thin plate. As the number of cracks increased, the thin plate demonstrated improved load-bearing capacity and enhanced ductility. To validate the aforementioned analysis and elucidate the impact of HPMC on TRLC thin plates at a microscopic level, two additional TRLC thin plates were prepared without epoxy resin impregnation on the textile to emphasize the influence of HPMC. Images of the interfaces of these two TRLC thin plates examined under an electron microscope are presented in Figures 12 and 13, respectively.

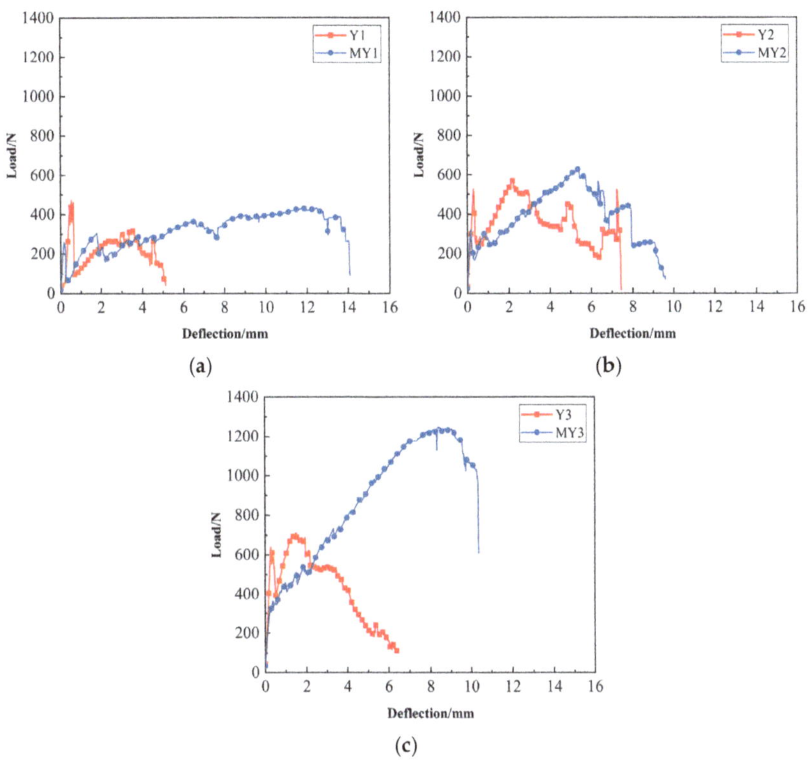

Figure 11. Effect of HPMC doping on load-deflection curves of TRLC thin plate. (**a**) Average curves of test groups Y1 and MY1, respectively, were selected for comparison of one layer of textile. (**b**) Average curves of test groups Y2 and MY2, respectively, were selected for comparison of two layers of textiles. (**c**) Average curves of test groups Y3 and MY3, respectively, were selected for comparison of three layers of textiles.

In Figure 12, it is evident that at the interface under 100 times magnification, the HPMC-doped specimen exhibits a notably higher presence of minute bubbles and holes compared to the undoped HPMC specimen. The internal structure appears loosely packed, leading to reduced stiffness at the initial loading stage, consequently causing a decrease in the initial cracking load in comparison to the undoped HPMC specimen.

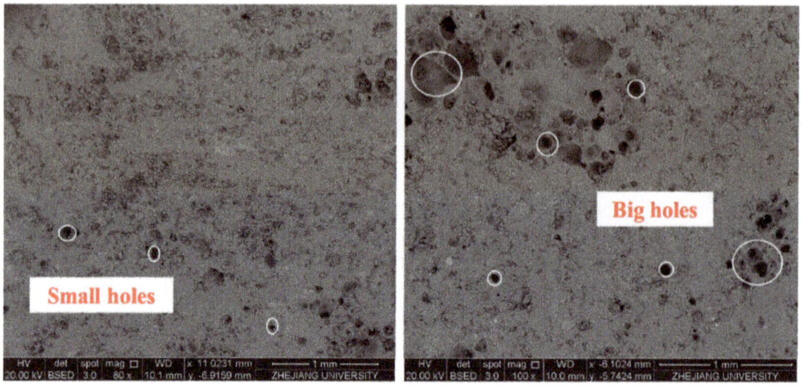

(**a**) (**b**)

Figure 12. Internal holes in (**a**) HPMC−undoped and (**b**) HPMC−doped specimens.

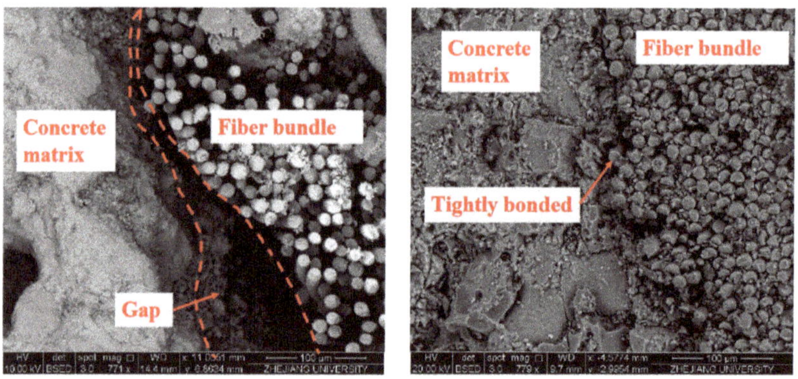

(**a**) (**b**)

Figure 13. Interface of (**a**) HPMC−undoped and (**b**) HPMC−doped specimens.

In Figure 13, it is apparent that the hydration reaction at the interface of the HPMC-doped specimens is more pronounced. This improved reaction could be attributed to the water retention properties of HPMC, facilitating better hydration of the cement at the interface that may have been inadequately wetted otherwise. As a result, the interfacial adhesion is bolstered to a certain degree. Additionally, the hydration products of cement penetrated into the fiber bundles, filling the gaps between the fiber filaments and augmenting the frictional forces.

As depicted in Figure 11, the ultimate loads of all test groups externally treated with HPMC exhibited an increase, except group Y1. The specimens in group Y demonstrated a sharp rise in the peak load at initial cracking, following which subsequent peak loads on the curve showed minimal increments. Conversely, the specimens in group MY exhibited a lower load at initial cracking, attributed to the presence of voids in the externally treated HPMC, resulting in reduced stiffness. However, as the textile progressively engaged during the loading process, the peak loads on the curve continued to rise. In the case of groups MY1 and MY2, where the initial cracking load experienced a significant decrease, a single or double layer of textile had a limited impact on the ultimate load of the specimens; thus, the ultimate loads in groups MY1 and MY2 remained unaffected by the HPMC treatment of the matrix. Conversely, in group MY3, the presence of multiple textile layers led to a noticeable enhancement in load, consequently increasing the ultimate load as well. For the group MY3 specimens, the effect of adding HPMC is very obvious, the ultimate load increased by 367.95 N with an increase of 40.0%. Although the addition of HPMC did not raise the ultimate loads of specimens in groups M1 and M2, it raised their ultimate deflections for all three sets of specimens. For one, two, and three textile layers, the ultimate deflection is increased by 150%, 24.6%, and 55.9%, respectively.

4. Flexural Toughness of TRLC Thin Plate

The flexural properties of the TRLC thin plate are related to both the ultimate deflection and ultimate load of the specimen. Therefore, the flexural toughness index is used to define the energy dissipation capacity of a material or structure from the onset of the load action to the cessation of damage. Various methodologies exist for determining the flexural toughness index, including those outlined by the American Concrete Institute (ACI 544) [23], the Japanese Society of Civil Engineering (JSCE G552) [24], and the American Society for Testing and Materials (ASTM C 1018) [25].

The simplified ASTM C 1018 standard is employed for the calculation in this test [25]. The flexural toughness index is defined as follows: the area under the curve corresponding to the initial cracking deflection is denoted as S_0, representing the initial cracking energy consumption, while the area under the curve corresponding to the ultimate deflection is denoted as S_1, representing the ultimate energy consumption. The calculation for S_0 or S_1 is conducted, as shown in Equation (2), with a visual representation provided in Figure 14.

$$S_0 = \int P_0 df \tag{2}$$

where P_0 is the corresponding load at initial cracking or ultimate, and f is the corresponding deflection.

The flexural toughness index (I) is expressed as the ratio of the ultimate energy consumption to the energy consumption at the first crack and is calculated as follows:

$$I = S_1/S_0 \tag{3}$$

where, S_0 and S_1 are the measured values of toughness (N·mm) when the deflection in the span is the initial crack and the limit, respectively.

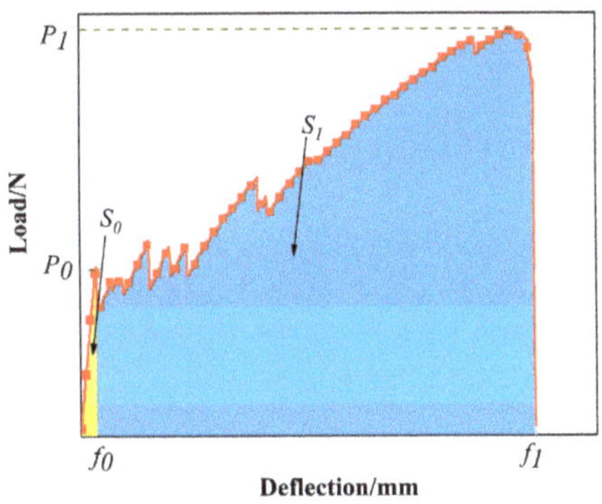

Figure 14. Schematic diagram of S_0 and S_1 calculation methods.

The flexural toughness index I of the TRLC thin plate under various conditions is presented in Table 7. From the data in Table 7, it can be seen that the flexural toughness index increases with the number of layers of the textile. The flexural toughness index of the Y2 and Y3 specimens was increased by a factor of 2.63 and 2.92, respectively, compared to the Y1 specimens. The flexural toughness index of the MY2 and MY3 specimens was increased by 60% and 65%, respectively, compared to the MY1 specimens. In order to further analyze the relationship between the number of textile layers and the flexural toughness index of the specimens, the best fit was conducted using linear regression analysis. The fitting results are shown in Figure 15. The coefficients of regression were 0.82 for the specimens in group Y and 0.80 for the specimens in group MY. The fitting results indicated a correlation between the flexural toughness index and the number of textile layers. When the number of textile layers was increased from one to two, the flexural toughness index increased by 263% and 60% for the group Y and group MY specimens, respectively. When the number of textile layers was increased from two to three, the flexural toughness index of the specimens in groups Y and MY increased by only 7.9% and 2.7%, respectively. This shows that when the number of textile layers is increased from one to two, it is effective in increasing the flexural toughness of the specimens. However, increasing the number of textile layers after reaching two textile layers has a very limited effect on the flexural toughness of the specimens. This may explain the small correlation coefficients in Figure 15.

Table 7. Calculated values of S_0, S_1, and I for TRLC thin plates.

Specimen	S_0 (N·mm)	S_1 (N·mm)	I
Y1	125.20 (35.13)	610.09 (92.31)	5.19 (1.99)
Y2	134.2 (12.73)	2527.68 (743.28)	18.85 (5.06)
Y3	147.32 (32.25)	2932.26 (535.56)	20.33 (4.57)
MY1	30.78 (3.79)	2179.45 (602.88)	64.38 (18.46)
MY2	30.02 (2.00)	3093.28 (254.45)	103.17 (7.9)
MY3	76.26 (4.64)	8086.29 (612.22)	105.97 (2.18)

Note: The data in the table are expressed as the means (standard deviation).

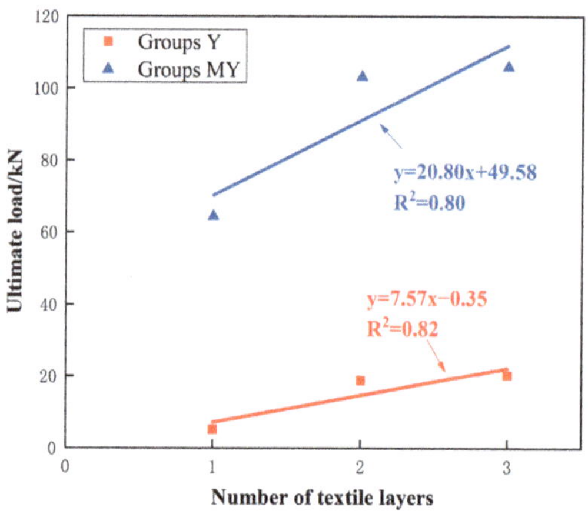

Figure 15. Linear relationship between the number of textile layers and the flexural toughness indices of the Y and MY groups.

The incorporation of HPMC is also an important factor affecting the flexural toughness index. On average, the initial cracking energy consumption of the HPMC-doped TRLC thin plate decreased by 67% compared to the non-HPMC-doped TRLC thin plate. This decline can be attributed to the presence of voids within the HPMC-doped thin plate itself, resulting in lower stiffness and an earlier onset of initial cracking. However, the flexural toughness index increased significantly by 11.40, 4.47, and 4.21 times for one, two, and three textile layers, respectively. It is noteworthy that the effect of HPMC doping in the matrix had a far more pronounced effect on the flexural toughness index of the specimens than the number of textile layers.

5. Conclusions and Recommendations

In this study, the synergistic effect between basalt textiles and lightweight aggregate concrete was investigated by employing four-point bending tests on six groups of 36 TRLC thin plates with and without the addition of HPMC to the matrix and the number of textile layers as parameters. In addition, the effect of the above parameters on the flexural properties of TRLC thin plates was quantitatively assessed using the flexural toughness index. Moreover, we focused on a single HPMC doping content and exclusively analyzed the flexural behavior of TRLC specimens. Despite these constraints, the following conclusions can be drawn.

(1) The increase in the number of textile layers elevates the initial cracking load, initial cracking deflection, residual stress after initial cracking, the possibility of multi-crack development after initial cracking, and the ultimate load and ductility of the TRLC thin plate.
(2) The mechanical properties of the HPMC-doped TRLC thin plate exhibited a reduction. While the initial cracking load decreased, the matrix consistency was enhanced. Furthermore, the increase in the ultimate load of the TRLC thin plate due to HPMC incorporation was more significant when a higher number of textile layers were present. For the group MY3 specimens, the effect of adding HPMC is very obvious, the ultimate load increased by 367.95 N with an increase of 40.0%.
(3) Multi-seam cracking represents a more favorable bending damage mode for TRLC thin plates. The quantity of cracks serves as a key indicator of bending performance, influenced by the number of textile layers and the presence of hydroxypropyl methylcellulose. In HPMC-doped TRLC thin plates, the number of cracks is notably higher compared to non-HPMC-doped plates.
(4) The impact of matrix doping with HPMC on the flexural toughness index of the specimens was significantly more pronounced compared to the influence of the number of textile layers. On average, the initial cracking energy consumption of TRLC thin plates doped with HPMC decreased by 67%, while the flexural toughness index showed consistent increases, with some reaching up to 11.40 times higher values.
(5) In this study, we focused solely on particular textile types and HPMC doping ratios, analyzing only the bending behavior of the specimens. Future research could explore the impact of various textile varieties and diverse doping ratios of HPMC on TRLC. Additionally, other dynamic properties of TRLC could be investigated beyond the bending characteristics.

Based on the experimental and theoretical results outlined in this paper, the multi-layer textile-reinforced lightweight aggregate concrete with HPMC exhibits characteristics of being lightweight and possessing high toughness. This material can be effectively utilized in the fabrication of assembly wall panels, fulfilling the criteria for lightweight assembly components while also enhancing the mechanical properties of the panels due to their high toughness. These attributes suggest a broad range of potential applications in civil engineering.

Author Contributions: J.W.: Funding acquisition, Investigation, Writing—reviewing, and supervision. D.Y.: Investigation, Data curation, Writing—original draft preparation, and editing. C.Z.: Investigation and Methodology. X.J.: Investigation and Validation. L.Y.: Validation. P.Z.: Funding acquisition and Conceptualization. S.Z.: Resources. All authors have read and agreed to the published version of the manuscript.

Funding: This research was funded by the Basic Public Welfare Research Project of Zhejiang Province, China (LGG22E080004), the National Key Research and Development Program of China (2019YFE0112600), the Science and Technology Plan Project of Zhoushan, China (2022C13025), and the Construction Research Project of Zhejiang Province, China (2021K128, 2023K141).

Data Availability Statement: The raw/processed data required to reproduce these findings cannot be shared at this time as the data also forms part of an ongoing study. The datasets generated during and/or analyzed during the current study are available from the corresponding author upon reasonable request.

Acknowledgments: The authors would like to express their gratitude for the invaluable assistance provided by Chenlin WAN and Yu PENG during the execution of the experiments. They would also like to acknowledge the support received from the Center for Balance Architecture and the Architectural Design & Research Institute Co., Ltd. of Zhejiang University in the research project.

Conflicts of Interest: Author Lingpeng Ye was employed by the company The Architectural Design & Research Institute of Zhejiang University Co., Ltd. Author Senlin Zhao was employed by the company Zhejiang Building—Tech Energy Dissipation Technology Co., Ltd. The remaining authors declare that the research was conducted in the absence of any commercial or financial relationships that could be construed as a potential conflict of interest.

References

1. Hassanpour, M.; Shafigh, P.; Bin Mahmud, H. Lightweight aggregate concrete fiber reinforcement—A review. *Constr. Build. Mater.* **2012**, *37*, 452–461. [CrossRef]
2. Gardiner, T.; Currie, B. Flexural behaviour of composite cement sheets using woven polypropylene mesh fabrics. *Int. J. Cem. Compos. Lightweight Concr.* **1983**, *5*, 193–197. [CrossRef]
3. Yin, S.P.; Xu, S.L. An experimental study on improved mechanical behavior of textile-reinforced concrete. *Adv. Mater. Res.* **2011**, *168–170*, 1850–1853. [CrossRef]
4. Wan, C.L.; Wang, J.Y.; Wang, S.B.; Ji, X.H.; Peng, Y.; Zhang, H.M. Tensile behavior of basalt textile reinforced concrete: Effect of test setups and textile ratios. *Materials* **2022**, *15*, 8975. [CrossRef] [PubMed]
5. Ghanem, H.; Chahal, S.; Khatib, J.; Elkordi, A. Experimental and Numerical Investigation of the Flexural Behavior of Mortar Beams Strengthened with Recycled Plastic Mesh. *Sustainability* **2023**, *15*, 5640. [CrossRef]
6. Hegger, J.; Voss, S. Investigations on the bearing behaviour and application potential of textile reinforced concrete. *Eng. Struct.* **2008**, *30*, 2050–2056. [CrossRef]
7. Peled, A.; Bentur, A.; Yankelevsky, D. Flexural performance of cementitious composites reinforced with woven fabrics. *J. Mater. Civ. Eng.* **1999**, *11*, 325–330. [CrossRef]
8. Peled, A.; Bentur, A. Fabric structure and its reinforcing efficiency in textile reinforced cement composites. *Compos. Part A Appl. Sci. Manuf.* **2003**, *34*, 107–118. [CrossRef]
9. Xu, S.L.; Li, H. Bond properties and experimental methods of textile reinforced concrete. *J. Wuhan Univ. Technol. Mater. Sci. Ed.* **2007**, *22*, 529–532. [CrossRef]
10. Yin, S.P.; Jing, L.; Yin, M.T.; Wang, B. Mechanical properties of textile reinforced concrete under chloride wet-dry and freeze-thaw cycle environments. *Cem. Concr. Compos.* **2019**, *96*, 118–127. [CrossRef]
11. Yin, S.P.; Xu, S.L.; Li, H.D. Improved mechanical properties of textile reinforced concrete thin plate. *J. Wuhan Univ. Technol. Mater. Sci. Ed.* **2013**, *28*, 92–98. [CrossRef]
12. Vastl, F. Desing of Coffee Table from Lightweight Concrete with Mesh Reinforcement. Bachelor's Thesis, CTU, Prague, Czechia, 2017.
13. Cibulka, T.; Musil, L.; Vodička, J. The lightweight textile reinforced concrete for thin-walled structures. *Acta Polytech. CTU Proc.* **2019**, *22*, 17–21. [CrossRef]
14. Rozycki, M.; Hager, I.; Zdeb, T.; Sitarz, M.; Mroz, K.; Zdeb, J.; Smoronska, N. Mechanical properties and water permeability of textile reinforced reactive powder concrete with lightweight aggregate. *Materials* **2023**, *16*, 7619. [CrossRef] [PubMed]
15. Naseri, H.; Danesh, M.; Naseri, M. An experimental investigation into fiber/textile-reinforced lightweight concrete precast panels for temporary housing. *Iran. J. Sci. Technol. Trans. Civ. Eng.* **2020**, *44*, 29–36. [CrossRef]
16. Chen, N.; Wang, P.; Zhao, L.; Zhang, G. Water Retention Mechanism of HPMC in Cement Mortar. *Materials* **2020**, *13*, 2918. [CrossRef] [PubMed]
17. Wang, J.Y.; Wan, C.L.; Zeng, Q.; Shen, L.H.; Malik, M.A.; Yan, D.M. Effect of eccentricity on retrofitting efficiency of basalt textile reinforced concrete on partially damaged masonry columns. *Compos. Struct.* **2020**, *232*, 111585. [CrossRef]
18. Wang, J.Y.; Wan, C.L.; Shen, L.H.; Zeng, Q.; Ji, X. Compressive behavior of masonry columns confined with basalt textile-reinforced concrete. *J. Build. Eng.* **2023**, *75*, 107019. [CrossRef]
19. Peled, A.; Mobasher, B. Tensile behavior of fabric cement-based composites: Pultruded and cast. *J. Mater. Civ. Eng.* **2017**, *19*, 340–348. [CrossRef]
20. Dai, Q.L.; Shen, L.H.; Xu, S.L.; Li, Q.H. Mechanical properties research of textile reinforced concrete. *J. Hydraul. Eng.* **2012**, *43*, 59–69. (In Chinese)
21. *GB/T 50081-2019*; Standard for Test Methods of Concrete Physical and Mechanical Properties. Ministry of Housing and Urban-Rural Development of the People's Republic of China: Beijing, China, 2019. (In Chinese)
22. *CECS 13: 2009*; Standard Test Methods for Fiber Reinforced Concrete. China Construction Engineering Association China Planning Press: Beijing, China, 2009. (In Chinese)
23. *ACI 544.2R-89*; Measurement of Properties of Fiber Reinforced Concrete. ASTM International: West Conshohocken, PA, USA, 1999.

24. *JSCE-G 552-1999;* Method of Test for Flexural Strength and Flexural Toughness of Steel-Fibre Reinforced Concrete. Test Methods and Specifications. Japan Society of Civil Engineers: Tokyo, Japan, 2005.
25. *ASTM C1018;* Standard Test Method for Flexural Toughness and First-Crack Strength of Fiber Reinforced Concrete (Using Beam with Third-Point Loading). ASTM International: West Conshohocken, PA, USA, 1997.

buildings

MDPI

Article

Study on the Stress Threshold of Preventing Interfacial Fatigue Debonding in Concrete Beams Strengthened with Externally-Bonded FRP Laminates

Xinzhe Min [1,*], Dong Yang [2], Shoutan Song [3] and Xing Li [4]

1 School of Civil Engineering and Architecture, Nanjing Institute of Technology, Nanjing 211167, China
2 School of Architectural Engineering, Jinling Institute of Technology, Nanjing 211169, China; yangdong@jit.edu.cn
3 Key Laboratory of Concrete and Prestressed Concrete Structures of Ministry of Education, National Engineering Research Center for Prestressing Technology, School of Civil Engineering, Southeast University, Nanjing 211189, China; songshoutan@seu.edu.cn
4 School of Civil Engineering, Suzhou University of Science and Technology, Suzhou 215009, China; 230189492@seu.edu.cn
* Correspondence: xinzhe_min@outlook.com

Abstract: Externally-bonded FRP laminate is widely used in structural strengthening due to the many advantages of FRP materials. Further enhancement of the strengthening effect can be achieved by inducing prestress into the FRP laminate. However, FRP debonding is still a main issue of this strengthening method, especially the Intermediate Crack-induced debonding (IC debonding). To better understand the impact of FRP debonding on the strengthening effect, a series of parameter analyses were conducted in this study based on the fatigue life prediction model proposed by the authors. The proposed model involves the fatigue damage accumulation of components of the beam, the mutual interaction between each component, and the impact of FRP fatigue debonding. As a result, a stress threshold for preventing FRP fatigue debonding in strengthening the concrete beam was proposed, which aimed to avoid safety hazards caused by IC debonding in practical engineering.

Keywords: fiber reinforced polymer; concrete beam; fatigue; debonding; prestress

1. Introduction

For engineering structures, an increase in service life, environmental erosion, and increased load can all cause the degradation of their performance and cause the structures to fail to meet the requirements of use. Demolishing and rebuilding all structures that fail to meet the requirements can not only cause a lot of resource waste but also seriously delay normal social production. Therefore, the strengthening of existing structures has become an economical and efficient method of structural repair and enhancement, which can enable engineering structures to achieve higher resistance and better performance on the existing basis. Since the 1980s, Fiber Reinforced Polymer (FRP), especially Carbon Fiber Reinforced Polymer (CFRP), has become a research hotspot in the field due to its high strength, lightweightness, corrosion resistance, fatigue resistance, and other advantages. Externally-bonded FRP laminates have also become a new method of strengthening and renovation that has received much attention. Compared with conventional strengthening methods such as increasing cross-sections, bonding steel plates, and adding extra supports, the externally-bonded FRP strengthening method has advantages such as convenient construction, manpower saving, the convenient on-site cutting of FRP sheets, no need for large construction equipment during the construction process, and minimal changes in the appearance of the structure after strengthening. Therefore, it has received increasing attention from engineering and academic circles. Swiss scholar Meier first used externally-bonded FRP technology in 1982 to reinforce the Ibach bridge with CFRP sheets [1]. After the Hanshin earthquake in 1995, CFRP sheet strengthening was also widely used in the strengthening of concrete structures such as bridges and buildings in Japan. By 1997, the use of CFRP sheets alone for concrete structure strengthening in Japan had reached over 1 million square meters [2]. Subsequently, FRP sheet strengthening technology has received great attention in developed countries such as Europe, the United States, and Japan, and a large amount of related scientific research has been carried out. The United States (ACI) [3], Europe (fib) [4], the United Kingdom (Concrete Society) [5], Japan (JSCE) [6], and Canada (ISIS Canada) [7] have also successively developed relevant codes or regulations for the FRP strengthening of concrete structures.

The externally-bonded FRP strengthening technology has also been widely applied in the strengthening engineering of bending members such as bridge girders (as shown in Figures 1 and 2). The bridge girders need to withstand millions of repeated loads during their service life, and the fatigue problem caused by repeated loads has always been one of the problems that cannot be ignored in bridge structures. The phenomenon of structural failure after multiple actions with internal forces lower than the static load-bearing capacity is called structural fatigue. The fatigue failure of a structure generally starts from the initial defects inside its constituent materials, such as microcracks, pores, impurities, etc. These initial defects will cause a stress concentration inside the material. Under the repeated action of fatigue loads, cracks continue to expand, and internal damage gradually accumulates to form macroscopic cracks. Ultimately, sudden structural failure occurs due to the penetration of cracks or insufficient effective stress areas. At present, there are relatively more experimental and theoretical studies on FRP-strengthened concrete flexural members under static load, and a relatively complete system has been formed. However, to better promote the application of externally bonded FRP in bridge strengthening, further exploration is needed on the performance of externally bonded and FRP-strengthened concrete structures under fatigue loading.

Figure 1. Externally-bonded FRP strengthening of Bridges along the Shuohuang Heavy Load Railway.

Figure 2. CFRP strengthening of Changsha Tuanjie Overpass.

2. FRP Laminates in Structural Strengthening

FRP is a composite material composed of continuous fibers and resin. The basic composition and the microstructure of the FRP cross-section under electron microscopy are shown in Figure 3. According to the different types of fibers contained in FRP, FRP can usually be divided into carbon fiber reinforced polymer (CFRP), glass fiber reinforced polymer (GFRP), basalt fiber reinforced polymer (BFRP), and aramid fiber reinforced polymer (AFRP).

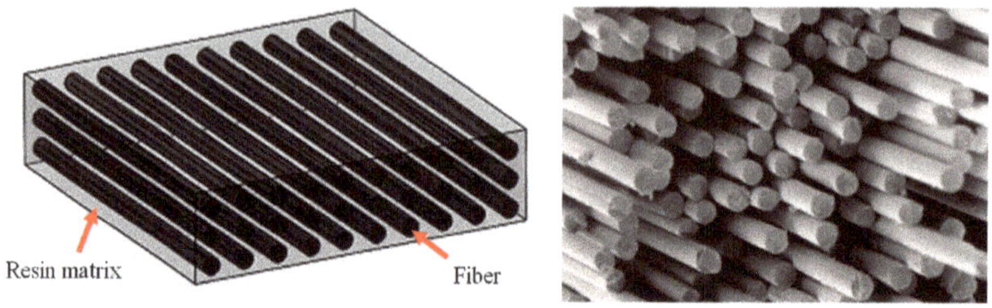

Figure 3. Composition and microstructure of FRP laminates.

In externally-bonded FRP strengthening, the applied FRP can be roughly divided into two categories based on the different forming processes: FRP sheet and FRP plate, which can also be collectively referred to as FRP laminates. FRP sheet, as shown in Figure 4a, is usually woven from bundles of fiber strands and requires resin infiltration and bonding at the construction site, known as the "wet bonding method" for construction molding. FRP plates, as shown in Figure 4b, are generally made in the factory through prefabricated molds or extrusion molding methods and are formed when leaving the factory. Both FRP laminates can be transported in rolls and are very lightweight.

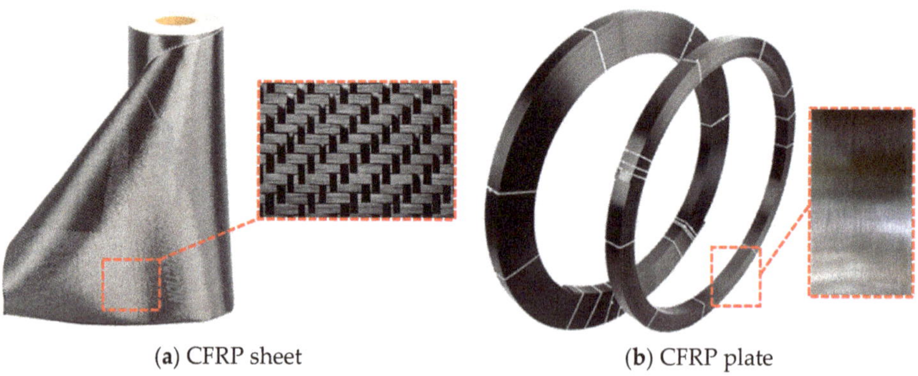

Figure 4. Two types of FRP laminates.

Compared to steel, FRP materials have some obvious advantages [8–12]: (1) Low density: the density of FRP materials is only about 1/8–1/4 of steel. As a result, FRP materials also significantly reduce the difficulty of construction and increase the length of continuously strengthened members; (2) High tensile strength: taking CFRP as an example, its tensile strength can reach 3–4 times that of steel, which further reflects the lightweight and high-strength characteristics of FRP materials; (3) Corrosion resistance: due to the fact that FRP materials are not corroded by acid, alkali, or chloride salts, they will not rust like steel. A large amount of FRP materials are used in the reinforcement engineering of coastal bridges; (4) Fatigue resistance: except for the slightly inferior fatigue performance of fiberglass materials, other common FRP materials exhibit good fatigue resistance, among which CFRP has the best fatigue resistance and is an ideal material for strengthening structures that need to withstand fatigue loads and is considered an ideal material to replace steel plates for externally-bonded strengthening.

Along with the many advantages mentioned above, FRP materials also have some undeniable drawbacks [13,14]: (1) Low material utilization: due to the premature debonding of externally-bonded FRP from the strengthened members [15], it loses its role in improving the load-bearing performance of the strengthened members; (2) High sensitivity to the performance of the bonded surface: due to the high strength of FRP material itself, the debonding of FRP normally initiates from the bonding surface. Especially when FRP material is bonded to concrete structures, debonding failure is an extremely common form of failure. Therefore, the strength of FRP materials depends heavily on the bonding quality or the stress performance of the substrate material; (3) Low elastic modulus: the elastic modulus of most FRP materials is lower than that of steel, so members strengthened with FRP materials often require greater deformation or deflection to enable the FRP material to exert its expected strength.

In structural strengthening, FRP laminates are generally bonded on the tensile side of the flexural member, and the tensile force is jointly borne by FRP and steel reinforcement to improve the flexural performance of the strengthened member. Furthermore, inducing prestress into FRP laminates can further improve the strengthening effect [16–21]. The prestress tensioning and anchoring system (PTA system) [22] development by the authors' research group is shown in Figure 5. The PTA system mainly consists of the following: (a) a CFRP plate tensioning end support, (b) a CFRP plate fixed end support, (c) a CFRP plate tensioning end anchorage, (d) a CFRP plate, (e) a CFRP plate fixed end anchorage, (f) a tensioning threaded rod, (g) a reaction steel plate, (h) a hydraulic jack, and (i) is composed of chemical anchor bolts and (j) anchor nuts. During tensioning, the hydraulic jack pulls the CFRP plate via the reaction steel plate and tensioning threaded rod. After reaching the controlling prestress, the fixed nuts on the CFRP plate tensioning end support are used to maintain the prestress. The schematic diagram of the tensioning process is shown in Figure 6.

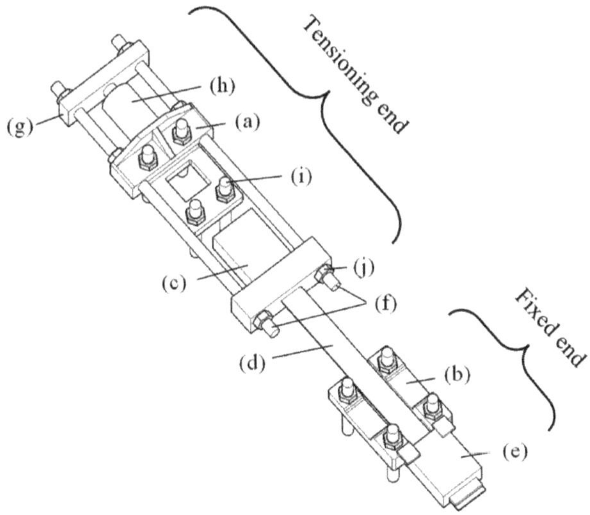

Figure 5. The Prestress Tensioning and Anchoring system (PTA system).

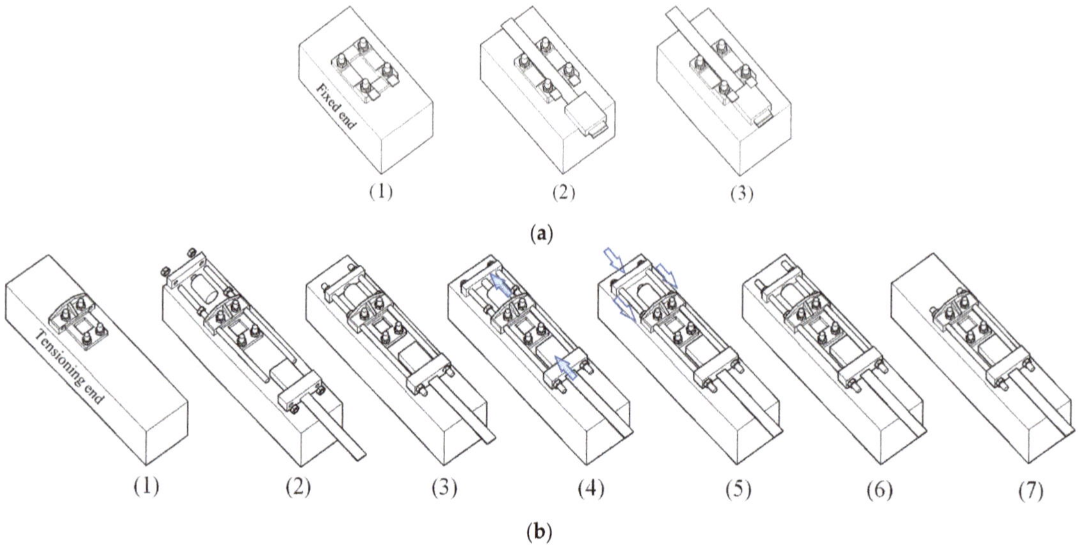

Figure 6. The process of applying prestress to the CFRP plate. (**a**) Anchoring process of the CFRP fixed end. ((1) Install the CFRP plate fixed-end support; (2) Install the CFRP plate fixed-end anchorage; (3) Fix the anchorage in position.); (**b**) Pre-tensioning and anchoring process of the CFRP tensioning end. ((1) Install the CFRP plate tensioning end support; (2)–(3) Install the accessories of the tensioning end; (4) Tension the CFRP plate; (5) Tighten the nuts and unload the jack for another tensioning process if needed; (6) CFRP plate reaches the controlling prestress; (7) remove the jack and the redundant length of tensioning threaded rods).

3. FRP Fatigue Debonding in Concrete Beam Strengthening

Externally-bonded FRP can bear tension on the tensile side of the strengthened member, thereby reducing the stress on the steel reinforcement inside the member, increasing fatigue life, reducing the deflection of the strengthened beam, and improving the crack resistance performance of the member. However, when subjected to external loads, the bonded FRP may endure debonding problems. After FRP debonding occurs, the tension originally borne by FRP in the debonding area will distribute to the steel rebars, increasing the stress on the steel rebars and seriously affecting the mechanical performance of the strengthened members [23–25]. In order to study the effects of FRP fatigue debonding in strengthened beams, the authors' group conducted a series of experimental studies [22,26]. The CFRP plate used in the tests was 1.4 mm in thickness and 100 mm in width. The tensile strength and longitudinal elastic modulus were 2758 MPa and 175 GPa, respectively. The details of the fatigue beam tests are shown in Figure 7. The test program and results are listed in Table 1. All experimental beams experienced FRP debonding during both the static and fatigue tests. The final fatigue failure modes of the strengthened beams were the same, which were the fatigue fracture of the tensile steel rebars at the main crack of the cross-section under one loading point. Research has shown that the debonding of FRP would affect the fatigue life of the strengthened beams by affecting the stress of the steel rebars [22].

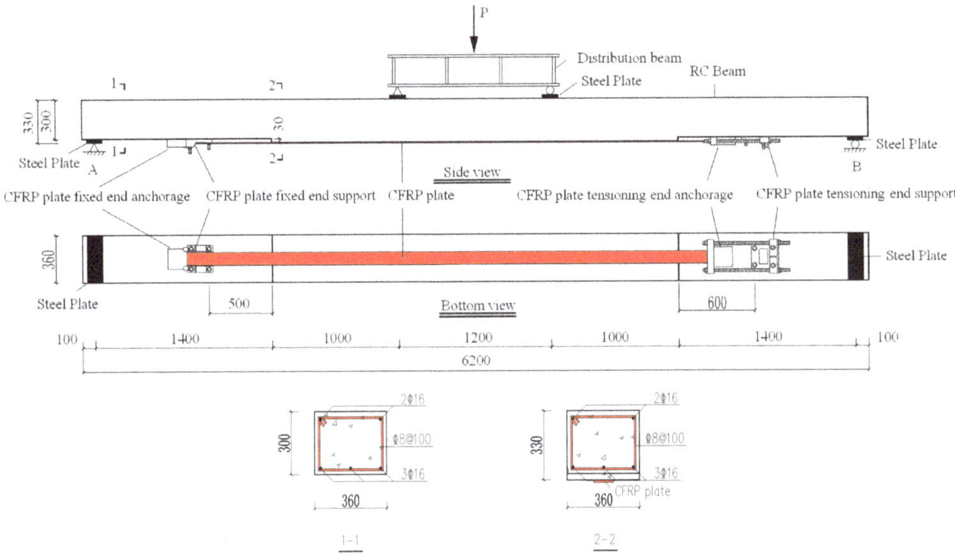

Figure 7. Details of the fatigue test specimens [22] (unit: mm).

Table 1. Summary of the test program and results of each fatigue specimen in [22].

Specimen	Designed Effective Prestress (MPa)	Maximum Load during Fatigue Test P_{max} (kN)	Minimum Load during Fatigue Test P_{min} (kN)	Fatigue Load Level $S_{max} = P_{max}/P_u$ *	Fatigue Life (N_f, Cycles)	Failure Mode
BF1-PS	1000	64	20	0.55	280,861	I *
BF2-PS	1000	54	20	0.47	883,645	I
BF3-PS	1000	48	20	0.41	>2,000,000	II *
BF4-S	0	48	20	0.46	351,761	I

* P_u—The ultimate load capacity of the specimen determined by monotonic testing. Naming of specimens: "F" = Fatigue test, "S" = Strengthened, "P" = Prestressed. Failure mode: I—fatigue fracture of steel reinforcement; II—did not fail during fatigue testing and was therefore monotonically loaded to failure after 2,000,000 loading cycles.

Effective interfacial stress transfer is the foundation of the collaboration between the bonded FRP and concrete soffit in externally bonded and FRP-strengthened concrete flexural members. However, under static or fatigue loads, there may be debonding between FRP and concrete, which greatly weakens the strengthening effect and accelerates the failure process of the strengthened members. It is generally believed that the interfacial stress between FRP and concrete is composed of interfacial shear stress and interfacial normal stress, and interfacial shear stress is the major part [27,28]. The interfacial normal stress only affects specific areas in certain special cases [29–31]. For example, at the plate end of externally-bonded FRP without mechanical anchoring, due to sudden changes in flexural stiffness near the FRP cut-off position, interfacial normal stress is generated between the FRP and beam, resulting in the problem of FRP plate-end debonding [32,33]. Moreover, when the beam on both sides of the inclined crack undergoes vertical relative displacement, normal stress will also be generated in the bonding interface between the FRP and the beam, thereby exacerbating the debonding of FRP. However, these bonding problems caused by interface normal stress can be solved by adding mechanical anchoring or pressure steel plates, and it is generally believed that normal stress is not the key to the debonding problem in externally bonded and FRP-strengthened members [27–31].

The Intermediate Crack-induced debonding (IC debonding) [27,30,34–36], which is dominated by interfacial shear stress, generally occurs in FRP-strengthened concrete flexural members with robust anchorages at the plate ends. The IC debonding commonly initiates from the major bending crack under the loading point and propagates toward the end of the CFRP plate. The IC debonding during the fatigue loading process will further lead to the formation of partially bonded prestresses in the strengthened beam, which is between bonded and unbonded prestressed members. That is, in the strengthened beam, the FRP and concrete beam in the bonded beam section maintain good adhesion and are bonded prestressed members in this section. However, in the beam section where IC debonding occurred, the FRP and the concrete bottom debonded, resulting in an approximate unbonded prestressed state. Therefore, FRP has completely different stress states in the bonded and debonded sections and has the stress transfer phenomenon between FRP and concrete in the transition zone of the two sections. The determination of the stress state of the strengthened beam plays a decisive role in the analysis, thereby affecting the fatigue life of the strengthened member. The FRP fatigue debonding is also coupled with the accumulation of fatigue damage in the concrete beam itself, as shown in Figure 8. For example, the stress amplitude of the steel rebars in specimen BF2-PS in [22] is lower than the fatigue requirements in the relevant National Design Code of China, "Code for the design of concrete structures (GB50010)", that is, the theoretical fatigue life should exceed 2 million cycles. However, with fatigue loading, debonding occurred in the beam, leading to an increase in stress amplitude in the steel rebars, and the beam ultimately failed at only 0.88 million cycles during the fatigue test. Therefore, the fatigue-checking calculation of externally bonded and FRP-strengthened concrete members cannot be simply completed by limiting the stress amplitude of the steel rebars.

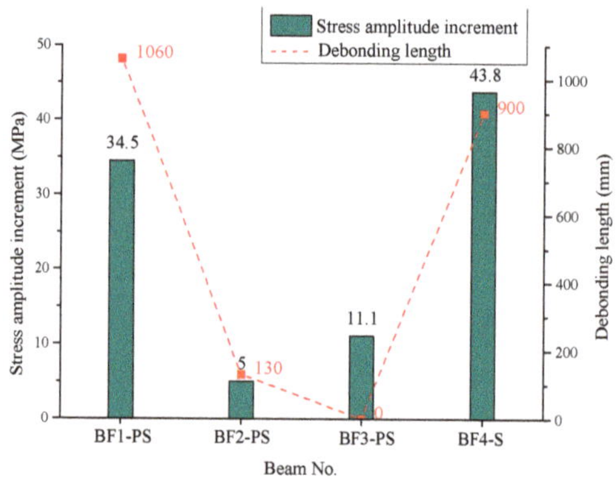

Figure 8. Steel reinforcement stress amplitudes and CFRP plate debonding lengths in the fatigue loading specimens after the 10,000th loading cycle [22].

In order to prevent the fatigue failure of flexural members with externally-bonded FRP laminates, some countries/regions have proposed requirements for the stress or strain limits of the FRP (as shown in Table 2). The ACI [3] in the United States believes that the creep and fatigue characteristics of FRP materials should be comprehensively considered, and it is stipulated that the stress in the FRP material on the strengthened beam should not exceed 55% of its ultimate strength regardless of static or fatigue loads. However, this regulation is not based on the impact of fatigue loading, as it does not take into account the average stress level of fatigue load. The Concrete Society in the UK [5] believes that, when bonded with FRP laminates, the stress limit of the FRP can reach 80% of the tensile strength. The European fib [4] believes that there is no essential difference in fatigue issues between FRP-strengthened beams and ordinary reinforced concrete beams. Therefore, it is recommended that the allowable fatigue stress range of longitudinal bars in FRP-strengthened beams and ordinary beams be consistent, which also omits the FRP debonding issue. The Canadian CSA S806 [37] stipulates that the strain of FRP should be less than 0.007 to prevent bonding failure. The Italian CNR [38] stipulates that the safety factor for all FRP materials is set at 0.5 to prevent potential FRP material failure. These existing provisions imply that the fatigue failure of FRP-strengthened beams may occur within the FRP materials, which means that the fatigue fracture of bonded FRP laminates may occur under fatigue loading. However, the fatigue failure of FRP-strengthened beams generally manifests as the fatigue fracture of steel rebars or the debonding of FRP. Brena [39,40] pointed out that FRP sheets can only use 15% to 25% of their ultimate strength in practical use, which is far from reaching the stress limit specified by the ACI. Many other literature [41–44] also hold the same view and believe that existing regulations or studies ignore the impact of FRP debonding on the stress of steel rebars in concrete beams. Therefore, it is necessary to propose an index that can comprehensively consider the influence of prestress, FRP stress and steel reinforcement stress. By limiting this index, it can be ensured that the externally bonded and FRP-strengthened concrete members can prevent FRP debonding issues and fatigue failure.

Table 2. Stress limits for externally-bonded FRP in various countries.

Country/Region	Code/Guideline	CFRP	GFRP	AFRP
United States	ACI 440.2R-02	$0.55f_{fu}$	$0.20f_{fu}$	$0.30f_{fu}$
UK	Technical Report 55	$0.80f_{fu}$	$0.30f_{fu}$	$0.70f_{fu}$
Europe	fib Biulletin 14	-	-	-
Italy	CNR-D 200/2004	$0.50f_{fu}$	$0.50f_{fu}$	$0.50f_{fu}$

Note: f_{fu} is the tensile strength of FRP laminate.

4. Fatigue Prediction Model for the FRP Strengthened Beams

Developing a model to analyze FRP-strengthened concrete beams under fatigue loading requires considering the influence of many parameters and variables, such as fatigue load level, loading rate, concrete strength, steel reinforcement ratio, FRP material properties, etc. Therefore, it is very difficult to make the model accurately reflect the actual situation of FRP debonding, concrete degradation, and damage accumulation of the steel reinforcement. To accurately predict the fatigue life of concrete beams with externally-bonded FRP while considering computational efficiency, the existing models have adopted different degrees of simplification, such as the assumption of plane section and the assumption of perfect bonding of FRP throughout the fatigue loading process. These assumptions greatly reduce the complexity of fatigue life prediction but also, to some extent, sacrifice accuracy. Existing experimental studies have shown that the interface between FRP and concrete will experience a certain degree of debonding under fatigue loading, and this issue could cause stress redistribution in the strengthened beam, that is, a decrease in FRP stress and an increase in rebar stress, resulting in a significantly lower fatigue life of the strengthened beam than expected. Meanwhile, in prestressed FRP-strengthened concrete beams, the debonding of FRP is related to the level of prestressing, considering the application of prestressing undoubtedly increases the complexity of the fatigue prediction model.

A full-range fatigue life prediction model for RC beams strengthened with prestressed CFRP plates was proposed by the authors [45], which can comprehensively consider the fatigue damage accumulation process of the beam, the FRP debonding growth, and the stress redistribution of the beam caused by FRP debonding. This model was based on the finite strip method and was analyzed by using Matlab programming. The model can obtain the tensile stress of steel rebars, the length of FRP debonding length, and the strain of CFRP plates under any number of fatigue loading cycles.

The key to this prediction model is to analyze the beam by dividing it into multiple segments based on whether FRP debonding occurs and whether there is shear force in the beam segment. The critical section is under one of the loading points. It was found that the debonding caused by fatigue loading started at the cross-section below the loading point and gradually propagated towards its adjacent support with the increase of loading cycles. Therefore, the beam was analyzed in three different segments along the beam length, namely (a) the pure bending bonded segment (PBBS), (b) the debonded segment in the shear span (DBS), and (c) the bonded segment in the shear span (BS), as shown in Figure 9. Due to the FRP debonding in segment DBS, there is no shear transfer between the FRP and concrete in this segment, and the deformations are not coordinated between the FRP and concrete bottom. The flowchart of the analysis process is shown in Figure 10, and the detailed process can be found in another paper written by the authors [45].

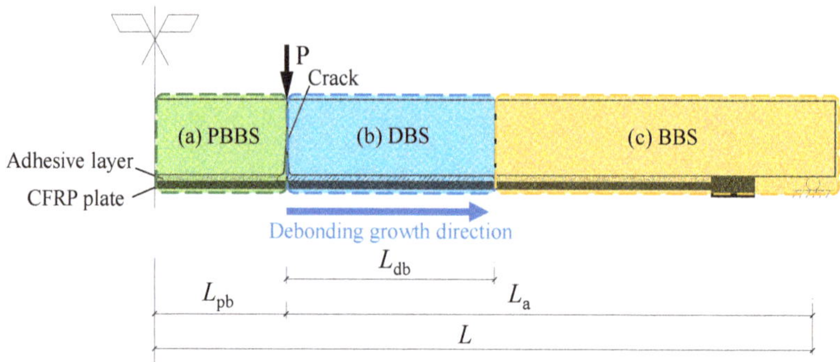

Figure 9. Beam segmentation in the prediction model (where L_{db} is the FRP debonding length, L_{pb} is the length of the pure bending region, L_a is the length of the shear span, and L is half of the beam span).

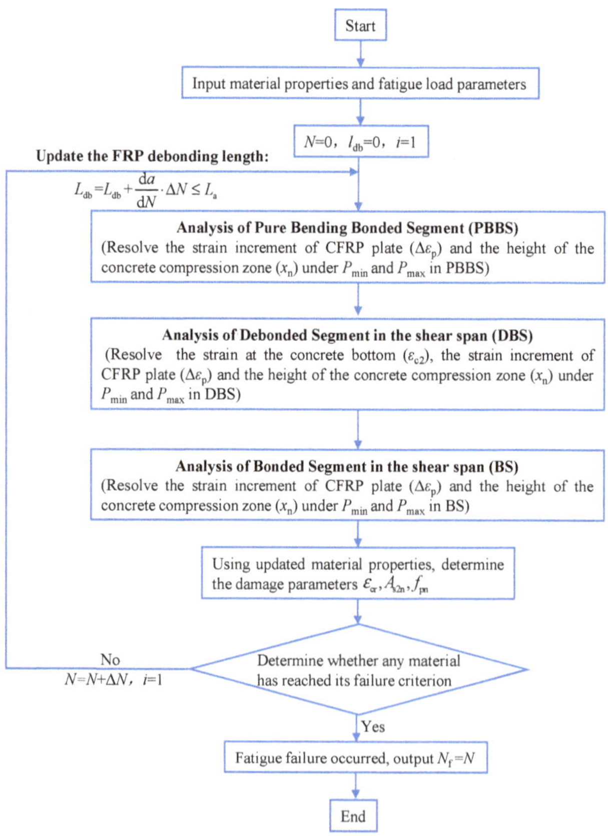

Figure 10. Flow chart for the full-range fatigue prediction model proposed by the authors [26].

5. The Proposed Stress Threshold for Preventing FRP Fatigue Debonding in the Strengthened Concrete Beam

There are many factors that can affect the fatigue life of the beams. Therefore, based on the proposed fatigue prediction model and according to the dimensions and reinforcement of the test beams in [22] as the benchmark (see Table 1), the main parameters affecting the fatigue life of CFRP plate reinforced concrete beams are quantitatively analyzed, and recommendations for the key parameters and the threshold of FRP debonding due to fatigue loading in the externally bonded and FRP-strengthened concrete beams are proposed.

Figure 11 shows the evolutions of critical FRP strain for prestressed strengthened beams under various fatigue loadings. For each beam, the effective prestress (σ_{pe}) is 1000 MPa, and the lower limit of fatigue load (P_{min}) is 20 kN. The upper load limit (P_{max}) varied from 45 kN to 64 kN. From Figure 11, it can be seen that the evolutions of FRP strain under the loading point show two trends as follows: Trend 1 (the downward-stable trend): the FRP strain decreases with the progress of fatigue loading and eventually tends to stabilize; and Trend 2 (the undulate trend): the FRP strain increases initially and then decreases with the fatigue loading cycles, and then reciprocates accordingly.

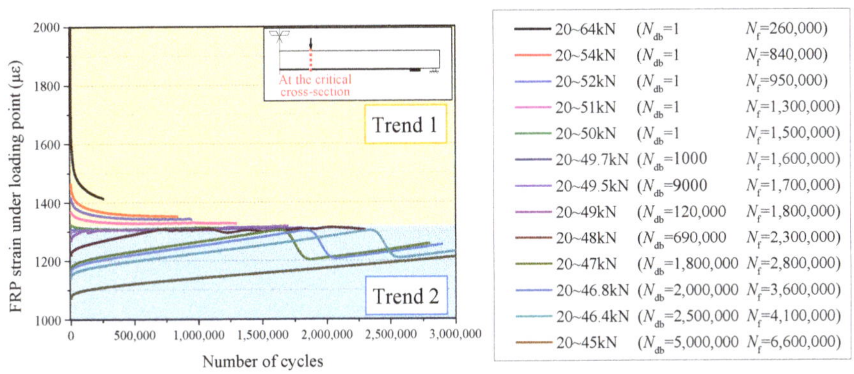

Figure 11. Evolutions of FRP strain at the critical cross-section for prestressed strengthened beams (σ_{pe} = 1000 MPa) under various fatigue loadings (Note: N_{db} is the loading cycle when the FRP debonding initiates, and N_f is the fatigue life).

For Trend 1, in the early stage of fatigue loading, the strain of FRP plates decreases continuously with the progress of fatigue loading because the fatigue load levels acting on these reinforced beams are relatively high, resulting in the FRP strain in the key section exceeding the FRP fatigue debonding strain threshold. That is, the fatigue load could lead to the debonding of FRP plates under the loading point from the beginning of tests. At the same time, even as debonding grows, the FRP strain will still remain at a high level during subsequent fatigue loading; that is, it will always exceed the debonding threshold at the interface. Debonding continues with the increase of fatigue loading cycles. FRP debonding causes the local area of the CFRP plate to be in an unbonded prestressed state. The strain in this debonded section is roughly homogenized, resulting in a continuous decrease in the strain of the FRP plate under the loading point. After the FRP debonded along the entire shear span, the strain of the FRP plate tends to stabilize, only slightly decreasing due to the deterioration of the concrete beam.

For Trend 2, in the early stage of fatigue loading, the strain of FRP plates increases continuously with the increasing number of loading cycles. The strain of FRP plates in key sections is lower than the debonding threshold, so the early fatigue loading could not lead to the initiation of FRP debonding. However, as fatigue loading continues, both concrete and steel rebars exhibit varying degrees of degradation. The degradation of rebars leads to a continuous decrease in the tensile force they bear, resulting in an increase in strain on the FRP plate. Eventually, after a certain number of fatigue loading cycles, the strain on the FRP plate exceeds the debonding threshold and initiates debonding. After the partial length of the FRP plate is debonded, the strain of the FRP plate in the debonded region decreases, which again returns to lower than the fatigue debonding threshold of the FRP plate, and the debonding growth pauses temporarily. Until the damage area of the steel reinforcement further increases due to the continuous fatigue loading, the strain of the FRP plate can increase to exceed the FRP fatigue debonding threshold at the front end of the debonding region, leading to the debonding growth of the FRP plate. Henceforth, the FRP fatigue debonding alternates between propagating and pausing.

5.1. The Impact of Prestress

Table 3 and Figure 12 show the impact of prestress in the FRP plate on the fatigue life of FRP-strengthened beams. It can be seen that when the fatigue load and FRP strengthening amount remain unchanged, the fatigue lives of the strengthened beams increase rapidly with the increase of the prestress level. When the FRP plate is not prestressed (σ_{pe} = 0), the fatigue life (N_f) of the strengthened beam under a fatigue loading of 20–54 kN is only about 170,000 cycles, which is also the shortest fatigue life among all beams. When prestressing the FRP plate and the effective prestressing reaches 500 MPa (approximately 21% f_p, where f_p is the FRP tensile strength), the fatigue life of the strengthened beam becomes 400,000 times, which is 2.35 times that of non-prestressed strengthened beams. By further increasing the effective prestress to 750 MPa, 1000 MPa, and 1250 MPa, the fatigue lives of the strengthened beams were increased to 3.29 times, 4.94 times, and 10 times higher than that of non-prestressed specimens, respectively. With effective prestress, σ_{pe} = 1250 MPa (about 52% f_p), the fatigue life of the strengthened beam achieves about 1.7 million cycles. Furthermore, when the FRP plate is prestressed to 1500 MPa (about 63% f_p), its fatigue life can reach 5.6 million cycles. It can be seen that if the amount of FRP plate remains unchanged, a much longer fatigue life can be achieved by inducing prestress to the FRP plate. In other words, the design threshold for preventing FRP fatigue debonding should account for the impact of prestress in the bonded FRP.

Table 3. Fatigue lives of prestressed FRP strengthened beams under different fatigue loadings.

No.	Effective Prestress, σ_{pe} (MPa)	Fatigue Loading (kN)	Fatigue Life, N_f (Cycles)
1	σ_{pe} = 0	P_{min} = 20, P_{max} = 54	170,000
2	σ_{pe} = 500	P_{min} = 20, P_{max} = 54	400,000
3	σ_{pe} = 750	P_{min} = 20, P_{max} = 54	560,000
4	σ_{pe} = 1000	P_{min} = 20, P_{max} = 54	840,000
5	σ_{pe} = 1250	P_{min} = 20, P_{max} = 54	1,700,000
6	σ_{pe} = 1500	P_{min} = 20, P_{max} = 54	5,600,000

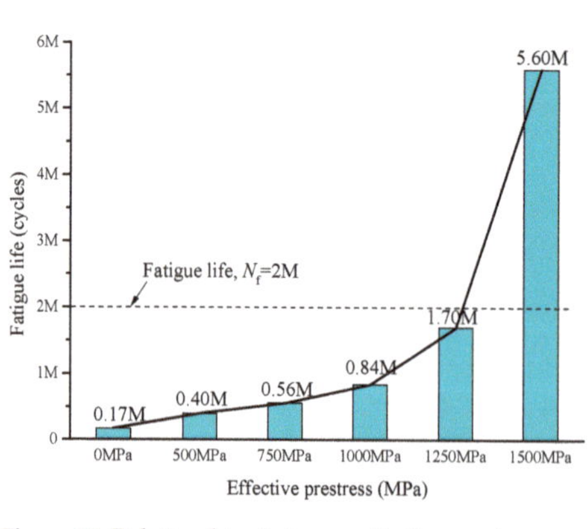

Figure 12. Relationships between effective prestresses and fatigue lives of FRP strengthened beams.

5.2. The Impact of FRP Fatigue Debonding

The fatigue debonding of the FRP plate has a significant impact on the mechanical performance of strengthened beams. In order to investigate the influence of this factor, a series of simulation analyses were carried out based on the full-range fatigue prediction model proposed by the authors [45], but without considering the FRP debonding. Specifically, throughout the entire analysis and calculation process, it is always believed that FRP was perfectly bonded to the bottom of the concrete beam, and the deformations of the FRP plate and the beam bottom were coordinated. The representative cross-sections of all elements can be analyzed by using the assumption of a flat section. Four test beams (BF1-PS, BF2-PS, BF3-PS, and BF4-S) in [22] were analyzed by this modified prediction model. The results of the analysis are shown in Table 4 and Figure 13. It can be seen that due to the neglect of FRP fatigue debonding, the stress on the steel rebars did not increase rapidly in the debonded region. When FRP debonding is not considered, the fatigue lives of the four test beams increased by 1.2 to 4.2 times compared to the results obtained by the fatigue prediction model that considers FRP fatigue debonding. Therefore, if the debonding is neglected in the analysis, it will greatly overestimate the fatigue life of the strengthened beam, resulting in significant safety hazards. Therefore, the impact of debonding also needs to be taken into consideration in strengthening design.

Table 4. Fatigue lives of FRP strengthened beams neglecting or accounting for the FRP debonding.

No.	Specimen	Fatigue Loading (kN)	Effective Prestress, σ_{pe}	Fatigue Life, N_f
1	BF1-PS, but neglect debonding	$P_{min} = 20, P_{max} = 64$	1000 MPa	573,000
2	BF1-PS	$P_{min} = 20, P_{max} = 64$	1000 MPa	260,000
3	BF2-PS, but neglect debonding	$P_{min} = 20, P_{max} = 54$	1000 MPa	1,925,000
4	BF2-PS	$P_{min} = 20, P_{max} = 54$	1000 MPa	840,000
5	BF3-PS, but neglect debonding	$P_{min} = 20, P_{max} = 48$	1000 MPa	5,300,000
6	BF3-PS	$P_{min} = 20, P_{max} = 48$	1000 MPa	2,300,000
7	BF4-S, but neglect debonding	$P_{min} = 20, P_{max} = 48$	0 MPa	1,267,000
8	BF4-S	$P_{min} = 20, P_{max} = 48$	0 MPa	300,000

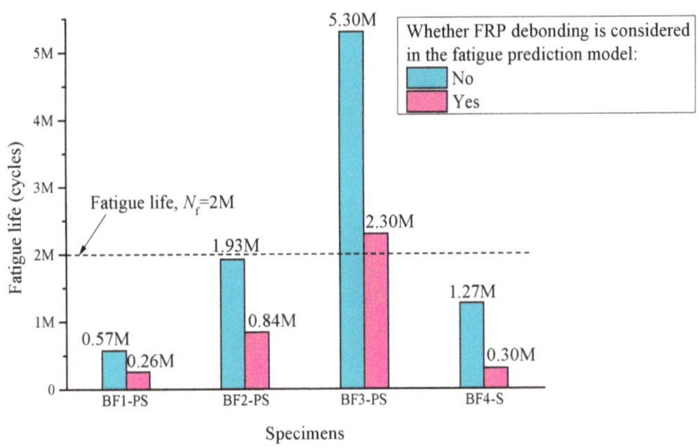

Figure 13. The impacts of FRP debonding on the predicted fatigue lives of FRP strengthened beams.

5.3. The Proposed Stress Threshold

For Trend 2 in Figure 11, which states that the debonding of the FRP plate does not occur during the initial stage of fatigue loading but only initiates after a certain number of fatigue cycles, the number of loading cycles, N, at this time is named "FRP debonding initiating a number of cycles, N_{db}," and the upper load limit corresponding to N_{db} is named $P_{db,N}$. The symbol "db" in the subscript indicates debonding, and "N" indicates the debonding initiating cycle number.

In order to be applicable to a wider range of cross-sectional dimensions and reinforcement of FRP-strengthened concrete beams, and to comprehensively consider the impacts of prestressing and FRP debonding, this paper proposes, "The tensile stress at the tensioning edge of the transformed section of the externally bonded and FRP-strengthened RC beams, $\sigma_{c0db,N}$". This design threshold refers to the calculation method of the cracking moment, M_{cr}, of prestressed concrete beams [46]. The rationale lies in the fact that IC debonding generally initiates from a major concrete crack, which is highly related to the calculation of M_{cr}. The "c0" in the subscript represents the transformed cross-section of concrete. The specific calculation method of $\sigma_{c0db,N}$ is as follows:

$$\sigma_{c0db,N} = \frac{M_{db,N}}{I_0} y_0 - \sigma_{pc} \tag{1}$$

where $M_{db,N}$ is the upper moment limit corresponding to N_{db}, σ_{pc} is the normal stress at the tensioning edge of the beam caused by prestress, I_0 is the second moment of area of the transformed section, and y_0 is the distance from the center of the transformed section to the calculated position. The definition and calculation methods of σ_{pc}, I_0, and y0 are all the same as the pertinent provisions 10.1.6 in the national standard of China "Code for the Design of Concrete Structures (GB 50010-2010) [46]".

For example, the FRP debonding did not initiate in the strengthened beam until the 1000th loading cycle (N_{db} = 1000); the corresponding fatigue load upper limit is named $P_{db,1000}$, and the relevant moment is $M_{db,1000}$. Consequently, the maximum tensile stress of the transformed cross-section, $\sigma_{c0db,1000}$, can be obtained via Equation (1). Table 5 shows the $\sigma_{c0db,1000}$ for FRP-strengthened beams with different effective prestress. It can be seen that even for beams without the occurrence of FRP debonding during the initial loading, it is still possible for the debonding to occur after a short period of loading cycles. Taking the strengthened beam with effective prestress of 1000 MPa as an example, when $\sigma_{c0db,1000}$ = 4.1 MPa, the FRP plate will begin to debond after 1000 loading cycles, resulting in an increase in tensile steel stress in the concrete beam, and the fatigue life of the beam will correspondingly decrease. Therefore, for FRP-strengthened beams that did not experience FRP debonding during the initial loading stage, the debonding verification should also be carried out. Otherwise, it may result in the failure to reach the design fatigue life. The analysis results also indicate that the fatigue debonding issue needs to be taken seriously in the design threshold.

Table 5. The maximum tensile stress of the transformed cross-section of the FRP-strengthened beam when FRP debonding initiates at the 1000th loading cycle: $\sigma_{c0db,1000}$.

Effective Prestress, σ_{pe} (MPa)	**FRP Debonding Initiating Number of Cycles, N_{db} (Cycles)**	**Upper Load Limit Corresponding to N_{db}, $P_{db,N}$ (kN)**	**Upper Moment Limit Corresponding to N_{db}, $M_{db,n}$ (kN·m)**	**Maximum Tensile Stress of the Transformed Cross-Section Corresponding to N_{db}, $\sigma_{c0db,1000}$ (MPa)**
0	1000	26.7	32.0	4.6
250	1000	32.4	38.9	4.5
500	1000	38.2	45.8	4.4
750	1000	43.8	52.6	4.2
1000	1000	49.6	59.5	4.1
1250	1000	55.4	66.5	4.0
1500	1000	61.2	73.4	3.9

Therefore, Table 6 shows the design threshold, $\sigma_{c0db,2E6}$, which presents the FRP-strengthened beams with different prestress levels that do not experience FRP debonding until reaching 2 million fatigue loading cycles. Taking the strengthened beam with an effective prestress of 1000 MPa as an example, when $\sigma_{c0db,2E6}$ = 3.6 MP, the FRP debonding will initiate after 2 million loading cycles. Therefore, to prevent FRP fatigue from debonding within 2 million loading cycles, the tensile stress at the tensioning edge of the transformed section of the strengthened RC beam should not exceed this design threshold.

Table 6. The maximum tensile stress of the transformed cross-section of the FRP-strengthened beam when FRP debonding initiates at the 2,000,000th loading cycle: $\sigma_{c0db,2E6}$.

Effective Prestress, σ_{pe} (MPa)	FRP Debonding Initiation Cycle Numbers, N_{db} (Cycles)	Upper Load Limit Corresponding to N_{db}, $P_{db,N}$ (kN)	Upper Moment Limit Corresponding to N_{db}, $M_{db,n}$ (kN·m)	Maximum Tensile Stress of the Transformed Cross-Section Corresponding to N_{db}, $\sigma_{c0db,2E6}$ (MPa)
0	2,000,000	25.4	30.5	4.4
250	2,000,000	31.0	37.2	4.2
500	2,000,000	36.6	43.9	4.1
750	2,000,000	41.7	49.8	3.8
1000	2,000,000	46.8	56.1	3.6
1250	2,000,000	51.6	61.9	3.3
1500	2,000,000	56.5	67.8	3.1

For the fatigue verification of reinforced concrete flexural members strengthened with externally-bonded FRP laminates, if it can be ensured that no debonding occurs during the 2 million fatigue loading process, then the conventional fatigue verification method of concrete flexural members can still be used [47]. Therefore, this paper suggests that the tensile stress at the tensioning edge of the transformed section of the externally bonded and FRP-strengthened RC beams needs to be lower than the design threshold $\sigma_{c0db,2E6}$ in Table 6, which is:

$$\sigma_{c0} = \frac{M}{I_0} y_0 - \sigma_{pc} < \sigma_{c0db,2E6} \tag{2}$$

where M is the applied moment on the member, σ_{c0} is the tensile stress at the tensioning edge of the transformed section of the strengthened beam, and $\sigma_{c0db,2E6}$ is the design stress threshold shown in Table 6. The stress threshold that is not listed in the table can be calculated by interpolation corresponding to the effective prestress level.

6. Conclusions

The fatigue problem of reinforced concrete beams strengthened with externally-bonded FRP laminates involves the fatigue damage accumulation inside the beam components and the interactions between different components. For externally bonded and FRP-strengthened concrete beams, even if there is no FRP debonding initiated during the initial loading stage, it may still be generated after a certain number of fatigue loading cycles. This study was based on the fatigue life prediction model proposed by the authors, and a series of parameter analyses were conducted. Furthermore, the stress threshold for preventing FRP debonding in the strengthened beams was also proposed. The following conclusions can be drawn:

(1) Through parameter analysis, it can be found that increasing the prestress level of FRP laminates can significantly increase the fatigue life of strengthened beams when the usage amount of FRP is constant. The inducing of prestress can not only improve the strengthening efficiency but also achieve a higher utilization rate of FRP material.

(2) The evolutions of FRP strain under the loading point for the strengthened beams under fatigue loadings can be categorized into two trends: the downward-stable trend and the undulate trend. The second trend can be used to guide the proposal of the stress threshold of FRP fatigue debonding in the strengthened beams.

(3) For the fatigue prediction of externally bonded and FRP-strengthened beams, the fatigue life of the beam could be greatly overestimated if the impact of FRP fatigue debonding is not taken into account, which can further lead to serious safety hazards.
(4) It can be found through parameter analysis and experimental observation that even if the FRP plate does not debond during the initial fatigue loading stage, the debonding can still initiate during the subsequent loading process. Therefore, based on the fatigue life prediction model proposed by the authors, a design stress threshold, $\sigma_{c0db,2E6}$, was proposed to prevent FRP fatigue debonding in the externally bonded and FRP-strengthened concrete beams. The tensile stress at the tensioning edge of the transformed beam section (σ_{c0}) needs to be lower than the stress threshold ($\sigma_{c0db,2E6}$) to ensure that there is no FRP debonding initiating during 2 million fatigue loading cycles. Further research is still desired to validate this suggestion.
(5) In general, to obtain a higher performance of the FRP-strengthened concrete beam, the FRP laminate has to be prestressed to maximize the utilization of its high tensile strength. In addition, robust mechanical anchorages need to be used at the ends of FRP laminate in order to prevent the debonding from its cut-off ends. Furthermore, the strengthened beams should be served within a reasonable load range and lower than the stress threshold to prevent IC debonding from the intermediate concrete crack opening. Thus, the advantages of FRP materials could be utilized to a greater extent.

Author Contributions: Methodology, X.M., D.Y. and S.S.; Software, X.M. and X.L.; Validation, X.M., D.Y. and X.L.; Formal analysis, X.M.; Investigation, X.M., D.Y. and X.L.; Writing—original draft, X.M.; Writing—review & editing, X.M., D.Y. and S.S.; Supervision, S.S. All authors have read and agreed to the published version of the manuscript.

Funding: This research was supported by the Natural Science Foundation of Jiangsu Province of China (BK20230703), the Scientific Research Foundation of Nanjing Institute of Technology (YKJ202122), the Natural Science Foundation of the Jiangsu Higher Education Institutions of China (23KJB560010).

Data Availability Statement: Data are contained within the article.

Conflicts of Interest: The authors declare no conflict of interest.

References

1. Meier, U. Proposal for a carbon fibre reinforced composite bridge across the Strait of Gibraltar at its narrowest site. *Proc. Inst. Mech. Eng. Part B Manag. Eng. Manuf.* **1987**, *201*, 73–78. [CrossRef]
2. Fukuyama, H.; Sugano, S. Japanese seismic rehabilitation of concrete buildings after the Hyogoken-Nanbu Earthquake. *Cem. Concr. Compos.* **2000**, *22*, 59–79. [CrossRef]
3. Committe, A. *Guide for the Design and Construction of Externally Bonded FRP Systems for Strengthening Concrete Structures*; 440.2R-02; American Concrete Institute: Farmington Hills, MI, USA, 1999.
4. Triantafillou, T.; Matthys, S.; Audenaert, K.; Balázs, G.; Blaschko, M.; Blontrock, H.; Czaderski, C.; David, E.; Di Tomasso, A.; Duckett, W.; et al. *Externally Bonded FRP Reinforcement for RC Structures*; Technical Report; International Federation for Structural Concrete: Lausanne, Switzerland, 2001; Volume 14.
5. Arya, C.; Clarke, J.L.; Kay, E.A.; O'regan, P.D. *Design Guidance for Strengthening Concrete Structures Using Fibre Composite Materials*; Concrete Society: Camberley, UK, 2012.
6. Japan Society of Civil Engineers (JSCE). Recommendations for upgrading of concrete structures with use of continuous fiber sheets. *JSCE Concr. Eng. Ser.* **2001**, *41*, 31–34.
7. ISIS Canada. *Strengthening Reinforced Concrete Structures with Externally Bonded Fibre Reinforced Polymers*; ISIS Canada: Winnipeg, MB, Canada, 2001.
8. Head, P. Advanced composites in civil engineering-a critical overview at this high interest, low use stage of development. In Proceedings of the 2nd International Conference on Advanced Composite Materials in Bridges and Structures, Acmbs-II, Montreal, QC, Canada, 11–14 August 1996.
9. Chole, A.; Tembhurne, A.; Bawanthade, A.; Bhadade, H.; Khan, H.A.; Shaw, S.K. Strengthening of reinforced concrete beams by using FRPs—An overview. *Mater. Today Proc.* **2023**, *in press*. [CrossRef]
10. Askar, M.K.; Hassan, A.F.; Al-Kamaki, Y.S. Flexural and shear strengthening of reinforced concrete beams using FRP composites: A state of the art. *Case Stud. Constr. Mater.* **2022**, *17*, e01189. [CrossRef]

11. Shi, J.-W.; Wu, Q.-Q.; Li, B.; Liu, Y.; Cao, W.-H.; Wang, H.-T. Fatigue bond behavior of FRP-to-concrete joints with various bonding adhesives. *Eng. Struct.* **2024**, *301*, 117311. [CrossRef]
12. Al-Saoudi, A.; Kalfat, R.; Al-Mahaidi, R.; Cervenka, J.; Pryl, D. Numerical and experimental investigation into the fatigue life of FRP bonded to concrete and anchored with bidirectional fabric patches. *Eng. Struct.* **2021**, *239*, 112335. [CrossRef]
13. Shabani, E.; Mostofinejad, D.; Saljoughian, A. Bond behavior of prestressed FRP sheet-concrete joints: Comparison of EBROG and EBR methods. *Constr. Build. Mater.* **2023**, *400*, 132752. [CrossRef]
14. Najaf, E.; Orouji, M.; Ghouchani, K. Finite element analysis of the effect of type, number, and installation angle of FRP sheets on improving the flexural strength of concrete beams. *Case Stud. Constr. Mater.* **2022**, *17*, e01670. [CrossRef]
15. Zhang, W.; Kang, S.; Lin, B.; Huang, Y. Mixed-Mode Debonding in CFRP-to-Steel Fiber–Reinforced Concrete Joints. *J. Compos. Constr.* **2024**, *28*, 04023069. [CrossRef]
16. Pino, V.; Hadad, H.A.; Basalo, F.D.C.Y.; Nanni, A.; Ebead, U.A.; El Refai, A. Performance of FRCM-Strengthened RC Beams Subject to Fatigue. *J. Bridge Eng.* **2017**, *22*, 04017079. [CrossRef]
17. Peng, H.; Zhang, J.; Shang, S.; Liu, Y.; Cai, C. Experimental study of flexural fatigue performance of reinforced concrete beams strengthened with prestressed CFRP plates. *Eng. Struct.* **2016**, *127*, 62–72. [CrossRef]
18. Mahal, M.; Täljsten, B.; Blanksvärd, T. Experimental performance of RC beams strengthened with FRP materials under monotonic and fatigue loads. *Constr. Build. Mater.* **2016**, *122*, 126–139. [CrossRef]
19. Charalambidi, B.G.; Rousakis, T.C.; Karabinis, A.I. Fatigue behavior of large-scale reinforced concrete beams strengthened in flexure with fiber-reinforced polymer laminates. *J. Compos. Constr.* **2016**, *20*, 04016035. [CrossRef]
20. Oudah, F.; El-Hacha, R. Research progress on the fatigue performance of RC beams strengthened in flexure using Fiber Reinforced Polymers. *Compos. Part B Eng.* **2013**, *47*, 82–95. [CrossRef]
21. Ferrier, E.; Bigaud, D.; Clément, J.C.; Hamelin, P. Fatigue-loading effect on RC beams strengthened with externally bonded FRP. *Constr. Build. Mater.* **2011**, *25*, 539–546. [CrossRef]
22. Min, X.; Zhang, J.; Li, X.; Wang, C.; Tu, Y.; Sas, G.; Elfgren, L. An experimental study on fatigue debonding growth of RC beams strengthened with prestressed CFRP plates. *Eng. Struct.* **2022**, *273*, 115081. [CrossRef]
23. Minnaugh, P.L.; Harries, K.A. Fatigue behavior of externally bonded steel fiber reinforced polymer (SFRP) for retrofit of reinforced concrete. *Mater. Struct.* **2009**, *42*, 271–278. [CrossRef]
24. Dai, J.; Ueda, T.; Sato, Y.; Ito, T. Flexural strengthening of RC beams using externally bonded FRP sheets through flexible adhesive bonding. In Proceedings of the International Symposium on Bond Behavior of FRP in Structures (BBFS 2005), Hong Kong, China, 7–9 December 2005.
25. Quattlebaum, J.B.; Harries, K.A.; Petrou, M.F. Comparison of three flexural retrofit systems under monotonic and fatigue loads. *J. Bridge Eng.* **2005**, *10*, 731–740. [CrossRef]
26. Min, X.; Zhang, J.; Li, X.; Wang, C.; Tu, Y.; Sas, G.; Elfgren, L. A nonlinear prediction model of the debonding process of an FRP-concrete interface under fatigue loading. *Constr. Build. Mater.* **2023**, *369*, 130583. [CrossRef]
27. Lu, X.Z.; Teng, J.G.; Ye, L.P.; Jiang, J.J. Intermediate crack debonding in FRP-strengthened RC beams: FE analysis and strength model. *J. Compos. Constr.* **2007**, *11*, 161–174. [CrossRef]
28. Lu, X.Z.; Teng, J.G.; Ye, L.P.; Jiang, J.J. Bond–slip models for FRP sheets/plates bonded to concrete. *Eng. Struct.* **2005**, *27*, 920–937. [CrossRef]
29. Smith, S.T.; Teng, J. Interfacial stresses in plated beams. *Eng. Struct.* **2001**, *23*, 857–871. [CrossRef]
30. Teng, J.; Smith, S.; Yao, J.; Chen, J. Intermediate crack-induced debonding in RC beams and slabs. *Constr. Build. Mater.* **2003**, *17*, 447–462. [CrossRef]
31. Lu, X.; Jiang, J.; Teng, J.; Ye, L. Finite element simulation of debonding in FRP-to-concrete bonded joints. *Constr. Build. Mater.* **2006**, *20*, 412–424. [CrossRef]
32. Yao, J.; Teng, J. Plate end debonding in FRP-plated RC beams—I: Experiments. *Eng. Struct.* **2007**, *29*, 2457–2471. [CrossRef]
33. Teng, J.; Yao, J. Plate end debonding in FRP-plated RC beams—II: Strength model. *Eng. Struct.* **2007**, *29*, 2472–2486. [CrossRef]
34. Chen, G.; Teng, J.; Chen, J. Finite-element modeling of intermediate crack debonding in FRP-plated RC beams. *J. Compos. Constr.* **2010**, *15*, 339–353. [CrossRef]
35. Liu, I.; Oehlers, D.; Seracino, R. Study of intermediate crack debonding in adhesively plated beams. *J. Compos. Constr.* **2007**, *11*, 175–183. [CrossRef]
36. Yao, J.; Teng, J.; Lam, L. Experimental study on intermediate crack debonding in FRP-strengthened RC flexural members. *Adv. Struct. Eng.* **2005**, *8*, 365–396. [CrossRef]
37. *CSA Standard S806-02*; Design and Construction of Building Components with Fibre-Reinforced Polymers. Canadial Standards Association: Toronto, ON, Canada, 2002.
38. *Guide for the Design and Construction of Externally Bonded FRP Systems for Strengthening Existing Structures*; CNR-DT200; National Research Council, Advisory, Committee on Technical Recommendations for Construction: Rome, Italy, 2004.
39. Brena, S.F.; Benouaich, M.A.; Kreger, M.E.; Wood, S.L. Fatigue tests of reinforced concrete beams strengthened using carbon fiber-reinforced polymer composites. *ACI Struct. J.* **2005**, *102*, 305.
40. Gussenhoven, R.; Breña, S.F. Fatigue Behavior of Reinforced Concrete Beams Strengthened with Different FRP Laminate Configurations. In *SP-230: 7th International Symposium on Fiber-Reinforced (FRP) Polymer Reinforcement for Concrete Structures*; American Concrete Institute: Farmington Hills, MI, USA, 2005.

41. Harries, K.A.; Aidoo, J. Debonding-and fatigue-related strain limits for externally bonded FRP. *J. Compos. Constr.* **2006**, *10*, 87–90. [CrossRef]
42. Aidoo, J.; Harries, K.A.; Petrou, M.F. Fatigue behavior of carbon fiber reinforced polymer-strengthened reinforced concrete bridge girders. *J. Compos. Constr.* **2004**, *8*, 501–509. [CrossRef]
43. Aidoo, J.; Harries, K.A.; Petrou, M.F. Full-scale experimental investigation of repair of reinforced concrete interstate bridge using CFRP materials. *J. Bridge Eng.* **2006**, *11*, 350–358. [CrossRef]
44. Harries, K.A.; Aidoo, J.; Zorn, A.; Quattlebaum, J. Deterioration of FRP-to-concrete bond under failure loading. *Adv. Struct. Eng.* **2006**, *9*, 779–789. [CrossRef]
45. Min, X.; Zhang, J.; Tu, Y.; Li, X.; Wang, C.; Sas, G.; Elfgren, L. A full-range fatigue life prediction model for RC beams strengthened with prestressed CFRP plates accounting for the impact of FRP debonding. *Eng. Struct.* **2024**, *301*, 117305. [CrossRef]
46. *GB 50010-2010*; Code for Design of Concrete Structures of the People's Republic of China. China Building Industry Press: Beijing, China, 2011.
47. Heffernan, C.P. Fatigue Behaviour of Reinforced Concrete Beams Strengthened with CFRP Laminates. Ph.D. Thesis, Royal Military College of Canada, Kingston, ON, Canada, 1997.

buildings

MDPI

Article

Parametric Study on Mechanical Properties of Basalt Fiber-Reinforced Pea Gravel Concrete

Jiming Li [1,2], Bu Wang [3,*], Peng Zhang [3], Zhenyu Wang [4] and Meng Wang [5]

1 School of Civil Engineering, Tsinghua University, Beijing 100084, China; li-jm22@mails.tsinghua.edu.cn
2 Shaanxi Academy of Building Research Co., Ltd., Xi'an 710082, China
3 School of Civil Engineering, Chang'an University, Xi'an 710064, China; zp1995@chd.edu.cn
4 SDEE Power Engineering Company of Operation and Maintenance Co., Ltd., Ji'nan 250000, China; wangzhenyu@sdee.cee-group.cn
5 China Xinxing Construction Engineering Co., Ltd., Beijing 100039, China; wangmeng7@xxg.gt.cn
* Correspondence: kingstep74@chd.edu.cn

Abstract: Basalt fiber-reinforced pea gravel concrete (BFRPGC) has remarkable potential for use as the retrofitting covers for masonry walls. However, a quantitative understanding of the mechanical properties of the BFRPGC material is still a perceived gap in the current literature. In this study, the role of basalt fibers in pea gravel concrete was evaluated by a comprehensive experimental investigation involving compressive strength tests and splitting tensile tests. Fiber length and volume fraction were selected as the key parameters. Two fiber lengths of 6 mm and 12 mm were considered, while the volume fraction corresponding to each of the fiber lengths was increased from 0.3% to 0.8%, with a step of 0.1%. The measured strengths were not simply proportional to the fiber volume fraction. The reason behind this phenomenon, i.e., the coupling effect of the bridging role of basalt fibers on concrete microcracks and the fiber agglomeration in concrete, was analyzed. The best performance of the BFRPGC material was achieved by incorporating 12-millimeter-length fibers with a volume fraction of 0.4%. Compared to that of the reference pea gravel concrete, a significant increase of up to 44.5% in compressive strength was recorded in this case. Furthermore, the failure mechanism of basalt fibers in pea gravel concrete was disclosed via the scanning electron microscope observations. In addition, the uniaxial compressive stress–strain model of the BFRPGC material was established.

Keywords: fiber-reinforced concrete; basalt fibers; pea gravel concrete; mechanical property

1. Introduction

Short-cut fibers are currently being considered as an approach for improving concrete, in terms of both strength and strain capacity [1–4], while unidirectional continuous fibers have been widely used for strengthening concrete structures [5–7]. Theoretically, the development of concrete microcracks can be prevented by the inclusion of short-cut fibers to some extent [8]. Such materials represent a new trend in the development of concrete.

Different kinds of short-cut fibers have been used to reinforce concrete, such as asbestos [3], cellulose [9], steel [10], polypropylene [11], polyvinyl alcohols [12], carbon [4], basalt [13], aramid [14], polyethylene [15], and glass fibers [16]. Among them, steel and glass fibers are the most widely used ones. The flexural toughness, flexural fatigue endurance, and impact resistance of concrete were upgraded by the use of steel fibers [17]. However, some problems still exist in the use of steel fibers, such as the reduced workability and vulnerability to corrosion [18]. Glass fiber-reinforced concrete is mainly used to produce lightweight building components. However, glass fibers suffer degradation in the alkaline environment of such concrete, which greatly limits their application [16]. In contrast, basalt fibers extruded from molten basalt represent a new type of inorganic fibers with high acid and alkali resistance [13,19]. Moreover, basalt fibers can bond well with concrete. In particular, the tensile strength of basalt fibers is several times higher than that of steel fibers. Compared to carbon fibers, basalt fibers offer a price advantage [20]. Therefore, basalt fibers can be used as a substitute for steel and glass fibers in fiber-reinforced concrete [21,22].

Dias et al. [23] evaluated the benefits of basalt fibers on the fracture toughness of concrete. Furthermore, the bridging effect of basalt fibers on concrete microcracks was revealed, and fiber length and volume fraction were confirmed as important parameters. Similar conclusions were obtained from the tests reported in Refs. [24–26]. In addition, the effect of basalt fibers on fly ash geopolymer concrete was investigated experimentally [13]. Basalt fibers were also used in combination with other highly elongated fibers to enhance concrete, e.g., glass [20], polypropylene [27], and steel fibers [28]. The highly elongated fibers were used to improve energy dissipation capacity during collisions.

Basalt fibers were also used to reinforce cement-based materials. During the experimental study performed by Xu et al. [22], unique strain-hardening and multiple-cracking behaviors were exhibited by basalt fiber-reinforced engineered cementitious composites. Zheng et al. [29] found that flexural strength, fracture energy, and crack resistance, although not compressive strength, were improved by the inclusion of short-cut basalt fibers to various extents. These results are not entirely consistent with those in the concrete case detailed in the previous paragraph, which implies that the effect of basalt fibers varies according to aggregate type.

Conversely, an excessive amount of basalt fibers can result in problems such as fiber agglomeration and mixing difficulties [30]. This indicates that the appropriate volume fraction should be determined [31,32].

Currently, the pea gravel concrete covers reinforced by steel wire mesh are used to improve the shear capacities of seismic resistance-deficient masonry walls [33,34]. The main difference between pea gravel concrete and normal concrete lies in aggregate size. Small-sized aggregate endows the pea gravel concrete with excellent impermeability and a relatively smooth surface, which is suitable for use as cladding material for brick masonry walls. In particular, some pea gravel aggregates are derived from solid-waste recycling [35,36]. It is important to note that the pea gravel aggregate considered in this study is not from solid waste but is instead commercial aggregate.

In view of the significant improvements gained from basalt fibers in the mechanical properties of normal concrete, as confirmed by the above-mentioned research, it is reasonable to believe that basalt fiber-reinforced pea gravel concrete (BFRPGC) has a remarkable potential for use as the retrofitting covers for brick masonry walls, as shown in Figure 1. In such cases, the BFRPGC covers themselves have sufficient resistance against brittle fracture. Hence, the cumbersome construction of the steel wire meshes for the conventional pea gravel concrete covers is replaced by the simple shear keys at the interfaces between brick masonry walls and the BFRPGC covers (i.e., the grooves in mortar joints and L-shaped steel connectors in the plot). This provides a low-cost and efficient technique for retrofitting masonry structures.

Considering the significant difference in the coarse aggregate between the BFRPGC material and normal concrete [37], the influence of basalt fibers on the mechanical properties of pea gravel concrete has its own particularity. As mentioned earlier, it is similar to the differences in performance seen between basalt fiber-reinforced cement-based material and basalt fiber-reinforced normal concrete. However, a quantitative understanding of the mechanical properties of the BFRPGC material is still a perceived gap in the current literature.

In the current study, the effects of the short-cut basalt fibers on the mechanical properties of pea gravel concrete were quantitatively evaluated by a comprehensive experimental investigation. To obtain the best mix proportion for the BFRPGC material, fiber length and volume fraction were selected as the key parameters. Two fiber lengths and six fiber volume fractions were considered. Both the compressive strength test and splitting tensile strength test were performed for each of the combinations at the different levels of the two parameters. Furthermore, scanning electron microscope observations were conducted to reveal the failure mechanism of the basalt fibers in the pea gravel concrete. Moreover, the compression stress–strain relationship of the BFRPGC material was established.

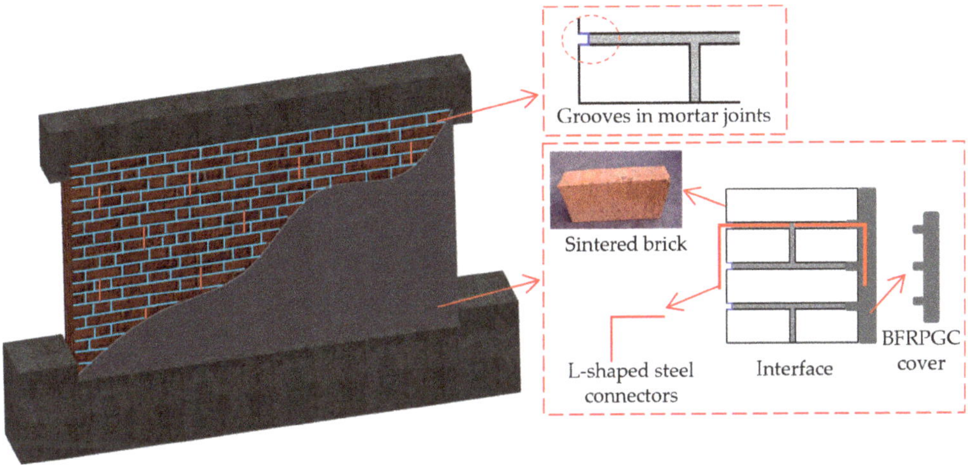

Figure 1. Masonry wall retrofitted with a BFRPGC cover.

2. Materials and Methods

2.1. Materials

A commercial composite silicate cement of Grade P.C32.5 [38] was chosen for this study. Pea gravel was used as a coarse aggregate (see Figure 2a), with a maximum particle size of 16 mm. The medium sand was used as the fine aggregate, with a fineness modulus of 2.5. The short-cut basalt fibers (see Figure 2b) were produced by Zhejiang Shijin Basalt Fibers Co., Ltd., Dongyang, China. The fiber properties provided by the manufacturer are listed in Table 1.

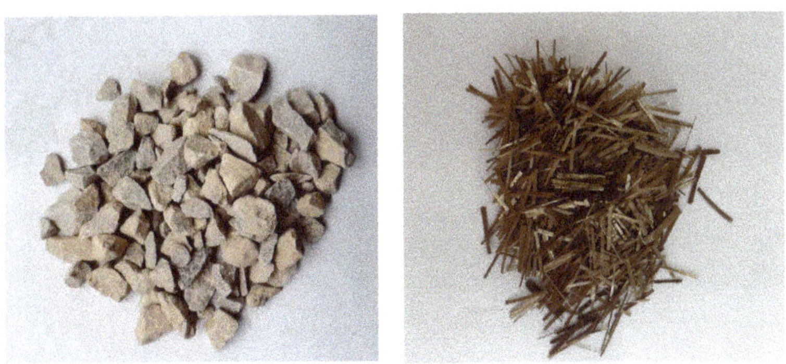

(**a**) (**b**)

Figure 2. Fibers and coarse aggregate (Photographed by B. Wang): (**a**) Pea gravel; (**b**) 12-millimeter-length basalt fibers.

Table 1. Physical and mechanical properties of basalt fibers.

Length (mm)	Diameter (μm)	Density (g·cm^{-3})	Elastic Modulus (GPa)	Tensile Strength (MPa)	Breaking Elongation (%)
6	16	2.65	81	4300	3.1
12	16	2.65	81	4300	3.1

2.2. Mix Proportions

The mix proportion for pea gravel concrete of Grade C30 [38] was used as the basis for the different BFRPGC test scenarios. In compliance with JGJ 55-2011 [38], the water–cement ratio and the sand ratio were designated as 0.42 and 0.50, respectively. The mix proportion for this reference pea gravel concrete are shown in Table 2. A total of twelve BFRPGC test scenarios were considered. Each of them was associated with a certain combination of fiber length and volume fraction. Two fiber lengths were selected, i.e., 6 mm and 12 mm. The volume fraction corresponding to each of the fiber lengths was increased from 0.3% to 0.8%, with a step of 0.1%, i.e., 0.3%, 0.4%, 0.5%, 0.6%, 0.7%, and 0.8%.

Table 2. Mix proportions of the reference pea gravel concrete.

Cement (kg·m^{-3})	Water (kg·m^{-3})	Sand (kg·m^{-3})	Pebble (kg·m^{-3})
485	204	785	771

2.3. Specimen Preparation

The specimens were the cubes with a side length of 100 mm. To avoid agglomeration of the short-cut basalt fibers as much as possible, a secondary mixing method was adopted. First, cement, sand, pea gravel, water, and half the amount of the short-cut basalt fibers were mixed for 100 s. Then, the other half of the basalt fibers was added, with a mixing time of 200 s. The mixed concrete material was cast into plastic molds treated with a pre-applied release agent. Working according to the provisions prescribed in GB/T 50081-2019 [39], the specimens were cured in a non-flow calcium hydroxide-saturated solution with a temperature of 20 °C ± 2 °C (see Figure 3).

Figure 3. Curing of specimens (Photographed by B. Wang).

A total of 255 specimens from 3 series were made. Among them, Series R consisted of the reference pea gravel concrete blocks without basalt fibers, while Series B6 and B12 corresponded to the BFRPGC specimens with fibers of 6 mm and 12 mm in length, respectively. There are one and six specimen groups in Series R and in each of the second two series, respectively. The six groups corresponded to the aforementioned six levels of the fiber volume fraction. Each of the groups consisted of 9 specimens. The groups in series B6 and B12 were labeled as GAB. In the label, "A" denoted fiber length; "B" denoted the name for the level of fiber volume fraction. The "A" values of 6 and X corresponded to fiber lengths of 6 mm and 12 mm, respectively. The "B" values of 3, 4, 5, 6, 7, and 8 corresponded to the fiber volume fractions of 0.3%, 0.4%, 0.5%, 0.6%, 0.7%, and 0.8%, respectively.

2.4. Experimental Program

2.4.1. Compressive Strength Test

The compressive strength tests were conducted in accordance with the provisions of GB/T 50081-2019 [39]. The WAW-1000 hydraulic universal testing machine was used, as shown in Figure 4. It was produced by Changchun Kexin Testing Instruments Co., Ltd. (Changchun, China). Its maximum loading capacity and accuracy are 1000 kN and ± 1%, respectively.

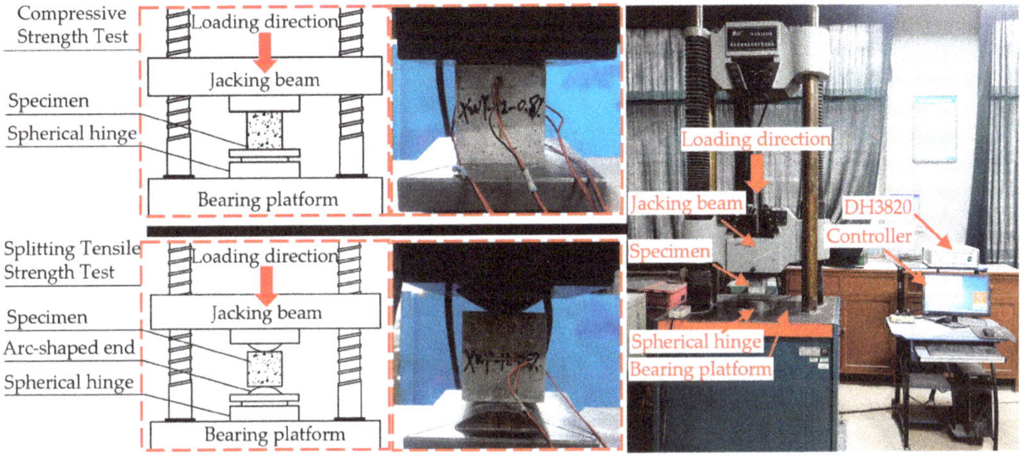

Figure 4. Test setup (Photographed by B. Wang).

During the tests, the loading speed was 5 kN/s. To disclose the strain–strain relationship of the BFRPGC material, a pair of strain gauges were set vertically on the centers of the two opposite sides of the concrete block (see Figure 5a), while a pair of strain gauges were applied to the centers of the other opposite sides along the transverse direction (see Figure 5b). The former and latter were installed to record the compressive strain and the tensile strain caused by the Poisson effect, respectively. The strain data were recorded with a DH3820 data acquisition instrument produced by Donghua Testing Instruments Co., Ltd., Jingjiang, China. The acquisition frequency used in this study was 100 Hz.

(a) (b) (c)

Figure 5. Strain measurements (Photographed by B. Wang): (**a**) Vertical strain gauge for the compressive strength test; (**b**) Transverse strain gauge for the compressive strength test; (**c**) Transverse strain gauge for the splitting tensile strength test.

2.4.2. Splitting Tensile Strength Test

The splitting tensile strength tests were also performed with the hydraulic universal testing machine and in compliance with GB/T 50081-2019 [39]. Except for the shapes of the loading and bearing ends, the specimens were loaded using a similar approach to that used for the compressive strength tests (as detailed in Section 2.4.1). The loading and bearing ends are arc-shaped (see Figure 4) and the radius of their cross-sections is 75 mm. Only two transverse strain gauges were set at the two opposite sides of the specimen (see Figure 5c).

2.4.3. Scanning Electron Microscope Observation

As shown in Figure 6, an EM-30 scanning electron microscope (SEM), manufactured by COXEM Co., Ltd., Daejeon, Republic of Korea, was used for clarifying the failure mechanism of the short-cut basalt fibers in the pea gravel concrete. With this instrument, the fragments from the BFRPGC blocks that remained after the tests were scanned by a narrowly focused high-energy electron beam. The various types of physical information were elucidated through interactions between the electron beam and the BFRPGC material. By collecting, amplifying, and re-imaging the information, the damage characteristics of the basalt fibers were presented. The magnification range and backscattered electron image resolution of the EM-30 SEM were from 20 to 100,000 times and 8.0 nm @30 kV SE, respectively. The morphologies of the BFRPGC material could be clearly observed at magnifications ranging from 500 times to 1000 times.

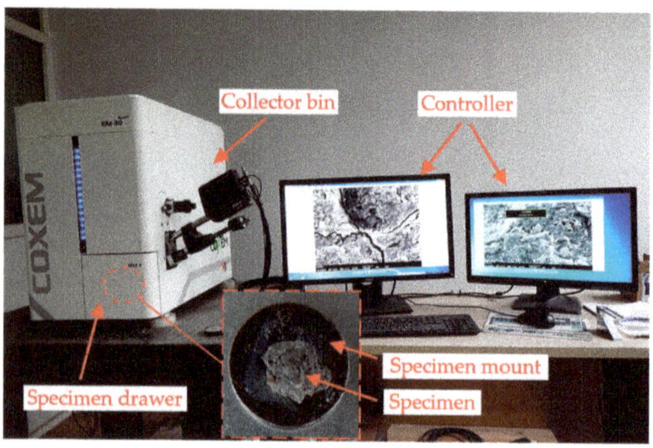

Figure 6. EM-30 SEM (Photographed by B. Wang).

3. Results and Discussion

3.1. Compressive Strength Test

3.1.1. Failure Modes

The typical compression failure of a reference pea gravel concrete block without basalt fibers is shown in Figure 7a. Two major penetrating cracks from the upper surface to the bottom surface of the specimen can be observed. As a result of these cracks, the specimen was divided into several vertical pieces that tended to fall apart. However, intensive but relatively short cracks appeared with the failure of the BFRPGC blocks, as shown in Figure 7b. However, the major penetrating cracks that can be seen in the reference pea gravel concrete block did not occur. This difference obviously came from the bridging effect of the basalt fibers on the pea gravel concrete. When microcracks developed in the concrete, the basalt fibers across the microcracks were pulled. As a result, the further development of the microcracks was restrained to a certain extent [19]. That is, a more uniform stress profile and microcrack distribution formed in the BFRPGC material. In this case, therefore, concrete failure was delayed, and the compressive strength was increased (as detailed in the next section).

(a)

(b)

Figure 7. Typical compression failures(Photographed by B. Wang): (**a**) Reference pea gravel concrete specimen; (**b**) BFRPGC specimen (Group GX4).

3.1.2. Compressive Strength

The compressive strength (f_{cp}), in MPa, was calculated in accordance with the provisions of GB/T 50081-2019 [39], as shown in Equation (1):

$$f_{cp} = F/A \quad (1)$$

where F is the maximum load in N; A is the bearing area in mm^2.

Each of the measured axial compressive strengths was multiplied by 0.95 to account for the size effect [39]. The average value and the coefficient of variation for the compressive strength obtained from each of the different specimen groups, the increase in the average value relative to the reference pea gravel concrete value, and the ratio between the coefficient of variation for the BFRPGC material and that of the reference pea gravel concrete are listed in Table 3. The increase in the average value is defined as $(f_{cp,ab} - f_{cp,a0})/f_{cp,a0} - 1$, where $f_{cp,ab}$ and $f_{cp,a0}$ are the average values of the BFRPGC material and the reference pea gravel concrete, respectively. As the volume fraction of basalt fibers was increased, there were changes in the average value and coefficient of variation in Series B6 and B12, as illustrated in Figures 8a and 8b, respectively.

Table 3. Results of the compressive strength tests.

Series	Group	Fiber Volume Fraction (%)	Average Compressive Strength (MPa)	Increase in Average Compressive Strength (%)	Coefficient of Variation (%)	Ratio for Coefficient of Variation Ratio (%)
R	/	0.00	31.9	/	7.03	/
B6	G63	0.30	30.5	−4.39	7.23	102.8
	G64	0.40	26.4	−17.2	5.44	77.38
	G65	0.50	28.4	−11.0	5.85	83.21
	G66	0.60	28.4	−11.0	5.96	84.78
	G67	0.70	30.5	−4.39	12.9	183.5
	G68	0.80	29.8	−6.58	5.93	84.35
B12	GX3	0.30	50.0	56.7	6.22	88.48
	GX4	0.40	46.1	44.5	6.24	88.76
	GX5	0.50	49.4	54.9	4.91	69.84
	GX6	0.60	37.4	17.2	5.37	76.39
	GX7	0.70	31.8	−0.31	6.70	95.31
	GX8	0.80	33.6	5.33	15.1	214.8

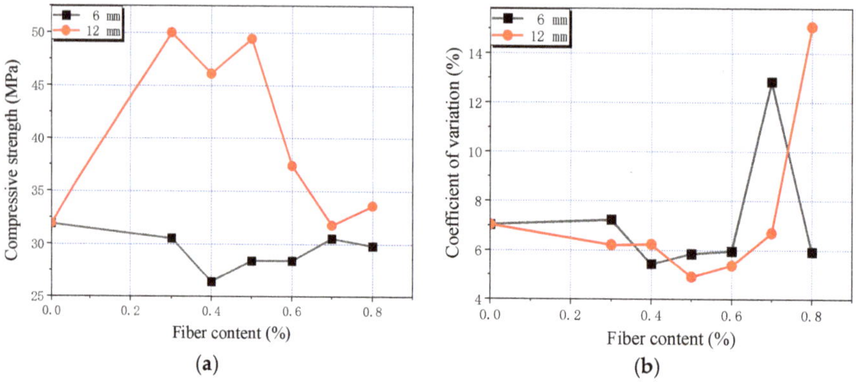

Figure 8. Effect of the fiber volume fraction on compressive strength: (**a**) Average value; (**b**) Coefficient of variation.

As shown in Table 3 and Figure 8, the average compressive strength and the corresponding coefficient of variation for the reference pea gravel concrete were 31.9 MPa and 7.03%, respectively. It is noteworthy that the value of the coefficient of variation was much lower than the lower limit of the recommended range for concrete given in GB 50010-2010 [40] (i.e., from 17.2% to 18.9%).

In Series B12, when the fiber volume fraction was increased, the average compressive strength first increased and then decreased (see Figure 8a). As the fiber volume fraction varied from 0.3% to 0.5%, the average compressive strength values remained in a range from 46.1 MPa to 50.0 MPa. These average compressive strength values were 44.5% to 56.7% higher than that of the reference pea gravel concrete. This clearly demonstrates that the short-cut basalt fibers rendered significant enhancement to the pea gravel concrete. However, the average compressive strength dropped dramatically to a range from 31.8 MPa to 33.6 MPa as the fiber volume fraction was further increased. The reason for the drop in strength is detailed in the next paragraph. There was no obvious improvement in comparison with the reference pea gravel concrete in such cases.

The above phenomenon is due to the variation in the compactness of concrete when the fiber volume fraction was increased. It is noteworthy that the compactness of concrete was not sensitive to relatively small amounts of basalt fibers when the fraction ranged from 0.3% to 0.5%. Hence, the compression failure of the BFRPGC material was delayed by the bridging effect of the basalt fibers (as discussed in Section 3.1.1), and the compressive strength was improved significantly. However, the excessive amounts of basalt fibers led to the marked degradation of the fluidity of the pea gravel concrete, in terms of obstruction due to the microfiber network composed of basalt fibers. In such cases, the negative effect of the agglomeration of basalt fibers in the pea gravel concrete could not be neglected. Concurrently, there were plenty of concrete pores that formed around the agglomerated basalt fibers, which resulted in a decrease in the compactness of the concrete. During compressive loading, this tended to induce stress concentrations in the concrete, where the premature crushing of the concrete was prone to occur. Consequently, the positive role played by the bridging effect from the basalt fibers was counteracted.

With the fiber volume fraction, the coefficient of variation for the compressive strength in Series B12 also fluctuated significantly, as demonstrated in Table 3 and Figure 8b. The coefficient of variation varied in a range from 4.91% to 6.70% and was lower than the value of 7.03% for the reference pea gravel concrete when the fraction was no more than 0.7%. However, as the fraction was increased to 0.8%, the coefficient of variation increased dramatically to 15.1%. The value was 114.7% higher than that of the reference pea gravel concrete but was still lower than the lower limit of the recommended range for concrete in GB 50010-2010 [40].

As presented in Table 3 and Figure 8a, in the different specimen groups in Series B6, the compressive strength of the BFRPGC material was slightly lower than that of the reference pea gravel concrete. This implies that the bridging effect of the basalt fibers on the concrete microcracks was not significant when the fibers were relatively short. Moreover, it is reasonable to believe that the negative effect of basalt fibers on the compactness of concrete still existed, to various extents. As the fiber volume fraction increased, the trend in the variation of the coefficient of variation in this series was similar to that in Series B12 (see Table 3 and Figure 8b). However, the maximum coefficient of variation, i.e., 12.9%, was recorded at a fiber volume fraction of 0.7%, which was slightly lower than the fiber volume fraction corresponding to the maximum coefficient of variation in Series B12.

From the above analyses, it is clear that the enhancement of the compressive strength was significant when the 12 millimeter-length fibers were used and when the fiber volume fraction was in a range from 0.3% to 0.5%. Furthermore, the discreteness of the compressive strength could be controlled effectively in such cases.

3.1.3. Strain Capacity

To further explore the mechanical behavior of the BFRPGC material, the typical axial stress–axial strain and axial stress–transverse strain relationship curves recorded in the compressive strength tests of the different specimen groups in Series B12 are compared to each other at the second and first quadrants in Figure 9, respectively.

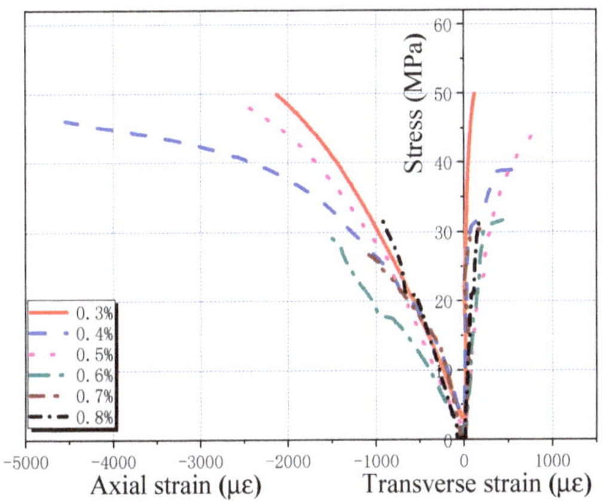

Figure 9. Stress–strain relationship curves obtained from the compressive strength tests in Series B12.

The initial linear portions of the axial stress–axial strain relationship curves associated with the different fiber volume fractions are very close to each other, as illustrated in Figure 9. It implies that the positive role of basalt fibers (i.e., the bridging effect) was not apparent before the development of microcracks in the concrete. However, an obvious discrete phenomenon exists among the nonlinear portions of the curves. It is the combined result of the bridging effect of basalt fibers on concrete microcracks and the negative effect of basalt fibers on the compactness of concrete. Furthermore, a direct relationship between the fiber volume fraction and the strain corresponding to the maximum compressive stress (hereafter referred to as "the maximum compressive strain") was observed. The recorded maximum compressive strains reached a range from 3000 με to 5000 με, while the fiber volume fraction varied from 0.3% to 0.5%. This clearly indicated delays in the compression failure. The phenomenon could only come from the effective bridging effect provided by the short-cut basalt fibers. However, the recorded maximum compressive strains associated with other levels of fiber volume fraction were generally no more than 2000 με. The above strain measurement results were consistent with the recorded compressive strengths.

It is also noted that the axial strain value was generally much higher than the corresponding transverse tensile value at the different compression stress levels. In particular, in the nonlinear response phase, significant strain-softening phenomena were observed in the axial stress–axial strain relationship curves associated with the fiber volume fractions of 0.3%, 0.4%, and 0.5%. However, the transverse stress–transverse strain relationship curves generally terminated at the ends of their respective initial linear portions. This was apparently due to the fracturing of the strain gauges that were spanned by the concrete cracks.

3.2. Splitting Tensile Strength Test

3.2.1. Failure Modes

A typical example of the splitting tensile failure of the reference pea gravel concrete specimens is presented in Figure 10a. The concrete block was split into two halves by the main crack formed at its middle and along the vertical direction. A different failure mode was observed in the BFRPGC specimen. Relatively intensive vertical cracks occurred in the BFRPGC block (see Figure 10b), which exhibited a certain degree of plasticity. Obviously, this is due to the bridging effect of the short-cut basalt fibers in the concrete.

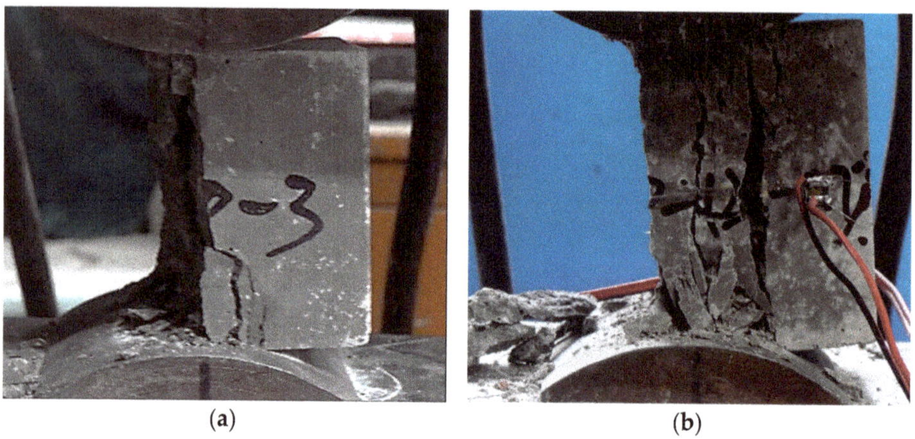

Figure 10. Typical splitting tensile failures (Photographed by B. Wang): (**a**) Reference pea gravel concrete specimen; (**b**) BFRPGC specimen (Group GX7).

3.2.2. Splitting Tensile Strength

The splitting tensile strength (f_{ts}), in MPa, was determined in accordance with the provisions of GB/T 50081-2019 [39], as shown in Equation (2):

$$f_{ts} = 2F/\pi A. \quad (2)$$

The size conversion coefficient for the measured tensile splitting strengths was 0.85 [39]. The average value and the coefficient of variation for the splitting tensile strength values obtained from each of the different specimen groups, the increase in the average value relative to the reference pea gravel concrete case, and the ratio between the coefficient of variation for the BFRPGC material and that of the reference pea gravel concrete are listed in Table 4. Along with the fiber volume fractions, the variations in the two values are illustrated in Figures 11a and 11b, respectively.

As shown in Table 4 and Figure 11a, the average splitting tensile strength in Series B12 oscillated substantially with the fiber volume fraction. An increasing trend in the average value was observed when the fiber volume fraction was increased from 0.3% to 0.4%. Hence, the maximum value for the average splitting tensile strength was reached at a fiber volume fraction of 0.4% and at 13.3 MPa. This was 27.9% higher than the maximum value for the reference pea gravel concrete. However, a dramatic drop in the average splitting tensile strength was recorded when the fiber volume fraction was further increased to 0.5%. In this case, the average splitting tensile strength was approximately equal to that of the reference pea gravel concrete. Thereafter, it rose again in accordance with the fiber volume fraction. At fiber volume fractions of 0.7% and 0.8%, the average splitting tensile strength was stable at a value of 12.5 MPa. It is noteworthy that this value was still 20.2% higher than that of the reference pea gravel concrete. From this phenomenon, relative to the compressive strength case, it appears that the splitting tensile strength was less sensitive to fiber agglomeration at a relatively high fiber volume fraction.

Table 4. Results of the splitting tensile strength tests.

Series	Group	Fiber Volume Fraction (%)	Average Splitting Tensile Strength (MPa)	Increase in Average Splitting Tensile Strength (%)	Coefficient of Variation (%)	Ratio for Coefficient of Variation Ratio (%)
R	/	0.00	10.4	/	13.0	/
B6	G63	0.30	10.7	2.88	18.0	138.5
	G64	0.40	9.80	−5.77	13.7	105.4
	G65	0.50	10.1	−2.88	7.06	54.31
	G66	0.60	11.2	7.69	8.43	64.85
	G67	0.70	12.3	18.3	7.15	55.00
	G68	0.80	11.4	9.62	12.7	97.69
B12	GX3	0.30	10.6	1.92	7.95	61.15
	GX4	0.40	13.3	27.9	14.1	108.5
	GX5	0.50	10.2	−1.92	11.0	84.62
	GX6	0.60	11.5	10.6	16.1	123.9
	GX7	0.70	12.5	20.2	11.4	87.69
	GX8	0.80	12.5	20.2	14.4	110.8

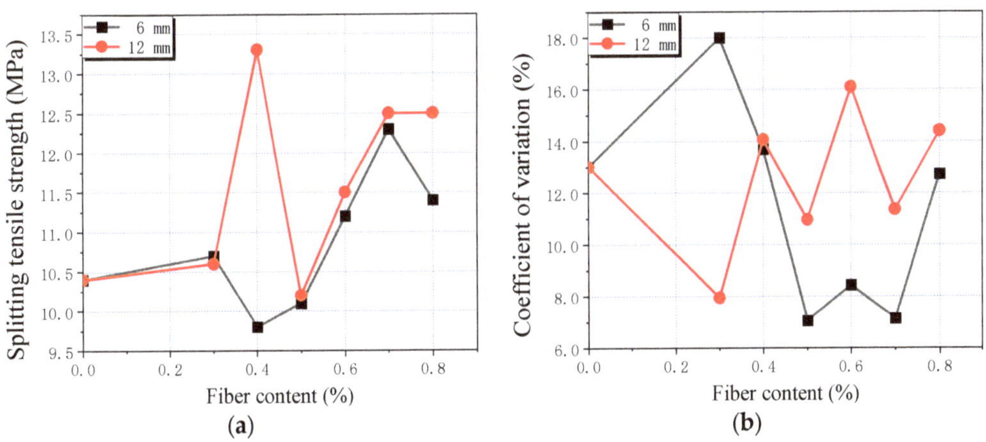

Figure 11. Effect of the fiber volume fraction on splitting tensile strength: (**a**) Average value; (**b**) Coefficient of variation.

The average splitting tensile strength in Series B6 also oscillated with the fiber volume fraction. At a fiber volume fraction of 0.7%, the maximum value for the average splitting tensile strength in this series was obtained. This was 12.3 MPa, 18.3% higher than that for Series R. However, at each of the fiber volume fractions of 0.4% and 0.5%, the average splitting tensile strength was slightly lower than that for Series R.

With the fiber volume fraction, the coefficients of variation for the splitting tensile strength in Series B12 oscillated in a range from 7.95% to 16.1%, as shown in Table 4 and Figure 11b. A similar trend existed in Series B6, with an oscillating range from 7.06% to 18.0%. Furthermore, it is noteworthy that the oscillations of the coefficient of variation in the case of the splitting tensile strength were more pronounced than those of compressive strength at the relatively low levels of fiber volume fraction ranging from 0.3% to 0.6%. This demonstrates the complexity of the effect of basalt fibers on the discreteness of the splitting tensile strength.

At a fiber volume fraction of 0.4% in Series B12, the maximum splitting tensile strength of this study was achieved, and the degree of dispersion was controlled at the same level as that of the reference pea gravel concrete. Hence, the mix proportion for the BFRPGC material in this test scenario is considered to be optimal for achieving the splitting tensile strength.

3.2.3. Strain Capacity

The typical axial stress–transverse strain relationship curves obtained from the splitting tensile strength tests of the different specimen groups in Series B12 are compared to each other in Figure 12. Apparently, each of the curves can be divided into two parts.

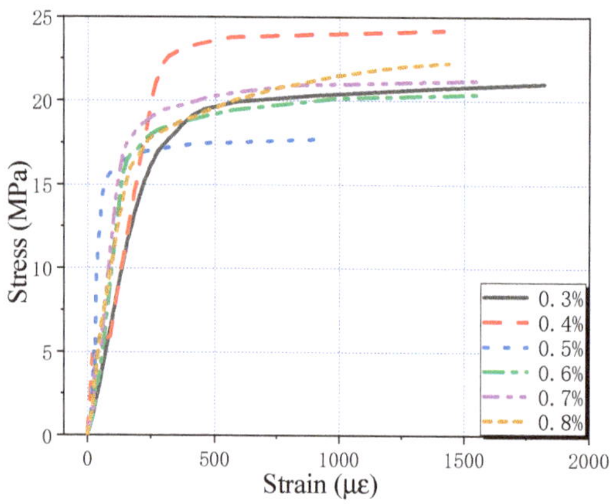

Figure 12. The stress–strain relationship curves obtained from the splitting tensile strength tests of Series B12.

The first part is characterized by a linear response, in which the transverse tensile strain is proportional to the axial compressive stress. The ends of the first parts for most of the curves are in a range from 250 με to 500 με. Obviously, these linear parts reflect the transverse expansion of the BFRPGC material, due to the Poisson's ratio effect.

The second part is characterized by a very slowly ascending trend. That is, the transverse tensile strain increases significantly, while the axial compressive stress changes a little. Most of the second parts terminate at approximately 1500 με, due to the fracturing of the transverse strain gauges. The characteristics of the second part reflect the development of concrete macrocracks. It is quite reasonable to suppose that the relatively high tensile strain levels reached in the second portion came from the bridging effect of the short-cut basalt fibers.

3.3. *Results of Scanning Electron Microscopy*

As shown in Figure 13a, the basalt fiber was a round rod and had a smooth surface. The measured fiber diameter was 14.32 μm. The typical failure characteristics of the short-cut basalt fibers during the compressive strength test are presented in Figure 13b–d. Generally, microscopic cracks existed within the pea gravel concrete itself and propagated as the compression was increased, which led to a reduction in the mechanical properties of the concrete. In the case of BFRPGC, the effective bridging role of basalt fibers in restraining the development of the microscopic cracks had the prerequisite that reliable interfacial bonds should be formed between the fibers and the concrete. This was confirmed by the phenomenon shown in Figure 13b, i.e., mortar debris adhered tightly to the fiber surface until the crushing of the BFRPGC material. The good bonding performance can be attributed to the large number of hydroxyl bonds formed on the fiber surfaces. During such a hydration process, a large amount of the hydration product continuously accumulated on the surfaces of the basalt fibers, which led to a strong adhesive force between the basalt fibers and the pea gravel concrete by forming shear keys at the interfaces [41,42].

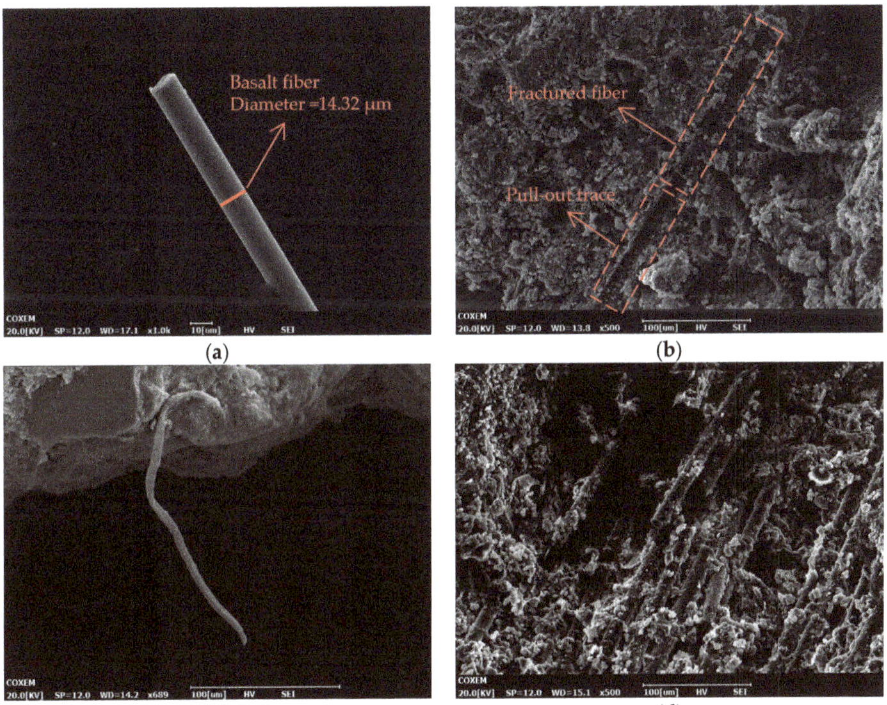

Figure 13. SEM images: (**a**) Basalt fiber; (**b**) Tensile fracture of basalt fibers in pea gravel concrete (Group GX4); (**c**) Pull-out part of basalt fiber (Group GX4); (**d**) Agglomeration of basalt fibers in pea gravel concrete (Group GX8).

As shown in Figure 13b, due to this good bonding performance, the short-cut basalt fibers across the macroscopic cracks suffered tensile fractures with the crushing of the BFRPGC block. The trace for the pull-out part of the fractured fiber is clearly present in the plot. As demonstrated in Figure 13c, the pull-out part of the fractured fiber shows characteristic tearing damage. This type of damage corresponds to a sheet-like fiber shape that was significantly different from the shape of the undamaged fiber (see Figure 13a). These phenomena confirmed the effectiveness of the bridging effect from the short-cut basalt fibers in the pea gravel concrete. Figure 13b,c was obtained with Series B12 and at a fiber volume fraction level of 0.4%. At the much higher fiber volume fraction of 0.8% found in Series B12, the short-cut basalt fibers agglomerated in the pea gravel concrete, as shown in Figure 13d. As a result, a larger number of voids are observed. In comparison with the 0.4% fiber volume fraction case (see Figure 13b), a significant deterioration in the compactness of the concrete can be observed. This provides intuitive evidence for explaining the oscillation in BFRPGC strength as the fiber volume fraction was increased.

3.4. Uniaxial Compressive Stress–Strain Model

As demonstrated by the test results, a significant enhancement to the mechanical behavior of the pea gravel concrete was achieved in the case of the fiber length of 12 mm, whereas the effect of the 6-millimeter-length fibers was not obvious. Therefore, the uniaxial compressive stress–strain model of the BFRPGC material was developed for the 12-millimeter-length scenario to facilitate the corresponding structural design and nonlinear analysis.

A reliable uniaxial compressive stress–strain model has been proposed for normal concrete with short-cut high-performance fibers [43–48], as presented in Table 5. In general, such a uniaxial compressive stress–strain model consists of two parts, i.e., the ascending and descending elements. This indicates that the general law for the constitutive relationship of the normal concrete still applies to the fiber-reinforced concrete [45,49]. Therefore, the classical uniaxial compressive constitutive relationship suggested by Guo et al. [46] was used as the basis for developing the model for the BFRPGC material. The model is shown in Equation (3):

$$y=\begin{cases} ax+(3-2a)x^2+(a-2)x^3 & (0\le x<1) \\ \frac{x}{b(x-1)^2+1} & (x\ge 1)\end{cases} \tag{3}$$

where x is the ratio of the compressive strain to its peak value and is defined as $\varepsilon/\varepsilon_{cr}$; ε and ε_{cr} are the compressive strain and its value corresponding to the maximum compressive stress, respectively; y denotes the ratio of the compressive stress to its maximum value and is defined as σ/f_{cr}; σ and f_{cr} are the compressive stress and its maximum value, respectively; a and b are the critical parameters associated with the rising and falling parts of the uniaxial compressive constitutive relationship, respectively.

Table 5. Proposed uniaxial compressive stress–strain models for fiber-reinforced concrete.

Researchers	Analytical Models
Wang et al. [43]	$y=\begin{cases}\alpha_1 x+(6-5\alpha_1)x^5+(4\alpha_1-5)x^6 & (0\le x<1) \\ \frac{x}{\alpha(x-1)^2+x} & (x\ge 1)\end{cases}$; α_1 = 1.417 + 0.697 V_{BF} − 6.699 V_{PPF}; here, α = 5.638 + 24.01 V_{BF} − 468.34 V_{PPF}; V_{BF} and V_{PPF} are the volume fractions of basalt fibers and polypropylene fibers, respectively.
Zhao et al. [44]	$y=\begin{cases}\frac{nx}{n-1+x^n} & (0\le x<1) \\ \frac{x}{\alpha(x-1)^b+x} & (x\ge 1)\end{cases}$; $n = E_{C\varepsilon c.r}/(E_{C\varepsilon c.r}+f_{c.r})$; $\alpha = 0.00022 f_{c.r}{}^{2.75} + 0.746$; here, $f_{c.r}$ and $E_{C\varepsilon c.r}$ are the peak stress and modulus of elasticity of steel fiber-reinforced expanded-shale lightweight concrete, respectively.
Chen et al. [45]	$y=\begin{cases}\frac{nx}{n-1+x^n} & (0\le x<1) \\ \frac{x}{\alpha(x-1)^2+x} & (x\ge 1)\end{cases}$; $n = E_c/(E_c+f_{cr})$; $\alpha = 0.157 f_{cr}{}^{0.785} - 0.905$; here, f_{cr} and E_c are the peak stress and modulus of elasticity of basalt fiber-reinforced rubberized recycled coarse aggregate concrete, respectively.
Ding et al. [46]	$y=\begin{cases}\left(\frac{E_c}{E_{c1}}x-x^2\right)\left[1+\left(\frac{E_c}{E_{c1}}-2\right)x\right]^{-1} & (0\le x<1) \\ \left[\left(\frac{1}{1.8}\right)^2\xi x(x_{lim}-1)+\frac{1}{0.9}(2x_{lim}-1)\right]^{-1} & (x\ge 1)\end{cases}$; $\xi=\frac{4[(x_{lim})^2(\frac{E_c}{E_{c1}}-2)+2x-\frac{E_c}{E_{c1}}]}{(x_{lim})(\frac{E_c}{E_{c1}}-2)+1}$; $x_{lim} = (0.05E_c/E_{c1}+0.9)+[(0.05E_c/E_{c1}+0.9)^2-0.9]^{0.5}$; here, E_c and E_{c1} are the tangent modulus and the secant modulus from the origin to the compressive strength, respectively; ξ and x_{lim} are the parameters related to the elastic modulus and the limited strain at a stress of $0.9f_c$, respectively.
Ou et al. [47]	$y=\frac{\beta x}{\beta-1+x^\beta}$; $\beta = 0.71(RI)^2 - 2(RI) + 3.05$; $f_{cr} = f_{cr.p} + 2.35RI$; $\varepsilon_{cr} = \varepsilon_{cr.p} + 0.0007RI$; $RI = V_f\, L_f/D_f$; here, V_f, L_f, D_f and $f_{cr.p}$ are the volume fraction, length, and diameter of the steel fibers and the compressive strength of plain concrete, respectively.
Abbass et al. [48]	$y=\frac{\beta x}{\beta-1+x^\beta}$; $\beta = 1.401(RI)^2 - 1.56(RI) + 2.42$; $f_{cr} = f_{cr.p} + 5.59RI$; $\varepsilon_{cr} = \varepsilon_{cr.p} + 0.000261RI$; $RI = V_f\, L_f/D_f$.

In this study, using the least squares method [43], the respective relationships of parameters a and b with the fiber volume fraction (V_{BF}) were established on the basis of the fitted uniaxial compressive stress–strain curves corresponding to the different fiber volume fractions. With the related test results, the two parts (i.e., the ascending and descending one) for each of the curves were fitted separately in the regression analyses. An example of these regression curves is shown in Figure 14 and corresponds to a fiber volume fraction of 0.4%. With regard to the fiber volume fraction range studied (i.e., from 0.3% to 0.8%), the proposed expressions for parameters a and b are shown in Equations (4) and (5), respectively.

$$a = 13.0V_{BF} - 0.2 \tag{4}$$

$$b = 1611.5V_{BF} - 2.5 \tag{5}$$

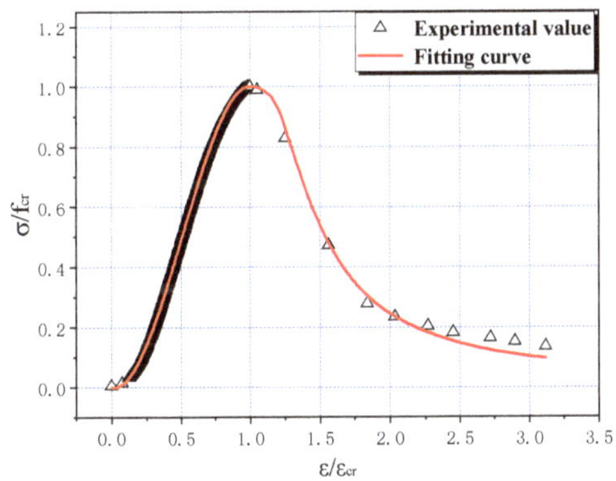

Figure 14. The fitted uniaxial compressive stress–strain curve in the case of 12-millimeter-length basalt fibers with a volume fraction of 0.4%.

Furthermore, as discussed in Sections 3.1 and 3.2, the best performance of the BFRPGC material was achieved by the 12-millimeter-length fibers with a volume fraction of 0.4%. This parameter combination is proposed for future engineering applications. In this case, the different stages in the mechanical behavior of this optimal BFRPGC material until compression failure is clearly reflected in the four parts of the fitted curve shown in Figure 14. Combined with the experimental observations in this study, the characteristics of the four parts can be summarized as follows.

The first stage corresponds to the initial near-linear part of the curve. At this stage, no significant changes to the initial microcracks inside the BFRPGC material commence. Hence, the mechanical behavior of the BFRPGC material is dependent on that of the matrix, i.e., the pea gravel concrete. The enhancement offered by the basalt fibers is not obvious.

In the second stage, the curve is characterized by a nonlinear ascending trend until a point of maximum compressive stress. During this process, the crushing of the concrete is delayed by the bridging effect provided by the basalt fibers across the microcracks, and higher compressive strength is reached.

This is followed by a stage dominated by the unstable expansion of cracks. After the maximum compressive stress is reached, the bridging effect degrades rapidly, while the fracture and pull-out phenomena in the basalt fibers occur across the board. Consequently, within the limited increase seen in the compressive strain, the curve dropped to an obvious inflection point that corresponds to approximately 30% of the maximum compressive stress.

The final stage reflects the residual behavior after intensive cracking. It is manifested by a relatively slowly descending trend beyond the inflection point. Correspondingly, the separation of the BFRPGC material commences. In such a severe cracking scenario, it is reasonable to consider that the bridging effect of basalt fibers is largely lost, and the residual compressive capacity comes from the damaged pea gravel concrete matrix.

4. Conclusions

In this study, the effect of short-cut basalt fibers on the mechanical properties of pea gravel concrete was quantitatively evaluated according to combinations of the different levels of the two key parameters. The main conclusions are summarized as follows:

- Fiber length is the most important parameter for the BFRPGC material. A significant bridging effect restraining the microcrack development in concrete was achieved by the 12-millimeter-length fibers, whereas no obvious role was observed in the 6-millimeter-length scenario.
- The mechanical properties of the BFRPGC material oscillated substantially with the fiber volume fraction. The phenomenon was the combined result of the bridging effect of basalt fibers on concrete microcracks and the negative effect of fiber agglomeration on the compactness of concrete.
- In both the compressive strength tests and splitting tensile tests, intensive but relatively short cracks appeared with the failure of the BFRPGC blocks. This specific failure mode demonstrates that the more uniform stress profile and microcrack distribution were formed in the BFRPGC material, in comparison with the reference concrete case.
- The parameter combination of a fiber length of 12 mm and a fiber volume fraction of 0.4% was found to be optimal. The average compressive and splitting tensile strengths in this case were 44.5% and 27.9% higher than those of the reference pea gravel concrete, respectively, and their respective dispersions were well controlled.
- The direct relationship between the fiber volume fraction and the strain corresponding to the maximum compressive stress was disclosed. The values of the strain reached a range from 3000 $\mu\varepsilon$ to 5000 $\mu\varepsilon$, while the fiber volume fraction was increased from 0.3% to 0.5%. This clearly indicated a delay in the compression failure.
- By scanning electron microscopy observation, not only was the failure mechanism of the tensile fracture disclosed for the basalt fibers in pea gravel concrete but a significant deterioration in the compactness of concrete was proved in the relatively high fiber–volume fraction scenario.
- The uniaxial compressive stress–strain model of the BFRPGC material was established on the basis of the test data. Four stages of mechanical behavior are clearly indicated by the model, namely, the elastic, crack development, unstable crack expansion, and residual behavior stages.

Based on the findings of this study, the following future research efforts are recommended, including the effects of the different shear keys at the interfaces between the BFRPGC covers and masonry walls, the combined enhancement effects of basalt fibers and high-elongation fibers on the pea gravel concrete, and the impact toughness of the BFRPGC material.

Author Contributions: Conceptualization, J.L.; methodology, J.L., B.W. and M.W.; formal analysis, B.W. and P.Z.; investigation, B.W., P.Z. and Z.W.; resources, J.L., B.W. and M.W.; writing—original draft, B.W. and P.Z.; writing—review and editing, B.W.; project administration, J.L. and B.W.; funding acquisition, J.L. and B.W. All authors have read and agreed to the published version of the manuscript.

Funding: This research was funded by the Shaanxi Academy of Building Research Co., Ltd. Research Project, grant number 2021-10-369.

Data Availability Statement: Data are contained within the article.

Conflicts of Interest: Author Jiming Li was employed by the company Shaanxi Academy of Building Research Co., Ltd. Author Zhenyu Wang was employed by the company SDEE Power Engineering Company of Operation and Maintenance Co., Ltd. Author Meng Wang was employed by the company China Xinxing Construction Engineering Co., Ltd. The remaining authors declare that the research was conducted in the absence of any commercial or financial relationships that could be construed as a potential conflict of interest.

References

1. Choi, Y.; Yuan, R.L. Experimental relationship between splitting tensile strength and compressive strength of GFRC and PFRC. *Cem. Concr. Res.* **2005**, *35*, 1587–1591. [CrossRef]
2. Nogueira, C.L. Anti-plane shear strength of plain concrete. *Mater. Today Commun.* **2020**, *24*, 101051. [CrossRef]
3. Ranaivomanana, H.; Leklou, N. Investigation of microstructural and mechanical properties of partially hydrated Asbestos-Free fiber cement waste (AFFC) based concretes: Experimental study and predictive modeling. *Constr. Build. Mater.* **2021**, *277*, 121943. [CrossRef]
4. Pirmohammad, S.; Shokorlou, Y.M.; Amani, B. Laboratory investigations on fracture toughness of asphalt concretes reinforced with carbon and kenaf fibers. *Eng. Fract. Mech.* **2020**, *226*, 106875. [CrossRef]
5. Wang, B.; Wu, X.H.; Liu, Q.; Wu, Y.Y.; Huang, F.; Xu, L.F.; Wu, X.; Deng, Y.X. Effectiveness and efficiency of externally bonded CFRP sheets for shear strengthening of RC beam-column joints. *Polymers* **2022**, *14*, 1347. [CrossRef]
6. Wang, B.; Wu, X.H.; Sun, Z.Y.; Liang, J.Q.; Mao, X.P.; Bi, T.C.; Song, Y.; Yang, W. Experimental investigation on low-velocity impact behavior of CFRP wraps in presence of concrete substrate. *Constr. Build. Mater.* **2021**, *301*, 124103. [CrossRef]
7. Wang, B.; Zhu, H.R.; Wu, X.H.; Zhang, N.Y.; Yan, B.Q. Numerical investigation on low-velocity impact response of CFRP wraps in presence of concrete substrate. *Compos. Struct.* **2020**, *231*, 111541. [CrossRef]
8. Camille, C.; Hewage, D.K.; Mirza, O.; Mashiri, F.; Kirkland, B.; Clarke, T. Performance behaviour of macro-synthetic fibre reinforced concrete subjected to static and dynamic loadings for sleeper applications. *Constr. Build. Mater.* **2021**, *270*, 121469. [CrossRef]
9. Aziz, M.A.; Zubair, M.; Saleem, M. Development and testing of cellulose nanocrystal-based concrete. *Case Stud. Constr. Mater.* **2021**, *15*, e00761. [CrossRef]
10. Iqbal, S.; Ali, I.; Room, S.; Khan, S.A.; Ali, A. Enhanced mechanical properties of fiber reinforced concrete using closed steel fibers. *Mater. Struct.* **2019**, *52*, 56. [CrossRef]
11. Liu, J.L.; Jia, Y.M.; Wang, J. Experimental Study on Mechanical and Durability Properties of Glass and Polypropylene Fiber Reinforced Concrete. *Fibers Polym.* **2019**, *20*, 1900–1908. [CrossRef]
12. Li, D.S.; Hai, C.; Ou, J.P. Fracture behavior and damage evaluation of polyvinyl alcohol fiber concrete using acoustic emission technique. *Mater. Des.* **2012**, *40*, 205–211. [CrossRef]
13. Wang, Y.; Hu, S.; He, Z. Mechanical and fracture properties of geopolymer concrete with basalt fiber using digital image correlation. *Theor. Appl. Fract. Mech.* **2021**, *112*, 102909. [CrossRef]
14. Moron, A.; Ferrandez, D.; Saiz, P.; Moron, C. Experimental Study with Cement Mortars Made with Recycled Concrete Aggregate and Reinforced with Aramid Fibers. *Appl. Sci.* **2021**, *11*, 7791. [CrossRef]
15. Simsek, B.; Uygunoglu, T. Multi-response optimization of polymer blended concrete: A TOPSIS based Taguchi application. *Constr. Build. Mater.* **2016**, *117*, 251–262. [CrossRef]
16. Wang, W.C.; Wang, H.Y.; Chang, K.H.; Wang, S.Y. Effect of high temperature on the strength and thermal conductivity of glass fiber concrete. *Constr. Build. Mater.* **2020**, *245*, 118387. [CrossRef]
17. Altun, F.; Haktanir, T.; Ari, K.; Karahan, O. A comparative experimental investigation of concrete, reinforced-concrete and steel-fibre concrete pipes under three-edge-bearing test. *Constr. Build. Mater.* **2007**, *21*, 1702–1708. [CrossRef]
18. Zhang, X.H.; Zhang, W.; Cao, C.D.; Xu, F.; Yang, C.Q. Positive effects of aligned steel fiber on bond behavior between steel rebar and concrete. *Cem. Concr. Compos.* **2020**, *114*, 103828. [CrossRef]
19. Chen, F.; Xu, B.; Jiao, H.; Chen, X.; Shi, Y.; Wang, J.; Li, Z. Triaxial mechanical properties and microstructure visualization of BFRC. *Constr. Build. Mater.* **2021**, *278*, 122275. [CrossRef]
20. Kizilkanat, A.B.; Kabay, N.; Akyuncu, V.; Chowdhury, S.; Akca, A.R. Mechanical properties and fracture behavior of basalt and glass fiber reinforced concrete: An experimental study. *Constr. Build. Mater.* **2015**, *100*, 218–224. [CrossRef]
21. Guo, Q.; Chen, Z.; Liu, P.; Li, Y.; Hu, J.; Gao, Y.; Li, X. Influence of basalt fiber on mode I and II fracture properties of asphalt mixture at medium and low temperatures. *Theor. Appl. Fract. Mech.* **2021**, *112*, 102884. [CrossRef]
22. Xu, M.F.; Song, S.; Feng, L.; Zhou, J.; Li, H.; Li, V.C. Development of basalt fiber engineered cementitious composites and its mechanical properties. *Constr. Build. Mater.* **2021**, *266*, 121173. [CrossRef]
23. Dias, D.P.; Thaumaturgo, C. Fracture toughness of geopolymeric concretes reinforced with basalt fibers. *Cem. Concr. Compos.* **2005**, *27*, 49–54. [CrossRef]
24. Sigruner, M.; Muscat, D.; Strubbe, N. Investigation on pull-out behavior and interface critical parameters of polymer fibers embedded in concrete and their correlation with particular fiber properties. *J. Appl. Polym. Sci.* **2021**, *138*, 50745. [CrossRef]
25. Jirawattanasomkul, T.; Minakawa, H.; Likitlersuang, S.; Ueda, T.; Dai, J.G.; Wuttiwannasak, N.; Kongwang, N. Use of water hyacinth waste to produce fibre-reinforced polymer composites for concrete confinement: Mechanical performance and environmental assessment. *J. Clean. Prod.* **2021**, *292*, 126041. [CrossRef]
26. Khan, M.; Cao, M.L.; Xie, C.P.; Ali, M. Effectiveness of hybrid steel-basalt fiber reinforced concrete under compression. *Case Stud. Constr. Mater.* **2022**, *16*, e00941. [CrossRef]

27. Fu, Q.; Bu, M.X.; Xu, W.R.; Chen, L.; Li, D.; He, J.Q.; Kou, H.L.; Li, H. Comparative analysis of dynamic constitutive response of hybrid fibre-reinforced concrete with different matrix strengths. *Int. J. Impact Eng.* **2021**, *148*, 103763. [CrossRef]
28. Dvorkin, L.; Bordiuzhenko, O.; Tekle, B.H.; Ribakov, Y. A Method for the Design of Concrete with Combined Steel and Basalt Fiber. *Appl. Sci.* **2021**, *11*, 8850. [CrossRef]
29. Zheng, Y.X.; Zhang, P.; Cai, Y.C.; Jin, Z.Q.; Moshtagh, E. Cracking resistance and mechanical properties of basalt fibers reinforced cement-stabilized macadam. *Compos. Part B Eng.* **2019**, *165*, 312–334. [CrossRef]
30. Branston, J.; Das, S.; Kenno, S.Y.; Taylor, C. Mechanical behaviour of basalt fibre reinforced concrete. *Constr. Build. Mater.* **2016**, *124*, 878–886. [CrossRef]
31. Khan, M.; Cao, M.L. Effect of hybrid basalt fibre length and content on properties of cementitious composites. *Mag. Concr. Res.* **2021**, *73*, 487–498. [CrossRef]
32. Khan, M.R.; Cao, M.L.; Xie, C.P.; Ali, M. Efficiency of basalt fiber length and content on mechanical and microstructural properties of hybrid fiber concrete. *Fatigue Fract. Eng. Mater. Struct.* **2021**, *44*, 2135–2152. [CrossRef]
33. Pul, S.; Arslan, M.E. Cyclic behaviors of different type of hollow brick infill walls: A hinged rigid frame approach. *Constr. Build. Mater.* **2019**, *211*, 899–908. [CrossRef]
34. Carrillo, J.; Pincheira, J.A.; Flores, L.E. Quasi-static cyclic tests of confined masonry walls retrofitted with mortar overlays reinforced with either welded-wire mesh or steel fibers. *J. Build. Eng.* **2020**, *27*, 100975. [CrossRef]
35. Shakir, Q.M.; Alghazali, A.F. New model of eco-friendly hybrid deep beams with wastes of crushed concrete. *J. Teknol.* **2023**, *85*, 145–154. [CrossRef]
36. Shakir, Q.M.; Alghazali, A.F. Effect of the arching action on the behavior of the RC precast concrete deep beams: Comparison between several hybrid models. *J. Build. Pathol. Rehabil.* **2024**, *9*, 23. [CrossRef]
37. Gaedicke, C.; Marines, A.; Miankodila, F. Assessing the abrasion resistance of cores in virgin and recycled aggregate pervious concrete. *Constr. Build. Mater.* **2014**, *68*, 701–708. [CrossRef]
38. *JGJ55-2011*; Specification for Mix Proportion Design of Ordinary Concrete. China Building Industry Press: Beijing, China, 2011. (In Chinese)
39. *GB/T50081-2019*; Standard for Test Method of Physical and Mechanical Properties of Concrete. China Building Industry Press: Beijing, China, 2019. (In Chinese)
40. *GB50010-2010*; Design Code for Concrete Structure. China Building Industry Press: Beijing, China, 2011. (In Chinese)
41. Xu, B.; Toutanji, H.A.; Gilbert, J. Impact resistance of poly(vinyl alcohol) fiber reinforced high-performance organic aggregate cementitious material. *Cem. Concr. Res.* **2010**, *40*, 347–351. [CrossRef]
42. He, J.; Shi, J.J.; Lu, Z.Y. Assessment of interfacial fracture energy between concrete and CFRP under water immersion conditions. *Constr. Build. Mater.* **2021**, *285*, 122942. [CrossRef]
43. Wang, D.; Ju, Y.; Shen, H.; Xu, L. Mechanical properties of high performance concrete reinforced with basalt fiber and polypropylene fiber. *Constr. Build. Mater.* **2019**, *197*, 464–473. [CrossRef]
44. Zhao, M.; Zhang, B.; Shang, P.; Fu, Y.; Zhang, X.; Zhao, S. Complete Stress-Strain Curves of Self-Compacting Steel Fiber Reinforced Expanded-Shale Lightweight Concrete under Uniaxial Compression. *Materials* **2019**, *12*, 2979. [CrossRef]
45. Chen, A.; Han, X.; Chen, M.; Wang, X.; Wang, Z.; Guo, T. Mechanical and stress-strain behavior of basalt fiber reinforced rubberized recycled coarse aggregate concrete. *Constr. Build. Mater.* **2020**, *260*, 119888. [CrossRef]
46. Ding, X.; Li, C.; Li, Y.; Lu, Y.; Song, C.; Zhao, S. Experimental and numerical study on stress-strain behavior of self-compacting SFRC under uniaxial compression. *Constr. Build. Mater.* **2018**, *185*, 30–38. [CrossRef]
47. Ou, Y.-C.; Tsai, M.-S.; Liu, K.-Y.; Chang, K.-C. Compressive Behavior of Steel-Fiber-Reinforced Concrete with a High Reinforcing Index. *J. Mater. Civ. Eng.* **2012**, *24*, 207–215. [CrossRef]
48. Abbass, W.; Khan, M.I.; Mourad, S. Evaluation of mechanical properties of steel fiber reinforced concrete with different strengths of concrete. *Constr. Build. Mater.* **2018**, *168*, 556–569. [CrossRef]
49. Guo, Z.; Zhang, X.; Zhang, D. Experimental Investigation of the Complete Stress-Strain Curve of Concrete. *J. Build. Struct.* **1982**, *122*, 81–89. [CrossRef]

buildings

MDPI

Article

Improved Bond Stress-Slip Relationships for Carbon Fibre-Reinforced Polymer-Strengthened Masonry Triplets

Seyyed Motasam Hashemi and Ashraf Ayoub *

School of Science & Technology, City, University of London, Northampton Square, London EC1V 0HB, UK; seyyed.hashemi.1@city.ac.uk
* Correspondence: ashraf.ayoub.1@city.ac.uk; Tel.: +44-(0)20-7040-8912

Abstract: Carbon fibre-reinforced polymer (CFRP) emerges as a viable solution for reinforcing unreinforced masonry (URM) walls subjected to shear loads. While masonry structures are straightforward to construct, the complexity of the construction materials, especially in terms of their mechanical properties, poses challenges for numerical studies of their structural behaviour. Walls, being fundamental components in masonry construction, play a crucial role in transferring both horizontal and vertical lateral forces. This study investigates the enhancement of masonry wall behaviour through the reinforcement of CFRP. CFRP reinforcement increases ductility and strength, reducing the risk of failure under shear conditions. Additionally, CFRP composites present a practical solution to strengthening masonry structures compared to traditional reinforcement. However, brick, mortar, and CFRP have not been thoroughly investigated. Experimental tests on the bond behaviour of different configurations of CFRP-retrofitted masonry triplets have not been performed before and are therefore presented in this paper. Triplet specimens, comprising three bricks and two mortar joints, both with and without CFRP strengthening, were subjected to bond testing. The study affirms that masonry triplets strengthened with CFRP under shear loads exhibit strength levels at least four to six times greater than those without CFRP. The experimental work was carried out with eight different CFRP configurations on triplet masonry, and each test was repeated four times. Further, the bond stress-slip relationship in the case of masonry triplets with and without CFRP was predicted with new mathematical equations based on the conducted test results. These equations were included in the commercial finite element software ANSYS and used to conduct simulations of CFRP-reinforced masonry triplets. The numerical results indicate good agreement between the finite element model and the test results. The outcome of this research improves the current knowledge on the use of CFRP to reinforce masonry walls with brick and mortar, which will contribute to the understanding of the effect of CFRP on masonry structures.

Keywords: unreinforced masonry; fibre-reinforced polymer; bond stress-slip; interface behaviour; masonry triplets; shear load; finite element modelling; bricks; mortar joints

1. Introduction

According to Ehsani and Saadatmanesh [1–3], a significant percentage of existing buildings are made from masonry, where the mechanical properties of brickwork are more complex than other structure materials. Failure in URM structures due to in-plane or out-of-plane loads is one of the major reasons for human injury. Strengthening and repairing the existing buildings is one of the main challenges in the field of civil engineering. The strengthening and reinforcement of masonry components can be carried out based on traditional materials such as steel and concrete. Reinforced concrete (RC) jackets are a technique for strengthening masonry constructions that involves placing jackets on one or both sides of masonry walls. The steel-reinforced brick masonry consists of brick masonry with steel reinforcement or mild steel iron mesh or bars embedded in the mortar. According to Ferretti et al. [4], from the mechanical point of view, masonry is a heterogeneous, nonlinear, and anisotropic material, consisting of masonry units, mortar, and masonry unit-to-mortar interfaces. Also, masonry is characterized by its very low tensile strength, substantial strength in compression, and weak performance in bending and shear. Indeed, it is well known that masonry buildings are weak against horizontal actions. Tensile and shear bond strengths are quite low and highly variable. Thus, masonry walls fail mostly through masonry unit-to-mortar interfaces, known as planes of weakness.

Ehsani and Saadatmanesh [5] concluded that traditional methods of strengthening often add extra load to structures, making them riskier. The interface behaviour of brick and mortar in masonry composites has been investigated by Mohammadipour [6]. Pavan and Nanjunda [7] studied the effectiveness of FRP composites in improving the brick–mortar interface. They studied two types of masonry assemblies to investigate the interfacial behaviour of brick–mortar joints in masonry triplets. The samples were bonded with glass and carbon FRP in strip form. According to their experimental tests, in the case of masonry triplets bonded with GFRP and CFRP composites, the percentage increases in shear strength covering the full triplet face and strips were approximately 165% and 185%, respectively. Many researchers have studied the behaviour of FRP-strengthened triplet masonry with one or two FRP configurations. Reinforcing masonry walls with fibre-reinforced polymers (FRP) was considered to increase strength and ductility and thus reduce the risk of failure under severe loading conditions. Fibre-reinforced polymer (FRP) composites may provide technically and economically viable solutions, as stated by Tumialan et al. [8–11]. El Malyh et al. [12] studied the effect of CFRP wrap on the strength and stiffness of URM walls. An experimental programme was set up to investigate the in-plane behaviour of the standardized specimens following ASTM E-519-02 [13]. The test included testing clay brick panels of 1200 × 1200 × 100 mm fixed at the ends. In all cases, the reinforced panels improved the shear strength and deformations of the samples before failure. According to their study, the retrofitting system restrains the URM wall from shear failure and shear sliding. It enhanced their strength by up to 319%, and all CFRP configurations significantly improved the load-bearing capacity of the specimens by 2.5 to 4.2 times. Babatunde [14] studied glass and carbon fibre-reinforced polymers (GFRP and CFRP) to reinforce masonry walls. The experimental tests showed that externally bonded FRP systems improve flexural strength and ductility and increase strain capacity and stability. Near-surface-mounted FRP systems improve out-of-plane bending and lateral resistance. The FRP strip strengthening technique for in-plane loading improves strength, ductility, and lateral load resistance by up to four times that of the unreinforced walls, and for out-of-plane loading, it minimizes displacement. According to Hamrat et al. [15], the restoration and reinforcement of cracked structural members with externally bonded (CFRP) sheets has become one of the greatest challenges confronting civil engineers. CFRP composites bonded to damaged structural members can significantly improve their mechanical behaviour and reduce crack propagation by increasing the load capacity. Luccioni and Rougier [16,17], indicated that CFRP improves the strength of bed joints under compression and increases ductility and failure mode. Depending on the dimensions and orientation, retrofitting with CFRP increases ductility under in-plane shear, prevents the joints from sliding, and increases the ultimate strength. According to Hernoune et al. [18], in order to extend the life of masonry structures, strengthening or repairing by employing new techniques has been developed. Many of these strengthening techniques, including the use of fibre-reinforced polymer (FRP) composites, have aided in reinforcing masonry structures. These composites are manufactured with different features depending on the fibre material type, such as carbon fibre (CFRP), glass (GFRP), and aramid (AFRP). FRP strips offer the possibility of application by glueing on externally bonded elements using the near-surface-mounted (NSM) technique. Hamid et al. [19] studied the shear behaviour of FRP-glued masonry composed of hollow concrete blocks. The study demonstrated the effectiveness of FRP in preventing the brittle failure modes exhibited by URM and improving the shear strength and deformation ability of masonry.

Saingam et al. [20] studied the flexural capacity of cement–clay interlocking (CCI) brick walls using ferrocement jackets with an expanded wire mesh. Their experimental results showed that the reinforced walls exhibited improved peak load capacity, notably in dual-sided reinforcement cases. With plastering on both sides and supplemented with anchors, an improvement of up to 221% in flexural capacity was obtained. The strengthened walls significantly improved CCI brick wall performance, increasing peak capacity by up to 46%. Specifically, one-sided strengthening enhanced peak capacity by 28.60%, while two-sided strengthening boosted it by 64.29%. Joyklad and Hussain [21] indicated that the diagonal compression capacity of cement–clay interlocking (CCI) brick walls was significantly improved by the addition of grout in the openings of CCI bricks. An increase of up to 418% in the diagonal compressive capacity was observed by the addition of grout, whereas the same improvement jumped to 567% after the addition of steel bars. Rashid et al. [22] investigated the bearing capacity performance of Kenaf fibre geotextile laid on inside the sand layer. A rigid footing was used to replicate a strip footing during the loading test, and sand was prepared based on 50% of the relative density in a rigid testing chamber for ground model preparation. In order to treat the soil, Kenaf fibre geotextile was laid at four different locations, which are on the soil surface and underneath the ground model surface at 50, 75, and 100 mm deep. It was found that the use of the Kenaf fibre geotextile has improved the bearing capacity of the sandy soil up to 414.9% as compared to untreated soil. Saidi et al. [23] developed an empirical model for predicting the bond strength between natural fibre-reinforced polymer (NFRP) sheets and concrete. The NFRP was produced from fabric using four different types of fibres, including jute, silk, and pineapple, as well as dense and non-dense abaca fibres. Moreover, three different types of bonding adhesives were applied to bond the NFRP composite produced from abaca fibre to the concrete surface, and these include standard epoxy resin, polyester resin, and thixotropic epoxy resin. The size of the specimens was 100 mm × 100 mm × 300 mm, with a single longitudinal reinforcing bar with a diameter of 10 mm in the centre of its cross-section and a pre-existing crack in the middle of its length on two opposite sides. NFRP composites with a size of 50 mm × 240 mm were subsequently applied to both sides containing a crack. They discovered that the fibre type used in producing NFRP affected the bonding strength between NFRP and concrete. Abaca was found to have the highest contribution, followed by jute and silk, while pineapple fibres have the smallest contribution. The use of two and three layers of fabrics enhanced the initial stiffness by 143–276% and 257–691%, respectively; bond strength by 118–350% and 312–400%, respectively; and toughness by 136–440% and 199–667%, respectively, depending on the fabric and bonding adhesive types. The behaviour analysis of strengthened concrete beams with natural fibre-reinforced polymer (NFRP) based on abaca fibre using the finite element method was investigated by Amalia et al. [24]. Based on the results of the study, the results of numerical analysis using ATENA V534 showed a lower maximum load value, higher stiffness, and lower deflection at maximum load compared to experimental results on all reinforced concrete beams analysed in their study. Based on the results of numerical analysis using ATENA V534 software (ATENA 3D Version 5.4.1) and experimental results in the laboratory, NFRP reinforcement with a vertical configuration was recommended because it can increase the maximum load, which is higher than the diagonal configuration.

Enhancing the resistance of masonry structures significantly relies on the integrity of the bond between the masonry unit and the mortar, impacting their shearing strength. Consequently, prioritizing improvements in the shear performance of these buildings becomes crucial. Typically, mortar joints constitute the weakest link within masonry, and their nonlinear response to tension and shear greatly influences the overall behaviour of the composite structure. Accurately predicting the load-bearing capacity of unreinforced masonry buildings when subjected to combined compression and shear demands a precise understanding of masonry strength and its broader response to bond stresses. Previous studies attempted to evaluate the bond stress-slip relationship of the brick–mortar interface for un-strengthened specimens. These relationships in the presence of FRP strengthening have not been fully evaluated yet. In this study, an extensive set of experiments was conducted for two main categories: 1—unreinforced masonry triplet specimens and 2—masonry triplets reinforced with carbon fibre-reinforced polymer (CFRP). The purpose of the tests is to evaluate the bond stress-slip behaviour of FRP-strengthened masonry triplets. These triplets were specifically built using mortar M4 and subjected to testing with varying configurations of CFRP. Based on the test results, mathematical models of the bond stress-slip were developed for each configuration and coded into a finite element (FE) platform. This FE model looked into bond-slip and normal separation occurring at the brick–mortar interface using a detailed micro-modelling approach. Furthermore, the numerical analysis was able to simulate the initiation and propagation of cracks within the mortar layer. Subsequently, the outcomes of these numerical simulations were compared with the results obtained from the experiments. A few researchers have studied the behaviour of FRP-strengthened triplet masonry with one or two FRP configurations. In this paper, the strength and bond stress-slip relationship of URM for eight different CFRP configurations were studied to fill this gap and provide accurate constitutive models for all practical cases.

The present research objectives are as follows:

1. To study the effect of different CFRP configurations on unreinforced triplet masonry (URTM) and investigate the behaviour of triplet specimens when failure occurs through the brick–mortar interface.
2. To evaluate experimentally the bond behaviour between brick and mortar in the presence of different configurations of CFRP, which has not been previously studied.
3. To develop mathematical equations for the bond stress-slip relationships identified in (2) to be used for accurate numerical simulation of FRP-strengthened masonry specimens.

2. Experimental Design, Material, and Methods

Small-scale specimens were used in an experimental test. This information was used to calibrate formulas for the design and computation of masonry retrofitted with composite materials. The mechanical properties of brick and mortar have an impact on the performance of masonry walls under stress. As a result, multiple tests were designed in the current study to assess the elastic modulus of brick, mortar, and masonry triplets. The elastic modulus and compressive strength of mortar cylinders with a diameter of 150 mm and a height of 300 mm were measured. The masonry and mortar samples were made using Portland cement and mortar M4 (1:4: Cement:Sand), which was produced in accordance with the UK National Annex Classification BS EN 998-2 (2003) [25] requirements. The mortar between the bricks was approximately 10 mm thick. Sand and cement were mixed with the appropriate amount of water (the water–cement ratio was 1:3).

2.1. Application of FRP

Rahman and Ueda [26] concluded that the shear strength of the reinforced walls is significantly improved by using FRP. According to Secondin [27], fibre-reinforced polymer (FRP) is a relatively new composite factory-made material from fibres and resins that is effective and economical for the repair of weak structures. Glass and carbon are connected by a matrix to form strong composite materials. The fibres provide stiffness and strength, while the matrix transfers the load between the fibres, stabilizing the composites. The results of the experimental tests showed an increase in the bending capacity due to out-of-plane loads and in-plane shear loads, as reported by Derias and El-Hacha [28].

In this study, the carbon fibre-reinforced polymer composites (CFRP) with a resin epoxy matrix were attached to the masonry triplets (Figures 1 and 2). In the current research, the use of these composites was based on the manufacturer's technical report. The thickness of CFRP was 0.3 mm. The characteristics of CFRP and resin are presented in Tables 1 and 2.

Figure 1. CFRP.

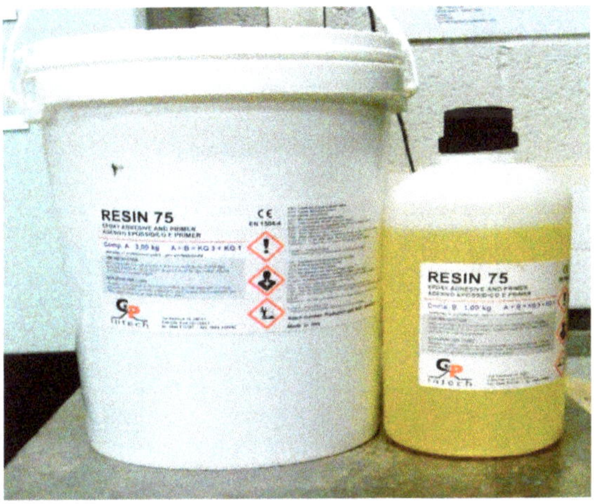

Figure 2. Resin.

Table 1. The characteristics of CFRP.

CFRP technical data	C-Sheet 240		
Tensile strength	4700 MPa		
Elongation	2%		
E-modulus	240 GPa		
Thickness, mm	0.22–0.35		
Weight, g/m^2	300	400	600
Standard width, cm	20	30	50
Standard length of rolls, m	50	50	50

Table 2. The characteristics of resin.

Technical data	Type 70–75
Sp. Gr	1.2 gr/cm^3
Hardening starts at 20 °C	6/7 h
Hardening completes at 20 °C	7 h
Resistance to compressive stress	83 MPa
Resistance flexural/tensile stress	45 MPa
Adhesion to concrete	>2.5 MPa
Minimum temperature for use	≥10 °C

2.2. Mechanical Properties of Brick, Mortar, and Triplets

The controlling factor of brick is its compressive strength. Bricks with various strengths are available, depending on architectural necessities. For low-rise buildings, bricks of 5.2 N/mm^2 should be sufficient. For damp-proof courses, low-absorption engineering bricks are usually required. For reinforced and pre-stressed brickwork, it is highly unlikely that brick with a strength lower than 20 N/mm^2 will be used in the United Kingdom. The sand for mortar must be clean, sharp, and free from salt and organic contamination. Most natural sand contains a small quantity of silt or clay. A small quantity of silt improves workability. Loam or clay is moisture-sensitive and, in large quantities, causes the shrinkage of mortar. Marine and estuarine sand should not be used unless washed completely to remove magnesium and sodium chloride salts, which are deliquescent and attract moisture. Masonry cement is a mixture of approximately 75% ordinary Portland cement, an inert mineral filler, and an air-entraining agent. The mineral filler is used to reduce the cement content, and the air-entraining agent is added to improve the workability. Mortar made from masonry cement will have lower strength compared to a normal cement mortar of a similar mix. The main elements influencing the performance of masonry buildings are their mechanical characteristics, such as the triplets, mortar, and bricks' elastic moduli (also known as Young's modulus). Experimental tests were created for this study in order to determine the masonry and mortar samples' elastic moduli. Furthermore, the samples' compressive strength was determined due to the strong association between them and the elastic modulus. Cylinder specimens were built in order to assess the compressive strength and elastic modulus of mortar. The masonry specimens were built using solid London bricks and mortar M4 with a thickness of 10 mm between each brick layer. The horizontal surfaces of all samples were checked for bubble levels. After the construction of each sample, it was cured, and its design capacity developed for more than 28 days. The axial compression test was conducted with an average load application rate of 0.6 kN/s, according to the compressing test machine manual, and each test lasted 15 to 20 min.

2.3. Measurement of the Elastic Modulus of Triplet Masonry Samples

Overall, three full-scale specimens were constructed with a nominal dimension of 215 mm × 104 mm × 65 mm (height, length, and thickness). The masonry triplets were built with solid clay bricks, and the thickness of mortar for each layer of the triplets was approximately 10 mm (Figure 3A). Due to the high water absorption capacity of bricks, the bricks were soaked before building the specimens so that the rapid absorption of mortar water, which leads to degradation of mortar and thus degradation in the joint (bonding) between bricks and mortars, could be prevented. When the brick wall construction was finished, for curing, the walls were wetted twice a day for three days. The horizontal surface of all specimens was checked for bubble level. The axial compressive load at a rate of 0.6 kN/s was applied to each sample, and after loading, the deformation was measured using the universal testing machine called Advantest 9, MCC 8—Multifunctional Control Console, and three mechanical strain gauges attached to the sample. This test was repeated six times (Figure 3B,C).

Figure 3. Masonry triplets before test (**A**), and Masonry triplets after compression tests (**B**,**C**).

2.4. Measurement of the Elastic Modulus of London Solid Brick Samples

In this study, the bricks were tested under compression to find the modulus of elasticity and compressive strength. Six bricks were tested with an average size of 220 mm × 104 mm × 65 mm. The specimens were tested under load control in the universal testing machine. Based on compressive strength, stress, and strain, the average modulus of elasticity of the bricks was calculated. To measure the elastic modulus of brick samples according to the testing machine manual, a compressive load with a rate of 0.6 kN/s was applied to the sample (Figure 4A,B). The distributed load was applied to the surface of the sample, and after loading, the deformation was measured by a mechanical strain gauge in the machine.

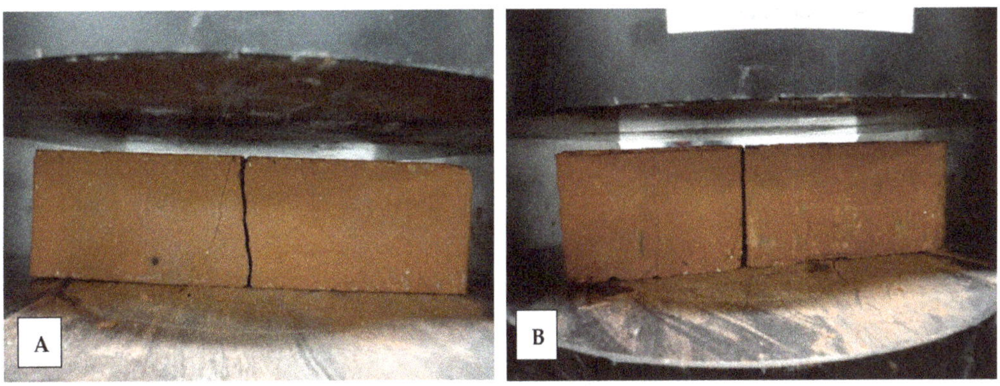

Figure 4. Compression tests on bricks after tests (**A**,**B**).

2.5. Measurement of the Elastic Modulus of Mortar M4 Samples

Twelve mortar cylinders were constructed with mortar M4 (1: ordinary Portland cement and 4: natural river sand, and the water–cement ratio was determined to be 1:3). To mix and prepare all the mortar specimens, regular tap water was used and mixed with an electric mixer. Mortar specimens were first cleaned of any loose material leftover from the previous test. To measure the elastic modulus of mortar samples, three strain gauges were attached to each sample, and an axial compressive load with a rate of 0.6 kN/s was applied to each sample, according to the testing machine manual (Figure 5A). All mortar specimens failed immediately after reaching the ultimate failure load. The failure mode of mortar specimens is presented in Figure 5B. All mortar specimens showed similar failure modes, expressed by crushing the specimen over the whole height at the end of loading. The deformation was measured using a mechanical strain gauge attached to the sample.

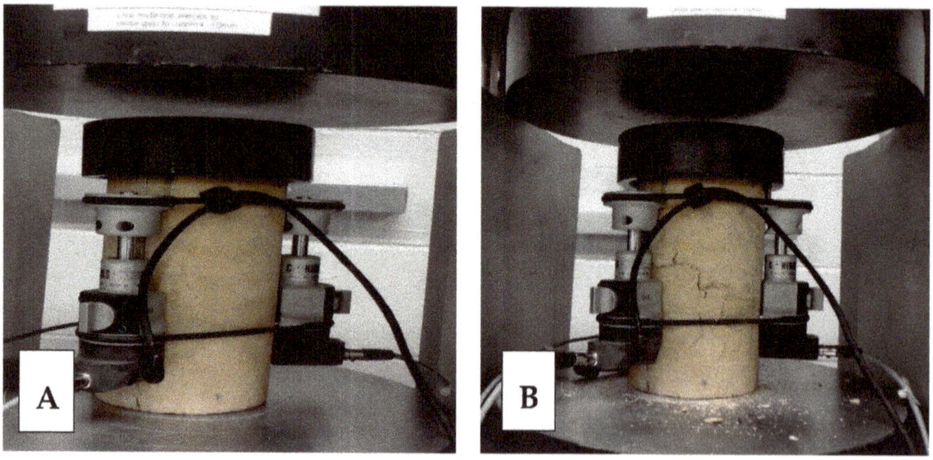

Figure 5. Compression tests on mortar before test (**A**), and after failure (**B**).

Figure 6 illustrates the comparative elastic modulus of brick, mortar M4, and triplet masonry, emphasizing variations in their resistance to deformation under stress. As expected, bricks exhibited a higher elastic modulus compared to mortar. Furthermore, Figure 7 displays the test outcomes of the stress-strain curves for triplet masonry, mortar, and bricks, including comparative curves. The stress-strain curve data indicates that the solid London brick exhibited a higher maximum compressive strength compared to mortar M4. This difference in strength between the brick and the weaker mortar joint contributed to a reduction in the overall compressive strength of the masonry.

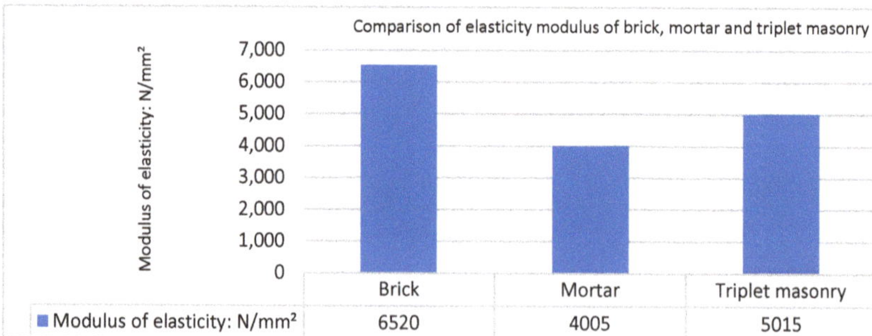

Figure 6. Comparison of elastic modulus for mortar, brick, and triplet masonry samples.

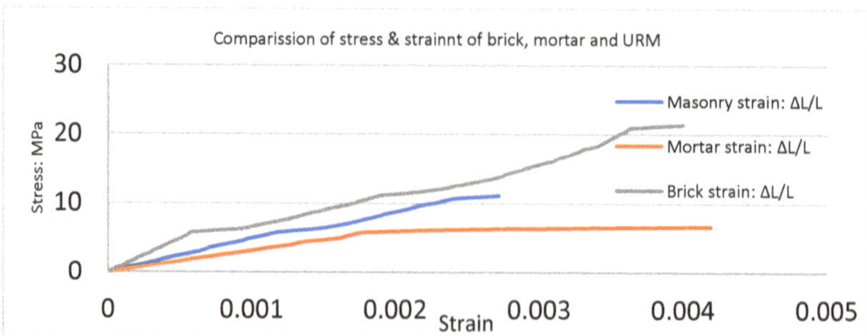

Figure 7. Comparison of stress-strain curves for mortar, brick, and URM samples.

3. Experimental Study on the Bond Behaviour of Strengthened Triplets under Shear Loading with Different CFRP and CFRP Net Configurations

Masonry can present different failure modes depending on the geometry, material, and load factors. According to Zimmermann et al. [29] and EN 1052-3 [30], the initial shear strength of masonry must be derived from the strength of small masonry specimens tested to destruction. The tests were conducted on specimens with dimensions of 220 × 103 × 215 mm prepared from solid clay bricks and mortar. All specimens had three brick units and a 10 mm joint thickness in between. The specimens were cured in the laboratory under controlled environmental conditions. Each specimen was tested at the age of 40 days. Bond stresses in the unit–mortar interface were introduced by gradually increasing the applied forces until failure. The experimental tests of masonry triplets were carried out with one-layer CFRP to evaluate the brick–mortar bed joints and bond stress-slip relationships of CFRP-strengthened triplet masonry. Additionally, the main goal of the present study was to evaluate how externally bonded CFRP composites improved the bond strength and deformation capacity of brick–mortar joints. For this reason, loads were applied to the triplet masonry assemblage in the current investigation, parallel to the bed joint. The loading has to be applied to the mortar joint interfaces between the centre brick and the side bricks in order to achieve the aforementioned goal. In order to do this, the side bricks were supported at the bottom, and the central brick was loaded at the top (Figure 8). Moreover, each experimental test was repeated on four specimens.

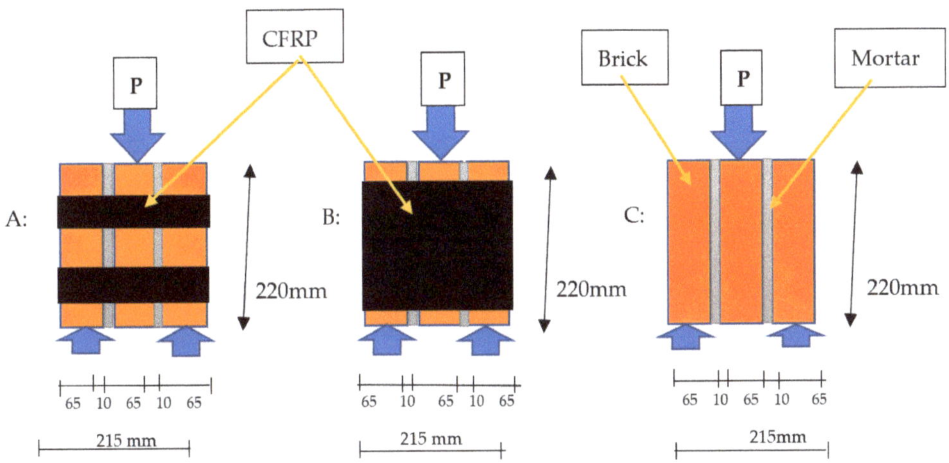

Figure 8. Modelling for shear analyses: (**A**) reinforced masonry prism (cover with CFRP on two sides), (**B**) reinforced masonry prism with two vertical CFRP on two sides, and (**C**) unreinforced masonry prism.

3.1. Triplet Masonry Test without CFRP

A pre-compression test was carried out on unreinforced masonry triplets constructed with mortar M4 to evaluate the main brick–mortar interface characteristics. These specimen prisms are placed longitudinally under the load that applies to the head of the middle brick masonry unit. This experiment can be done with or without a lateral pre-compression load. Four URTM specimens were constructed without FRP as control samples. The samples suddenly failed at 10 kN, 10 kN, 12 kN, and 13 kN. The mode of failure under in-plane loading was sudden and brittle in all specimens. The specimens failed due to the development of cracks along the interface between brick and mortar, followed by complete collapse, as shown in Figure 9A,B. The failure (cracking) in the mortar joints was either confined to one of the interfaces (Figure 9A) or to the inner-to-outer interface (as viewed in the direction of loading), as shown in Figure 9B (the upper interface of the specimen on the right side cracked horizontally). As expected, low bond strength of masonry specimens was obtained, and all triplets failed in the brick–mortar interface by shear sliding. This failure is one of the valid failures established by EN 1052-3 [30]. The stiffness of the stress-slip behaviour increases in the early portion of the curve, as seen in Figure 10. This phenomenon can be attributed to the effect of dilatation; the control triplet specimens' average ultimate displacement was 2.0 mm. The maximum load and final displacement of the control triplets in this study and the ones tested by Mojsilovic [31] are comparable. Failures in the mortar layer were the primary source of the failure modes. As the load increased, a diagonal crack appeared in the mortar layer and spread throughout the mortar. As a result, the specimen split into two separate parts that came from either the outer or inner brick, or from both. No considerable damage was seen on the brick sides. A similar failure mode was similar to that reported by Prakash [32] and Bompa et al. [33]. All triplet specimens failed similarly, and failure occurred mainly in the mortar joints. The specimens demonstrated failure in the joint with the relative movement of the middle brick.

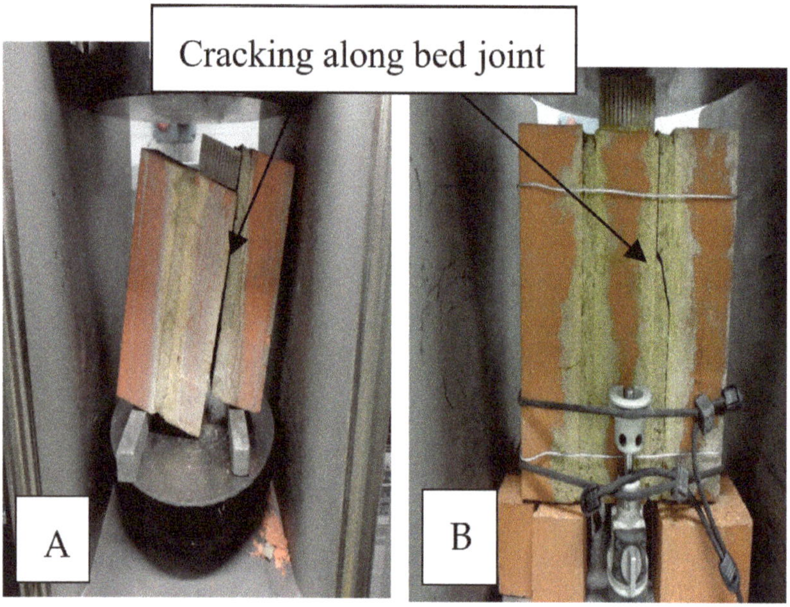

Figure 9. The failure (cracking) in the mortar joints (**A**,**B**).

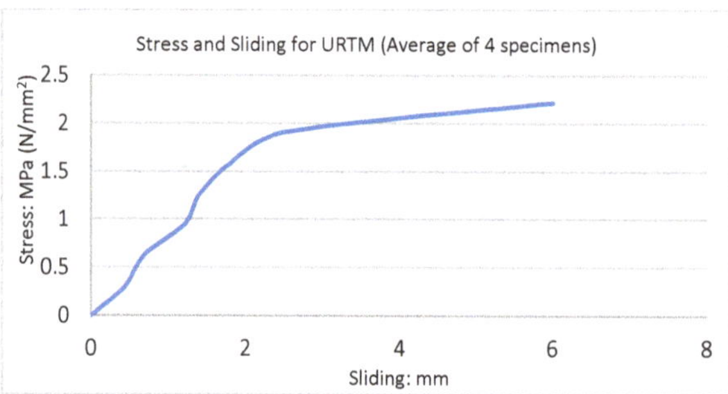

Figure 10. Stress-strain curve for the pre-compression test on URTM.

3.2. Reinforcement with One Layer of CFRP on Two Sides

Four samples were constructed and strengthened with one layer of CFRP sheet on two sides of triplet masonry. CFRP composites were attached to masonry triplets. Before applying CFRP, minor imperfections, such as small pieces of mortar, were removed to gain a flat surface by scrubbing the specimens with sandpaper. Epoxy resin was mixed with a hardener, according to the manufacturer's manual. The resin was applied to the prism's surface, followed by the application of FRP to the masonry triplets. The sliding was measured by a mechanical gauge in the testing machine. The ultimate load was approximately 84 kN, and the mode of failure was ductile. Debonding also occurred just before failure; the FRP ruptured at mid-span, followed by shearing of the sheets at the bottom. Debonding occurs when stresses transferred by the CFRP composite exceed the tensile strength of the substrate material. This causes the CFRP to peel off from the brick. The failure in the unit–mortar interface is shown in Figure 11. Figure 12 illustrates the stress-sliding curve. In this test, the triplet masonry with one layer of CFRP sheets on two sides was seven times stronger than the URM samples.

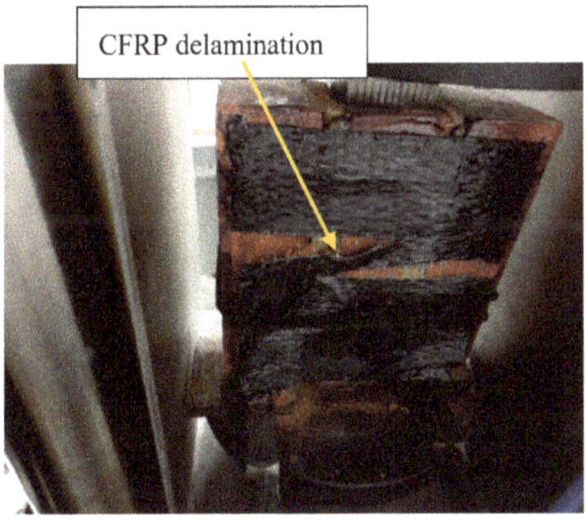

Figure 11. CFRP on two sides.

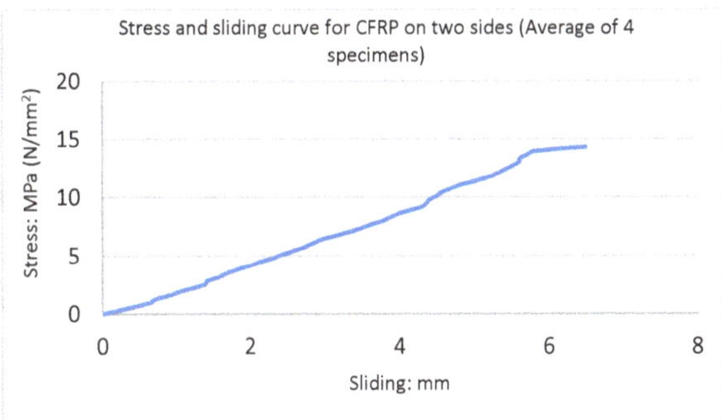

Figure 12. Stress-slip curve for CFRP on two sides.

The CFRP reinforcement stopped the micro-cracks from expanding through the mortar layer and prevented the early shear failure of the triplet. This demonstrated how the CFRP reinforcement supported the bed joints' tensile load. The primary cause of the failure of CFRP-strengthened triplet masonry was the debonding of CFRP from triplet; this failure generally happens when the bonding between the reinforcement composite and masonry interface is lost. Reinforced triplets, as compared with URM walls, remained stable after failure. Before the masonry cracked and the stress dropped, it was found that the triplet had a nearly linear stress-slip relationship.

3.3. Strengthening with One Layer of CFRP Net on Two Sides of Triplet

These samples were strengthened with one layer of CFRP net on two sides of the triplets. The mode of failure was observed to be ductile, and the ultimate load was approximately 71 kN. These triplets also displayed the propagation of cracks, as did the previous test. Delamination before failure and failure in the unit–mortar interface also occurred, as shown in Figure 13. Figure 14 illustrates the stress-slip curve; the initial slope of the load-displacement curves for a CFRP net on four sides and CFRP on two sides of triplet specimens were almost equal. In this test, the triplet masonry with one layer of CFRP net on two sides was six times stronger than the URM samples, and the ultimate load capacity of CFRP net glued on four sides and two sides was almost equal. Shear cracks in the mortar layer began to appear at the brick–mortar interface, which is where the cracking in the specimen reinforced with CFRP net on two sides occurred. Before the mortar cracked and the stress dropped, it was found that the triplet had an almost linear stress-sliding behaviour.

Figure 13. CFRP net on two sides.

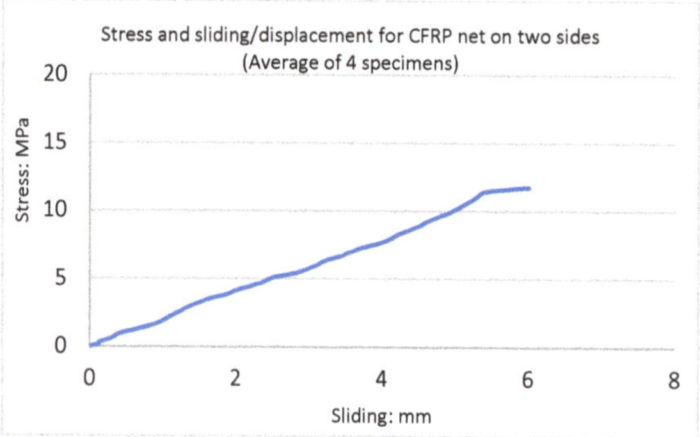

Figure 14. Stress-slip curve for strengthening with CFRP net on two sides.

3.4. Strengthening with One Layer of CFRP Strip on Two Sides of a Triplet

The samples were reinforced with one layer of CFRP, a 50 mm wide strip on two sides of triplets. The mode of failure was ductile, and the ultimate load was approximately 57 kN. These triplets also displayed the propagation of cracks, as did the previous test.

Delamination also occurred before failure. The failure in the unit–mortar interface is shown in Figure 15. Figure 16 shows the stress-sliding curve; the triplet masonry with one layer of CFRP strip on two sides was four times stronger than the URM samples. The presence of the CFRP improved the performance by changing the brittle failure to a ductile failure. In the strengthened triplets, the failure occurred due to the loss of bonding between the CFRP and the masonry units. The crack was formed by the debonding of the brick from the mortar until the debonding of the epoxy from the masonry units. At this point, the triplet failed because the tensile stresses were no longer transferred to the CFRP.

Figure 15. CFRP strips on two sides.

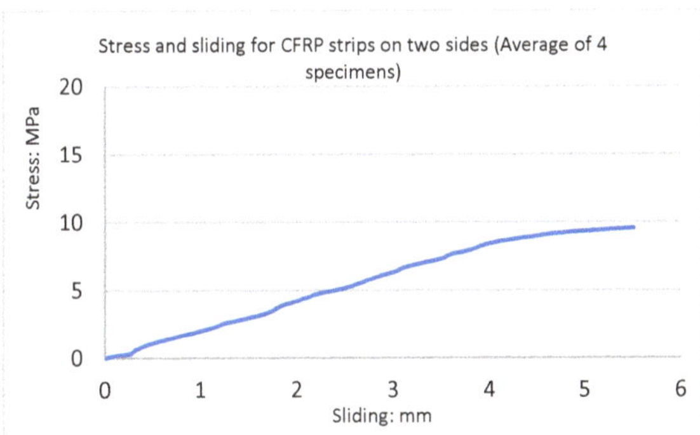

Figure 16. Stress-slip curve for strengthening with CFRP strips on two sides.

3.5. Strengthening with One Layer of CFRP Net Strip on Two Sides

These samples were strengthened with one layer of two CFRP net strips, 50 mm wide, on two sides of the triplet. The mode of failure was observed to be ductile, and the ultimate load was approximately 52 kN. The triplets also displayed the propagation of cracks as per the previous test, and delamination occurred before failure. Figure 17 illustrates failure in the unit–mortar interface, and Figure 18 shows the stress-sliding curve. In this test, the triplet masonry with one layer of CFRP net strip on two sides was four times stronger than the URM samples. The debonding of CFRP from triplet was the main reason for the failure of the CFRP-strengthened triplet masonry.

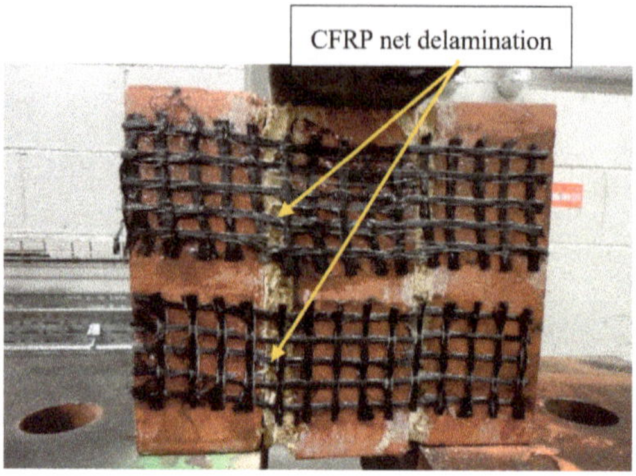

Figure 17. CFRP net strips on two sides.

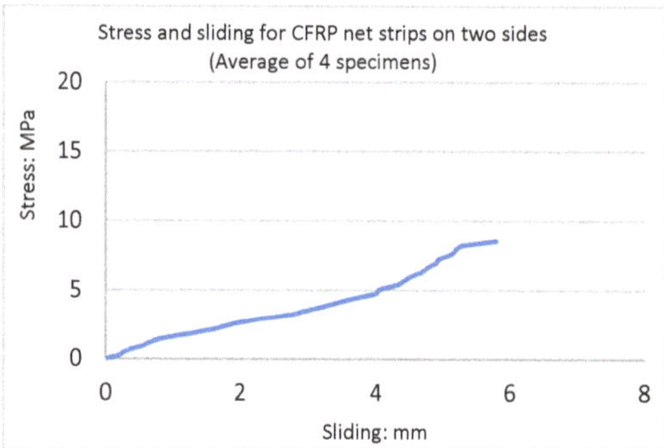

Figure 18. Stress-slip curve with CFRP net strips on two sides.

3.6. Strengthening with One Layer of CFRP Wrapping on Four Sides

Samples were strengthened with one layer of CFRP on four sides of the triplet. The mode of failure was observed to be ductile, and the ultimate load was about 88 kN. These samples also displayed the propagation of cracks as did the previous test, and delamination also occurred before failure. The shear failure in the unit–mortar interface is shown in Figure 19. Figure 20 shows the stress-sliding curve. In this test, the triplet masonry with one layer of CFRP on four sides was six times stronger than the URM samples. The maximum load capacity of CFRP sheets on four sides and two sides is nearly the same, but reinforced triplets on four sides showed less displacement.

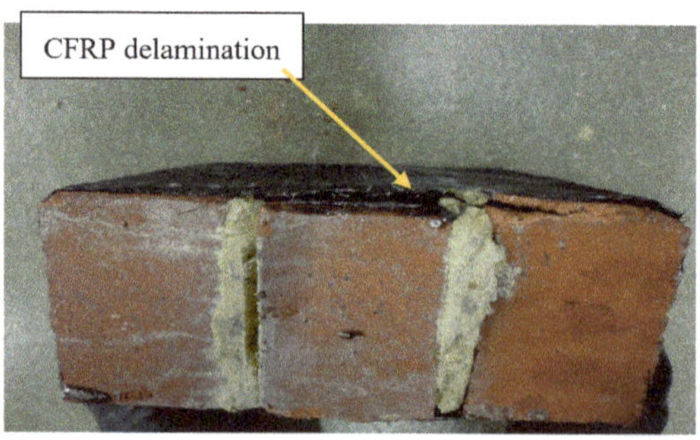

Figure 19. Strengthening with CFRP wrapping on four sides.

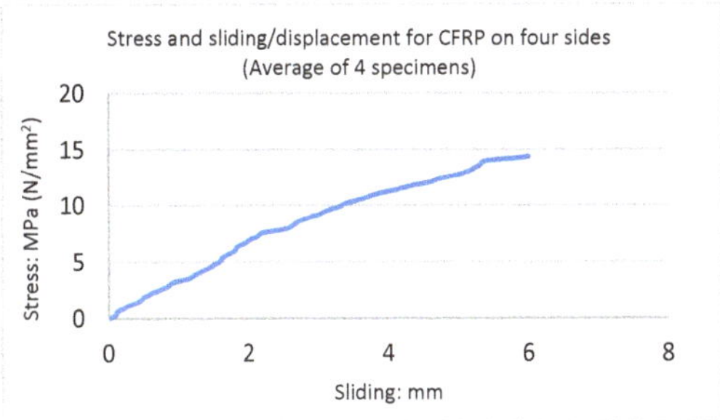

Figure 20. Stress-slip curve for strengthening with CFRP on four sides.

3.7. Strengthening with One Layer of CFRP Net on Four Sides of Triplet

These samples were strengthened with one layer of CFRP net on four sides of triplet masonry. The mode of failure was observed to be ductile, and the ultimate load was approximately 78 kN. Debonding occurs when stresses transferred by the CFRP composite exceed the tensile strength of the substrate material. This causes the CFRP net to peel off and cause damage to the triplet specimen. The shear failure in the unit–mortar interface is shown in Figure 21. Figure 22 displays the stress-sliding curve, and in this test, the triplet masonry was six times stronger than the URM samples. The failure in the specimens occurred through the development of sliding on the brick–mortar interface. The debonding of CFRP from triplet was the main reason for the failure of CFRP-strengthened triplet masonry, which mainly occurs when the bonding between the reinforcement composite and masonry interface is lost. In contrast to URM walls, reinforced triplets were stable after failure.

Figure 21. CFRP net on four sides.

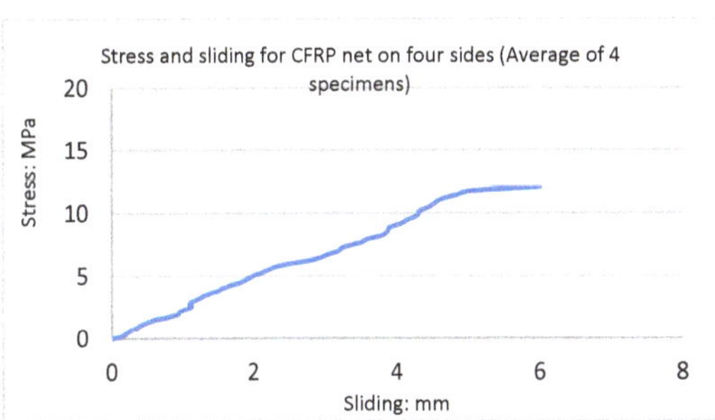

Figure 22. Stress-slip curve for CFRP net on four sides of a triplet.

3.8. Strengthening with One Layer of CFRP Strip on the Four Sides of a Triplet

These samples were strengthened with one 50 mm wide layer of CFRP strip on four sides of triplet masonry. The mode of failure was ductile, and the ultimate load was approximately 57 kN, which was slightly stronger than CFRP strips on two sides of the triplet. The propagation of cracks and delamination also occurred before failure. Figure 23 presents the shear failure in the unit–mortar interface. Figure 24 displays the stress-sliding curve; triplet masonry with one layer of CFRP strip on four sides was five times stronger than the URM samples. The failure in the specimens occurred through the development of sliding on the brick–mortar interface; debonding of CFRP from the triplet was the reason for the failure of the CFRP-strengthened triplet.

Figure 23. CFRP strips on four sides.

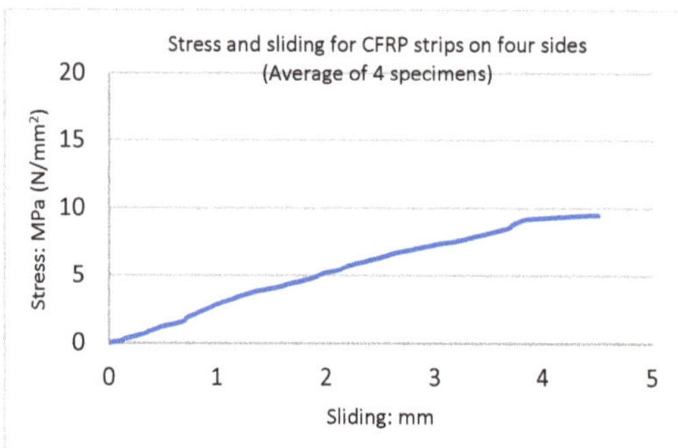

Figure 24. Stress-slip curve for a triplet with one layer of CFRP strip on four sides.

3.9. Strengthening with One Layer of CFRP Net Strip on Four Sides

The samples were strengthened with one layer of two CFRP net strips, 50 mm wide, on four sides of the triplet. The mode of failure was ductile, and the ultimate load was approximately 60 kN. These triplets displayed the propagation of cracks and delamination that occurred prior to failure. The shear failure in the unit–mortar interface and brick is shown in Figures 25 and 26, which illustrate the stress-sliding curves. Furthermore, it was observed that the shear triplet presented an almost linear shear stress–strain relationship before cracking of mortar, followed by a drop of shear strength. In this test, the triplet masonry with one layer of CFRP net strip on four sides was five times stronger than the URM samples.

Figure 25. CFRP net strips on four sides.

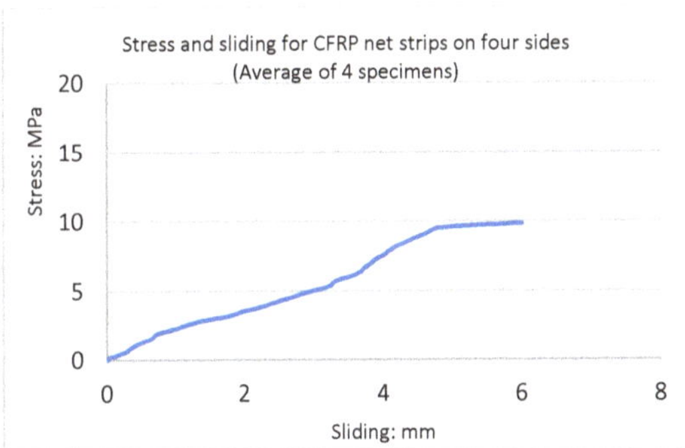

Figure 26. Stress-slip curve of CFRP net strips on four sides.

Table 3 illustrates a comparison of bond stress and sliding from the experimental tests at failure points for the different CFRP configurations. The tests demonstrated that four sides covered with FRP and FRP strips resulted in stronger bonds and less displacement in triplet masonry. In addition, the FRP wrap performed better than the FRP strips.

Table 3. Comparison of bond stress and sliding from the experimental test.

No	CFRP Configurations on Triplet Masonry (One Layer)	Load: KN	Stress: MPa	Maximum Sliding in Experimental Tests at Failure Point: mm
1	URTM	10	1.59	2.5
2	CFRP on two sides	90	14.33	6.2
3	CFRP on four sides	90	14.33	5.6
4	CFRP strips on two sides	62	9.85	5.3
5	CFRP strips on four sides	62	9.85	4.4
6	CFRP net on two sides	76	12.06	6
7	CFRP net on four sides	78	12.38	5.4
8	CFRP net strips on two sides	54	8.57	5.8
9	CFRP net strips on four sides	64	10.16	5.2

4. Equation for Shear Behaviour of Triplet Masonry

4.1. Empirical Equations for the Shear Strength of URM

The results of the standard shear tests on triplets are presented in this section to derive empirical equations for the shear behaviour of unreinforced masonry elements. According to EN 1052-3 [30], the shear stresses can be assumed to be uniformly distributed over the two-bed joints during the triplet test. The following method can be used to evaluate the maximum value of the experimental shear stress τ_{max} on the bed joints:

$$\tau_{max} = \frac{F_{max}}{2A} \quad (1)$$

where τ_{max} is the maximum value of the experimental shear stress, F_{max} is the maximum shear force at failure, and A is the total area of the bed joint. Based on the experimental tests in the current study, the above equation demonstrated extremely low shear stress compared to experimental tests and was therefore unreliable. Thus, this equation was improved for maximum shear strength of triplet masonry with solid brick and mortar M4 as follows:

$$\tau_{max} = \frac{0.70 \times F_{max}}{A \times \rho} \quad (2)$$

where ρ = 0.20 for mortar M4, A is the total area of the bed joint, and F_{max} is the maximum shear force at failure. The coefficient of 0.70 was obtained based on experimental tests through trial and error. In this new equation, τ can be calculated at any shear load point. Table 4 and Figure 27 show the comparison of the experimental tests and the results from Equation (2).

Table 4. Comparison of the experimental shear tests and the results from Equations (2) and (3) and EN 1052-3:200 [30].

Test	Maximum Load: KN	$\tau_{max}=\frac{F_{max}}{2A}$ (MPa) EN 1052-3:2002	$\tau_{max}=\frac{0.70\times F_{max}}{A\times\rho}$ (MPa) New Equation	Experimental Shear Tests (MPa)
1	4	0.087	0.612	0.64
2	6	0.13	0.918	0.95
3	8	0.18	1.23	1.27
4	10	0.22	1.53	1.59
5	12	0.26	1.84	1.88
6	14	0.31	2.1	2.15

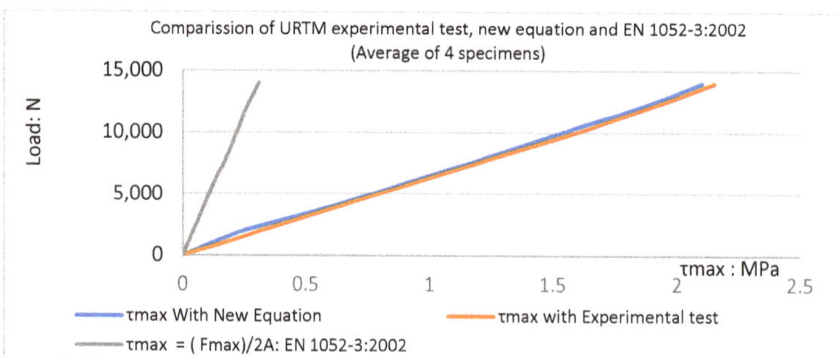

Figure 27. Comparison between results from EN 1052-3:2002 [30], the experimental test, and the new equation (URTM).

4.2. *New Equations for Strengthened Triplet Masonry with CFRP*

As per codes of practice, information about shear tests on unreinforced and reinforced triplet masonry with CFRP is limited.

Therefore, the combined resistance offered by the FRP and the brick–mortar joint can be used to express the shear strength of masonry that has been strengthened with FRP.

$$S_m = S_{m0} + S_{frp} \tag{3}$$

where

S_m = reinforced masonry shear strength with FRP

S_{m0} = shear resistance of the unreinforced masonry

S_{frp} = additional resistance offered by FRP confinement

Sm_0 can be calculated using formulas found in design guidelines such as IS: 1905 (BIS 2002a). In this research, S_{m0} for solid London brick and mortar M4 was obtained from the shear tests performed on control triplet specimens (Equation (4)).

$$Sm_0 = \tau_{max} = \frac{0.70 \times Fmax}{A \times \rho} \tag{4}$$

The expressions that quantify S_{frp} needed to be developed. Below, the results of the experimental tests used to calibrate new expressions for S_{frp} are presented.

Tensile stress in CFRP was caused by the centre brick slipping during the loading process. As the middle brick slipped further, the compressive stress on the bed joint grew until the specimen broke as a result of the FRP composite rupturing or debonding from the masonry. Consequently, a new equation (Equation (5)) could be used to express the maximum shear strength of CFRP:

$$S_{frp} = \sigma frp \times Af//At \times K \times \mu f \tag{5}$$

where σfrp is the tensile strength of the CFRP, which is given by the factory. The average tensile strength of the CFRP and CFRP net was 3500 MPa.

S_{frp} is the maximum shear strength of masonry with CFRP; At is the area of triplet masonry; Af is the area of triplet masonry covered by CFRP; k = 1, for one layer of CFRP on two sides; k = 0.7, for one layer of CFRP on four sides; µf is the coefficient for CFRP; and µf = 0.007 for CFRP.

Table 5 shows a comparison of the maximum shear strength of different configurations of the experimental triplet masonry strengthened with CFRP and the results from the new Equations (3) and (5).

Table 5. Comparison of the shear strength of different configurations of experimental triplets with CFRP and results from the new equation.

Test	CFRP Configurations on Triplet Masonry	A_f/A_t	$S_{frp} = \sigma_{frp} \times A_f/A_t \times K \times \mu f$ $S_m = S_{m0} + S_{frp}$ (MPa)	Experimental Shear Tests (MPa)	Error %
1	One-layer CFRP on two sides	0.51	14.10	14.33	2
2	One-layer CFRP on four sides	0.75	14.45	14.33	1
3	One-layer CFRP strips on two sides	0.23	7.3	9.53	25
4	One-layer CFRP strips on four sides	0.35	10.1	9.53	6
5	One-layer CFRP net on two sides	0.51	14	11.75	16
6	One-layer CFRP net on four sides	0.75	14.45	12.06	17
7	One-layer CFRP net strips on two sides	0.23	7.3	8.57	17
8	One-layer CFRP net strips on four sides	0.35	7.6	9.57	20
	$S_{m0} = \tau_{max} = 1.59$ (MPa)				

4.3. *New Equations for Bond Stress-Slip of Strengthened Triplets with CFRP*

In this study and based on the database of the experimental results, if the sliding/slipping distance is known, the bond stress for all CFRP configurations of triplet masonry can be calculated with the newly developed equations shown in Table 6. Moreover, bond stress can be calculated at any sliding point, and the failure point is predictable. To find a reliable equation, a quadratic formula with a trial-and-error coefficient was used to develop equations that were validated against experimental tests. The bond stresses predicted by the equations were compared with the experimental tests. They are shown in Figures 28–36. The bond stress and sliding curves are in good agreement with the experimental results. In all diagrams, L denotes the CFRP layer, and S represents the side of the triplet. By using the new equations, it was possible to predict the correct bond stress at any sliding distance for solid London brick and mortar M4 reinforced with eight different CFRP configurations as well as for unreinforced triplet masonry. The equations in the present study are an advancement directly benefiting engineers in predicting the correct bond stress-slip resistance. (L represents layer, and S denotes side).

Table 6. The newly developed formula for predicting bond stress.

	CFRP Configurations on Triplet Masonry (One Layer)	Equation Number	Bond Stress $\sigma_b = f(S)$
1	URTM	(6)	$\sigma b = 0.075S^2 - 0.65S + 0.075S\sqrt{S^2 - 20S + 73.33}$
2	CFRP on two sides	(7)	$\sigma b = 0.14S^2 - 2.1S + 0.14S\sqrt{S^2 - 30S + 210.71}$
3	CFRP on four sides	(8)	$\sigma b = 0.14S^2 - 2.1S + 0.14S\sqrt{S^2 - 30S + 210.71}$
4	CFRP strips on two sides	(9)	$\sigma b = 0.10S^2 - 1.5S + 0.14S\sqrt{S^2 - 30S + 205}$
5	CFRP strips on four sides	(10)	$\sigma b = 0.11S^2 - 1.65S + 0.11S\sqrt{S^2 - 30S + 206.68}$
6	CFRP net on two sides	(11)	$\sigma b = 0.12S^2 - 2.3S + 0.12S\sqrt{S^2 - 30S + 208.33}$
7	CFRP net on four sides	(12)	$\sigma b = 0.125S^2 - 1.875S + 0.125S\sqrt{S^2 - 30S + 209}$
8	CFRP net strips on two sides	(13)	$\sigma b = 0.075S^2 - 1.125S + 0.075S\sqrt{S^2 - 30S + 198.33}$
9	CFRP net strips on four sides	(14)	$\sigma b = 0.105S^2 - 1.575S + 0.105S\sqrt{S^2 - 30S + 205.95}$

S represents sliding (mm), and σ_b is bond stress (MPa) for any sliding.

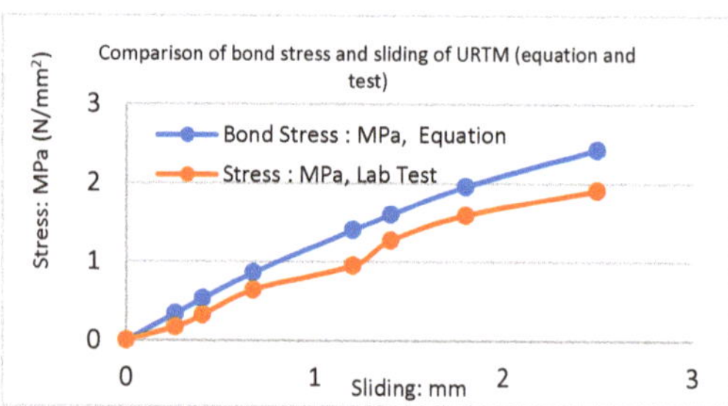

Figure 28. Comparison of bond stress and sliding of URTM (equation and test).

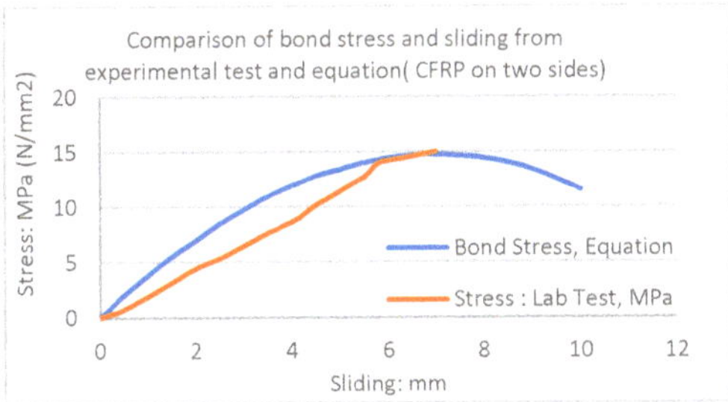

Figure 29. Comparison of bond stress and sliding from the experimental test and equation (CFRP on two sides).

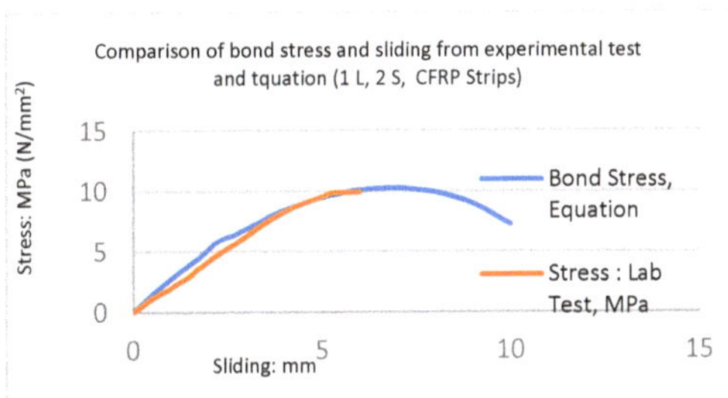

Figure 30. Bond stress and sliding from the experimental test and equation (1 L, 2 S, and CFRP strips).

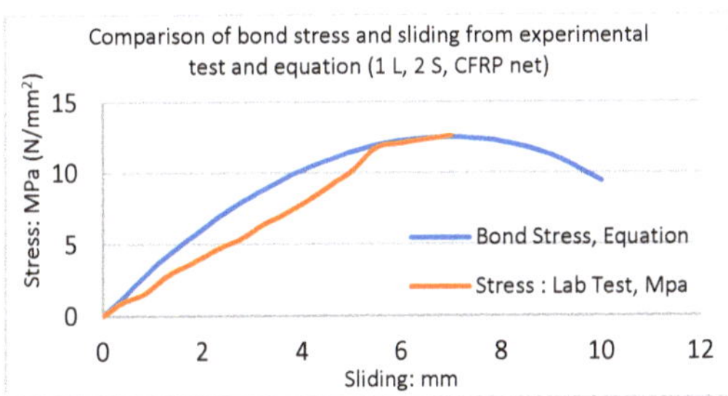

Figure 31. Comparison of bond stress and sliding from the experimental test and equation (1 L, 2 S, and CFRP net).

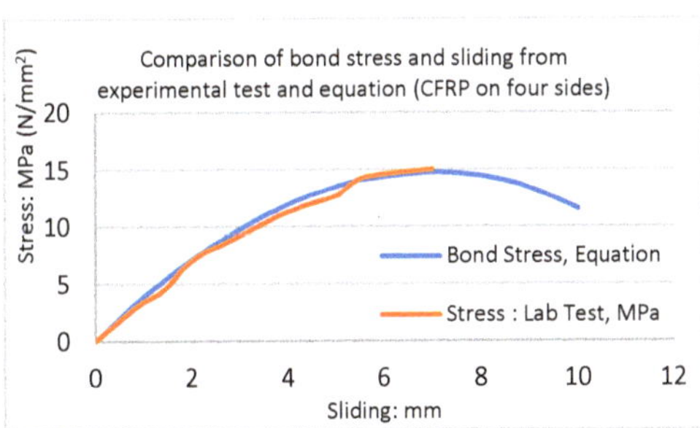

Figure 32. Comparison of bond stress and sliding from the experimental test and equation (CFRP on four sides).

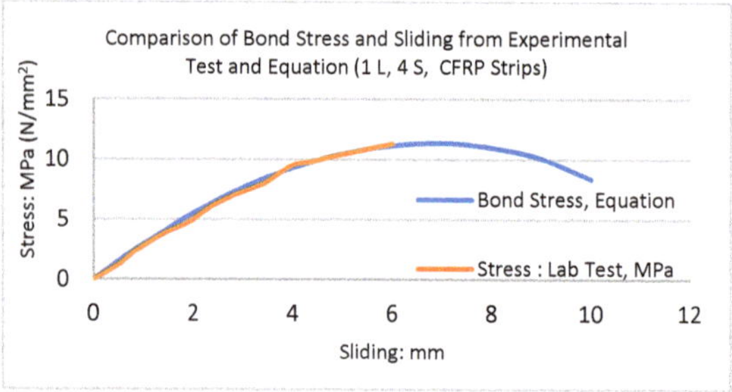

Figure 33. Comparison of bond stress and sliding from the experimental test and equation (1 L, 4 S, and CFRP strips).

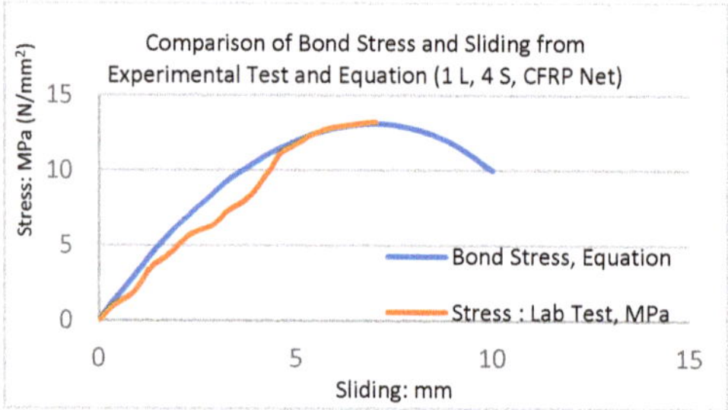

Figure 34. Comparison of bond stress and sliding from the experimental test and equation (1 L, 4 S, and CFRP net).

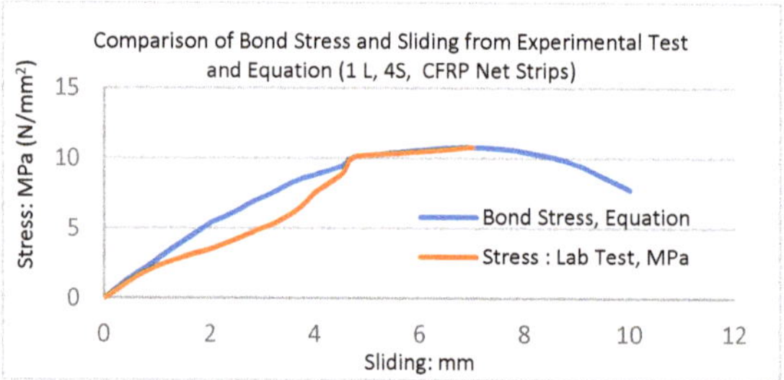

Figure 35. Comparison of bond stress and sliding from the experimental test and equation (1 L, 4 S, and CFRP net strips).

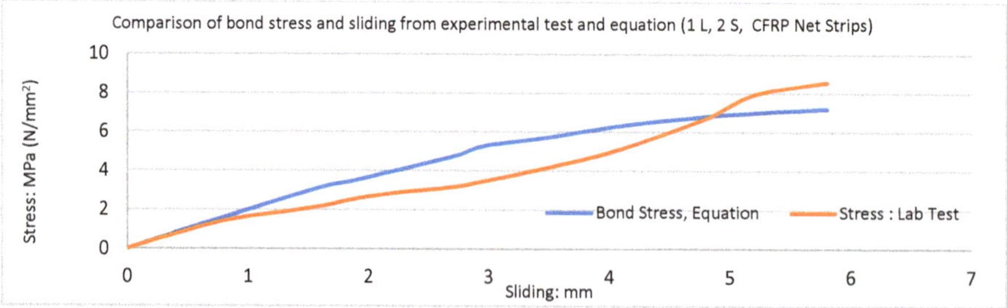

Figure 36. Comparison of bond stress and sliding from the experimental test and equation (1 L, 2 S, and CFRP net strips).

5. Finite Element Simulation and Verification via Results from an Experimental Study

The interaction of bricks and mortar is critical in the nonlinear analysis of masonry structures. The use of interface elements in the micro- and simplified micro-models enables the identification and analysis of potential failure mechanisms occurring within the mortar joint. These failure mechanisms encompass brick/mortar decohesion and cracking across the mortar layer. As a result, interface elements have a wide range of applications in the analysis of masonry structures, and ongoing research is being conducted to further their development. Page [34] was the pioneer in endeavouring to incorporate interface elements within masonry walls. D'Altri et al. [35,36] introduced a new 3D micro-model that depicts the response of masonry walls under both in-plane and out-of-plane loading scenarios. This model comprises textured units that are composed of a single brick and mortar layer. To address the limitations of the detailed micro-modelling approach, a simplified micro-modelling technique was developed. The use of simplified micro-modelling is prevalent in the numerical analysis of masonry due to its straightforwardness and substantial decrease in analysis duration in comparison to intricate micro-modelling, as evidenced by various studies, such as those by Doran et al. [37–39]. Micro-models, according to Lourenco et al. [40–43], are probably the best tools available for understanding masonry behaviour. The advantage of using this method is that it takes into account all of the different failure mechanisms, such as (a) cracking and sliding along the bed or head joints, (b) cracking of the units in direct tension, and (c) masonry crushing, which is commonly identified as the splitting of units in tension as a result of mortar at high normal stress values. Ravi et al. [44] carried out a numerical investigation that provided detailed information on masonry behaviour that could be compared to the actual behaviour of masonry structures. Masonry is anisotropic due to the presence of nonlinear and non-homogeneous mortar joints. The finite element method can be used to perform nonlinear analysis of unreinforced brick masonry. Prakash and Alagusundaramoorthy [45] have demonstrated that micro-modelling of triplet masonry is an effective approach to identifying the behaviour and shear transfer mechanism across the mortar joint. The shear transfer mechanisms at joints can be categorized into three types, namely sliding, interlocking, and overriding.

5.1. Modelling by ANSYS and Verification via Experimental Studies

In this study, the shear and bond behaviour of FRP-strengthened triplet masonry was investigated analytically through the use of the commercial software ANSYS versions 18 and 19.2 [46,47]. Different strengthened specimens were analysed, and the behaviour of the samples was compared with results from the experimental studies. The analytical modelling process was based on the following set of assumptions:

1. The cross-sectional area and the modulus of elasticity remain unchanged throughout the loading process.
2. Masonry samples are regarded as consisting of a material that is both non-homogeneous and isotropic in nature.

Researchers prefer employing homogeneous modelling and linear elastic behaviour in order to simulate masonry walls. However, in this study, non-homogeneous micro-modelling and nonlinear elastic behaviour were assumed for the URM triplets. The compression and shear behaviour of strengthened masonry could be compared together, particularly since the displacement of triplet masonry was one of the main objectives of the present research.

5.2. Formulation of the Model (Micro-Modelling of the Brick Masonry)

Brickwork strength is reliant on the physical features of the masonry unit, the mortar, and the bond among them. In the current study, models with two different materials are presented, which are brick units and mortar joints. The brick unit was modelled using solid 185. Solid 185 is used for the 3-D modelling of structures and is defined by eight nodes. Each node has three degrees of freedom: translations in the nodal x, y, and z directions. The element considers large deflection, large strain, plasticity, creep, and stress stiffening capabilities. Deformations of nearly incompressible elastoplastic materials and fully incompressible hyperelastic materials can also be simulated, as it has mixed formulation capabilities. This is available as a homogenous structural solid element and a layered structural solid element. The mortar joint was modelled using solid 65; solid 65 is used for 3-D modelling of solids with or without reinforcing bars (rebar). Solid 65 is capable of cracking in tension and crushing in compression. The element is defined by eight nodes having three degrees of freedom at each node: translations in the nodal x, y, and z directions. Up to three different rebar specifications may be defined. The most important aspect of this element is the treatment of nonlinear material properties.

To predict the mechanical performance of masonry, finite element models of specimens have been developed. The current study attempted to develop a finite element model to predict the shear strength of masonry prisms subjected to shear loading. The micro-level models consider the units and mortar joints separately. Thus, the structural analysis was performed, taking into consideration each element of the masonry material. The stack-bonded brick masonry prism was built with three solid London clay bricks measuring 220 mm × 104 mm × 65 mm in 10 mm thick cement mortar with a 1:4 (M4) ratio. Separate layers of elements represented the bricks and mortar joints. Each element type was represented by its properties as well as its initial modulus of elasticity. The loading was applied to the model's top surface. The micro-modelling approach was used to represent brick and mortar joints as continuum elements. The bond between brick and mortar was modelled using contact elements. The constitutive behaviour of the contact elements was based on the aforementioned Equations (6)–(14) given in Table 6.

5.3. CFRP Configurations for Analytical Study by ANSYS

This study on finite element analysis (FEA) provided evidence that the enhancement in resistance to triplet load was a result of the augmentation in fibre-reinforced polymer (FRP) composite material. The experimental results showed that retrofitting CFRP composites increased the load resistance of masonry triplets. CFRP retrofitting increased the strength and deformation capacity of triplets. Until the point of failure, the bond between the CFRP and the triplet surface remained unbroken. In the present investigation, modelling was conducted on triplet specimens to find the strength of the brick masonry. The study also examined the performance of three sets of triplet samples that were retrofitted using various arrangements of CFRP composites. The retrofitted triplets' test results were used to validate the micro-modelling. The FE modelling consisted of nine different triplet masonry configurations with one-layer CFRP and CFRP net as follows: (1) CFRP on two sides, (2) CFRP on four sides, (3) CFRP strips on two sides, (4) CFRP strips on four sides, (5) CFRP net on two sides, (6) CFRP net on four sides, (7) CFRP net strips on two sides, and (8) CFRP net strips on four sides of triplet masonry and unreinforced triplet masonry (URTM); they were analysed under shear loading.

5.4. Loading, Boundary Conditions, and Load-Deformation Behaviour

It was assumed that an ideal connection existed between the surface of the triplet and the CFRP composite. The load was applied to the upper part of the central brick, while the lower parts of the side bricks were considered to be fixed, allowing no movement in any direction. The software employed a surface-to-surface contact algorithm to simulate the interface between brick and mortar. In this simulation, the brick surface was designated as the target surface owing to its higher modulus of elasticity, while the mortar surface was identified as the contact surface. According to the experimental data, significant stress was transferred to the FRP around the cracked elements, and the CFRP retrofitted triplet exhibited nonlinear load resistance-displacement behaviour. After cracking through the bed joints, the CFRP began to resist the load.

6. Finite Element Analyses for Masonry Triplets with and without CFRP

The FE modelling demonstrated that the effect of CFRP increased the load resistance of the samples. The experimental results showed that retrofitting masonry triplets with CFRP composites increased their load resistance. The bricks and mortar were represented as separate elements with their properties using the initial modulus of elasticity from the experimental tests, and the brick and mortar joints were modelled using the micro-modelling approach. Figures 37a, 38a, 39a, 40a, 41a, 42a, 43a, 44a and 45a present a comparison of the experimental tests and developed finite element simulations for all CFRP configurations. Figures 37b, 38b, 39b, 40b, 41b, 42b, 43b, 44b and 45b depict the localized increase in stress at the interfaces of the brick-and-mortar surfaces. The failure started at the peripheral edges of the interface layer, gradually progressing towards the middle of the specimen. The mortar with a lower modulus of elasticity experienced higher stress and cracks compared to the brick. The deformation experienced by the entire mortar was relatively high, pointing towards the possibility of interface delamination. Based on the FE analysis of masonry triplets, it was observed that the stress is effectively transferred to the CFRP composite, leading to a stronger connection between the brick and mortar. As shown in Figures 37b, 38b, 39b, 40b, 41b, 42b, 43b, 44b and 45b, the stress transferred from mortar and bricks to CFRP and the failure of all triplets began in the brick-and-mortar interface layer, sweeping to the CFRP on the specimen. The cracks observed in both the mortar layer and the brick during the experimental test were similar to the ones anticipated by the suggested model. In experimental tests and also in finite element analyses, the debonding of CFRP from triplets was the main reason for failure; the loss of bonding between the reinforcement composite and masonry interface is primarily responsible for the occurrence of this failure. The occurrence of failure indicated that the crack originated from the separation of the masonry units from the mortar and propagated across the entire thickness until there was the detachment of the epoxy and CFRP from the masonry units. At this point, the triplet failed because the tensile stresses were no longer transferred to the CFRP, confirming that the developed numerical model can simulate the failure modes of triplets with sufficient accuracy.

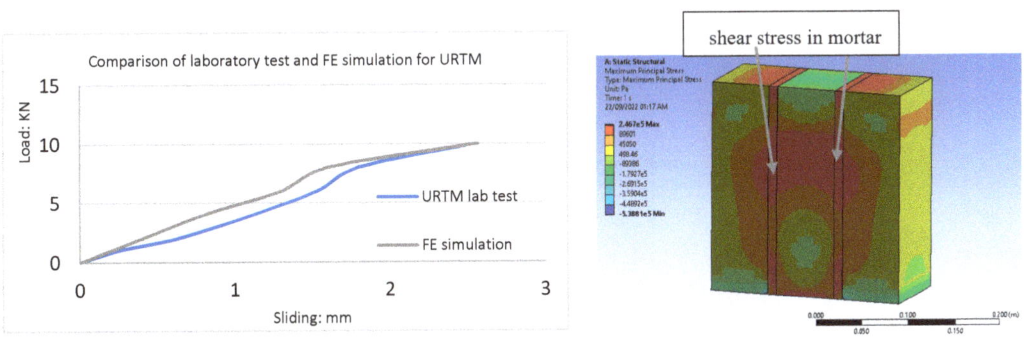

(a) (b)

Figure 37. (**a**) Comparison of laboratory test and FE simulation. (**b**) Stress in FE simulation.

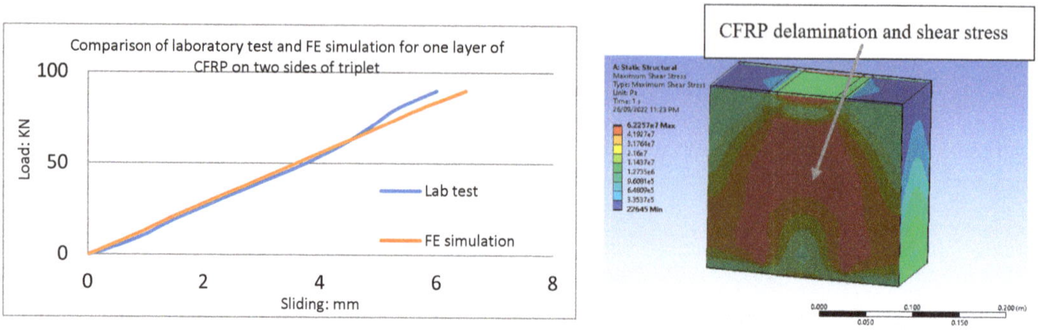

(a) (b)

Figure 38. (**a**) Comparison of laboratory test FE simulation. (**b**) Shear stress at the failure point.

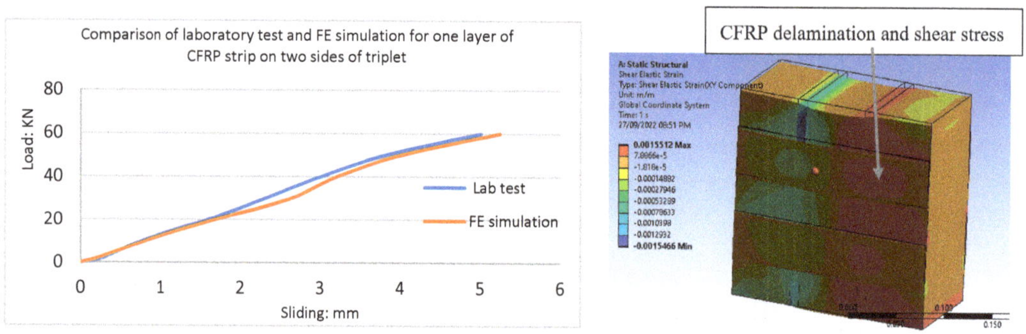

(a) (b)

Figure 39. (**a**) Comparison of laboratory test and FE simulation. (**b**) Shear stress at the failure point.

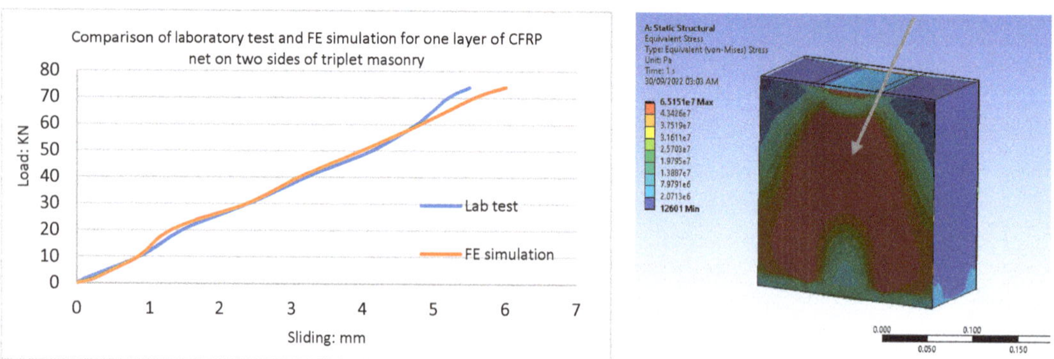

(a) (b)

Figure 40. (**a**) Comparison of laboratory test and FE simulation. (**b**) Shear stress in FE simulation.

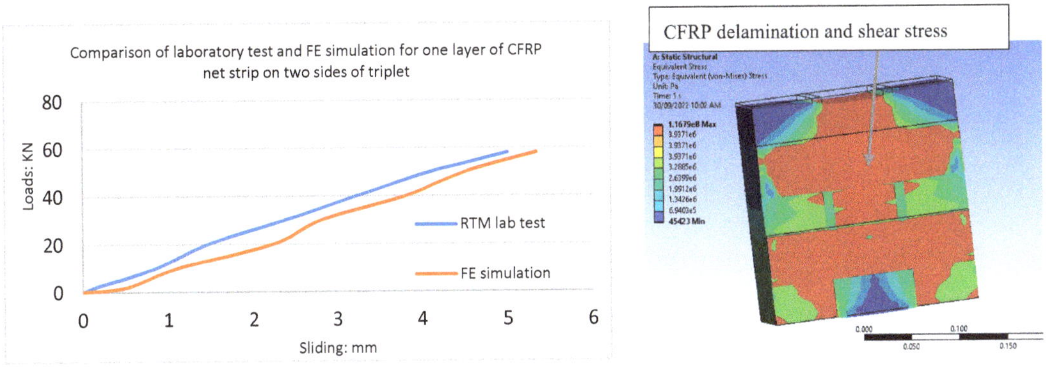

(**a**) (**b**)

Figure 41. (**a**) Comparison of laboratory test and FE simulation. (**b**) Shear stress at the failure point.

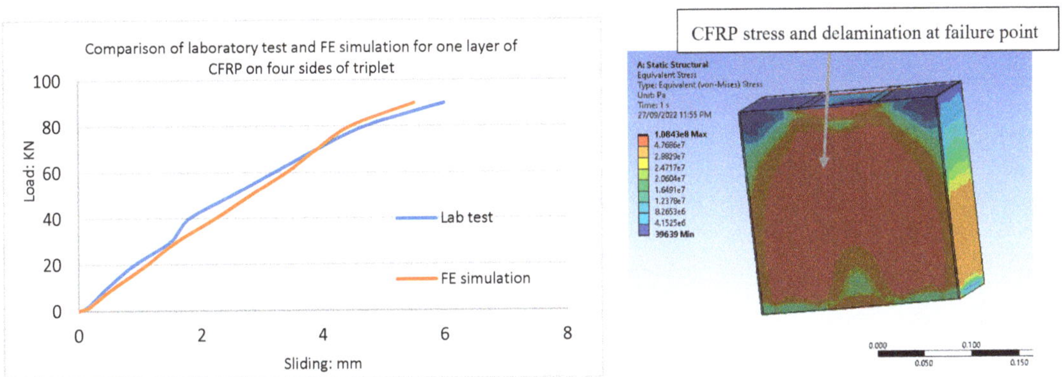

(**a**) (**b**)

Figure 42. (**a**) Comparison of lab test FE simulation (**b**) Shear stress at the failure point.

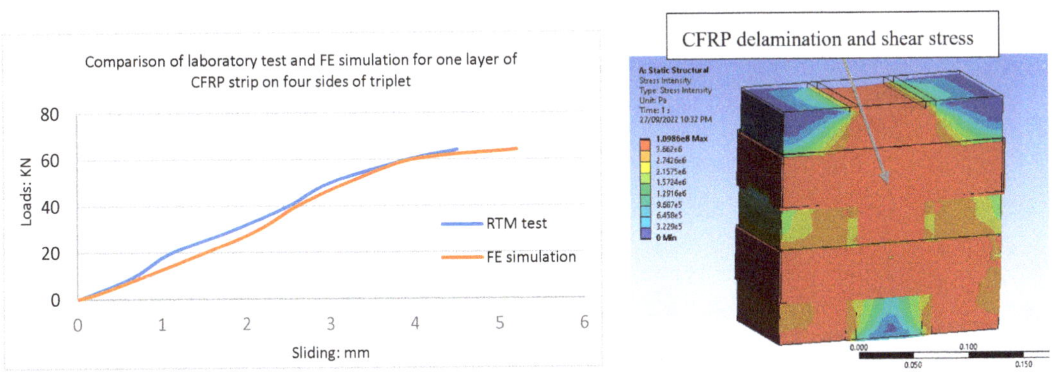

(**a**) (**b**)

Figure 43. (**a**) Comparison of laboratory test and FE simulation. (**b**) Shear stress in FE simulation.

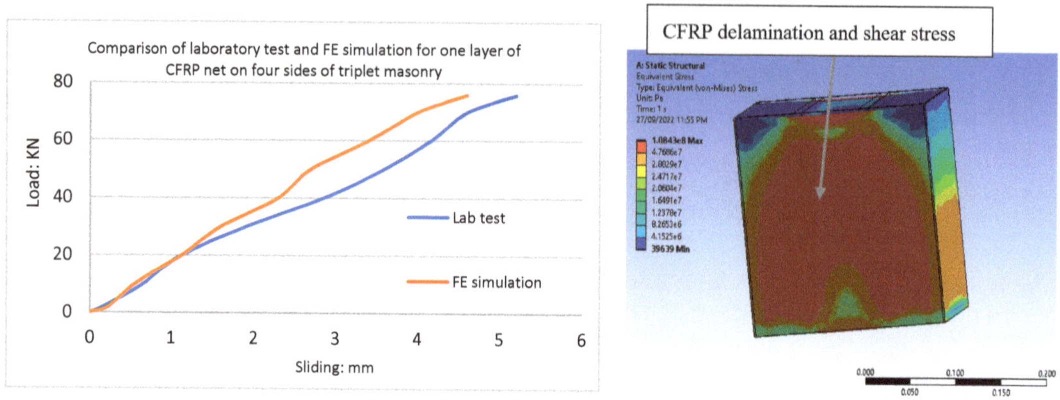

(a) (b)

Figure 44. (**a**) Comparison of laboratory test and FE simulation. (**b**) Shear stress at the failure point.

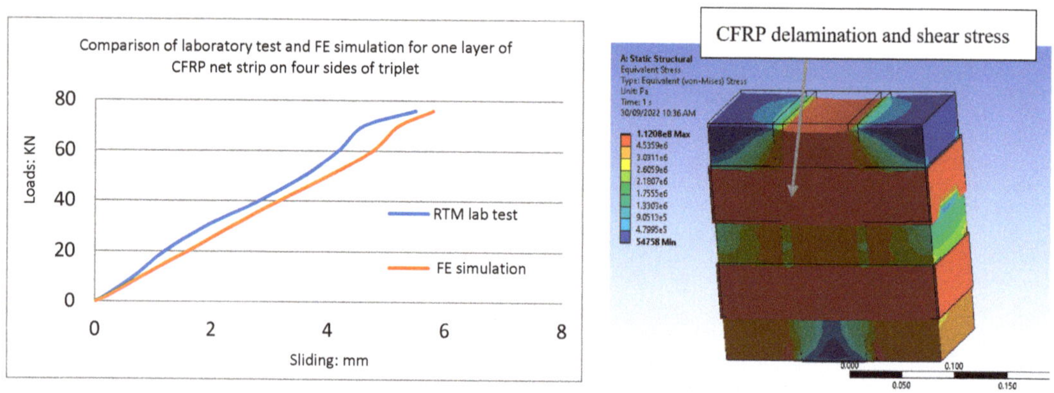

(a) (b)

Figure 45. (**a**) Comparison of laboratory test and FE simulation. (**b**) Shear stress at the failure point.

The ultimate displacement predicted by the numerical modelling for all CFRP configurations on triplet masonry was much lower compared with the results obtained in the experimental test. It is normal to find disagreement between analytical finite element modelling and experimental tests. In this study, the comparison of the load-displacement graphs from the theoretical modelling with the results from the experimental tests showed significantly low displacement and were not accurate, thus the user-defined functions for finite element modelling were developed.

The reason for lower sliding in FE modelling than in the experimental tests is that the finite element program in Ansys assumes a perfect bond between brick and mortar. The perfect homogeneous material is the other assumption in finite element analysis. Therefore, this error can be corrected by using the developed functions in Table 6 as a user-defined function (UDF) to create correct sliding and 3D modelling for eight CFRP configurations. By using the UDF functions, it is possible to predict correct sliding in 3D modelling at any load or stress point. The FEM stress in mortar and sliding of the middle brick in this study is similar to the results obtained by Hernoune, et al. [18].

According to the diagrams from the FE modelling, interface stresses and sliding failure propagation initiated at the brick-and-mortar interface layer spread to the CFRP and brick interface on the specimens. The failure of interface elements commenced at the middle brick interfaces, which might have been due to the boundary conditions on the left and right ends of the specimen's outer bricks and the direction of loading. Full cracking was only observed at the inner interfaces, starting at the maximum load.

7. Discussion and Conclusions

Overall, reinforced triplet masonry shear tests play a significant role in ensuring the safety, reliability, and longevity of masonry constructions by providing valuable insights into the behaviour of materials under shear stress. Engineers and construction professionals can use the test results to select the most effective reinforcement method for masonry walls in earthquake-prone regions. The findings can influence building codes and construction practices to enhance the seismic resilience of structures.

In the strengthened triplets, the failure was caused by the loss of bonding between the CFRP and the masonry units. The mode of failure for all configurations was ductile, and delamination occurred just before failure. FRP ruptured at mid-span, followed by shearing of the sheets in the middle of the sample being observed. It was also observed that the presence of the CFRP improved the triplets' performance by changing the failure from brittle to ductile.

In summary, the experimental data yielded the following observations:

1. All types of reinforced triplets increased bond strength by up to four to six times compared with triplets without CFRP, and the extent of reinforcement effectiveness varied based on the configuration employed.
2. The shear failure of brickwork commenced with cracking and sliding through the mortar joints. The failure occurred along the interface between the brick-and-mortar joints.
3. The highest ductility in terms of maximum sliding at the ultimate load was observed for all samples.
4. The retrofitting of shear triplets with CFRP composites reduced the interfacial shear stress concentrations between the brick-and-mortar surfaces, and in contrast with URM triplets, the strengthened samples were stable after failure.
5. The bond strength in reinforced triplets with the same amount of CFRP and CFRP net was found to be almost equal. The main reason for this could be comparable values of the modulus of elasticity.
6. A debonding failure pattern was observed in reinforced triplets with CFRP sheet and CFRP strip configurations.
7. Comparison of the finite element analyses using the newly developed equations with experimental tests for reinforced triplet masonry with different CFRP configurations showed good agreement, resulting in a high level of accuracy.

8. Suggestions for Further Investigations

The effectiveness of the FRP composites for reinforcement of the masonry structures has been proven in this study, but further investigations are needed to clarify the behaviour of FRP-strengthened masonry walls. Some of the areas that need to be investigated are indicated below:

1. Further information about the durability of composite materials used in the reinforcement of URM walls under different environmental conditions would be very beneficial.
2. Investigation of surface preparation methods and the amount of impregnating resins are also needed.
3. An investigation of the coupled behaviour of compression forces and FRP strengthening of masonry specimens is needed.
4. Since debonding of FRP from the masonry surface is the main reason for failure and has a direct relationship with the porosity of the masonry unit and mortar, the different types of masonry units and mortar should be investigated.
5. A numerical investigation of the different parameters affecting the behaviour (e.g., masonry properties, CFRP properties, and thickness) needs to be performed.

6. Investigation of strengthened masonry specimens under high temperatures.
7. Investigation of large-scale strengthened masonry walls with different CFRP configurations under combinations of in-plane and out-of-plane loadings.

Author Contributions: Conceptualization, methodology, software, validation, formal analysis, investigation, data curation and writing—original draft preparation, S.M.H.; supervision, project administration, writing—review and editing, A.A. All authors have read and agreed to the published version of the manuscript.

Funding: This research did not receive any specific grant from funding agencies in the public, commercial, or not-for-profit sectors.

Data Availability Statement: Data are contained within the article.

Acknowledgments: The authors are grateful for the support of the technical staff of the structural laboratory at City, University of London, in particular to the Brett McKinley.

Conflicts of Interest: The authors declare no conflict of interest.

References

1. Ehsani, M.R.; Saadatmanesh, H. Shear behaviour of URM retrofitted with FRP Overlays. *J. Compos. Constr.* **1997**, *1*, 17–25. [CrossRef]
2. Ehsani, M. Strengthening of Concrete and Masonry Structures with Fiber Reinforced Polymers (FRP). In Proceedings of the 30th Conference on Our World in Concrete & Structures, Singapore, 23–24 August 2005.
3. Saadatmanesh, H. Extending service life of concrete and masonry structures with fibre composites. *Constr. Build. Mater.* **1997**, *11*, 327–335. [CrossRef]
4. Ferretti, F.; Ferracuti, B.; Mazzotti, C.; Savoia, M. Destructive and minor destructive tests on masonry buildings: Experimental results and comparison between shear failure criteria. *Constr. Build. Mater.* **2019**, *199*, 12–29. [CrossRef]
5. Ehsani, M.; Saadatmanesh, H. Fibre Composites: An Economical Alternative for Retrofitting Earthquake-Damaged Precast-Concrete Walls. *Earthq. Spectra* **1997**, *13*, 225–241. [CrossRef]
6. Mohammadipour, A.; Willam, K. Lattice simulations for evaluating interface fracture of masonry composites. *Theor. Appl. Fract. Mech.* **2016**, *82*, 152–168. [CrossRef]
7. Pavan, G.S.; Nanjunda Rao, K.S. The behaviour of Brick–Mortar Interfaces in FRP-Strengthened Masonry Assemblages under Normal Loading and Shear Loading. *Am. Soc. Civ. Eng.* **2015**, *28*, 04015120. [CrossRef]
8. Tumialan, G.J.; Nanni, A. *In-Plane and Out-of-Plane Behaviour of Masonry Walls Strengthened with FRP Systems*; Department of Civil Engineering, University of Missouri: Rolla, MO, USA, 2001.
9. Tumialan, G.; Micelli, F.; Nanni, A. A Structural Engineering Odyssey. In *Proceedings of Structures Congress & Exposition*; Chang, P.C., Ed.; ASCE: New York, NY, USA, 2001.
10. Tumialan, G.; Vatovec, M.; Kelley, P.L. FRP Composites for Masonry Retrofitting. *Structure Magazine*, 12 May 2009; pp. 12–14. Available online: https://www.structuremag.org/wp-content/uploads/2014/08/C-BuildingBlocks-Tumialan-May091.pdf (accessed on 8 January 2024).
11. Tumialan, J.G. *In-Plane and Out-of-Plane Behaviour of Masonry Walls Strengthened with FRP Systems*; CIES Research Report; University of Missouri: Rolla, MO, USA, 2003; pp. 1–4.
12. El Malyh, S.; Bouyahyaoui1, A.; Cherradi1, T.; Rotaru, A.; Mihai, P. In-Plane Shear Behaviour of Unreinforced Masonry Walls Strengthened with Fibre Reinforced Polymer Composites. Advances in Science. *Technol. Eng. Syst. J.* **2020**, *5*, 360–367.
13. *ASTM E-519-02*; Standard Test Method for Diagonal Tension (Shear) in Masonry Assemblages. ASTM International: West Conshohocken, PA, USA, 2002.
14. Babatunde, S.A. Review of strengthening techniques for masonry using fibre reinforced polymers. *Compos. Struct.* **2017**, *161*, 246–255. [CrossRef]
15. Hamrat, M.; Bouziadi, F.; Boulekbache, B.; Daouadji, T.H.; Chergui, S.; Labed, A.; Amziane, S. Experimental and Numerical investigation on the deflection behaviour of pre-cracked and repaired reinforced concrete beams with fibre-reinforced polymer. *Constr. Build. Mater.* **2020**, *249*, 118745. [CrossRef]
16. Luccioni, B.; Rougier, V.C. In-plane retrofitting of masonry panels with fibre reinforced composite materials. *Constr. Build. Mater.* **2010**, *25*, 1772–1788. [CrossRef]
17. Luccioni, B.; Rougier, V.C. Shear behaviour of brick–mortar interface in CFRP retrofitted or repaired the masonry. *Int. J. Mech. Sci.* **2010**, *52*, 602–611. [CrossRef]
18. Hernoune, H.; Benabed, B.; Abousnina, R.; Alajmi, A.; Alfadhili, A.M.G.; Shalwan, A. Experimental Research and Numerical Analysis of CFRP Retrofitted Masonry Triplets under Shear Loading. *Polymers* **2022**, *14*, 3707. [CrossRef]
19. Hamid, A.A.; El-Dakhakhni, W.W.; Hakam, Z.H.; Elgaaly, M. Behavior of composite unreinforced masonry– fiber-reinforced polymer wall assemblages under in-plane loading. *J. Compos. Constr.* **2005**, *9*, 73–83. [CrossRef]

20. Saingam, P.; Hlaing, H.H.; Suwannatrai, R.; Ejaz, A.; Hussain, Q.; Khan, K.; Joyklad, P. Enhancing the flexural behavior of brick masonry walls with ferrocement overlays and low-cost anchors. *Case Stud. Constr. Mater.* **2023**, *19*, e02558. [CrossRef]
21. Joyklad, P.; Hussain, Q. Development of strength models for brick walls: Experimental and theoretical study. *Results Eng.* **2023**, *18*, 101103. [CrossRef]
22. Rashid, A.S.A.; Shirazi, M.G.; Mohamad, H.; Sahdi, F. Bearing capacity of sandy soil treated by Kenaf fibre geotextile. *Environ. Earth Sci.* **2017**, *76*, 431. [CrossRef]
23. Saidi, T.; Hasan, M.; Amalia, Z.; Salsabila, S. The analysis of the bond strength between natural fiber reinforced polymer (NFRP) sheets and concrete. *Result Eng.* **2023**, *18*, 101124. [CrossRef]
24. Amalia, Z.; Hasan, M.; Saidi, T.; Nadia, S. Bonding Behavior of Pineapple and Silk Fiber-based NFRP on Strenghthened Reinforced Concrete Beams Using Direct Tensile Test. *AIP. Conf. Proc.* **2023**, *2711*, 060005. [CrossRef]
25. *EN 998-2*; Specification for Mortar for Masonry-Part 2: Masonry Mortar. European Committee for Standardization: Brussels, Belgium, 2003.
26. Rahman, A.; Ueda, T. In-Plane shear performance of masonry walls after strengthening by two different FRPs. *J. Compos. Constr.* **2016**, *20*, 04016019. [CrossRef]
27. Stefano, S. Masonry Reinforced with FRP. In *Centre for Infrastructure Engineering Studies (CIES)*; University of Missouri, Engineering Research Lab: Rolla, MO, USA; Available online: https://quakewrap.com/frp%20papers/MasonryReinforcedwithFRPSystems.pdf (accessed on 8 January 2024).
28. Derias, M.; El-Hacha, R. Flexural and Shear Strengthening of Masonry Walls with FRP Composite Materials. In Proceedings of the 2007 International Institute for FRP in Construction, Hong Kong, China, 12–14 December 2007.
29. Zimmermann, T.; Strauss, A.; Bergmeister, K. Structural Behaviour of Low- and Normal-Strength Interface Mortar of Masonry. 8 November 2011. Available online: https://www.researchgate.net/publication/257896156_Structural_behavior_of_low_and_normalstrength_interface_mortar_of_masonry (accessed on 8 November 2011).
30. *EN 1052-3*; Methods of Test for Masonry–Part 3: Determination of Initial Shear Strength. European Committee for Standardization: Brussels, Belgium, 2007.
31. Mojsilovi'c, N. Masonry elements with damp-proof course membrane: Assessment of shear strength parameters. *Constr. Build. Mater.* **2012**, *35*, 1002–1012. [CrossRef]
32. Prakash, S.S.; Alagusundaramoorthy, P. Load resistance of masonry wallets and shear triplets retrofitted with GFRP composites. *Cem. Concr. Compos.* **2007**, *30*, 745–761. [CrossRef]
33. Bompa, D.; Elghazouli, A. Experimental and numerical assessment of the shear behaviour of lime mortar clay brick masonry triplets. *Constr. Build. Materials. Constr. Build. Mater.* **2020**, *262*, 120571. [CrossRef]
34. Page, A.W. Finite element model for masonry. *J. Struc. Div. ASCE* **1978**, *104*, 1267–1285. [CrossRef]
35. D'Altri, A.M.; de Miranda, S.; Castellazzi, G.; Sarhosis, V. A 3D detailed micro-model for the in-plane and out-of-plane numerical analysis of masonry panels. *Comput. Struct.* **2018**, *206*, 18–30. [CrossRef]
36. D'Altri, A.M.; Sarhosis, V.; Milani, G.; Rots, J.; Cattari, S.; Lagomarsino, S.; Sacco, E.; Tralli, A.; Castellazzi, G.; de Miranda, S. Modeling strategies for the computational analysis of unreinforced masonry structures: Review and classification. *Arch. Comput. Methods Eng.* **2020**, *27*, 1153–1185. [CrossRef]
37. Doran, B.; Koksal, H.O.; Aktan, S.; Ulukaya, S.; Oktay, D.; Yuzer, N. In-plane shear behaviour of traditional masonry walls. *Int. J. Archit. Herit.* **2017**, *11*, 278–291. [CrossRef]
38. Doran, B.; Yuzer, N.; Aktan, S.; Oktay, D.; Ulukaya, S. Numerical modelling of traditional masonry walls strengthened with grout injection. *Int. J. Archit. Herit.* **2020**, *14*, 1517–1532. [CrossRef]
39. Doran, B.; Karslioglu, M.; Unsal Aslan, Z.; Vatansever, C. Experimental and Numerical Investigation of Unreinforced Masonry Walls with and without Opening. *Int. J. Archit. Herit.* **2022**, *17*, 1833–1854. [CrossRef]
40. Lourenço, P.B.; Barros, J.O.; Oliveira, J.T. Shear testing of stack bonded masonry. *Constr. Build. Mater.* **2004**, *18*, 125–132. [CrossRef]
41. Lourenço, P.B. Multisurface interface model for analysis of masonry structures. *J. Eng. Mech.* **1997**, *123*, 660–668. [CrossRef]
42. Lourenço, P.B.; De Borst, R.; Rots, J.G. A plane stress softening plasticity model for orthotropic materials. *Int. J. Numer. Methods Eng.* **1997**, *40*, 4033–4057. [CrossRef]
43. Lourenço, P.B.; Rots, J.; Blaauwendraad, J. Continuum model for masonry: Parameter estimation and validation. *J. Struct. Eng.* **1998**, *124*, 642–652. [CrossRef]
44. Ravil, S.; Viswanathan, S.; Nagarajan, T.; Srinivas, V.; Narayanan, P. Experimental and Numerical Investigations on Material Behaviour of Brick Masonry. In Proceedings of the 2nd International Conference on Research in Science, Engineering and Technology (ICRSET), Paris, France, 22–24 November 2019.
45. Prakash, S.S.; Alagusundaramoorthy, P. Study on Brick Masonry Infill Walls with Air Gap. In Proceedings of the Ninth Canadian Conference on Earthquake Engineering, Ottawa, ON, Canada, 26–29 June 2007.

46. ANSYS Inc. *ANSYS User Manuals: Released 18.0*; ANSYS Inc.: Canonsburg, PA, USA, January 2017; Available online: https://users.abo.fi/rzevenho/ansys%20fluent%2018%20tutorial%20guide.pdf (accessed on 8 January 2024).
47. ANSYS User Manual. Element Reference and Structural Analysis Guide, Release 19.2. August 2018. Available online: https://forum.ansys.com/uploads/534/5I3WFK94SATN.pdf (accessed on 8 January 2024).

buildings

MDPI

Article

Nonlinear Coupled Vibration Behavior of BFRP Cables on Long-Span Cable-Stayed Bridges under Parametric Excitation

Yaqiang Yang [1,*], Zixian Zhou [1], Yanlin Guan [1], Jianzhe Shi [2,*], Qiwei Zhan [1], Mohamed F. M. Fahmy [3] and Bitao Wu [4]

1 School of Architecture and Civil Engineering, Jiangsu University of Science and Technology, Zhenjiang 212100, China; 211110901104@stu.just.edu.cn (Z.Z.); 221210901109@stu.just.edu.cn (Y.G.); zhanqw@just.edu.cn (Q.Z.)
2 College of Civil and Transportation Engineering, Hohai University, Nanjing 210098, China
3 Department of Civil Engineering, Faculty of Engineering, Assiut University, Assiut 71515, Egypt; m.fahmy@aun.edu.eg
4 School of Civil Engineering and Architecture, East China Jiao Tong University, Nanchang 330013, China; wubitao@ecjtu.edu.cn
* Correspondence: yangyq@just.edu.cn (Y.Y.); sjz@hhu.edu.cn (J.S.)

Abstract: Based on the cable-stayed beam model, this paper studies the nonlinear coupled vibration behavior of basalt fiber-reinforced polymer (BFRP) cables on long-span cable-stayed bridges under parametric excitation. Considering the sag, damping of BFRP cables, and the coupled interactions between stayed cables and the main girder, the nonlinear coupling vibration model of the BFRP cable–beam composite structure has been established. Taking the longest cable of Sutong Bridge as a case study, the nonlinear coupled vibration behavior of BFRP cables under parametric excitation has been numerically analyzed using the finite difference method. The analysis results indicate that (1) under parametric excitation, the large amplitude nonlinear vibration of the BFRP cable will be induced with an evident "beat" phenomenon. (2) Under the same parametric excitation, the nonlinear coupling vibration response and the beta frequency of the BFRP cable were both smaller than that of the traditional steel cable. (3) The nonlinear coupling vibration response of the BFRP cable increased with an increment in excitation amplitude and a decrement in cable force. With the increase in the excitation frequency, weight per unit length, and axial stiffness, the nonlinear vibration response of the BFRP cable increased first and then decreased. Meanwhile, the damping ratio of the BFRP cable had no significant influence on the nonlinear coupling vibration.

Keywords: BFRP cable; cable–beam composite structure; nonlinear coupled vibration; parametric excitation; long-span cable-stayed bridge

1. Introduction

With the increase in the span capacity of cable-stayed bridges, the disadvantages of traditional steel cables, including large self-weight, pronounced sag effect, and low carrying efficiency, have become more serious [1]. Due to the limitation of construction technology and economic performance, a main span of 1300 m has become the bottleneck of the long-span cable-stayed bridge [2]. In addition, the severe air pollution and increasing traffic load have aggravated the fatigue degradation and corrosion damage of the traditional steel cable, which cannot meet the requirements of the long-span cable-stayed bridge for long spans, long life, and high durability. It is necessary to develop a new material cable with superior mechanical and chemical properties and economic performance to apply on the long-span cable-stayed bridge to overcome the challenges in span and durability of bridge engineering [3]. Fiber-reinforced polymer (FRP for short) materials have been widely used in repairing existing structures and reinforcing new structures due to their high strength-to-weight ratio, good anti-fatigue properties, and superior corrosion resistance during the past 30 years [4]. Manufacturing stayed cables with FRP materials can not only take full advantage of FRP materials such as high strength and lightweight qualities, good fatigue resistance, and superior anti-corrosion properties, but it also avoids the FRP material shortcomings of poor shearing performance. Therefore, the application of FRP cable on long-span cable-stayed bridges can increase span and service life significantly, which will provide an effective way to improve the short- and long-term performance of long-span cable-stayed bridges. Due to the superior mechanical and chemical properties, carbon fiber-reinforced polymer (CFRP for short) material has become the first choice for making FRP cables for long-span cable-stayed bridges [5]. However, the high cost and significant wind sensitivity of CFRP cable restrict its application on long-span cable-stayed bridges. Taking mechanical properties and economic performance into consideration, the feasibility of the other FRP materials used to manufacture FRP cables has been studied. Compared with the other FRP materials, basalt fiber-reinforced polymer (BFRP for short) has shown its mechanical and economic advantages, such as high strength, significant elongation, good fatigue resistance, superior anti-corrosion property, good chemical stability, and low cost [6]. With no emissions of harmful gases during manufacture, the BFRP cable has been regarded as one of the green inorganic fiber materials. With its outstanding advantages, theoretical and experimental studies on the mechanical behavior and structural performance of BFRP cables on long-span cable-stayed bridges have been carried out [7]. According to the existing studies [8], anchorage of BFRP cables with multiple tendons and large tonnage can be achieved using the anchorage method with a continuous fiber-reinforced load transfer component. Applying BFRP cable on long-span cable-stayed bridges can not only meet the requirement of bridge structure design but also reduce the axial force of the girder. In addition, BFRP cables can help to suppress the significant amplitude vibration of cables and optimize the economic performance of long-span cable-stayed bridges [9]. With the improvement in manufacturing and the accumulation of engineering experience, BFRP cables have broad prospects of application on long-span cable-stayed bridges.

With a quarter of the weight per unit length of the traditional steel cable, the BFRP cable is more sensitive to external excitation [10]. Under the same load excitation, the amplitude of vibration acceleration will be significantly larger than that of the steel cable [11], which shows the high sensitivity of the FRP cable under external excitation. According to existing studies [12,13], the large-amplitude vibration of the stayed cable will be induced under parametric excitation, whose excitation frequency is twice the natural frequency of the stayed cable. However, the results of the analysis for the stayed cable do not agree with the actual vibration behavior on long-span cable-stayed bridges due to the overlooking of the coupled vibration interaction between the cable and the deck of the long-span cable-stayed bridge. It is necessary to analyze the parametric vibration behavior of the FRP cable on a long-span cable-stayed bridge considering the coupled vibration interaction. For the cable-stayed bridges, masts, and suspended roofs, the cable-stayed beam model has been widely adopted in coupled vibration analysis. Through a mass block [14], simply supported beam [15], or cantilever beam [16] simulating the bridge deck, reduced cable–beam models were employed to study the nonlinear coupled behavior of the stayed cable on a long-span cable-stayed bridge. Based on the cable-stayed beam model, the coupled vibration responses were constructed with their vibration characteristics, influence factors, and the coupling effects on the cable. With the reduced cable–beam model (cable–mass block model), Guo [17,18] investigated the cable's forced nonlinear vibration that was excited by beam oscillation and numerically analyzed the influence of the deck/cable mass ratio, cable inclination, boundary damping, and internal detuning on the nonlinear vibration responses. According to the cable–simple support beam model, Kang [19] and Cong [20] explored the spatial dynamic theory and analyzed the multimodal interaction of the system, which deduced a wealth of behavior from the results, including the saddle-node bifurcation, Hopf bifurcation, and veering phenomenon. Based on the cable–cantilever beam, Cong [21] clarified that the cable experienced a more significant response than the beam, whether excitations were applied to the beam or the cable. In addition, Mohammad [22] discussed the effect of bending stiffness and sag of the cable on the system dynamics through theoretical analysis and vibration experiments. With regard to the numerical simulation methods for reproducing the nonlinear behavior of cable elements, there are various numerical strategies such as the multi-truss element approach and the catenary element model. Using the multi-truss element approach, Lonetti [23] proposed a numerical model for predicting the structural integrity of a self-anchored cable-stayed suspension bridge considering both geometric and material nonlinearities. With the catenary element model, Zhou [24] numerically assessed the load effects on the prototype of a long-span cable-stayed bridge and revealed that the design live load from AASHTO may be unsafe for use as the live load. Although phase achievements in the nonlinear vibration behavior of the cable–beam composite structure have been made, the coupled effect between the stayed cable and the beam has not been taken into full consideration, resulting in an inaccurate description of the parametric vibration behavior of the stayed cables. Meanwhile, the existing studies of parametric vibration cannot be directly applied to BFRP cables due to the changing conditions of nonlinear vibration and the complex mechanism of damping the dissipation energy of BFRP cables. It is necessary to establish the nonlinear coupling vibration model of the BFRP cable–beam composite structure to study the nonlinear vibration behaviors of BFRP cables on the long-span cable-stayed bridges under parametric excitation. Taking the nonlinear factors of BFRP cables and coupling effects between the stayed cable and beam into consideration, this paper proposed the nonlinear coupled vibration model. It carried out the theoretical and numerical studies of BFRP cables on the long-span cable-stayed bridges under parametric excitation. To better understand the nonlinear parametric vibration of BFRP cables, the key influencing factors of the nonlinear vibration behavior of BFRP cables on long-span cable-stayed bridges under parametric excitation have been discussed in this paper.

2. Nonlinear Coupling Vibration Model of the BFRP Cable–Beam

In cable-stayed bridges, the vibration behaviors of stayed cables are usually influenced by the girder and the tower. Compared with the girder, the impact of the tower on stayed cables can be negligible due to the significant stiffness of the tower. As shown in Figure 1, the cable–beam model has been employed to study the nonlinear coupling vibration behavior of the BFRP cables on long-span cable-stayed bridges [25]. According to the Cartesian coordinate system, the vibration of the girder and stayed cable can be described by $O_cX_cZ_c$ and $O_bX_bZ_b$ in the coordinates of the cable–beam model, where w_c and u_c indicate the in-plane and axial displacement of BFRP cable, respectively, and w_b and u_b imply the in-plane and axial displacement of the girder, respectively. By ignoring the secondary factors, the following basic assumptions were proposed to study the nonlinear vibration of BFRP cables on long-span cable-stayed bridges:

(1) During the whole vibration process of the cable–beam composite structure, the BFRP cable and girder are in a linear elastic state;

(2) Under the effect of gravity, the structural configuration of the BFRP cable can be described by the parabolic equation. Taking the sag into consideration, the structural shape of the BFRP cable can be indicated as in Equation (1), in which mc and H_c represent the mass per unit length and cable force of the BFRP cable, respectively; x_c and z_c represent the axial and transverse coordinates of the BFRP cable in the $O_cX_cZ_c$ Cartesian coordinate system; g denotes the gravity acceleration; θ indicates the angle between the girder and the stayed cable; l_c and l_b imply the length of the girder and of the BFRP cable, respectively;

$$z_c(x_c) = \frac{m_c g \cos\theta}{2H_c} x_c(l_c - x_c) \tag{1}$$

(3) Compared with transverse acceleration and axial stiffness, the axial acceleration, shear stiffness, torsional stiffness, and flexural stiffness of the BFRP cable have no

significant influence on the nonlinear vibration behaviors, which can be ignored during the analysis;

(4) Due to the small axial and shear deformation, only the flexural deformation of the girder has been taken into consideration.

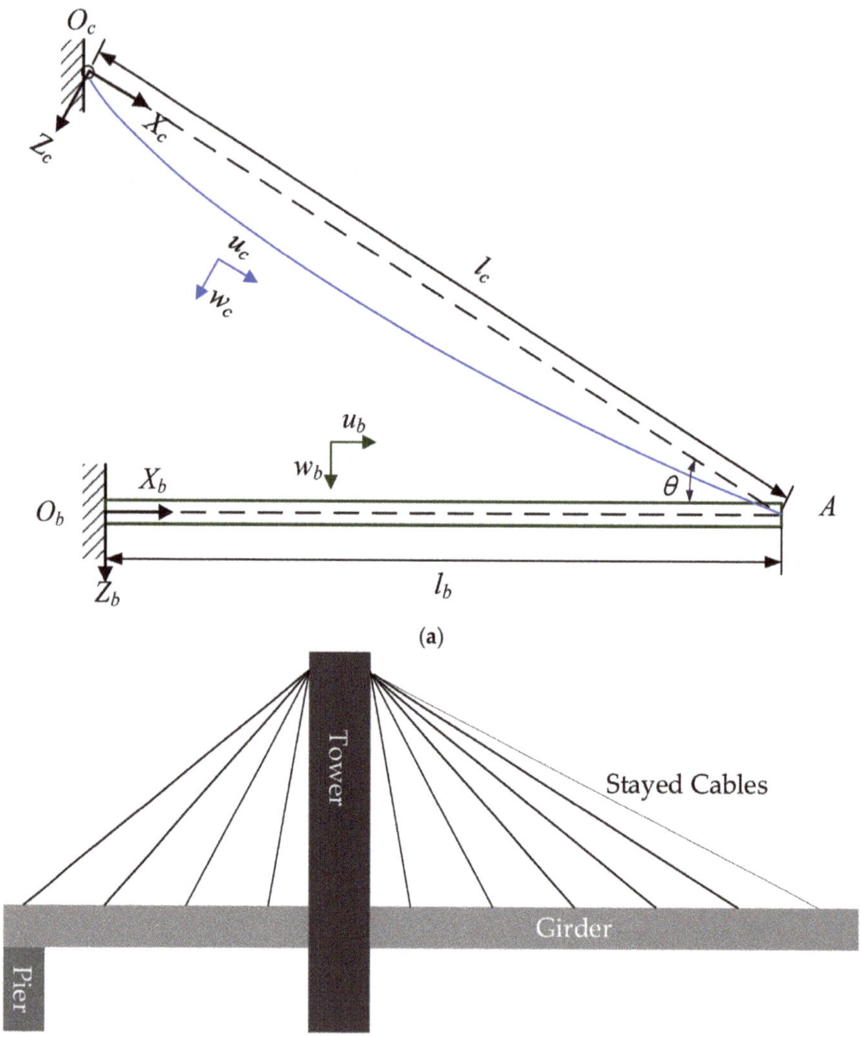

Figure 1. Cable–beam composite structure: (**a**) model structure; (**b**) actual structure.

Based on the above assumptions, the nonlinear vibration differential equation of the BFRP cable and girder can be established according to the D'Alembert principle. As shown in Equations (2) and (3), $w_c(x_c,t)$ and w_b denote the transverse vibration displacement of the BFRP cable and the vibration displacement of the girder, respectively; E_cA_c and E_bI_b represent the axial stiffness of the BFRP cable and the flexural stiffness of the girder; f_c and f_b represent the distribution load of the BFRP cable and the girder, respectively; c_c and c_b indicate the damping coefficients of the BFRP cable and the girder, respectively; m_b and

x_b imply the mass per unit length and transverse coordinates the in $O_bX_bZ_b$ system of the girder, respectively; L_e states the virtual length of the BFRP cable.

$$\begin{gathered} m_c\frac{\partial^2 w_c}{\partial t^2} + c_c\frac{\partial w_c}{\partial t} - H_c\frac{\partial^2 w_c}{\partial {x_c}^2} + \frac{E_cA_c}{L_e}\left(\frac{m_cg\cos\theta}{H_c}\right)^2\int_0^{l_c} w_c dx_c + \frac{E_cA_c}{L_e}\frac{m_cg\cos\theta}{H_c}\cdot \\ \int_0^{l_c}\left(\frac{\partial w_c}{x_c}\right)^2 dx_c - \frac{E_cA_c}{L_e}\left[u_c(l_c) - \frac{m_cgl_c\cos\theta}{2H_c}w_c(l_c) + \frac{m_cg\cos\theta}{H_c}\int_0^{l_c} w_c dx\right]\frac{\partial^2 w_c}{\partial {x_c}^2} \\ = f_c - \frac{E_cA_c}{L_e}\frac{m_cg\cos\theta}{H_c}\left[u_c(l_c) - \frac{m_cgl_c\cos\theta}{2H_c}w_c(l_c)\right] \end{gathered} \tag{2}$$

$$m_b\frac{\partial^2 w_b}{\partial t^2} + c_b\frac{\partial w_b}{\partial t} + H_c\cos\theta\frac{\partial^2 w_b}{\partial {x_b}^2} + E_bI_b\frac{\partial^4 w_b}{\partial {x_b}^4} = f_b \tag{3}$$

To obtain the nonlinear vibration responses, the boundary condition and displacement continuous condition of the BFRP cable and the girder should be proposed based on the cable–beam composite model.

Compared with only considering the vibration behavior of the BFRP cable, the coupling interaction between the BFRP cable and the girder has a significant influence on the nonlinear vibration responses of the BFRP cable–beam composite structure, which can result in changes in the cable force and in the displacement/rotation between the girder and BFRP cable end. Therefore, the nonlinear vibration differential equation of the BFRP cable–beam composite structure should also satisfy the dynamic equilibrium condition induced by the above coupling interaction, expressed in Equation (4), in which h_c represents the additional cable force generated during the nonlinear vibration of BFRP cable, and φ_{c0} denotes the initial rotation of the BFRP cable end.

$$E_bI_b\frac{\partial^3 w_b}{\partial {x_b}^3}(l_b) = h_c(\sin\theta - \cos\theta\tan\varphi_{c0}) - H_c\frac{\partial w_b}{\partial x_b}\cos\theta + H_c\frac{\partial w_c}{\partial x_c}\cos\theta \tag{4}$$

To simplify the nonlinear vibration analysis, dimensionless parameters were introduced. Based on the finite difference method, the numerical algorithm of the nonlinear coupling vibration response of the cable–beam composite structure was proposed and is shown in Figure 2.

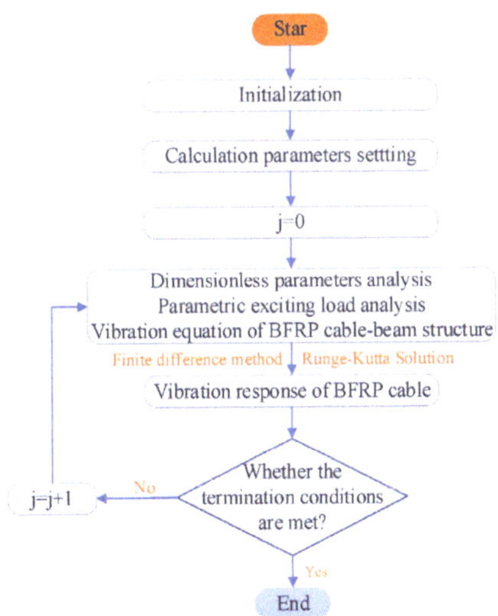

Figure 2. Algorithmic program of nonlinear coupling vibration response for BFRP cable–beam model.

3. Nonlinear Vibration Study of BFRP Cable–Beam Composite Structures under Parametric Excitation

Based on the above nonlinear coupling vibration model, the longest stayed cable on the Sutong Bridge in China, shown in Figure 3, was selected as an example to discuss the nonlinear coupling vibration behavior of the BFRP cable on a long-span cable-stayed bridge under parametric excitation [26]. Compared to short cables, the nonlinear coupling vibration of the long cable has significant effects on the dynamic performance of the long-span cable-stayed bridge. Therefore, the longest stayed cable of Sutong Bridge was taken as the research object. In Table 1, the structural parameters of the prototype cable have been listed. Through considering mechanical properties and economic performance [27], the section parameter of the BFRP cable was designed in terms of the strength principle, as shown in Equation (5), in which $F_{d,S}$ and $F_{d,B}$ denote the design cable force of the steel cable and the BFRP cable, respectively; $f_{s,B}$ represents the tensile strength of the material design strength of the cable; S_B and A_B indicate safety factor and sectional area of the BFRP cable, respectively. The structural parameters of the BFRP cable are also listed in Table 1.

$$\{F_{d,B} = F_{d,S}; f_{d,B} = \frac{f_{s,B}}{S_B}; A_B = \frac{F_{d,B}}{f_{d,B}}\} \Rightarrow A_B = \frac{F_{d,B} \cdot S_B}{f_{s,B}} \tag{5}$$

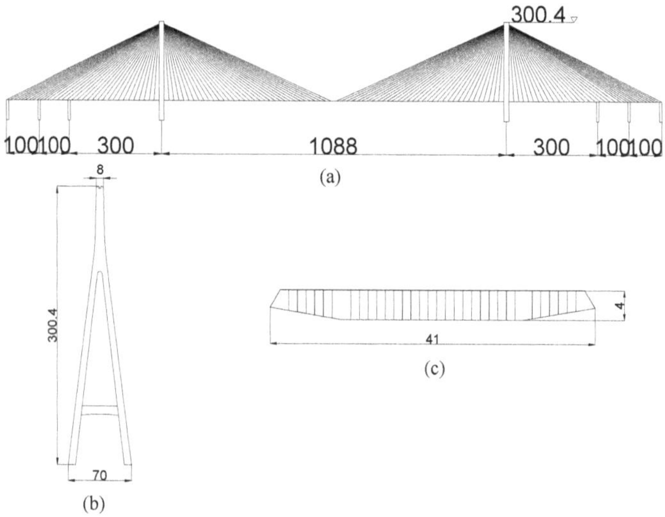

Figure 3. Sutong Bridge structure diagram: (**a**) layout of the whole bridge; (**b**) tower; (**c**) girder.

Table 1. Parameter of material and structure of steel cable and BFRP cable.

Cable Type	Density (kg/m^3)	Length (m)	Inclination (°)	Elastic Modulus (GPa)	Tensile Strength (MPa)	Safety Factor	Design Cable Force (10^7 N)	Sectional Area (10^{-2} m^2)	Damping Ratio (%)	Natural Frequency (Hz)
HS	7850	576	23	210	1860	2.5	1.19	1.89	0.17	0.25
BFRP	1739	576	23	58.23	1470	4.0	1.19	3.24	0.78	0.40

In Table 1, the natural frequency (ω) of the stayed cable can be derived based on the theoretical formula for string vibration, as shown in Equation (6). Meanwhile, the experimental results of the damping ratios of the steel cable and the BFRP cable were adopted in this analysis [25]. As shown in Figure 3, the girder of the Sutong Bridge is the steel box beam made of Q345 and Q370 grade steel, whose sectional parameters are listed in Table 1.

$$\omega = \frac{1}{2l_c}\sqrt{\frac{H_c}{m_c}} \tag{6}$$

Once the exciting frequency from the external environment achieves twice the natural frequency of the stayed cable, the large-amplitude parametric vibration of the cable will be induced. Therefore, the nonlinear vibration behavior of the BFRP cable on the long-span cable-stayed bridge under parametric excitation was studied by applying harmonic loads with the twice the natural frequency of the stayed cable, which can be expressed using dimensionless parameters such as $f_c = 0, f_b = 10^{-5}\, sin\, (\pi x_b)\, cos\, (2\, t)$. Based on the proposed numerical algorithm, the nonlinear vibration response of the steel cable and the BFRP cable has been analyzed. As shown in Figure 4, the displacement–time curves of the midpoint and endpoint of the BFRP cable under parametric excitation are compared with that of the traditional steel cable. It indicates that a significant amplitude vibration can be induced by parametric excitation both for the steel cable and the BFRP cable. In addition, there are apparent beating vibration phenomena during the nonlinear vibration process of the steel cable and the BFRP cable. Furthermore, the vibration response of the midpoint is more significant than that of the endpoint both for the steel cable and the BFRP cable under parametric excitation. However, the parametric vibration amplitudes of the midpoint and endpoint of the BFRP cable are about 44% and 8% less than that of the steel cable, respectively, as shown in Table 2. Furthermore, the frequency of the beat vibration of the BFRP cable is lower than that of the steel cable. In addition, the beat vibration of the steel cable maintained a presence and had no fundamental amplitude decay or frequency decay under the parametric excitation. However, the beat frequency of the BFRP cable decreased gradually with the increase in parametric excitation. With the reduction in the vibration amplitude, the parametric vibration of the BFRP cable decayed to a steady amplitude. This is because the BFRP cable as a composite material is endowed with viscoelastic properties and vibration absorption. To sum up, applying BFRP cables on long-span cable-stayed bridges can effectively reduce the parametric vibration response and improve the vibration stability of the stayed cable under the parametric excitation.

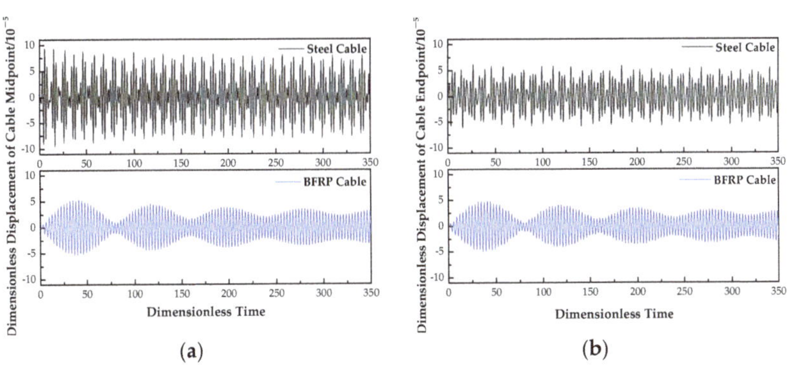

Figure 4. Nonlinear coupled vibration response of steel cable and BFRP cable under parametric excitation: (**a**) displacement–time curves of midpoint of stayed cable; (**b**) displacement–time curves of endpoint of stayed cable.

Table 2. Maximum amplitude of steel cable and BFRP cable under parametric excitation.

Cable Type	Midpoint ($\times10^{-5}$)	Endpoint ($\times10^{-5}$)
Steel Cable	9.46	5.34
BFRP Cable	5.34	4.91

4. Analysis of Influence Factors

According to the nonlinear coupling vibration model of the BFRP cable–beam, the factors that influence the parametric vibration behavior of the BFRP cable include the frequency and amplitude of the external excitation, the mass per unit length, the damping ratio, the axial stiffness, and the cable force of the BFRP cable. To explore the critical factors that influence the regularity of the parametric vibration behavior of the BFRP cable, the nonlinear coupling vibration behavior of the BFRP cable under parametric excitation has been studied by changing the influencing factors and conducting numerical analysis.

4.1. Excitation Frequency

Keeping the other parameters unchanged, the nonlinear coupling vibration responses of the midpoint and endpoint of the BFRP cable was obtained under three different excitation frequencies, which were applied on the BFRP cable–beam composite structure and expressed using dimensionless parameters such as $f_c = 0, f_b = 10^{-5}\ sin\ (\pi x_b)\ cos\ (t); f_c = 0$, $f_b = 10^{-5}\ sin\ (\pi x_b)\ cos\ (2\ t); f_c = 0, f_b = 10^{-5}\ sin\ (\pi x_b)\ cos\ (3\ t)$. As shown in Figure 5, the nonlinear coupling vibration responses of the BFRP cable under different excitation frequencies are compared. It can be seen from Figure 5 that the vibration amplitude of the BFRP cable first increases and then decreases with the increase in the excitation frequency. In particular, the vibration amplitude achieves the maximum value (5.34×10^{-5} or 4.91×10^{-5}) with the most obvious beat vibration when the excitation frequency is twice the natural frequency of the BFRP cable, as shown in Figure 5 and Table 3. This is because the external excitation will continuously input vibration energy and focus on the cable vibration when the excitation frequency is twice the natural frequency. In addition, the maximum vibration amplitude of the BFRP cable exciting at three times the natural frequency was about 30% less than that excited by the natural frequency, as shown in Table 3. Moreover, there was no significant beat vibration once the BFRP cable was excited by natural frequency and triple natural frequency, excluding the vibration responses of the midpoint of the BFRP cable under triple natural frequency. In conclusion, it is prone to induce a large-amplitude parametric vibration for the BFRP cable when the excitation frequency reaches twice the natural frequency of the BFRP cable. Therefore, effective measures should be taken to suppress the large-amplitude parametric vibration.

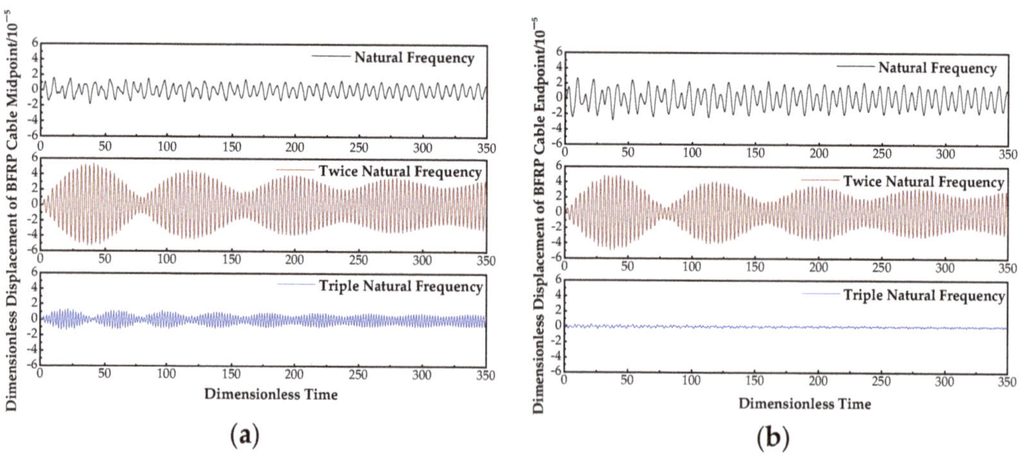

Figure 5. Nonlinear coupled vibration response of BFRP cable under parametric excitation with different frequencies: (**a**) nonlinear vibration response of midpoint; (**b**) nonlinear vibration response of endpoint.

Table 3. Maximum amplitude of BFRP cable with different frequencies.

Frequency	Midpoint ($\times 10^{-5}$)	Endpoint ($\times 10^{-5}$)
$f_c = 0, f_b = 10^{-5}\ sin\ (\pi x_b)\ cos\ (t)$	1.79	2.80
$f_c = 0, f_b = 10^{-5}\ sin\ (\pi x_b)\ cos\ (2\ t)$	5.34	4.91
$f_c = 0, f_b = 10^{-5}\ sin\ (\pi x_b)\ cos\ (3\ t)$	1.37	2.97

4.2. Excitation Amplitude

At the parametric excitation frequency (twice the natural frequency), the excitations with different amplitudes were applied to the BFRP cable–beam composite structure as the input loads to discuss the influence of the excitation amplitude on the parametric vibration response of the BFRP cable. There were three—large, medium, and small—amplitude parametric excitations separately acting on the BFRP cable–beam model, which can be referred to, respectively, as $f_c = 0$, $f_b = 5 \times 10^{-6}\ sin\ (\pi x_b)\ cos\ (2\ t)$; $f_c = 0$, $f_b = 10^{-5}\ sin\ (\pi x_b)\ cos\ (2\ t)$; and $f_c = 0$, $f_b = 2 \times 10^{-5}\ sin\ (\pi x_b)\ cos\ (2\ t)$ according to dimensionless parameters. The parametric vibration responses of the midpoint and endpoint on the BFRP cable under different vibration amplitudes are compared in Figure 6 and Table 4. As shown in Figure 6, the parametric vibration responses of the BFRP cable increased with the increase in the excitation amplitude. In addition, an increment in the maximum amplitude of the BFRP cable was nearly the same as that of the excitation amplitude as shown in Table 4. Furthermore, it would take more time for the BFRP cable to achieve a stable vibration amplitude when the excitation amplitude increased. Consequently, more attention should be paid to the large-amplitude parametric vibration of the BFRP cable used on long-span cable-stayed bridges.

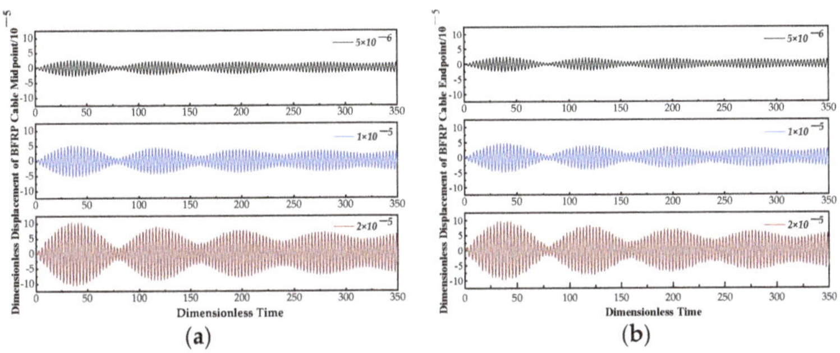

Figure 6. Nonlinear coupled vibration response of BFRP cable under parametric excitation with different vibration amplitudes: (**a**) nonlinear vibration response of the midpoint; (**b**) nonlinear vibration response of the endpoint.

Table 4. Maximum amplitude of BFRP cable with different vibration amplitudes.

Vibration Amplitude	Midpoint ($\times 10^{-5}$)	Endpoint ($\times 10^{-5}$)
$f_c = 0, f_b = 5 \times 10^{-6}\ sin\ (\pi x_b)\ cos\ (2\ t)$	2.67	2.46
$f_c = 0, f_b = 10^{-5}\ sin\ (\pi x_b)\ cos\ (2\ t)$	5.34	4.91
$f_c = 0, f_b = 2 \times 10^{-5}\ sin\ (\pi x_b)\ cos\ (2\ t)$	10.69	9.82

4.3. Section Parameters

For stayed cable, the section parameter, such as mass per unit length or axial stiffness, changes with the section area of the cable. Proportional to the section area, the mass per unit length and axial stiffness of the BFRP cable increase according to the section area of the BFRP cable. By changing the section area of the BFRP cable, the effects of mass per unit length and axial stiffness on the nonlinear coupling vibration behavior of the BFRP cable can be studied. Three representative sections of the BFRP cable were used to discuss the vibration response of the BFRP cable under parametric excitation, which involved 2.43×10^{-2} m^2, 3.24×10^{-2} m^2, and 4.05×10^{-2} m^2. The parametric vibration–time history curves of the BFRP cable with different section areas are shown in Figure 7 and Table 5. It is indicated from Figure 7 and Table 5 that the vibration amplitude of the BFRP cable under parametric excitation increased and then decreased with the increase in the mass per unit length and axial stiffness of the cable. However, with a rise in the mass per unit length and axial stiffness of the cable, the frequency of the beat vibration of the BFRP cable under parametric excitation decreased firstly and then increased. As the ratio of the mass per unit length and axial stiffness of the BFRP cable approximates 1:1.3:1.7, the ratio of the beat vibration frequency of the BFRP cable under parametric excitation approaches 4:1:2.

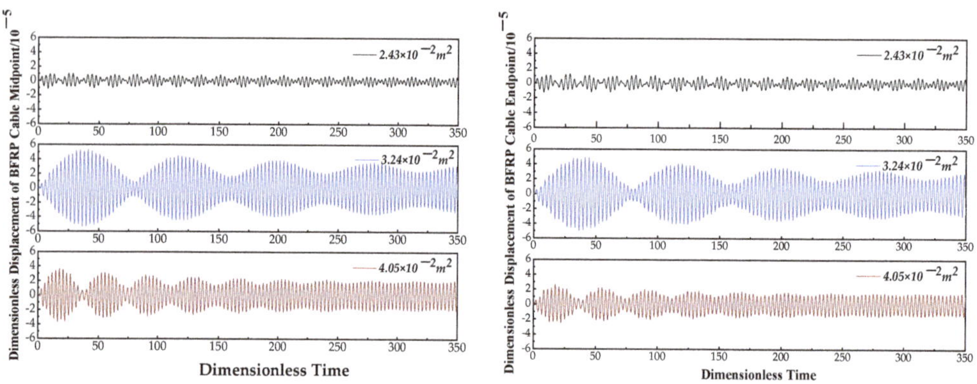

Figure 7. Nonlinear coupled vibration response of BFRP cable under parametric excitation at different section areas: (**a**) nonlinear vibration response of the midpoint; (**b**) nonlinear vibration response of the endpoint.

Table 5. Maximum amplitude of BFRP cable at different section areas.

Section Area	Midpoint ($\times 10^{-5}$)	Endpoint ($\times 10^{-5}$)
2.43×10^{-2} m^2	1.11	1.28
3.24×10^{-2} m^2	5.34	4.91
4.05×10^{-2} m^2	3.66	2.61

4.4. Cable Force

As one of the critical influencing factors, variation in the cable force would have significant effects on the vibration behavior of the BFRP cable. Under the condition of constant parametric excitation, the nonlinear vibration responses of the BFRP cable were analyzed by applying large, medium, and small cable forces, which were 1.01×10^7 N, 1.19×10^7 N, and 1.37×10^7 N, respectively. The results of the analysis of the midpoint and endpoint of the BFRP cable with different cable forces are compared in Figure 8 and Table 6. As shown in Figure 8 and Table 6, the nonlinear parametric vibration response of the midpoint of the BFRP decreased with the increase in the cable force. For the parametric vibration of the endpoint of the BFRP cable when applying different cable forces, it increased slightly initially and decreased obviously afterward as shown in Table 6. However, the beat frequency of the parametric vibration of the BFRP cable decreased at first and then increased with the increase in the cable force both for the midpoint and the endpoint of the cable, as shown in Figure 8.

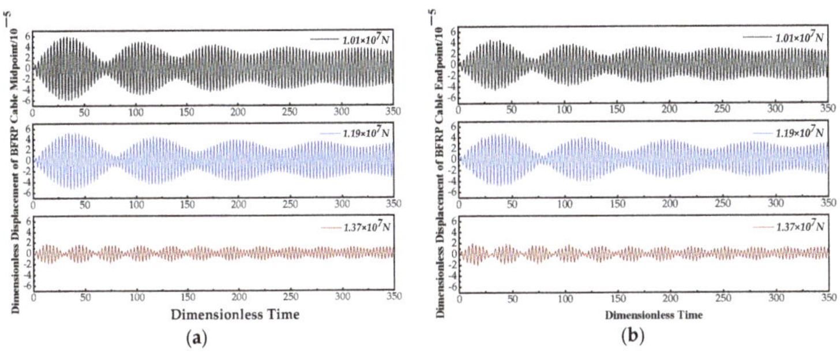

Figure 8. Nonlinear coupled vibration response of BFRP cable under parametric excitation with different cable forces: (**a**) nonlinear vibration response of the midpoint; (**b**) nonlinear vibration response of the endpoint.

Table 6. Maximum amplitude of BFRP cable with different cable forces.

Cable Force	Midpoint ($\times 10^{-5}$)	Endpoint ($\times 10^{-5}$)
1.01×10^7 N	5.94	4.61
1.19×10^7 N	5.34	4.91
1.37×10^7 N	1.84	1.99

4.5. Damping Ratio

As a key factor in vibration attenuation and energy dissipation, damping ratio has an essential influence on the nonlinear vibration behavior of the stayed cable on long-span cable-stayed bridges. To study the influence regularity of damping ratio on the parametric vibration behavior, BFRP cables with three different damping ratios relating to 7.8×10^{-4}, 7.8×10^{-3}, and 7.8×10^{-2} were employed to carry out the nonlinear vibration analysis. Under similar parametric excitation, the vibration time–history responses of the BFRP cable with different damping ratios are compared in Figure 9. It can be seen from Figure 9 that there were no significant differences in the parametric vibration amplitude and frequency of the midpoint and endpoint of the BFRP cable with different inherent damping ratios as shown in Table 7. Once the large amplitude parametric vibration is induced, the intrinsic damping ratio has no significant effect on suppressing the nonlinear vibration of the BFRP cable under parametric excitation. It is necessary to install additional dampers to control the large-amplitude parametric vibration for the BFRP cable on long-span cable-stayed bridges.

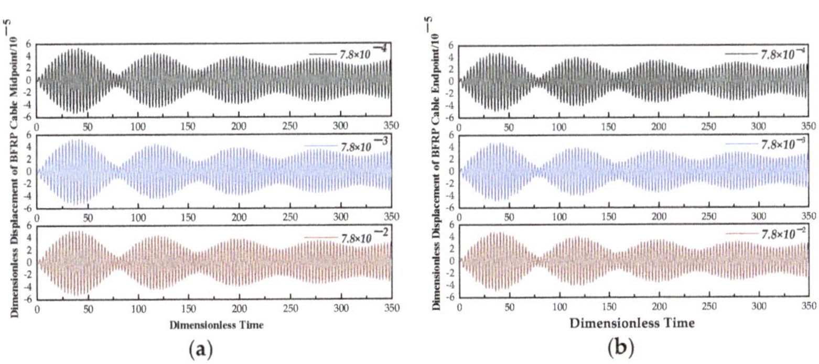

Figure 9. Nonlinear coupled vibration response of BFRP cable under parametric excitation with different damping ratios: (**a**) nonlinear vibration response of the midpoint; (**b**) nonlinear vibration response of the endpoint.

Table 7. Maximum amplitude of BFRP cable with different damping ratios.

Damping Ratio	Midpoint ($\times 10^{-5}$)	Endpoint ($\times 10^{-5}$)
7.8×10^{-4}	5.35	4.91
7.8×10^{-3}	5.34	4.91
7.8×10^{-2}	5.29	4.90

Based on the analysis of the influencing factors of the parametric vibration behavior of the BFRP cable under parametric excitation, the external excitation and mechanical characteristics of BFRP cables have significant effects on the nonlinear coupled vibration of BFRP cables on long-span cable-stayed bridges. Therefore, effective measures should be taken to suppress the nonlinear coupled vibration of BFRP cables once applied on long-span cable-stayed bridges in the future. On the one hand, the girder section and restriction damping of the cable-stayed bridges should be properly designed to change the natural frequency and amplitude of the girder, which could help BFRP cables avoid the natural frequency and amplitude of parametric vibration. On the other hand, the external damper and cross-tie cable should be applied on BFRP cables to improve the mechanical performance including the free length, natural frequency, and stiffness of the cable system, which could reduce the response of the parametric vibration.

5. Conclusions

Based on the cable–beam composite structure, the nonlinear coupling vibration model of BFRP cables on long-span cable-stayed bridges was proposed. The longest-stayed cable of the Sutong Bridge was taken as a case study to investigate the nonlinear coupling vibration behavior of the BFRP cable under the parametric excitation according to the BFRP cable–beam model. Through influence parameter analysis, the vibration regulations of the BFRP cable under parametric excitation were studied. Based on these analyses, the following major conclusions were drawn:

(1) Under parametric excitation, large-amplitude and significant beat vibration would be induced both for the steel cable and the BFRP cable. However, the frequency of the parametric vibration of the BFRP cable was considerably less than that of the steel cable. Furthermore, the maximum amplitude of the BFRP cable was up to 44% less than that of the steel cable due to superior mechanical performance.

(2) The maximum vibration amplitude with evident beat vibration would be obtained once the excitation frequency is twice the natural frequency of the BFRP cable. In addition, the larger the excitation amplitude was, the greater the nonlinear parametric vibration responses of the BFRP cable became. The increment in the maximum amplitude of the BFRP cable was nearly the same as that of the excitation amplitude.

(3) With an increase in the mass per unit length and axial stiffness, the vibration amplitude of the BFRP cable under parametric excitation increased firstly and then decreased, while the frequency of the beat vibration decreased firstly and then increased.

(4) The nonlinear parametric vibration response of the BFRP decreased with an increase in the cable force. The inherent damping ratio of the BFRP cable had no significant influence on the parametric vibration behavior.

(5) Parametric vibration control has become the key issue for the application of BFRP cables on long-span cable-stayed bridges. Establishing parametric vibration control theory and developing parametric vibration control methods will be the development direction of future research.

Author Contributions: Conceptualization, Y.Y. and Z.Z.; methodology, Y.Y. and B.W.; software, Y.Y., Q.Z. and Y.G.; validation, Y.Y., Z.Z. and M.F.M.F.; formal analysis, Y.Y.; investigation, Y.Y. and B.W.; resources, Y.Y. and Q.Z.; data curation, Y.Y.; writing—original draft preparation, Y.Y.; writing—review and editing, Y.Y. and M.F.M.F.; visualization, Y.Y. and J.S.; supervision, Y.Y.; project administration, Y.Y.; funding acquisition, Y.Y., Q.Z. and J.S. All authors have read and agreed to the published version of the manuscript.

Funding: This research was funded by National Natural Science Foundation of China (Grant Number 51808265); Natural Science Foundation of the Jiangsu Higher Education Institutions of China (Grant Number 18KJB560005); Key Research and Development of Shandong Province (Grant Number 2019GSF111013); Postgraduate Research & Practice Innovation Program of Jiangsu Province (Grant Number KYCX22_3831); and Key Research & Development Project of Zhenjiang City (Industry Foresight and Common Key Technologies) (Grant Number GY2023028).

Data Availability Statement: The data presented in this study are available in the article.

Conflicts of Interest: The authors declare no conflict of interest.

References

1. Wang, Z.; Xian, G. Impact performances of fiber reinforced polymer composites and cables: A review. *Compos. Struct.* **2023**, *319*, 117128. [CrossRef]
2. Wu, Z.; Liu, J.; Zou, D.; Wang, X.; Shi, J. Status and Development Trend of Light-Weight, High-Strength, and Durable Structural Materials Applied in Marine Bridge Engineering. *Strateg. Study CAE* **2019**, *21*, 31–40. [CrossRef]
3. Guo, R.; Li, C.; Xian, G. Water absorption and long-term thermal and mechanical properties of carbon/glass hybrid rod for bridge cable. *Eng. Struct.* **2023**, *274*, 601–617. [CrossRef]
4. Wei, Y.; Bai, J.; Zhang, Y.; Miao, K.; Zheng, K. Compressive performance of high-strength seawater and sea sand concrete-filled circular FRP-steel composite tube columns. *Eng. Struct.* **2021**, *240*, 112357. [CrossRef]
5. Zhou, J.; Wang, X.; Ding, L.; Liu, S.; Wu, Z. Numerical and Experimental Study on Large-Diameter FRP Cable Anchoring System with Dispersed Tendons. *Buildings* **2023**, *13*, 92. [CrossRef]
6. Macro, D.; Nicola, N. A Split-Wedge Anchorage for CFRP Cables: Numerical Model vs. Experimental Results. *Polymers* **2022**, *14*, 2675. [CrossRef]
7. Yang, Y.; Fahmy, M.F.M.; Guan, S.; Pan, Z.; Zhan, Y.; Zhao, T. Properties and applications of FRP cable on long-span cable-supported bridges: A review. *Compos. Part B Eng.* **2020**, *190*, 107934. [CrossRef]
8. Shi, J.; Wang, X.; Zhang, L.; Wu, Z.; Zhu, Z. Composite-Wedge Anchorage for Fiber-Reinforced Polymer Tendons. *J. Compos. Constr.* **2022**, *26*, 04022005. [CrossRef]
9. Zhou, J.; Wang, X.; Peng, Z.; Wu, Z.; Wei, X. Enhancement of FRP Cable Anchor System: Optimization of Load Transfer Component and Full-Scale Cable Experiment. *J. Compos. Constr.* **2022**, *26*, 801–815. [CrossRef]
10. Kim, T.; Jung, W. Improvement of Anchorage Performance of Carbon Fiber-Reinforced Polymer Cables. *Polymer* **2022**, *14*, 1239. [CrossRef]
11. Yang, Y.; Fahmy, M.F.M.; Pan, Z.; Guan, S.; Zhan, Y. Analytical estimation on damping behaviors of the Self-Damping fiber reinforced polymer (FRP) cable. *Structures* **2020**, *25*, 774–784. [CrossRef]
12. Wang, X.; Wu, Z. Vibration control of different FRP cables in long-span cable-stayed bridge under indirect excitations. *J. Earthq. Tsunami* **2011**, *5*, 167–188. [CrossRef]
13. Mei, K.H.; Lu, Z.T.; Sun, S.J. Property of Nonlinear Parametric Vibration of CFRP Cables. *China J. Highw. Transp.* **2007**, *17*, 43–45.
14. Guo, T.; Kang, H.; Wang, L.; Zhao, Y. An asymptotic expansion of cable–flexible support coupled nonlinear vibrations using boundary modulations. *Nonlinear Dyn.* **2017**, *88*, 33–59. [CrossRef]
15. Su, X.; Kang, H.; Guo, T.; Cong, Y. Dynamic analysis of the in-plane free vibration of a multi-cable-stayed beam with transfer matrix method. *Arch. Appl. Mech.* **2019**, *89*, 31–48. [CrossRef]
16. Wang, L.; Zhang, X.; He, K.; Peng, J. Revisited dynamic modeling and eigenvalue analysis of the cable-stayed beam. *Acta Mech. Sin.* **2020**, *36*, 950–963. [CrossRef]
17. Guo, T.; Kang, H.; Wang, L.; Zhao, Y. An inclined cable excited by a non-ideal massive moving deck: An asymptotic formulation. *Nonlinear Dyn.* **2019**, *95*, 749–767. [CrossRef]
18. Guo, T.; Kang, H.; Wang, L.; Zhao, Y. Cable dynamics under non-ideal support excitations: Nonlinear dynamic interactions and asymptotic modelling. *J. Sound Vib.* **2016**, *384*, 253–272. [CrossRef]
19. Kang, H.; Guo, T.; Zhu, W. Multimodal interaction analysis of a cable-stayed bridge with consideration of spatial motion of cables. *Nonlinear Dyn.* **2020**, *99*, 123–147. [CrossRef]
20. Cong, Y.; Kang, H.; Guo, T. Planar multimodal 1:2:2 internal resonance analysis of cable-stayed bridge. *Mech. Syst. Signal Process.* **2019**, *120*, 505–523. [CrossRef]
21. Cong, Y.; Kang, H.; Yan, G. Investigation of dynamic behavior of a cable-stayed cantilever beam under two-frequency excitations. *Int. J. Non-Linear Mech.* **2021**, *129*, 103670. [CrossRef]

22. Jalali, M.H.; Rideout, G. Analytical and experimental investigation of cable–beam system dynamics. *J. Vib. Control* **2019**, *25*, 2678–2691. [CrossRef]
23. Paolo, L.; Arturo, P. A numerical study on the structural integrity of self-anchored cable-stayed suspension bridges. *Frat. Integrità Strutt.* **2016**, *38*, 359–376. [CrossRef]
24. Zhou, Y.; Chen, S. Investigation of the Live-Load Effects on Long-Span Bridges under Traffic Flows. *J. Bridge Eng.* **2018**, *23*, 101–118. [CrossRef]
25. Fujino, Y.; Warnitchai, P.; Pacheco, B.M. Experimental and analytical study of autoparametric resonance in a 3DOF model of cable-stayed-beam. *Nonlinear Dyn.* **1993**, *4*, 111–138. [CrossRef]
26. Ma, C.M.; Duan, Q.S.; Liao, H.L. Experimental Investigation on Aerodynamic Behavior of a Long Span Cable-stayed Bridge Under Construction. *KSCE J. Civ. Eng.* **2018**, *22*, 2492–2501. [CrossRef]
27. Yang, Y.; Wang, X.; Wu, Z. Experimental Study of Vibration Characteristics of FRP Cables for Long-Span Cable-Stayed Bridges. *J. Bridge Eng.* **2015**, *20*, 04014074. [CrossRef]

buildings

MDPI

Review

A Review on Research Advances and Applications of Basalt Fiber-Reinforced Polymer in the Construction Industry

Sheng-Jie Duan [1], Ru-Ming Feng [2], Xin-Yan Yuan [2], Liang-Tao Song [1], Gen-Shu Tong [1,*] and Jing-Zhong Tong [1,*]

[1] Institute of Advanced Engineering Structures, Zhejiang University, Hangzhou 310058, China; sjduan@zju.edu.cn (S.-J.D.); ltsong@zju.edu.cn (L.-T.S.)
[2] Red Bay Laboratory, Shanwei 516600, China; fengrm@hhwlab.com (R.-M.F.); yuanxy@hhwlab.com (X.-Y.Y.)
* Correspondence: tonggs@zju.edu.cn (G.-S.T.); tongjz@zju.edu.cn (J.-Z.T.)

Abstract: Compared to glass fiber-reinforced polymer (GFRP) and carbon fiber-reinforced polymer (CFRP), basalt fiber-reinforced polymer (BFRP) offers distinct advantages, including the relatively lower cost and superior creep resistance. As a result, its application in the construction industry has been gaining growing attention. This paper begins by providing an overview of the fundamental background, as well as the mechanical and microscopic properties, of BFs. By exploring various application types, including one-dimensional (e.g., bars, cables), two-dimensional (e.g., grids, sheets), and three-dimensional (e.g., profiles) applications, the research progress of BFRP products in the construction industry is comprehensively summarized. Research has demonstrated the effectiveness of BFRP in a variety of structural applications, such as reinforcing existing structures (e.g., concrete or masonry) using BFRP bars, grids, or sheets, and the development of novel design concepts that integrate BFRP products with existing structural systems. Furthermore, this paper identifies unresolved challenges and proposes potential research directions, intending to promote BFRP's broader adoption as a standardized and innovative material in the construction industry.

Keywords: basalt fiber; fiber-reinforced polymer; bar; cable; grid; sheet; profile

1. Introduction

In the course of human societal development, the exploration and application of materials stand as vital substance bases and driving forces for social progress [1,2]. The discovery and utilization of each significant material signify the advancement of an era, enhancing not only humanity's capacity to transform the natural world but also greatly elevating societal productivity and quality of life. Against the backdrop of rapid modern technological advancement, traditional single materials no longer suffice to meet complex performance requirements [3–5]. Consequently, global attention toward research in the field of composite materials has been escalating. The level of a country's or region's composite material industry has become a crucial indicator for assessing its economic and technological prowess. Generally, a composite material comprises two main components: the reinforcement material and the matrix material. While the reinforcement material does not form the continuous phase of the composite material, it predominantly serves as the carrier of its mechanical properties. On the other hand, the matrix material forms the continuous phase within the composite material and is typically a single material. Fiber-reinforced polymer (FRP), as a category of high-performance composites, is formed through the combination of two or more materials with differing properties. The fibers, serving as the reinforcement, commonly include carbon fiber (CF), glass fiber (GF), and basalt fiber (BF), among others. Currently, FRP has gradually found applications in various fields, such as construction, advanced manufacturing, new energy, low-carbon environmental protection, and military industries [6–9], as illustrated in Figure 1.

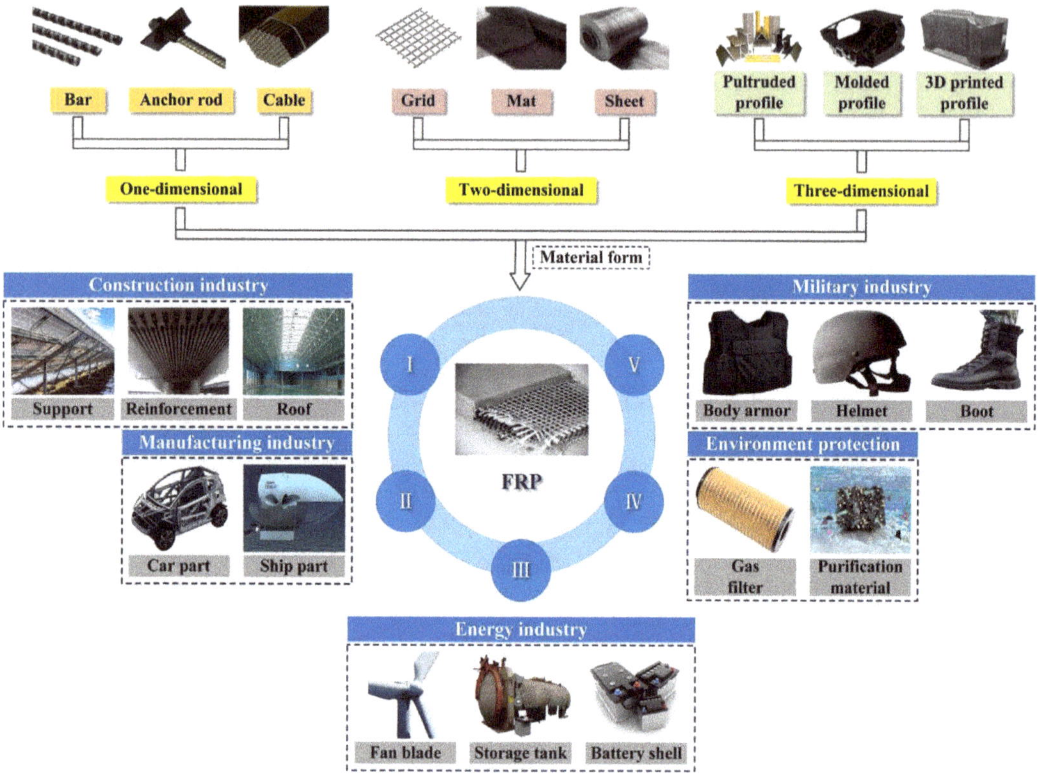

Figure 1. Applications of FRP.

Infrastructure serves as the cornerstone of a country's economic development, directly impacting its competitiveness and the quality of life of its citizens. Particularly in a globalized context, the construction of modern infrastructure holds significant importance for economic growth, social stability, and national security. Traditional building materials such as steel and concrete exhibit issues like poor corrosion resistance and high self-weight [10–14]. In comparison, fiber and its composite materials offer a light weight, high strength, excellent corrosion resistance, and fatigue resistance, making them an ideal choice for achieving long-lasting, lightweight, corrosion-resistant structures and reducing the overall life cycle costs in construction [15–17].

Basalt, widely present in nature, possesses stable chemical and mineral compositions. It exhibits extremely low reactivity with air and water, is non-flammable and non-toxic, and features explosion-proof characteristics. Typically characterized by high ecological compatibility and recyclability, basalt is regarded as a high-performance green inorganic material [18,19]. Not being a new material, basalt has been utilized since the Roman era for cobblestone roads and building materials. However, its innovative applications and practical value in modern industry and structural fields have brought significant convenience and progress to numerous industries.

Basalt fiber (BF), as a high-performance novel composite material reinforcement, has garnered widespread attention in recent years. It is deemed a pollution-free, recyclable, "green industrial raw material", refs. [20,21] with energy consumption of only 1/16th of CF, priced at 1/6th of CF, exhibiting a creep rate merely 1/4th of aramid fiber (AF), and boasting superior corrosion resistance compared to GF [22]. Therefore, BF is considered a highly cost-effective fiber material in the realm of building materials. BF can maintain excellent strength and stiffness even in high-temperature environments, showcasing outstanding acid resistance, solvent resistance, low water absorption, and significant thermal and acoustic insulation properties. Furthermore, BF demonstrates good processability and relatively low production costs [18].

It is worth mentioning that BF can seamlessly integrate with various resins, such as epoxy resin, polyurethane, and phenolic resin, among others, to produce basalt fiber-reinforced polymer (BFRP). Throughout the entire production process, BFRP does not generate particulate matter (such as asbestos) that is harmful to human health. Similar to traditional FRP (carbon fiber-reinforced polymer (CFRP), glass fiber-reinforced polymer (GFRP)), BFRP finds a wide range of applications, particularly gaining increasing usage in the construction industry in recent years. BFRP, for instance, finds applications in the reinforcement and retrofitting of existing structures, such as bars, grids, and sheets; as cables in large-span bridges; as load-bearing profiles in structural components; and as insulating and soundproofing panels in buildings. The growing use of BFRP in the construction industry compels us to deepen our understanding of the mechanical properties of this material and to devise effective design methods. However, up to this point, many countries' design standards have primarily catered to traditional composite materials (such as CFRP and GFRP), prompting extensive research efforts in that direction [18]. Consequently, the study of BFRP products and the development of corresponding design standards are indispensable.

In the research on BFRP, understanding the unique characteristics of BF is critical. Therefore, this paper first introduces the background, mechanical properties, and microscopic properties of BF. Subsequently, BFRP products are classified based on their applications, with the reviewed literature primarily focusing on studies published within the past 15 years. Various product types, including bars, cables, grids, sheets, and profiles, are examined. The research primarily encompasses the mechanical properties of the products themselves, the bond performance with concrete structures, the mechanical behavior of reinforced concrete structures, the anchoring performance of cables, and the connection performance of profile joints. Furthermore, the advancements and limitations in the application of BFRP in the construction industry are comprehensively summarized, and the future development trends in this field are outlined. The objective is to provide potential research directions for scholars in this area.

2. Basalt Fiber

2.1. Background

Basalt is a type of rock formed by the cooling of basaltic magma erupted from the Earth's surface, which is widely distributed among all igneous rocks. From a mineralogical perspective, basalt primarily consists of plagioclase and pyroxene. In terms of its chemical composition, its main components include SiO_2, Al_2O_3, and Fe_xO_y, where the SiO_2 content is approximately 45–52%, Al_2O_3 ranges from 6 to 15%, and Fe_xO_y (the total of Fe_2O_3 and FeO) accounts for about 5–13% [23]. Currently, the basalt used for BF production requires an SiO_2 content of no less than 46%. This is because an increase in the SiO_2 content within a certain range helps improve spinning processes and enhance fiber performance [24]. Simultaneously, the Al_2O_3 content should be kept within 18% to maintain the high melt viscosity, ensuring the fiber possesses good chemical corrosion resistance. The Fe_xO_y content also has specific requirements to guarantee the thermal, crystalline, mechanical, and optical properties of the fiber [25]. Additionally, the essential addition of MgO and TiO_2 in appropriate amounts is indispensable to enhance fiber waterproofing and corrosion resistance [26].

Basalt is one of the most abundant types of rocks in the Earth's crust [27], making BF one of the most sustainable reinforcement fibers. Furthermore, BF offers significant advantages, such as its non-toxic nature, high tensile strength, strong durability, excellent corrosion resistance, and lower production costs [18,19]. In today's society, with an increasing awareness of ecosystem conservation and sustainable use of environmental resources, various industries are adopting or developing eco-friendly materials and processes. The construction industry, in particular, should actively promote sustainable materials with low carbon footprints. According to market forecasts [28], the market size of BF is expected to grow from USD 227 million in 2019 to USD 397 million in 2024, with a compound annual growth rate of 11.8%. Particularly in the sectors of the construction industry, there is a significant growth trend in the market demand. The increasing demand in the construction industry for non-corrosive materials is one of the key factors driving the growth in the BF demand. This not only helps with cost reduction but also effectively extends the lifespan of structures.

2.2. Comparison of GF, CF and BF

GF, CF, and BF, as common composite material reinforcements, have been widely used in the construction industry. Table 1 summarizes the mechanical properties and sustainability characteristics of GF, CF, and BF. BF exhibits high strength, effectively meeting the design requirements of structural elements in buildings. Its lower elastic modulus can reduce the prestress losses when used as prestressed tendons in concrete structures. Additionally, BF has an elongation rate of up to 3.1%, providing structures with better ductility. Of particular note is the excellent recyclability of BF and its minimal harm to human health. Compared to other reinforcement fibers, BF demonstrates clear advantages and holds broader application prospects in various fields, particularly in the construction industry.

Table 1. Comparison of GF, CF, and BF [29–31].

Fiber	Density (g/cm^3)	Elongation (%)	Elastic Modulus (MPa)	Tensile Strength (MPa)	Renewability	Health Risks
GF	2.5	2.5	65–72	1700–3500	No	Yes
CF	1.8	1.5–2.0	200–600	2000–5000	No	Yes
BF	1.4	3.1	80–90	2800–3100	Yes	No

2.3. Microscopic Property

In recent years, BF has been widely utilized as a reinforcement material for matrices such as epoxy resin and polyurethane, enabling the production of composite materials for various applications. Since the interfacial bonding performance between BF and matrix materials plays a crucial role in determining the mechanical properties of BFRP, the analysis of its microstructure has become particularly essential and a noteworthy area of research. By applying appropriate surface treatments to the fibers, the bonding performance with the matrix can be effectively enhanced, thereby ensuring the stable transfer of loads [32]. The common surface treatment methods include fiber surface roughening [33], enhancing interactions with nanoparticles [34], chemical functionalization [35], and plasma treatment [36], as shown in Figure 2.

(a) (b) (c)

Figure 2. SEM images of the BF surface: (**a**) surface roughening [33]; (**b**) enhancing interactions with nanoparticles [34]; and (**c**) plasma treatment [36].

Halasová et al. (2019) [37] analyzed the microstructural changes in BFRP at temperatures of 650 °C and 700 °C using SEM, TEM, and spectroscopy. Fe^0 crystals were identified within the BF, and the BCC lattice structure of the Fe^0 particles was confirmed. Additionally, it was observed that BFRP exhibited high fracture resistance at 650 °C. However, at 700 °C, due to fiber crystallization and matrix transformations, the mechanical properties of BFRP deteriorated.

Zhao et al. (2020) [38] investigated the effects of different resin matrices on the performance of BFRP through static and fatigue tests, supplemented by in situ SEM observations. The experiments revealed that an increase in matrix ductility significantly reduced the extent of crack propagation. For BFRP with a vinyl resin matrix, the fatigue strength after 10 million cycles improved from 69.89% to 75.56% following toughening treatment.

The interfacial fracture resistance of BFRP with two different matrix materials was explored by Chlup et al. (2020) [39]. Compared to epoxy resin and cured polysiloxane matrices, the partially pyrolyzed polysiloxane matrix exhibited a fracture resistance level that was reduced by an order of magnitude. Moreover, the cured matrices demonstrated significant plastic deformation and crack-arresting capabilities.

Lilli et al. (2021) [40] modified the surface of basalt fibers (BFs) using plasma polymerization technology (PECVD), resulting in the formation of polymer films composed of tetravinyl silane or its mixture with oxygen on the fiber surface. The performance of the modified composites was evaluated through short beam shear tests, which demonstrated that the interlaminar shear strength of the modified basalt fibers increased by at least 180% compared to untreated fibers. In the same year, Lilli et al. [41] also utilized plasma-enhanced chemical vapor deposition to deposit a layer of tetravinyl silane (TVS) or its mixture on the BF surface. While the strength of the modified BF remained unchanged, the interfacial shear strength between the BF and the epoxy matrix improved by 79%, with SEM analysis confirming the strong bonding between the fibers and the matrix.

Pai et al. (2022) [42] investigated the effects of different aging conditions—environmental aging (25 °C), low-temperature aging (−10 °C), and humid aging (40 °C with 60% relative humidity)—on the mechanical properties of hybrid basalt–aramid/epoxy composites. It was revealed that low-temperature aging had the least impact on the mechanical properties, followed by humid aging, while environmental aging caused the most significant deterioration. FTIR and SEM analyses identified matrix decomposition, matrix cracking, and interfacial debonding as the primary failure mechanisms in the aged materials.

The surface of BF was modified by Zhang et al. (2024) [43] using a silane coupling agent (SCA). Analysis through SEM, FTIR, and XPS revealed the presence of hydrophilic NH_2 groups on the modified BF surface. This modification partially increased the surface roughness of the fibers and effectively protected them from erosion, thereby enhancing the interfacial bonding between the fibers and the matrix.

3. Research and Application of BFRP in the Construction Industry

To promote the application of BF and its composites in the construction industry, a series of high-performance BFRP products, such as BFRP bars, cables, grids, sheets, profiles, etc., have been developed, as illustrated in Figure 3. These products exhibit excellent mechanical properties, durability, and construction feasibility, meeting the high requirements for strength, stability, and corrosion resistance in building structures. The latest research progress related to the classification of different product types of BFRP in applications has been introduced.

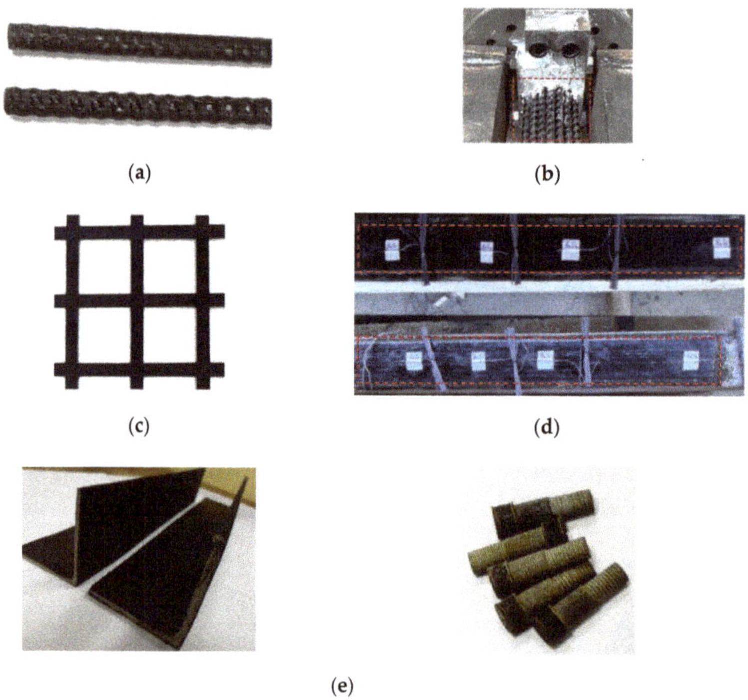

Figure 3. BFRP products: (**a**) bar [44]; (**b**) cable [45]; (**c**) grid [46]; (**d**) sheet [47]; and (**e**) profile [48,49].

3.1. One-Dimensional Level

3.1.1. Bar/Tendon

BFRP bars are recognized for their exceptional performance characteristics, including their high tensile strength, low weight, non-magnetic properties, superior corrosion resistance, enhanced fatigue resistance, and thermal expansion coefficient closely matching that of concrete. These attributes position them as a viable alternative to traditional steel reinforcement, providing a promising solution to the persistent challenge of insufficient durability in concrete structures.

Revealing the mechanical properties of BFRP bars is meaningful. In an early study by El Refai (2013) [50], static and fatigue performance tests were conducted on BF bars submerged in saltwater and alkaline solutions. It was observed that the tensile strength of the BF bars decreased by 7% and 9%, respectively, following immersion. The fatigue life of the anchorage system was primarily influenced by the stress amplitudes, with alkaline solutions found to reduce the likelihood of premature failure in the anchorage zone, exhibiting a fatigue limit of 0.04 times the ultimate load capacity. In a subsequent study by Serbescu et al. (2015) [51], tensile tests were performed under various temperature and pH conditions to collect material property data and develop an improved method for predicting the long-term strength of BF bars, as depicted in Figure 4. After 100 years of exposure to concrete and mortar environments, the residual strength of BF bars was approximately 70%. Further investigation by Lu et al. (2020) [52] involved immersing both cement mortar-coated and bare BF bars in alkaline solutions and distilled water, followed by mechanical property assessments. The alkaline environment was the primary factor contributing to the degradation of BF bars. Additionally, an alkali–aggregate reaction (AAR) was observed at the interface between the cement mortar and the BF bars, which negatively impacted the overall durability of the bars.

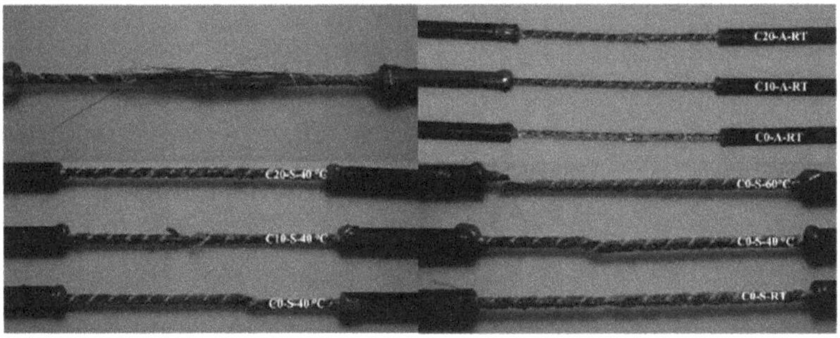

Figure 4. Failure mode of BFRP bars under different conditions [52].

The bond performance of BFRP bars was studied by Dai et al. (2021) [53]. Pull-out tests were carried out to investigate the bonding properties between BFRP bars and coral reef sand concrete (CRSC). Smaller-diameter BFRP bars with short anchorage lengths were prone to pull-out failure, while larger-diameter bars with longer anchorage lengths tended to experience splitting failure. Based on existing bond-slip models, a bond-slip model suitable for BFRP bars and CRSC was proposed. The durability of carbon nanotube (CNT)-modified BFRP bars in seawater sand concrete (SWSSC) beams was examined by Su et al. (2022) [54]. The CNTs significantly enhanced the resin matrix's fracture toughness, effectively mitigating the degradation of the tensile strength and bond strength of BFRP bars in marine environments. After a 9-month corrosion test, the bond strength of the BFRP bars remained at 87.1%. To further investigate the bonding performance between BFRP bars and reactive powder concrete (RPC), Xiao et al. (2023) [55] conducted 27 pull-out tests to explore the influence of the steel fiber volume content, protective layer thickness, and anchorage length on the bonding performance. It was revealed that there was good bonding performance between BFRP bars and RPC. Steel fiber effectively filled cracks, enhancing the bond strength, and increased the protective layer thickness, contributing to improved bond strength. Moreover, increasing the anchorage length showed an initial increase followed by a decrease in the bond strength, indicating the existence of an optimal anchorage length.

The impact of two alkaline levels of sea sand concrete (SSC) and the surface characteristics of BFRP bars (sand-coated (SC), helically wound (HWo), helically wrapped (HWr)) on the degradation at the BFRP bar and bond interface were investigated by Feng et al. (2024) [56] through shear, tension, and pull-out tests. Due to the hindrance of the sand layer, after 180 days of immersion in 55 °C seawater, the tensile strength retention of the SC-BFRP bars was 13.4% and 26.4% higher than that of the HWo-BFRP bars and HWr-BFRP bars, respectively. The bond strength of BFRP bars with low-alkalinity SSC was less affected by the environment, with over 75% bond strength remaining after 180 days. However, the bond strength of BFRP bars with normal SSC significantly decreased, with less than 50% bond strength remaining after 180 days.

The bond performance between ultra-high-performance sea sand concrete (UHP-SSC) and BFRP bars was investigated by Jiang et al. (2024) [44] to explore the effects of surface treatment methods and the anchorage length through pull-out tests. It was found that the inclusion of fibers notably enhanced the bond strength between the materials, with BFRP bars demonstrating superior reinforcement effects compared to steel fibers at identical volume fractions. The bond strength was observed to exhibit a positive correlation with the bar height and a negative correlation with the anchorage length, attributed to the nonlinear distribution of the bond stress. Additionally, a modified formula was proposed to predict the bond strength more accurately.

In a separate study, Jin et al. (2024) [57] performed numerical simulations to examine the bond performance of BFRP bars and concrete under dynamic loading conditions and elevated temperatures. The bond strength increased nonlinearly with the strain rate at various temperatures. However, at excessively high temperatures, the bond strength became largely independent of the strain rate and began to decrease progressively with the temperature rise. A semi-empirical model, which accounted for the effects of both the strain rate and temperature variations on the bond performance, was developed and validated.

Exploring the performance of components reinforced with BFRP bars was deemed crucial. Abed and Alhafiz (2019) [58] studied the influence of different types of fiber on the bending performance of concrete beams reinforced with BFRP bars. For this purpose, a four-point bending test on the beams was carried out, and the failure mode is shown in Figure 5. All the specimens first experienced concrete crushing, followed by the rupture of the BFRP bars. Compared with plain concrete beams, both BFs and synthetic fibers significantly enhanced the bending load-bearing capacity of the beams, increasing it by 12% and 19%, respectively. Additionally, BFs exhibited superior performance in terms of crack control, leading to a 21% enhancement in the cracking load of plain concrete beams.

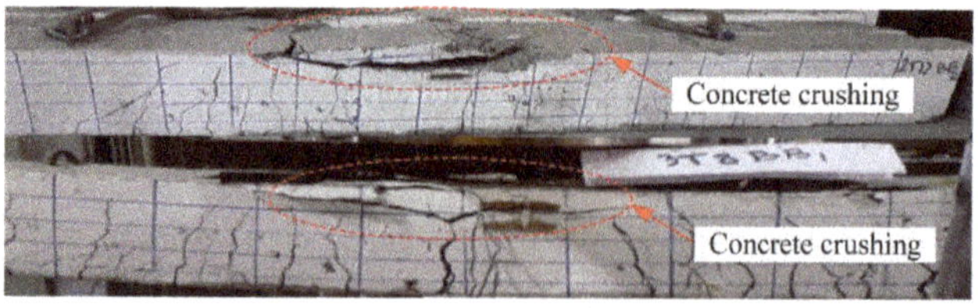

Figure 5. Failure mode of RC beams strengthened with BFRP bars [58].

Attia et al. (2019) [59] carried out experimental research on the bending performance of basalt fiber-reinforced concrete slabs with BFRP and GFRP bars and predicted their bending load-bearing capacity using current design standards. The cracking load of concrete incorporating BMF increased by 46% to 93%. During the tests, the slabs exhibited concrete compression failure, with no degradation observed in the FRP bars. Furthermore, the formulas in the ACI 440.1R-15 [60] and ACI 544.4R-88 [61] standards effectively predicted the bending load-bearing capacity of the slabs. The influence of basalt macro-fiber (BMF) on the shear performance of BFRP bar-reinforced concrete beams was studied by Muhammad and Yousif (2023) [62] through experiments. A substantial improvement in the shear performance of high-strength concrete beams incorporating BMF was induced, particularly in terms of the increased stiffness following the cracking of the concrete beam. Additionally, a formula for calculating the shear capacity of such beams was proposed.

Experimental research on and theoretical analysis of the eccentric compressive performance of slender BFRP bar-reinforced concrete columns were conducted by Liu et al. (2023) [63], as shown in Figure 6. No buckling or fracture of the BFRP bars was observed during the experiments, and it was found that the axial force contribution of the BFRP bars was lower than that of steel bars. An increase in eccentricity resulted in a marked decrease in the load-bearing capacity and lateral stiffness of the columns. Additionally, a formula for calculating the cross-sectional strength of BFRP bar-reinforced concrete columns was established, and the performance of existing design methods was evaluated. Mostafa et al. (2023) [64] performed four-point bending tests on BFRP bar-reinforced concrete beams, assessing the effects of the reinforcement ratio, BFRP bar dimensions, concrete strength, and bar type on the bending performance. The concrete beams reinforced with BFRP and GFRP bars exhibited a 77% higher bending load-bearing capacity than steel-reinforced concrete beams at the same reinforcement ratio, although the cracking moment was lower. The bending load-bearing capacity and mid-span deflection of the specimens were predicted using the ACI 440.1R-15 [60] standard formula, and a modified formula was proposed and validated.

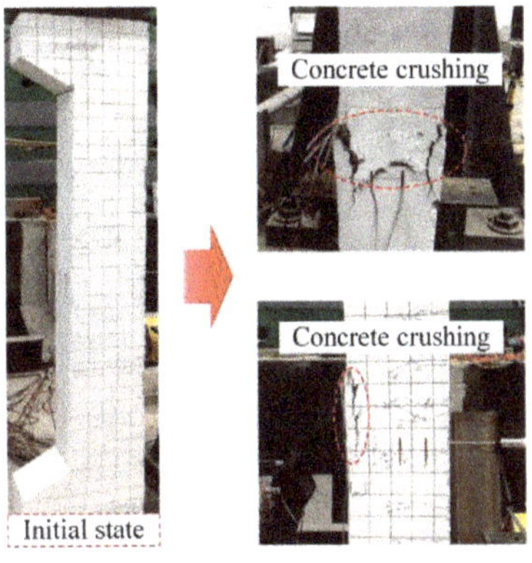

Figure 6. Failure mode of slender rectangular columns strengthened with BFRP bars [63].

To address the inadequate reinforcement effects in large-span structures and the underutilization of the mechanical properties of FRP materials, the emergence of FRP prestressed tendons became crucial, representing one of the significant means of achieving large-span and lightweight structures. The creep characteristics of BFRP prestressed tendons were identified as crucial performance indicators. Wang et al. (2014) [65] conducted experimental research on and mechanism analysis of key factors such as the creep rate and residual strength, as shown in Figure 7. There was a close correlation between the creep rate and the stress levels. Moreover, even after 1000 h of sustained loading, BFRP prestressed tendons could retain approximately 95% of their residual strength. The component performance of concrete beams externally prestressed with BFRP tendons was investigated by Wang et al.

(2015) [66], as shown in Figure 8. The experimental findings showed that all the specimens exhibited concrete crushing failure. The study also indicated that utilizing epoxy resin bonding anchorage could effectively utilize the strength of BFRP tendons. Additionally, an appropriate deviation design could efficiently offset the second-order effects of the beams. Existing design methods were able to predict the reinforcement effects quite effectively.

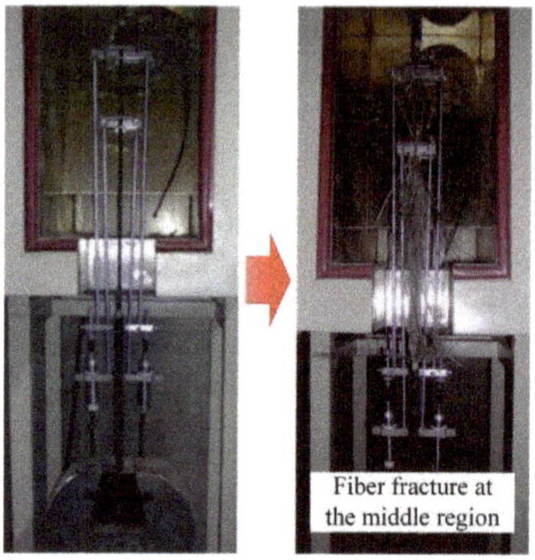

Figure 7. Failure mode of BFRP tendons [65].

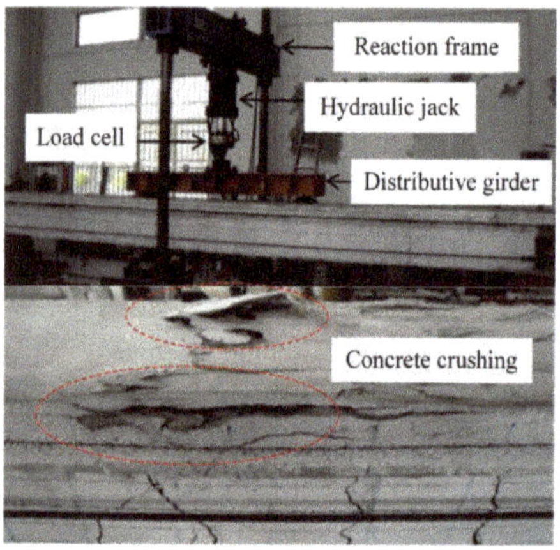

Figure 8. Failure mode of an RC beam strengthened with external BFRP tendons [66].

An effective wrapping fiber plate anchorage method for testing the fatigue performance of prestressed BFRP tendons was introduced by Wang et al. (2016) [67]. The fatigue failure of BFRP tendons primarily stemmed from debonding between the outer fiber and the matrix interface, with the fatigue life being highly sensitive to the fatigue stress range. Recommended stress ranges (0.04 fu) and maximum stress (0.53 fu) for BFRP tendons were provided in the study. As shown in Figure 9, Dal Lago et al. (2017) [68] proposed a BFRP prestressed concrete slab and provided a detailed description of the associated manufacturing processes. Three-point loading tests were conducted on the slabs, and the bending performance of the slabs was predicted using nonlinear viscoelastic and elastoplastic models. The predicted results aligned with the experimental results, demonstrating good accuracy. Atutis et al. (2018) [69] investigated the stress relaxation phenomena in

BFRP prestressed concrete beams and proposed a recommended stress relaxation curve that accounts for the prestress losses. The stress relaxation rate of BFRP tendons was 6.67% when the initial stress level was 55% of the ultimate strength.

Figure 9. Failure mode of a concrete slab strengthened with BFRP tendons [68].

Motwani et al. (2022) [70] conducted experimental research on the application of BFRP prestressed tendons in concrete beams. It revealed a prestress transfer length between 250 and 500 mm, an end slip range of 1.6–2.7 mm, and an average bonding stress ranging from 1.8 to 2.6 MPa. Ji et al. (2024) [71] applied BFRP prestressed tendons in concrete cylinder pipes and concluded through experimental research and theoretical analysis that as the bending radius increased, the degree of strength degradation of the BFRP tendons decreased, especially in large-diameter pipes where the BFRP tendons demonstrated higher cost-effectiveness.

A bending test on and numerical simulation analysis of RC beams reinforced with BFRP bars were conducted by Erfan et al. (2019) [72]. The application of BFRP bars not only increased the ductility and load-bearing capacity of the elements but also reduced the width and length of the cracks. It was noted that a brittle failure mode was exhibited by the BFRP bars when the specimens reached their ultimate load-bearing capacity. Experimental research on and numerical simulation analysis of the bending performance of BFRP bars and steel-reinforced concrete slabs were carried out by Erfan et al. (2021) [73], as shown in Figure 10. A significant improvement in the bending load-bearing capacity of BFRP bar-reinforced concrete slabs was noticed compared to steel-reinforced concrete slabs. An inverse relationship between the deflection of the slabs and the content of BFRP bars was revealed, with BFRP bar-reinforced concrete slabs exhibiting bilinear elastic behavior before failure. The bending performance of concrete beams reinforced with a combination of steel and BFRP bars through experimental and numerical simulation studies was investigated by Hussein et al. (2022) [74]. The contribution of BFRP bars to the bending capacity of the beams mainly depended on their bond strength with the concrete after steel yielding. The roughening of the surface of the BFRP bars had no significant effect on the bending performance before steel yielding. To meet the deflection limit requirements, the ratio of mixed reinforcement with BFRP bars to steel bars should be less than 1.9. Shi et al. (2022) [75,76] introduced a novel wedge anchor design for prestressed BFRP tendons. Through finite element analysis, optimization was conducted on the wedge segment length and elastic modulus. The experimental results demonstrated that the efficiency coefficient of this new anchoring system reached 91%, significantly outperforming traditional steel wedge anchors. Additionally, the fatigue life of the tendon–anchor system was found to be capable of reaching 200 million cycles when subjected to a load range of 0.05 to 0.5 times the tensile capacity.

Figure 10. Failure mode of concrete slabs strengthened with BFRP bars [73].

3.1.2. Cable

Traditional steel cables are often challenged by issues such as excessive self-weight and insufficient load-bearing capacity, which render them inadequate for fulfilling the design requirements of large-span bridges. Additionally, steel cables are characterized by a limited lifespan and are prone to corrosion and fatigue damage, leading to the necessity of frequent replacements and ongoing maintenance. This not only disrupts the traffic flow but also results in substantial maintenance costs. In contrast, BFRP cables, with their lightweight nature, high strength, and extended durability, have emerged as a key solution for achieving lightweight and highly durable cable structures, offering a more dependable alternative for bridge design.

The material properties of hybrid basalt and carbon (B/C) FRP cables and various cable bridges were analyzed by Wang and Wu (2010) [77], indicating that superior mechanical performance, durability, and fatigue resistance are exhibited by B/C FRP cables compared to steel cables. Excellent mechanical properties were found by Wang et al. (2013) [78] in composites made by blending CFs or steel fibers with BFs. A significant improvement in the fatigue strength of mixed FRP composites compared to pure BFRP was shown by the experimental results. Lower and upper limits for the design of large-span cable-stayed bridges using FRP cables in terms of the strength and stiffness were proposed in the study. Yang et al. (2016) [79] conducted dynamic experiments and theoretical derivations on BFRP and CFRP cables, leading to the development and validation of a modified modal damping ratio equation. It was discovered that the CRFID model effectively captures the out-of-plane modal damping ratio. Zhou et al. (2022) [80] enhanced the compressive performance of the load transfer component (LTC) in BFRP cables by incorporating GFs and CFs. Experimental studies revealed that the optimal fiber lengths were 75, 150, and 900 μm. During testing, the primary failure modes of the LTC were identified as fiber debonding, pull-out, and fracture. A fitting formula was proposed to predict the compressive performance of the LTC, and the anchoring efficiency of the BFRP cables reached an impressive 101%.

As shown in Figure 11, the design of and experimental research on the anchoring system of high-capacity BFRP cables were conducted by Zhou et al. (2022) [45], revealing an average anchoring efficiency of 95% for BFRP cables. With increasing loads, the maximum axial strain of the BFRP cables gradually increased, while the shear stress exhibited a trend of initial increase, subsequent decrease, re-increase, and final decrease, with these two parameters, respectively, following linear and cubic function relationships. Fatigue performance tests on BFRP cables were conducted by Zhou et al. (2023) [81], considering factors such as the manufacturing processes, cable surface shapes, and anchoring methods. Ribbed cable failure was initiated by transverse cracking of the resin matrix, leading to continuous debonding at the interface. The surface shape of the cable was a key factor determining the performance of BFRP cables, while the influence of the anchoring method remains unclear.

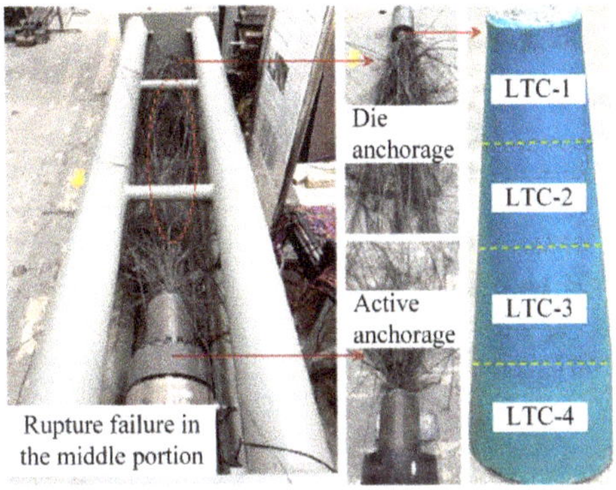

Figure 11. Failure mode of BFRP cables [45].

Feng et al. (2019) [82] carried out finite element analysis of the fatigue life of CFRP, BFRP, and steel cables, proposing an appropriate method for fatigue life estimation. It was revealed that the fatigue life of CFRP cables and BFRP cables was approximately 3 times and 1.4 times that of steel cables, respectively. Furthermore, to meet the design requirement of a 100-year service life, the safety factors for CFRP, BFRP, and steel cables should not be less than 2.55, 3.27, and 3.64, respectively. Wang et al. (2020) [83] proposed a novel anchoring system for BFRP cables and conducted static tensile tests, numerical simulations, and theoretical analyses. The results demonstrated that the average anchoring efficiency of the BFRP cables reached as high as 99.2%, effectively preventing shear failure at the ends. Furthermore, the tensile strain in the middle section of the BFRP cables exhibited a linear relationship with the applied external load, and the cables showed almost no shear lag effect. The nonlinear coupled vibration characteristics of BFRP cables were investigated by Yang et al. (2023) [84] through numerical analysis. Under the same excitation, the vibration frequency of BFRP cables was found to be lower than that of steel cables, while their vibration amplitude was reduced by 44% compared to steel cables. An increase in cable tension was observed to decrease the vibration response. Moreover, the increment in the excitation amplitude was consistent with the increment in the maximum amplitude of the BFRP cables. Zhou et al. (2023) [85] developed a novel anchoring system for BFRP cables and evaluated its static performance through full-scale experiments and numerical simulations. It was demonstrated that the load transfer component (LTC) remained intact, achieving an anchoring efficiency of 93%. Additionally, it was found that using LTCs with variable stiffness effectively eliminated the stress concentration in the anchoring zone of the BFRP cables. However, current research on BFRP cables is relatively limited, both domestically and internationally, primarily focusing on other types of FRP cables [86–94].

3.2. Two-Dimensional Level

3.2.1. Grid

Traditional fiber cloth reinforcement techniques have certain limitations in terms of the peel resistance, fire resistance, and durability, and they are challenging to widely apply in special environments such as underwater structures and tunnels. Meanwhile, CF grids are costly, and GFs exhibit poor durability, with existing production processes being relatively inefficient. To address these issues, BFRP grids have emerged as an ideal choice combining high performance with lower costs. Ali et al. (2015) [95] conducted experimental research on the tensile properties of BFRP grids, their bond performance with concrete, and the flexural performance of reinforced steel–concrete beams. It was shown that the elastic modulus of BFRP grids ranged from 40 to 43 GPa, and the tensile strength ranged between 815 and 931 MPa. The bond performance between BFRP grids and concrete exhibited a ductile failure mode, with no instances of debonding failure in any specimens. Zheng et al. (2016) [96] combined BFRP grids with ECC for reinforcing steel–concrete beams, which is depicted in Figure 12. Through bending tests, good bond performance between the reinforcement layer and the concrete was found, with the beam primarily experiencing concrete crushing or BFRP grid fracture during loading. Finally, an analytical model was proposed to predict the beam deflection. The compressive performance of rubble walls reinforced with BFRP grids was investigated by Huang et al. (2018) [97] through experimental testing. It was shown that the load-bearing capacity of the walls could be increased by up to 2.5 times with BFRP grid reinforcement, while the out-of-plane displacement was significantly reduced.

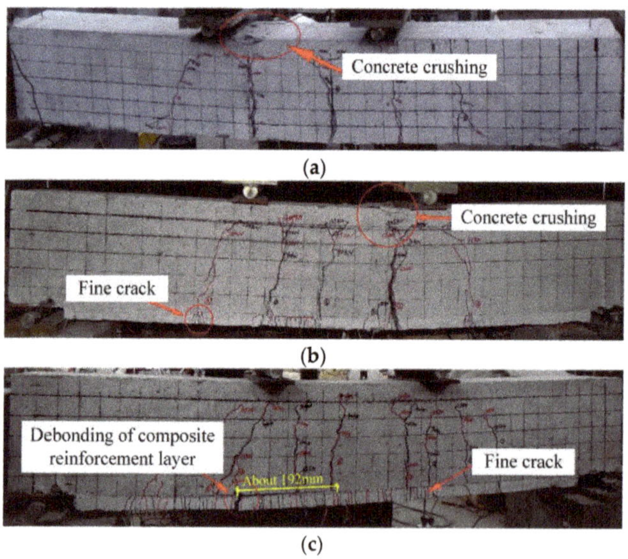

Figure 12. Failure mode of an RC beam strengthened with a BFRP grid [96]: (**a**) concrete crushing; (**b**) concrete crushing after the rupture of the BFRP grid; and (**c**) partial debonding of the CRL after the rupture of the BFRP grid.

Dong et al. (2021) [98] used BFRP grids as shear reinforcement for a new type of seawater–sea sand concrete beams (SWSSC) and conducted four-point bending tests. A significant improvement in the bending stiffness when using BFRP grids was noticed, with smaller crack widths. Wen and Wan [99] carried out shear fatigue tests on the interface between BFRP grids and concrete, analyzing the fatigue failure modes and proposing a theoretical model of bond-slip fatigue degradation based on the experimental results. Uniaxial tensile tests on high-ductility fiber-reinforced concrete embedded with BFRP grids were conducted by Zhang et al. (2022) [100]. The tensile capacity of concrete was significantly enhanced by BFRP grids. Based on the experimental results, corresponding formulas for calculating the ultimate tensile strength capacity were proposed. He et al. (2024) [101] conducted experimental research on the fatigue performance of BFRP grid-reinforced steel–concrete beams. The beam's fatigue life was significantly extended by the BFRP grids, with the effect being more pronounced when using prestressed BFRP grids. Additionally, the bond performance between BFRP grids and concrete was good and reliable. Finally, a fatigue life prediction model was proposed.

Bending tests on tailings sand ultra-high ductile concrete beams reinforced with BFRP grids were carried out by Song et al. (2024) [102]. It was noticed that there was a significant improvement in the bending load-bearing capacity of the beams due to the BFRP grids. Based on existing formulas, the research team further derived the corresponding formulas for the bending load-bearing capacity. The shear performance of steel–concrete beams reinforced with prestressed BFRP grids was investigated and a shear capacity calculation formula was derived by Zhou et al. (2024) [46]. The shear performance of steel–concrete beams reinforced with BFRP grids was revealed by Zhou et al. (2025) [103] through experimental research. A significant improvement in the initial stiffness and bending load-bearing capacity of the reinforced beams was found. Moreover, the ACI 549.4R-13 [104] standard was suitable for predicting the contribution of BFRP grids to the shear load-bearing capacity of beams.

Zheng et al. (2020) [105] reinforced RC beams using BFRP grids and engineered cementitious composites (ECCs), uncovering the changes in the shear performance before and after reinforcement. The bond between the BFRP grid and the concrete was strong, effectively suppressing diagonal cracks in the beams. Based on the finite element simulation results, a corresponding formula for calculating the shear load-carrying capacity was proposed. Zhang et al. (2023) [106] investigated the flexural performance of corroded RC beams reinforced with BFRP grids and ECC. Through a combination of experimental testing and numerical simulation, it was found that the tensile stress in the BFRP grid at failure was only 50–70% of its ultimate tensile strength. Furthermore, a theoretical model to predict the moment values at critical points was successfully developed.

3.2.2. Sheet/Strip

Due to their excellent corrosion resistance, substantial lightweight advantages, and good processability, BFRPs are deemed suitable for various construction and repair projects. Their environmental friendliness and durability make them stand out in harsh environments, where the structural self-weight can be effectively reduced. In recent years, significant interest in BFRP sheets has been shown by scholars.

As presented in Figure 13, experimental research was conducted by Zhou et al. (2013) [107] using BFRP sheets to reinforce unreinforced masonry walls, followed by detailed testing of their in-plane seismic performance. The enhancement effect of BFRP reinforcement on the seismic performance of unreinforced masonry walls was evaluated, and the corresponding design models were proposed based on the results. Earthquake-damaged reinforced concrete circular bridge columns were repaired by Jiang et al. (2016) [108] using BFRP sheets, as depicted in Figure 14, followed by cyclic tests under lateral loads. A significant increase in the flexural load-bearing capacity of the repaired columns was induced, with some restoration of ductility achieved. A study conducted by

Qin et al. (2019) [109] revealed that after reinforcing cracked concrete beams with BFRP sheets, there was a significant increase in both the flexural stiffness and the load-bearing capacity. The beam failed because the concrete in the compression zone was crushed. Additionally, it was noted that increasing the number of BFRP sheets helped further enhance the flexural load-bearing capacity, albeit with certain limitations in terms of the improvement.

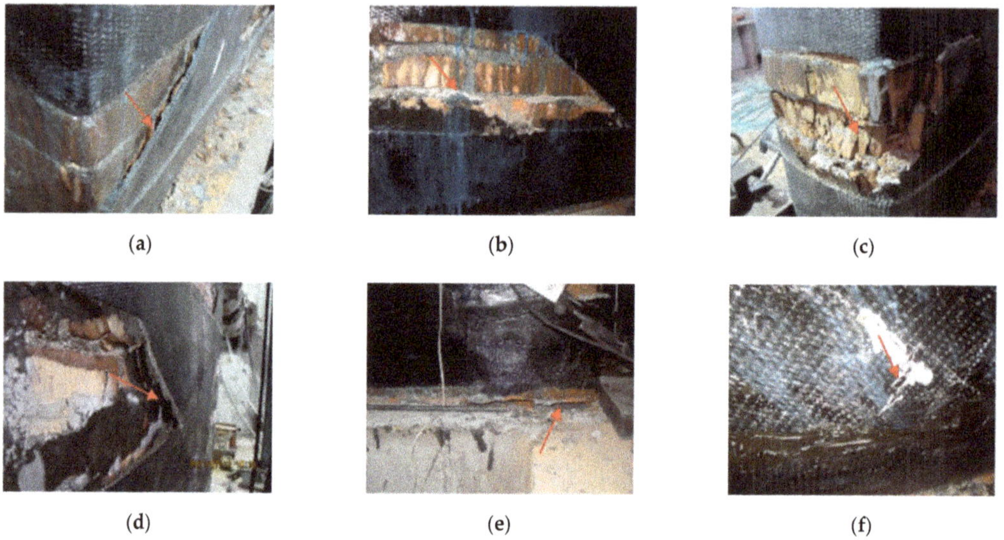

Figure 13. Failure of a masonry wall strengthened with BFRP sheets [107]: (**a**) FRP debonding; (**b**) slight shear sliding crack; (**c**) masonry crushing; (**d**) FRP buckling; (**e**) rockling behavior; and (**f**) FRP rupture.

Figure 14. Failure of an RC column strengthened with BFRP sheets [108]: (**a**) interface failure; and (**b**) fracture of NSM-BFRP bars.

To study the dynamic bond strength between BFRP sheets and steel fiber-reinforced concrete (SFRC), shear tests were carried out by Yuan et al. (2019) [110]. The bond strength and bond slip were highly sensitive to the loading rates, with the failure mode transitioning from concrete matrix failure to interface failure with increasing loading rates. A bond-slip model considering the influence of the loading rates was established based on the experimental data. As illustrated in Figure 15, Pham et al. (2020) [111] investigated the impact performance of BFRP sheet-reinforced lightweight rubberized concrete beams. Rubberized concrete beams slowed down the stress wave velocity and exhibited reduced displacement and peak impact force. A novel epoxy anchor rod for enhancing the bond performance between BFRP sheets and reinforced concrete beams was proposed by Yuan et al. (2021) [47].

Figure 15. Failure of a lightweight rubberized beam strengthened with BFRP sheets [111].

As shown in Figure 16, the repair of modified reinforced concrete beam–column joints with BFRP sheets was carried out by Shen et al. [112], followed by a study on their seismic performance. The joints' seismic performance was effectively improved by the BFRP sheets, leading to a delay in concrete spalling and enhancement of the bond performance between the steel reinforcement and the concrete. Reinforced concrete columns with BFRP sheets were strengthened by Zeng et al. (2022) [113], followed by lateral push and seismic tests. It was determined that externally bonded BFRP improved the strength, ductility, and energy dissipation capacity while also enhancing the recoverability.

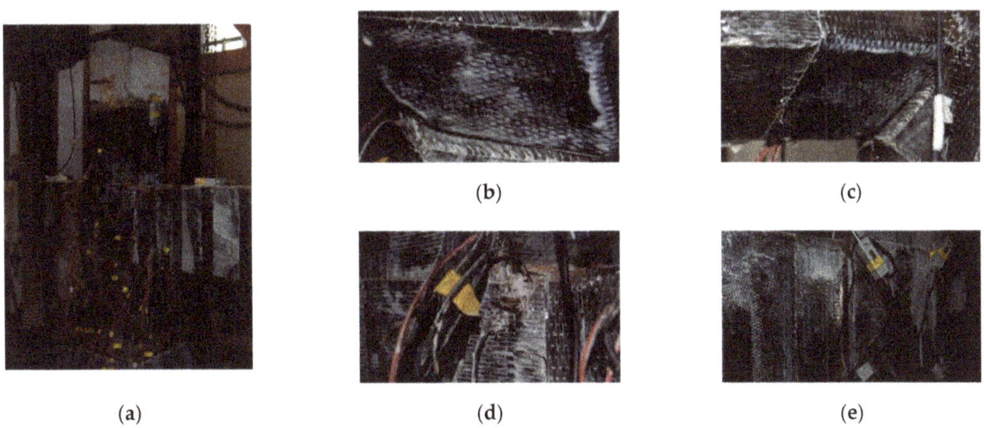

Figure 16. Failure of RC beam–column joints strengthened with BFRP sheets [112]: (**a**) initial state; (**b**) epoxy crack; (**c**) local bulging; (**d**) rupture of BFRP sheet; and (**e**) crushing of core concrete.

The interface bond performance and flexural strength of steel-reinforced concrete beams strengthened with BFRP sheets were studied by He et al. (2023) [114,115]. The maximum stress at the interface occurred at the end of the bonding region, primarily resulting in debonding failure at that location. Additionally, after reinforcement with BFRP sheets, the flexural load-bearing capacity of the steel-reinforced concrete beams increased by 78.56%. It was suggested that BFRP sheets should be applied for reinforcement before the load-bearing capacity of the steel-reinforced concrete beams reaches 90% to enhance their repairability. Evaluation of the bending failure process of steel-reinforced concrete beams reinforced with BFRP sheets was conducted by Liu et al. [116] using non-destructive testing techniques, studying the damage process and failure mechanisms. Kang et al. [117] found through experimental research that when damaged high-performance concrete (HPC) composite beams were reinforced with BFRP sheets and subjected to secondary chloride corrosion, most specimens exhibited the failure mode of BFRP sheet peeling. However, with an increase in the number of BFRP sheets, the cracking load and flexural load-bearing capacity of the specimens increased by 27.9% and 35.8%, respectively.

Bending tests on concrete T-beams reinforced with BFRP sheets were conducted by Nayak et al. [118], demonstrating that the BFRP sheets effectively increased the bending load-bearing capacity and reduced the deflection. When the BFRP plates are installed at a 45-degree angle, the reinforcement effectiveness surpassed that achieved with a 90-degree installation or in the absence of BFRP plates altogether. Based on the experimental result, a refined finite element model was established and validated.

3.3. Three-Dimensional Level

FRP profiles, with their excellent properties, such as the light weight, high strength, and corrosion resistance, have unmatched advantages in reducing the structural self-weight, enhancing construction convenience, and improving the structural durability compared to other FRP products. Currently, there is more research on CFRP profiles and GFRP profiles [119–124]. GFRP profiles, with their lower cost advantage, have gained a certain market share compared to CFRP profiles. However, GFRP profiles face issues such as the lower creep fracture stress, higher creep rate, and significant long-term deformation. BFRP profiles, while ensuring a higher cost–performance ratio, have successfully overcome the drawbacks of GFRP, better meeting the demands of the construction industry for lightweight, large-span, and highly corrosion-resistant structures.

Lu et al. (2014) [125–127] studied the water absorption and mechanical property changes of pultruded BFRP sheets after thermal aging, as well as the impact of continuous loading and high-temperature environments on the mechanical performance, proposing relevant recommendations. As presented in Figure 17, the axial compression and axial tensile mechanical properties of pultruded BFRP round tubes were revealed by Ding et al. (2018) [48], with an analysis of the influence of the slenderness ratio on the failure mode and the derivation and verification of a stable load-bearing capacity calculation formula. Experimental research on the tensile performance changes of BFRP profiles in acidic environments was carried out by Wang et al. (2018) [128]. BFRP profiles have excellent durability, with almost no change in the tensile modulus and only an 18.4% decrease in the strength. A strength degradation model was established considering the temperature and pH values. The impact of BFRP bolts, stainless steel bolts, and hybrid steel–FRP bolts on the static and fatigue performance of pultruded sheets was studied by Abdelkerim et al. (2019) [49]. HSFRP bolts outperformed the other types significantly, exhibiting ductile failure. The specimens maintained their rigidity throughout the entire fatigue life, and under the same fatigue load range, BFRP bolts could replace SS bolts.

The impact of different connecting materials on the static and fatigue performance of pultruded BFRP bolt connections was revealed by Ding et al. (2021) [129] through experiments. All the bolt connections experienced shear failure, independent of the connecting material. Nodes composed of BFRP bolts and BFRP cover plates exhibited better fatigue performance than steel bolt-connected nodes. Stability analysis of pultruded BFRP circular and rectangular tubes, including fitting relevant stable load-bearing capacity calculation formulas based on the experimental results, was conducted by Chen and Zhang (2024) [130] and the on-site test is shown in Figure 18. The tensile performance of pultruded BFRP sheets under multi-bolt connections was studied by Liang et al. (2024) [131,132]. The nodes exhibited brittle shear failure when the end-to-diameter ratio and pitch-to-diameter were small, while they displayed ductile pin-bearing compressive failure when the end-to-diameter ratio was not less than 4 and pitch-to-diameter was not less than 4.5. Ultimately, the study provided a formula for calculating the ultimate load-bearing capacity of the bolt connections in BFRP sheets.

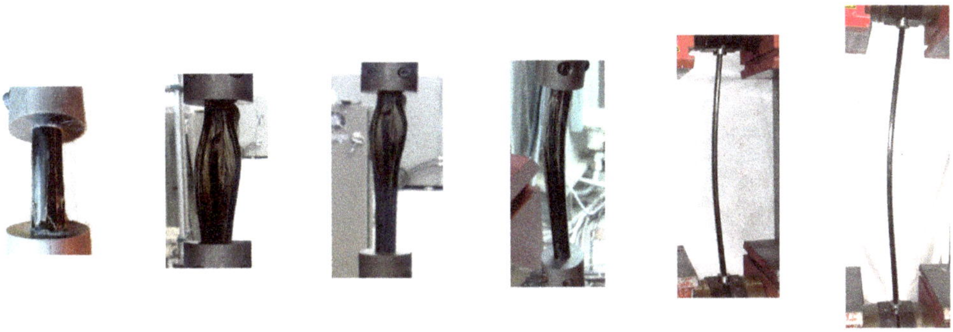

(a) (b) (c) (d) (e) (f)

Figure 17. Failure mode of a BFRP tube [48]: (**a**) $\lambda = 6$; (**b**) $\lambda = 10$; (**c**) $\lambda = 30$; (**d**) $\lambda = 50$; (**e**) $\lambda = 70$; and (**f**) $\lambda = 90$.

(a) (b) (c)

Figure 18. Failure mode of a BFRP profile [130]: (**a**) circular tube; (**b**) rectangular tube; and (**c**) box tube.

Liu et al. (2020) [133] proposed a bolt connection form for FRP profiles, conducted experimental research on different BFRP ply angles and sequences, and combined numerical simulations to analyze the failure process of nodes. The optimal arrangement and proportion of 0-degree, 45-degree, and 90-degree ply layers were derived, noting that the Hashin–Rotem failure criterion was not entirely applicable.

4. Discussion

A comprehensive review of the research on BFRP products in the construction industry over the past 15 years has been conducted. These studies are categorized based on the different application types, primarily including the mechanical properties of BFRP products themselves, the bond performance between BFRP products and concrete structures, the mechanical behavior of reinforced concrete structures, the anchoring performance of cables, and the connection performance of profile joints. The distribution of the relevant literature is shown in Figure 19.

4.1. One-Dimensional Level

4.1.1. Bar/Tendon

BFRP bars exhibit exceptional durability in saline solutions and alkaline environments, and under high-temperature conditions, particularly in marine engineering, humid and hot environments, and alkaline concrete, where their mechanical property retention is relatively high. At the same time, they significantly improve the structural ductility, flexural capacity, and crack control ability. However, alkaline environments can induce alkali–aggregate reactions (AARs) at the interface, adversely affecting the durability. The bond performance between BFRP bars and concrete is excellent, with notable effectiveness in ordinary concrete, ultra-high-performance concrete (UHP-SSC), and sea sand concrete (SSC) used in special environments. The bond performance is significantly influenced by the surface treatments (such as sand coating), anchorage length, and cover thickness, with optimization of the design further enhancing the bond strength.

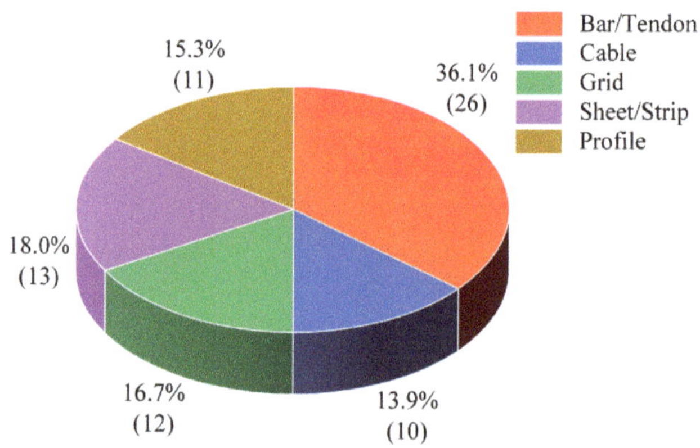

Figure 19. The proportion of research work on different BFRP products.

To address the inherent brittleness of BFRP bars, modification measures such as the incorporation of carbon nanotubes (CNTs) and steel fibers have been investigated, resulting in improved toughness and interfacial strength, while also delaying corrosion and degradation. Under flexural, shear, and eccentric compressive conditions, structures reinforced with BFRP bars demonstrate superior bending and shear performance, surpassing that of reinforced concrete structures. However, brittle failure limits their application in scenarios requiring high ductility.

As a significant development direction, BFRP tendons have played an important role in long-span and lightweight structures. Research on their creep performance and fatigue life has shown that high residual strength is maintained even under long-term loads, while the development of novel wedge-shaped anchorage designs has significantly enhanced the anchorage efficiency and reliability.

4.1.2. Cable

BFRP cables, owing to their light weight, high strength, corrosion resistance, and fatigue resistance, have emerged as a critical technological approach in long-span bridge design. Compared to traditional steel cables, they significantly reduce the self-weight while enhancing the durability. Studies have demonstrated that hybrid FRP cables, composed of basalt fibers combined with carbon fibers or steel fibers, exhibit superior mechanical properties and fatigue strength compared to pure BFRP cables. Furthermore, an optimized anchoring system can achieve an anchoring efficiency of up to 95%, maintaining stable deformation under long-term cyclic loading and enduring over two million fatigue cycles. The surface morphology of the cables (e.g., ribbed cables) has been shown to significantly influence the interfacial bonding performance and fatigue lifespan. Transverse cracking and interfacial debonding are regarded as the primary modes of fatigue failure.

Compared to steel, BFRP cables exhibit superior fatigue resistance, reduced vibration amplitudes, and a safety factor that guarantees a service life of 100 years. Despite advancements in anchoring systems, load transfer components, and dynamic performance, the current research remains limited, primarily focusing on other types of FRP materials. This highlights the need for further exploration of BFRP cables in practical applications.

4.2. *Two-Dimensional Level*

4.2.1. Grid

Studies have indicated that the tensile strength of BFRP grids can reach 815–931 MPa, with an elastic modulus ranging between 40 and 43 GPa. The interface between the BFRP grid and the concrete exhibits a ductile failure mode, without significant debonding, thereby significantly enhancing the flexural, shear, and fatigue performance of the concrete components. When applied to the reinforcement of reinforced concrete beams, BFRP grids not only improve the initial stiffness and flexural capacity of the components but also markedly extend the fatigue life of the beams. The use of prestressed BFRP grids has been shown to yield even more pronounced effects. Furthermore, it has been observed that concrete components reinforced with BFRP grids primarily fail due to concrete crushing or grid rupture during the loading process. A bond-slip degradation model and load-bearing capacity calculation formula, proposed based on experimental results, provide theoretical support for engineering applications.

In specific applications, the combination of BFRP grids with novel concrete materials—such as ECCs (engineered cementitious composites), high-ductility fiber-reinforced concrete, and ultra-high-toughness tailings sand concrete—has been demonstrated to further enhance the overall tensile capacity, flexural stiffness, and crack control capabilities, offering reliable solutions for reinforcement design in complex environments. In seawater–sea sand concrete beams (SWSSC), the use of BFRP grids as stirrups has been found to significantly improve the flexural stiffness and durability of the beams. Additionally, research has revealed the failure modes in shear fatigue and shear resistance, demonstrating that the shear capacity of BFRP grids is significantly superior to that of traditional reinforcement materials. The applicability of the ACI 549.4R [104] guidelines in predicting the contribution of BFRP grids to shear resistance in components has also been validated.

4.2.2. Sheet/Strip

In specific applications, the reinforcement of unreinforced masonry walls with BFRP sheets has been found to significantly improve their in-plane seismic performance. When applied to the repair of earthquake-damaged bridge piers, BFRP sheets have demonstrated a remarkable enhancement of the flexural capacity and ductility. For cracked concrete beams, the use of BFRP sheets effectively improves the flexural capacity and stiffness, although the extent of the improvement is limited by the number of sheet layers applied. Additionally, dynamic bond tests have revealed that the bonding performance between BFRP sheets and concrete is highly sensitive to the loading rates, with higher loading rates being more likely to result in interfacial failure rather than the failure of the concrete substrate. By introducing innovative epoxy anchoring techniques and optimizing the laying angles (e.g., 45° orientation), the bonding performance and flexural effectiveness of BFRP sheets have been further enhanced.

The application of BFRP sheets in complex structures and specialized materials has also shown promising results. For instance, the impact performance of lightweight rubberized concrete beams has been improved, with BFRP sheets shown to mitigate the stress wave propagation and peak impact forces, thereby significantly enhancing the impact resistance of the structure. In high-performance concrete composite beams subjected to chloride corrosion, the use of BFRP sheets effectively increased the cracking loads and flexural capacity, although multi-layer BFRP sheet reinforcement may lead to a peeling failure mode. Tests on the lateral loading and seismic performance of reinforced concrete columns have demonstrated that externally bonded BFRP sheets significantly improve the strength, ductility, and energy dissipation capacity, while also enhancing the recoverability of the columns to some extent.

Moreover, experimental results have indicated that the flexural capacity of reinforced concrete beams strengthened with BFRP sheets can be increased by up to 78.56%, with interfacial debonding failure typically occurring at the ends of the bonded region. Consequently, it is recommended that BFRP sheets be applied to beams before they reach 90% of their load-bearing capacity to enhance their reparability. The monitoring and damage assessment of flexural failure processes using non-destructive testing techniques has further revealed the underlying failure mechanisms, providing critical insights for practical engineering applications.

4.3. Three-Dimensional Level

BFRP profiles have demonstrated exceptional durability in complex environments. For instance, in acidic conditions, the tensile elastic modulus remains almost unaffected, while the strength decreases by only 18.4%. The long-term performance can be evaluated through the establishment of strength degradation models. The mechanical properties of BFRP profiles vary depending on factors such as the cross-sectional shape, slenderness ratio, ply design, and connection method. For example, in pultruded circular and rectangular tubes, the compressive strength and buckling capacity are significantly influenced by the slenderness ratio, and relevant calculation formulas have been derived.

In terms of the structural connections, the bolted connection performance of BFRP profiles is a key area of research. Experiments have shown that appropriate end distances and bolt spacing can significantly enhance the joint performance and prevent brittle failure. Furthermore, BFRP bolted connections exhibit superior fatigue performance compared to steel bolted joints, with higher ductility and stiffness observed. Additionally, studies on the connections of BFRP profiles under various environmental and complex loading conditions, including optimized designs for different ply angles and stacking sequences, have significantly improved the overall load-bearing performance.

4.4. Limitations of Current Research

Although BFRP products have garnered widespread attention in the construction industry due to their excellent properties, such as the high strength, light weight, corrosion resistance, and eco-friendliness, the current research still faces limitations in certain aspects. These limitations are primarily manifested in the incompleteness of design codes, insufficient studies on long-term performance, and lack of research on extreme environmental adaptability.

Currently, the application of BFRP in actual engineering lacks comprehensive design theories and codes. Standardization efforts in terms of the strength design, deformation design, fatigue life design, and fire resistance design are still at the stage of improvement. Moreover, most existing design codes are based on traditional materials such as CFRP and GFRP, failing to fully consider the uniqueness of BFRP in terms of the mechanical properties, durability, and construction techniques, thereby limiting its application in complex engineering projects to a certain extent.

Despite the superior corrosion resistance and fatigue performance of BFRP, systematic studies on its performance degradation under long-term service conditions (such as high-temperature, high-humidity, and alkaline environments) have not been conducted. For instance, in marine environments, the degradation of the bond strength between BFRP and the mortar interface and the impact of alkali–aggregate reactions still require further exploration. Additionally, there is a lack of sufficient data support for the changes in mechanical properties of BFRP exposed to extreme environments (such as salt spray, ultraviolet radiation, and high temperature), thus limiting its use in high-demand environments to some extent.

There is limited research on the mechanical behavior of BFRP products under extreme loads (such as impact loads and fires), particularly in critical applications like bridge cables and prefabricated components. This uncertainty in performance significantly affects the safety and reliability of engineering design. Furthermore, there is insufficient experimental and numerical analysis data to support the structural response characteristics of BFRP under complex dynamic actions such as earthquakes and wind loads, further restricting its widespread application in special environments.

5. Conclusions

This paper aims to categorize the applications of BFRP in the construction industry. On that basis, the advances and limitations in terms of the research of BFRP products are summarized in detail. The major conclusions and outlook of this study are listed in the following.

BF, derived from natural basalt rock, possesses significant advantages compared to traditional fiber materials such as CF and GF, including being non-toxic, exhibiting high tensile strength, exceptional durability, and outstanding corrosion resistance, and having lower production costs. Additionally, the production and utilization of BF are environmentally friendly, which has led to its widespread application in fields such as the construction industry. To enhance the interfacial bonding performance between BF and the matrix, effective strategies include fiber surface roughening, strengthening interactions with nanoparticles, chemical functionalization, and plasma treatment.

This characteristic holds substantial value, as BF can be combined with resins such as epoxy or polyurethane to produce BFRP. This innovative composite material has garnered extensive attention from researchers. In the field of the construction industry, BFRP products are categorized by their application types into three groups: one-dimensional (e.g., bars, cables), two-dimensional (e.g., grids, sheets), and three-dimensional (e.g., profiles) applications. In recent years, research findings related to BFRP have been extensively summarized and explored, focusing primarily on the mechanical properties of BFRP products, the bonding performance between BFRP bars, grids, or sheets and concrete structures, the mechanical behavior of concrete structures strengthened with BFRP, the anchoring performance of BFRP cables, and the connection performance of BFRP profile joints.

The existing design methods and standards are often not fully applicable or precise when accounting for BFRP. While the research on various applications of BFRP has been extensive, there is still a lack of comprehensive design methods for the component strength, deformation, fatigue life, fire resistance, and other aspects. The design standards are worthy of improvement. Future studies could integrate machine learning techniques to further investigate these domains, leveraging data-driven analysis and optimization to enhance the accuracy and efficiency of related design systems. Moreover, integrating sensing technology with BFRP products to obtain real-time data on the stress, strain, and environmental effects would enhance the maintenance strategies, improve safety, and extend the service life of structures. Innovative applications, such as BFRP prestressed modular structures and full BFRP profile systems, could revolutionize construction, especially in large-scale infrastructure projects. Combining BFRP with advanced techniques like 3D printing could also lead to lightweight and durable structures with complex designs.

It was valuable to explore BFRP's long-term performance in harsh environments, including high humidity, alkalinity, extreme temperatures, and UV exposure, to better predict its durability. Additionally, studies on its adaptability to extreme conditions, such as fire, seismic loads, and impact forces, are essential for expanding its use in critical infrastructure. Developing hybrid systems that combine BFRP with materials like CFRP or steel fibers could further enhance its mechanical properties. Furthermore, exploring eco-friendly resins, recycling methods, and life cycle sustainability will help reduce costs and make BFRP a more environmentally friendly alternative.

Author Contributions: Conceptualization, S.-J.D., G.-S.T. and J.-Z.T.; methodology, R.-M.F., G.-S.T. and J.-Z.T.; validation, X.-Y.Y.; formal analysis, S.-J.D.; investigation, S.-J.D. and L.-T.S.; resources, G.-S.T. and J.-Z.T.; writing—original draft preparation, S.-J.D. and R.-M.F.; writing—review and editing, S.-J.D. and X.-Y.Y.; supervision, G.-S.T. and J.-Z.T.; funding acquisition, G.-S.T. and J.-Z.T. All authors have read and agreed to the published version of the manuscript.

Funding: This research was funded by the Zhejiang Provincial Natural Science Foundation of China, grant number LR24E080002, and the Natural Science Foundation of China, grant number 52478219.

Data Availability Statement: The original contributions presented in this study are included in the article. Further inquiries can be directed to the corresponding author.

Conflicts of Interest: The authors declare no conflicts of interest.

References

1. Chen, Y.; Tong, J.; Li, Q.; Xu, S.; Shen, L. Application of high-performance cementitious composites in steel–concrete composite bridge deck systems: A review. *J. Intell. Constr.* **2024**, *2*, 9180012. [CrossRef]
2. Tong, J.; Wang, L.; Wu, R.; Hou, J.; Li, Q.; Xu, S. Cyclic test and analysis of UHTCC-enhanced buckling-restrained steel plate shear walls. *Earthq. Eng. Struct. Dyn.* **2024**, *53*, 4006–4031. [CrossRef]
3. Yu, C.; Tong, J.; Zhou, S.; Zhang, J.; Shen, J.; Zhang, L.; Tong, G.; Li, Q.; Xu, S. State-of-the-art review on steel-concrete composite walls. *Sustain. Struct.* **2024**, *4*, 35. [CrossRef]
4. Yu, C.; Tong, J.; Zhang, J.; Tong, G.; Chen, M.; Xu, S.; Gao, W. Axial compressive behavior of multi-celled corrugated-plate CFST walls: Tests and numerical simulations. *Eng. Struct.* **2025**, *322*, 119033. [CrossRef]
5. Tong, J.; Zhang, J.; Yu, C.; Tong, G.; Li, Q.; Xu, S. Seismic experiments and shear resistance prediction of multi-celled corrugated-plate CFST walls. *Earthq. Eng. Struct. Dyn.* **2024**, *53*, 1681–1704. [CrossRef]
6. Fiore, V.; Scalici, T.; Di Bella, G.; Valenza, A. A review on basalt fibre and its composites. *Compos. Part B Eng.* **2015**, *74*, 74–94. [CrossRef]
7. Saleem, A.; Medina, L.; Skrifvars, M.; Berglin, L. Hybrid Polymer Composites of Bio-Based Bast Fibers with Glass, Carbon and Basalt Fibers for Automotive Applications—A Review. *Molecules* **2020**, *25*, 4933. [CrossRef] [PubMed]
8. Selcuk, S.; Ahmetoglu, U.; Gokce, E. Basalt Fiber Reinforced Polymer Composites (BFRP) other than rebars: A review. *Mater. Today Commun.* **2023**, *37*, 107359. [CrossRef]
9. Ke, L.; Li, Y.; Li, C.; Cheng, Z.; Ma, K.; Zeng, J. Bond behavior of CFRP-strengthened steel structures and its environmental influence factors: A critical review. *Sustain. Struct.* **2024**, *4*, 38. [CrossRef]

10. Tong, J.; Wu, R.; Wang, L. Experimental and numerical investigations on seismic behavior of stiffened corrugated steel plate shear walls. *Earthq. Eng. Struct. Dyn.* **2023**, *52*, 3551–3574. [CrossRef]
11. Wu, R.; Yu, C.; Wang, L.; Tong, J. Shear elastic buckling of corrugated steel plate shear walls with stiffeners considering torsional rigidity. *Thin-Walled Struct.* **2025**, *206*, 112646. [CrossRef]
12. Tong, J.; Chen, Y.; Li, Q.; Dai, J.; Wang, G.; Shen, J.; Gao, W.; Xu, S. Flexural performance and crack width prediction of steel-UHTCC composite bridge decks with wet joints. *Eng. Struct.* **2025**, *323*, 119264. [CrossRef]
13. Wang, Z.; Gong, F.; Maekawa, K. Multi-scale and multi-chemo–physics lifecycle evaluation of structural concrete under environmental and mechanical impacts. *J. Intell. Constr.* **2023**, *1*, 9180003. [CrossRef]
14. Zhu, X.; Abe, H.; Hayashi, D.; Tanaka, H. Behavioral characteristics of RC beams with non-uniform corrosion along the reinforcement. *J. Intell. Constr.* **2023**, *1*, 9180019. [CrossRef]
15. Lopresto, V.; Leone, C.; De Iorio, I. Mechanical characterisation of basalt fibre reinforced plastic. *Compos. Part B Eng.* **2011**, *42*, 717–723. [CrossRef]
16. Wei, B.; Cao, H.; Song, S. Degradation of basalt fibre and glass fibre/epoxy resin composites in seawater. *Corros. Sci.* **2011**, *53*, 426–431. [CrossRef]
17. Fiore, V.; Di Bella, G.; Valenza, A. Glass–basalt/epoxy hybrid composites for marine applications. *Mater. Des.* **2011**, *32*, 2091–2099. [CrossRef]
18. Monaldo, E.; Nerilli, F.; Vairo, G. Basalt-based fiber-reinforced materials and structural applications in civil engineering. *Compos. Struct.* **2019**, *214*, 246–263. [CrossRef]
19. Perevozchikova, B.; Pisciotta, A.; Osovetsky, B.; Menshikov, E.; Kazymov, K. Quality Evaluation of the Kuluevskaya Basalt Outcrop for the Production of Mineral Fiber, Southern Urals, Russia. *Energy Procedia* **2014**, *59*, 309–314. [CrossRef]
20. Wu, Z.; Wang, X.; Shi, J. Advancement of basalt fiber-reinforced polymers (BFRPs) and the novel structures reinforced with BFRPs. *Eng. Mech.* **2020**, *37*, 1–14.
21. Jain, N.; Verma, A.; Singh, V. Dynamic Mechanical Analysis and Creep-recovery behaviour of Polyvinyl Alcohol based cross-linked Biocomposite reinforced with Basalt fiber. *Mater. Res. Express* **2019**, *6*, 105373. [CrossRef]
22. Wang, X.; Shi, J.; Wu, Z.; Zhu, Z. Creep strain control by pretension for basalt fiber-reinforced polymer tendon in civil applications. *Mater. Des.* **2016**, *89*, 1270–1277. [CrossRef]
23. Dhand, V.; Mittal, G.; Rhee, K.; Park, S.; Hui, D. A short review on basalt fiber reinforced polymer composites. *Compos. Part B Eng.* **2015**, *73*, 166–180. [CrossRef]
24. Chen, X.; Zhang, Y.; Huo, H.; Wu, Z. Study of high tensile strength of natural continuous basalt fibers. *J. Nat. Fibers* **2020**, *17*, 214–222. [CrossRef]
25. Chen, M.; Liu, J.; Wu, Z. Effect of Fe_2O_3Concentration on the Properties of Basalt Glasses. *J. Nat. Fibers* **2022**, *19*, 575–585. [CrossRef]
26. Liu, J.; Chen, M.; Yang, J.; Wu, Z. Study on Mechanical Properties of Basalt Fibers Superior to E-glass Fibers. *J. Nat. Fibers* **2020**, *19*, 882–894. [CrossRef]
27. Colombo, C.; Vergani, L.; Burman, M. Static and fatigue characterisation of new basalt fibre reinforced composites. *Compos. Struct.* **2012**, *94*, 1165–1174. [CrossRef]
28. Al-Rousan, E.; Khalid, H.; Rahman, M. Fresh, mechanical, and durability properties of basalt fiber-reinforced concrete (BFRC): A review. *Dev. Built Environ.* **2023**, *14*, 100155. [CrossRef]
29. Saheb, D.; Jog, J. Natural fiber polymer composites: A review. *Adv. Polym. Technol.* **1999**, *18*, 351–363. [CrossRef]
30. Sathishkumar, T.; Naveen, J.; Satheeshkumar, S. Hybrid fiber reinforced polymer composites—A review. *J. Reinf. Plast. Compos.* **2014**, *33*, 454–471. [CrossRef]
31. Santulli, C. Mechanical and Impact Damage Analysis on Carbon/Natural Fibers Hybrid Composites: A Review. *Materials* **2019**, *12*, 517. [CrossRef] [PubMed]
32. Varley, R.; Tian, W.; Leong, K.; Leong, A.; Fredo, F.; Quaresimin, M. The effect of surface treatments on the mechanical properties of basalt-reinforced epoxy composites. *Polym. Compos.* **2013**, *34*, 320–329. [CrossRef]
33. Wei, B.; Cao, H.; Song, S. Surface modification and characterization of basalt fibers with hybrid sizings. *Compos. Part A Appl. Sci. Manuf.* **2011**, *42*, 22–29. [CrossRef]
34. Wei, B.; Song, S.; Cao, H. Strengthening of basalt fibers with nano-SiO_2–epoxy composite coating. *Mater. Des.* **2011**, *32*, 4180–4186. [CrossRef]
35. Lee, J.; Nam, I.; Kim, H. Thermal stability and physical properties of epoxy composite reinforced with silane treated basalt fiber. *Fibers Polym.* **2017**, *18*, 140–147. [CrossRef]
36. Wang, G.; Liu, Y.; Guo, Y.; Zhang, Z.; Xu, M.; Yang, Z. Surface modification and characterizations of basalt fibers with non-thermal plasma. *Surf. Coat. Technol.* **2007**, *201*, 6565–6568. [CrossRef]

37. Halasová, M.; Kuběna, I.; Roupcová, P.; Černý, M.; Strachota, A.; Chlup, Z. Iron precipitation in basalt fibres embedded in partially pyrolysed methylsiloxane matrix. *Compos. Part A Appl. Sci. Manuf.* **2019**, *123*, 286–292. [CrossRef]
38. Zhao, X.; Wang, X.; Wu, Z.; Wu, J. Experimental study on effect of resin matrix in basalt fiber reinforced polymer composites under static and fatigue loading. *Constr. Build. Mater.* **2020**, *242*, 118121. [CrossRef]
39. Chlup, Z.; Černý, M.; Kácha, P.; Hadraba, H.; Strachota, A. Fracture resistance of partially pyrolysed polysiloxane preceramic polymer matrix composites reinforced by unidirectional basalt fibres. *J. Eur. Ceram. Soc.* **2020**, *40*, 4879–4885. [CrossRef]
40. Lilli, M.; Jurko, M.; Sirjovova, V.; Zvonek, M.; Cech, V.; Scheffler, C.; Rogero, C.; Ilyn, M.; Tirillò, J.; Sarasini, F. Basalt fibre surface modification via plasma polymerization of tetravinylsilane/oxygen mixtures for improved interfacial adhesion with unsaturated polyester matrix. *Mater. Chem. Phys.* **2021**, *274*, 125106. [CrossRef]
41. Lilli, M.; Zvonek, M.; Cech, V.; Scheffler, C.; Tirillò, J.; Sarasini, F. Low temperature plasma polymerization: An effective process to enhance the basalt fibre/matrix interfacial adhesion. *Compos. Commun.* **2021**, *27*, 100769. [CrossRef]
42. Pai, Y.; Pai, K.D.; Kini, M. Experimental investigations on the moisture absorption and mechanical behaviour of basalt-aramid/epoxy hybrid interply composites under different ageing environments. *Cogent Eng.* **2022**, *9*, 2080354. [CrossRef]
43. Zhang, T.; Wang, K.; Lin, B.; Yao, Y. The enhancement mechanism of modified basalt fiber on the performance of geopolymer concrete. *Constr. Build. Mater.* **2024**, *417*, 135123. [CrossRef]
44. Jiang, K.; Wang, X.; Ben, Q.; Ding, L.; Chen, Z.; Liew, J.; Du, H.; Wu, Z. Bond behavior of BFRP bars in ultra-high performance seawater sea-sand concrete reinforced by non-metallic fibers. *Eng. Struct.* **2024**, *318*, 118699. [CrossRef]
45. Zhou, J.; Wang, X.; Peng, Z.; Wu, Z.; Zhu, Z. Evaluation of a large-tonnage FRP cable anchor system: Anchorage design and full-scale experiment. *Eng. Struct.* **2022**, *251*, 113551. [CrossRef]
46. Zhou, J.; Wang, X.; Ding, L.; Tao, J.; Liang, X.; Wu, Z. Evaluation of the shear behavior of RC beams strengthened with prestressing BFRP grids. *Constr. Build. Mater.* **2024**, *452*, 138978. [CrossRef]
47. Yuan, C.; Chen, W.; Pham, T.; Hao, H.; Chen, L.; Wang, J. Experimental and analytical study of flexural behaviour of BFRP sheets strengthened RC beams with new epoxy anchors. *Eng. Struct.* **2021**, *241*, 112441. [CrossRef]
48. Ding, L.; Liu, X.; Wang, X.; Huang, H.; Wu, Z. Mechanical properties of pultruded basalt fiber-reinforced polymer tube under axial tension and compression. *Constr. Build. Mater.* **2018**, *176*, 629–637. [CrossRef]
49. Abdelkerim, D.; Wang, X.; Ibrahim, H.; Wu, Z. Static and fatigue behavior of pultruded FRP multi-bolted joints with basalt FRP and hybrid steel-FRP bolts. *Compos. Struct.* **2019**, *220*, 324–337. [CrossRef]
50. El Refai, A. Durability and Fatigue of Basalt Fiber-Reinforced Polymer Bars Gripped with Steel Wedge Anchors. *J. Compos. Constr.* **2013**, *17*, 4013006. [CrossRef]
51. Serbescu, A.; Guadagnini, M.; Pilakoutas, K. Mechanical Characterization of Basalt FRP Rebars and Long-Term Strength Predictive Model. *J. Compos. Constr.* **2015**, *19*, 4014037. [CrossRef]
52. Lu, Z.; Tan, S.; Huang, P.; Lei, Z.; Liu, F.; Xie, J. Durability of cement mortar-covered BFRP bars in simulated seawater environment. *Constr. Build. Mater.* **2020**, *234*, 117803. [CrossRef]
53. Dai, J.; Yin, S.; Hu, C. Analysis of key influencing factors of the bond performance between BFRP bars and coral reef and sand concrete. *Constr. Build. Mater.* **2021**, *269*, 121248.
54. Su, C.; Wang, X.; Ding, L.; Wu, Z.; Ma, X. Durability of seawater sea sand concrete beams reinforced with carbon nanotube-modified BFRP bars in a marine environment. *Compos. Struct.* **2022**, *292*, 115642. [CrossRef]
55. Xiao, J.; Murong, Y.; Chen, X.; Liu, L.; Zhai, K.; Jiang, H.; Huang, L.; Wang, G. Research on the Bonding Performance of BFRP Bars with Reactive Powder Concrete. *Buildings* **2023**, *13*, 2083. [CrossRef]
56. Feng, G.; Guo, S.; Zhou, L.; Luo, W.; Guo, X.; Jin, Z.; Zhu, D. Effects of surface characteristics and alkalinity on the deterioration of BFRP bars and BFRP-SSC interface in seawater environment. *Compos. Part B Eng.* **2024**, *268*, 111072. [CrossRef]
57. Jin, L.; Wu, J.; Zhang, R.; Du, X. Dynamic bond behavior of deformed BFRP bars embedded in concrete subjected to high-temperatures. *Int. J. Mech. Sci.* **2024**, *262*, 108748. [CrossRef]
58. Abed, F.; Alhafiz, A. Effect of basalt fibers on the flexural behavior of concrete beams reinforced with BFRP bars. *Compos. Struct.* **2019**, *215*, 23–34. [CrossRef]
59. Attia, K.; Alnahhal, W.; Elrefai, A.; Rihan, Y. Flexural behavior of basalt fiber-reinforced concrete slab strips reinforced with BFRP and GFRP bars. *Compos. Struct.* **2019**, *211*, 1–12. [CrossRef]
60. *ACI 440.1R-15*; Guide for the Design and Construction of Structural Concrete Reinforced with Fibre-Reinforced Polymer (FRP) Bars. American Concrete Institute (ACI) Committee 440: Farmington Hills, MI, USA, 2015.
61. *ACI 544.4R-88*; Design Considerations for Steel Fiber Reinforced Concrete. American Concrete Institute (ACI) Committee 544: Farmington Hills, MI, USA, 1988.
62. Muhammad, J.; Yousif, A. Shear behavior of normal and high-strength concrete beams reinforced with BFRP bars and basalt macro fibers. *Constr. Build. Mater.* **2023**, *408*, 133516. [CrossRef]

63. Liu, S.; Wang, X.; Ali, Y.; Huang, H.; Zhou, J.; Wu, Z. Experimental study on eccentric compression behavior of slender rectangular concrete columns reinforced with steel and BFRP bars. *Eng. Struct.* **2023**, *293*, 116626. [CrossRef]
64. Mostafa, O.; Rahman, M.; Al-Zahrani, M.; Adekunle, S.; Al-Osta, M.; Najamuddin, S. Flexural behavior and bond coefficient of BFRP bar reinforced normal and high strength concrete beams. *Constr. Build. Mater.* **2023**, *401*, 132896. [CrossRef]
65. Wang, X.; Shi, J.; Liu, J.; Yang, L.; Wu, Z. Creep behavior of basalt fiber reinforced polymer tendons for prestressing application. *Mater. Des.* **2014**, *59*, 558–564. [CrossRef]
66. Wang, X.; Shi, J.; Wu, G.; Yang, L.; Wu, Z. Effectiveness of basalt FRP tendons for strengthening of RC beams through the external prestressing technique. *Eng. Struct.* **2015**, *101*, 34–44. [CrossRef]
67. Wang, X.; Shi, J.; Wu, Z.; Zhu, Z. Fatigue Behavior of Basalt Fiber-Reinforced Polymer Tendons for Prestressing Applications. *J. Compos. Constr.* **2016**, *20*, 4015079. [CrossRef]
68. Dal Lago, B.; Taylor, S.; Deegan, P.; Ferrara, L.; Sonebi, M.; Crosset, P.; Pattarini, A. Full-scale testing and numerical analysis of a precast fibre reinforced self-compacting concrete slab pre-stressed with basalt fibre reinforced polymer bars. *Compos. Part B Eng.* **2017**, *128*, 120–133. [CrossRef]
69. Atutis, M.; Valivonis, J.; Atutis, E. Experimental study of concrete beams prestressed with basalt fiber reinforced polymers. Part II: Stress relaxation phenomenon. *Compos. Struct.* **2018**, *202*, 344–354. [CrossRef]
70. Motwani, P.; Rather, A.; Laskar, A. Transfer stage parameters for concrete beams prestressed with BFRP bars: Experimental and finite element studies. *Constr. Build. Mater.* **2022**, *315*, 125639. [CrossRef]
71. Ji, J.; Wang, X.; Dong, Z.; Wang, J.; Li, J.; Li, K. Study on mechanical properties of curved prestressed BFRP bars for future application in PCCP. *Constr. Build. Mater.* **2024**, *451*, 138873. [CrossRef]
72. Erfan, A.; Algash, Y.; El-Sayed, T. Experimental & analytical flexural behavior of concrete beams reinforced with basalt fiber reinforced polymers bars. *Int. J. Sci. Eng. Res.* **2019**, *10*, 297–315.
73. Erfan, A.; Elnaby, R.; Badr, A.; El-sayed, T. Flexural behavior of HSC one way slabs reinforced with basalt FRP bars. *Case Stud. Constr. Mater.* **2021**, *14*, e00513. [CrossRef]
74. Hussein, A.; Huang, H.; Okuno, Y.; Wu, Z. Experimental and numerical parametric study on flexural behavior of concrete beams reinforced with hybrid combinations of steel and BFRP bars. *Compos. Struct.* **2022**, *302*, 116230. [CrossRef]
75. Shi, J.; Wang, X.; Zhang, L.; Wu, Z.; Zhu, Z. Composite-Wedge Anchorage for Fiber-Reinforced Polymer Tendons. *J. Compos. Constr.* **2022**, *26*, 4022005. [CrossRef]
76. Shi, J.; Wang, X.; Wu, Z.; Zhu, Z. Optimization of anchorage and deviator for concrete beams prestressed with external fiber-reinforced polymer tendons. *Compos. Struct.* **2022**, *297*, 115970. [CrossRef]
77. Wang, X.; Wu, Z. Integrated high-performance thousand-metre scale cable-stayed bridge with hybrid FRP cables. *Compos. Part B Eng.* **2010**, *41*, 166–175. [CrossRef]
78. Wang, X.; Wu, Z.; Wu, G.; Zhu, H.; Zen, F. Enhancement of basalt FRP by hybridization for long-span cable-stayed bridge. *Compos. Part B Eng.* **2013**, *44*, 184–192. [CrossRef]
79. Yang, Y.; Wang, X.; Wu, Z.; Peng, C. Damping properties of FRP cables for long-span cable-stayed bridges. *Mater. Struct.* **2016**, *49*, 2701–2713. [CrossRef]
80. Zhou, J.; Wang, X.; Peng, Z.; Wu, Z.; Wei, X. Enhancement of FRP Cable Anchor System: Optimization of Load Transfer Component and Full-Scale Cable Experiment. *J. Compos. Constr.* **2022**, *26*, 4022008. [CrossRef]
81. Zhou, J.; Wang, X.; Su, C.; Huang, H.; Liu, X.; Wu, Z. Experimental investigation on the fatigue performance of multi-tendon basalt fiber reinforced polymer cable. *Eng. Struct.* **2023**, *297*, 116963. [CrossRef]
82. Feng, B.; Wang, X.; Wu, Z. Fatigue life assessment of FRP cable for long-span cable-stayed bridge. *Compos. Struct.* **2019**, *210*, 159–166. [CrossRef]
83. Wang, X.; Zhou, J.; Ding, L.; Song, J.; Wu, Z. Static Behavior of Circumferential Stress-Releasing Anchor for Large-Capacity FRP Cable. *J. Bridge Eng.* **2020**, *25*, 4019127. [CrossRef]
84. Yang, Y.; Zhou, Z.; Guan, Y.; Shi, J.; Zhan, Q.; Fahmy, M.F.M.; Wu, B. Nonlinear Coupled Vibration Behavior of BFRP Cables on Long-Span Cable-Stayed Bridges under Parametric Excitation. *Buildings* **2023**, *13*, 3082. [CrossRef]
85. Zhou, J.; Wang, X.; Ding, L.; Liu, S.; Wu, Z. Numerical and Experimental Study on Large-Diameter FRP Cable Anchoring System with Dispersed Tendons. *Buildings* **2023**, *13*, 92. [CrossRef]
86. Wang, X.; Xu, P.; Wu, Z.; Shi, J. A Novel Anchor Method for Multitendon FRP Cable: Manufacturing and Experimental Study. *J. Compos. Constr.* **2015**, *19*, 4015010. [CrossRef]
87. Yang, Y.; Fahmy, M.; Guan, S.; Pan, Z.; Zhan, Y.; Zhao, T. Properties and applications of FRP cable on long-span cable-supported bridges: A review. *Compos. Part B Eng.* **2020**, *190*, 107934. [CrossRef]
88. Górski, P.; Tatara, M.; Stankiewicz, B. Vibration serviceability of all-GFRP cable-stayed footbridge under various service excitations. *Measurement* **2021**, *183*, 109822. [CrossRef]

89. Zhou, J.; Wang, X.; Peng, Z.; Wu, Z. Optimization of load transfer component for FRP cable anchor system. *Compos. Struct.* **2022**, *282*, 115009. [CrossRef]
90. Peng, Z.; Wang, X.; Wu, Z. Statistical modeling and reliability analysis of large-scale multi-tendon fiber-reinforced polymer cables. *Eng. Struct.* **2024**, *300*, 117226. [CrossRef]
91. Fang, Y.; Xiang, Y.; Fang, Z. Impact Behaviour of Carbon Fibre-Reinforced Polymer (CFRP) Cables with Protective Sheaths. *Constr. Build. Mater.* **2024**, *450*, 138599. [CrossRef]
92. Kim, T.; Hwang, S.; Kim, J.; Jung, W.; Yoon, J. Analysis of bond strength of CFRP cables with concrete using random forest model. *J. Build. Eng.* **2024**, *96*, 110658. [CrossRef]
93. Kim, T.; Kim, J.; Hwang, S.; Yoon, J. Comprehensive investigation of bond properties of CFRP cables with concrete and mortar. *Structures* **2024**, *66*, 106861. [CrossRef]
94. Qi, L.; Bai, J.; Wu, H.; Xu, G.; Xiong, H.; Yang, Y. The first engineering application of 10MN CFRP cables in cable-stayed bridge in China. *Structures* **2024**, *68*, 107199. [CrossRef]
95. Ali, N.; Wang, X.; Wu, Z.; Hassanein, A. Basalt fiber reinforced polymer grids as an external reinforcement for reinforced concrete structures. *J. Reinf. Plast. Compos.* **2015**, *34*, 1615–1627. [CrossRef]
96. Zheng, Y.; Wang, W.; Brigham, J. Flexural behaviour of reinforced concrete beams strengthened with a composite reinforcement layer: BFRP grid and ECC. *Constr. Build. Mater.* **2016**, *115*, 424–437. [CrossRef]
97. Huang, H.; Jia, B.; Li, W.; Liu, X.; Yang, D.; Deng, C. An Experimental Study on Compressive Behavior of Rubble Stone Walls Retrofitted with BFRP Grids. In *Advances in Energy Science and Environment Engineering II*; AIP Publishing: Melville, NY, USA, 2018; Volume 1944.
98. Dong, Z.; Sun, Y.; Wu, G.; Zhu, H.; Zhao, X.; Wei, Y.; Zhang, P. Flexural behavior of seawater sea-sand concrete beams reinforced with BFRP bars/grids and BFRP-wrapped steel tubes. *Compos. Struct.* **2021**, *268*, 113956. [CrossRef]
99. Wen, B.; Wan, C. Experimental study on the bond degradation of basalt fiber reinforced polymer grid-concrete interface under fatigue loading. *Constr. Build. Mater.* **2021**, *270*, 121383. [CrossRef]
100. Zhang, W.; Deng, M.; Han, Y.; Li, R.; Yang, S. Uniaxial tensile performance of high ductile fiber-reinforced concrete with built-in basalt textile grids. *Constr. Build. Mater.* **2022**, *315*, 125716. [CrossRef]
101. He, W.; Cai, F.; Wang, X.; Wu, Z.; Sun, Z.; Chen, C. Fatigue performance of RC beams strengthened with non-prestressed and prestressed BFRP grids. *Eng. Struct.* **2024**, *304*, 117594. [CrossRef]
102. Song, L.; Wei, M.; Zhang, X.; Li, C.; Dong, Z.; Wang, Y. Flexural performance of reinforced concrete beams repaired with BFRP grid-TSUHDC. *Constr. Build. Mater.* **2024**, *443*, 137748. [CrossRef]
103. Zhou, J.; Wang, X.; Liu, X.; Tao, J.; Wu, Z. Investigation on the shear behavior of RC beams strengthened with BFRP grids and PCM. *Eng. Struct.* **2025**, *322*, 119173. [CrossRef]
104. *ACI 549.4R-13*; Guide to Design and Construction of Externally Bonded FRCM Systems for Repair and Strengthening Concrete and Masonry Structures. American Concrete Institute (ACI) Committee 549: Farmington Hills, MI, USA, 2013.
105. Zheng, Y.; Wang, W.; Mosalam, K.M.; Fang, Q.; Chen, L.; Zhu, Z. Experimental investigation and numerical analysis of RC beams shear strengthened with FRP/ECC composite layer. *Compos. Struct.* **2020**, *246*, 112436. [CrossRef]
106. Zhang, Z.; Guo, X.; Yang, Z.; Ji, J.; Sun, Q.; Tian, P.; Wu, X. Flexural behavior of the corroded RC beams strengthened with BFRP grid-reinforced ECC. *Structures* **2023**, *58*, 105541. [CrossRef]
107. Zhou, D.; Lei, Z.; Wang, J. In-plane behavior of seismically damaged masonry walls repaired with external BFRP. *Compos. Struct.* **2013**, *102*, 9–19. [CrossRef]
108. Jiang, S.; Zeng, X.; Shen, S.; Xu, X. Experimental studies on the seismic behavior of earthquake-damaged circular bridge columns repaired by using combination of near-surface-mounted BFRP bars with external BFRP sheets jacketing. *Eng. Struct.* **2016**, *106*, 317–331. [CrossRef]
109. Qin, Z.; Tian, Y.; Li, G.; Liu, L. Study on bending behaviors of severely pre-cracked RC beams strengthened by BFRP sheets and steel plates. *Constr. Build. Mater.* **2019**, *219*, 131–143. [CrossRef]
110. Yuan, C.; Chen, W.; Pham, T.; Hao, H.; Cui, J.; Shi, Y. Strain rate effect on interfacial bond behaviour between BFRP sheets and steel fibre reinforced concrete. *Compos. Part B Eng.* **2019**, *174*, 107032. [CrossRef]
111. Pham, T.; Chen, W.; Elchalakani, M.; Karrech, A.; Hao, H. Experimental investigation on lightweight rubberized concrete beams strengthened with BFRP sheets subjected to impact loads. *Eng. Struct.* **2020**, *205*, 110095. [CrossRef]
112. Shen, D.; Li, M.; Liu, C.; Kang, J.; Li, C.; Yang, J. Seismic performance of corroded reinforced concrete beam-column joints repaired with BFRP sheets. *Constr. Build. Mater.* **2021**, *307*, 124731. [CrossRef]
113. Zeng, X.; Jiang, S.; Deng, K.; Huang, H.; Cui, E. Seismic performance of circular RC columns strengthened in flexure using NSM reinforcement and externally bonded BFRP sheets. *Eng. Struct.* **2022**, *256*, 114033. [CrossRef]

114. He, J.; Lei, D.; She, Z.; Xi, B. Investigation on bonding behavior of basalt fiber reinforced polymer (BFRP) sheet reinforced concrete beam. *J. Build. Eng.* **2023**, *75*, 106963. [CrossRef]
115. He, J.; Lei, D.; She, Z.; Xi, B. Flexural performance and damage evaluation on basalt fiber reinforced polymer (BFRP) sheet reinforced concrete. *Constr. Build. Mater.* **2023**, *395*, 132321. [CrossRef]
116. Liu, K.; Wulan, T.; Yao, Y.; Bian, M.; Bao, Y. Assessment of damage evolution of concrete beams strengthened with BFRP sheets with acoustic emission and unsupervised machine learning. *Eng. Struct.* **2024**, *300*, 117228. [CrossRef]
117. Kang, W.; Mei, K.; Lu, D.; Sun, S. Study of the flexural behavior of UHPC-HPC composite beams strengthened with BFRP sheet after chloride secondary erosion. *Eng. Struct.* **2024**, *316*, 118536. [CrossRef]
118. Nayak, C.; Narule, G.; Surwase, H. Structural and cracking behaviour of RC T-beams strengthened with BFRP sheets by experimental and analytical investigation. *J. King Saud Univ. Eng. Sci.* **2022**, *34*, 398–405. [CrossRef]
119. Hashem, Z.; Yuan, R. Short vs. long column behavior of pultruded glass-fiber reinforced polymer composites. *Constr. Build. Mater.* **2001**, *15*, 369–378. [CrossRef]
120. Bai, Y.; Keller, T. Shear Failure of Pultruded Fiber-Reinforced Polymer Composites under Axial Compression. *J. Compos. Constr.* **2009**, *13*, 234–242. [CrossRef]
121. Cardoso, D.; Harries, K.; Batista, E. Compressive strength equation for GFRP square tube columns. *Compos. Part B Eng.* **2014**, *59*, 1–11. [CrossRef]
122. Almeida, J.H.; Tonatto, M.; Ribeiro, M.; Tita, V.; Amico, S. Buckling and post-buckling of filament wound composite tubes under axial compression: Linear, nonlinear, damage and experimental analyses. *Compos. Part B Eng.* **2018**, *149*, 227–239. [CrossRef]
123. Al-saadi, A.; Aravinthan, T.; Lokuge, W. Effects of fibre orientation and layup on the mechanical properties of the pultruded glass fibre reinforced polymer tubes. *Eng. Struct.* **2019**, *198*, 109448. [CrossRef]
124. Alajarmeh, O.; Zeng, X.; Aravinthan, T.; Shelley, T.; Alhawamdeh, M.; Mohammed, A.; Nicol, L.; Vedernikov, A.; Safonov, A.; Schubel, P. Compressive behaviour of hollow box pultruded FRP columns with continuous-wound fibres. *Thin Walled Struct.* **2021**, *168*, 108300. [CrossRef]
125. Lu, Z.; Xian, G.; Li, H. Effects of thermal aging on the water uptake behavior of pultruded BFRP plates. *Polym. Degrad. Stab.* **2014**, *110*, 216–224. [CrossRef]
126. Lu, Z.; Xian, G.; Li, H. Effects of exposure to elevated temperatures and subsequent immersion in water or alkaline solution on the mechanical properties of pultruded BFRP plates. *Compos. Part B Eng.* **2015**, *77*, 421–430. [CrossRef]
127. Lu, Z.; Xian, G. Combined effects of sustained tensile loading and elevated temperatures on the mechanical properties of pultruded BFRP plates. *Constr. Build. Mater.* **2017**, *150*, 310–320. [CrossRef]
128. Wang, X.; Jiang, L.; Shen, H.; Wu, Z. Long-Term Performance of Pultruded Basalt Fiber Reinforced Polymer Profiles under Acidic Conditions. *J. Mater. Civ. Eng.* **2018**, *30*, 4018096. [CrossRef]
129. Ding, L.; Liu, L.; Wang, X.; Shen, H.; Wu, Z. Effects of connecting materials on the static and fatigue behavior of pultruded basalt fiber-reinforced polymer bolted joints. *Constr. Build. Mater.* **2021**, *273*, 121683. [CrossRef]
130. Chen, Y.; Zhang, C. Stability analysis of pultruded basalt fiber-reinforced polymer (BFRP) tube under axial compression. *Compos. Struct.* **2024**, *327*, 117660. [CrossRef]
131. Liang, M.; Wang, X.; Wang, Y.; Ding, L.; Jiang, Z.; Huang, H.; Liu, X.; Wu, Z. Tensile properties and load distribution of multi-bolted pultruded basalt FRP joints with multi-directional fiber lay-ups. *Structures* **2024**, *70*, 107637. [CrossRef]
132. Liang, M.; Wang, X.; Zhong, L.; Jiang, Z.; Huang, H.; Wu, Z. Effects of off-axis load direction on tensile properties of pultruded fiber-reinforced polymer bolted joints with multi-directional fiber lay-ups. *Polym Compos.* **2024**, *45*, 10172–10189. [CrossRef]
133. Liu, L.; Wang, X.; Wu, Z.; Keller, T. Optimization of multi-directional fiber architecture for resistance and ductility of bolted FRP profile joints. *Compos. Struct.* **2020**, *248*, 112535. [CrossRef]

MDPI AG
Grosspeteranlage 5
4052 Basel
Switzerland
Tel.: +41 61 683 77 34

Buildings Editorial Office
E-mail: buildings@mdpi.com
www.mdpi.com/journal/buildings

www.ingramcontent.com/pod-product-compliance
Lightning Source LLC
Chambersburg PA
CBHW061609220926
51850CB00005B/464
* 9 7 8 3 7 2 5 8 4 6 9 7 9 *